Catalysis of Organic Reactions

CHEMICAL INDUSTRIES

A Series of Reference Books and Textbooks

Founding Editor

HEINZ HEINEMANN
Berkeley, California

Series Editor

JAMES G. SPEIGHT
University of Trinidad and Tobago
O'Meara Campus, Trinidad

Catalysis of Organic Reactions

Edited by

Michael L. Prunier

Eli Lilly & Company
Indianapolis, Indiana, U.S.A.

CRC Press
Taylor & Francis Group
Boca Raton   London   New York

CRC Press is an imprint of the
Taylor & Francis Group, an **informa** business

CRC Press
Taylor & Francis Group
6000 Broken Sound Parkway NW, Suite 300
Boca Raton, FL 33487-2742

© 2009 by Taylor & Francis Group, LLC
CRC Press is an imprint of Taylor & Francis Group, an Informa business

Library of Congress Cataloging-in-Publication Data

Conference on the Catalysis of Organic Reactions (22nd : 2008 : Richmond, Virginia)
   Catalysis of organic reactions / editor, Michael L. Prunier.
      p. cm. -- (Chemical industries ; no. 123)
   Includes bibliographical references and index.
   ISBN 978-1-4200-7076-7 (hardcover : alk. paper)
   1. Organic compounds--Synthesis--Congresses. 2. Catalysis--Congresses. I. Prunier, Michael L. II. Title. III. Series.

QD262.C563 2008
660'.2995--dc22   2008042113

Visit the Taylor & Francis Web site at
http://www.taylorandfrancis.com

and the CRC Press Web site at
http://www.crcpress.com

Contents

[1]Department of Chemical Engineering and [2]Department of Chemistry, Imperial College London SW7 2AZ, United Kingdom, [3]Process Studies Group, Syngenta, Huddersfield HD2 1FF, United Kingdom

Xiaoying Ouyang[1], Jun Li[2,3], Ram Seshadri[1,2,3] and Susannah L. Scott[1,4]
[1]Department of Chemistry and Biochemistry, [2]Materials Department, [3]Materials Research Laboratory, and [4]Department of Chemical Engineering University of California, Santa Barbara, CA 93106

Steven D. Dietz, Claire M. Ohman, Trudy A. Scholten, Steven Gebhard and Girish Srinivas
TDA Research, Inc., 12345 W. 52nd Ave., Wheat Ridge, CO 80033

Andreas Bernas[1], Johan Ahlkvist[1], Johan Wärnå[1], Päivi Mäki-Arvela[1], Juha Lehtonen[2], Tapio Salmi[1] and Dmitry Yu Murzin[1]
[1]Åbo Akademi, Process Chemistry Centre, Laboratory of Industrial Chemistry, FI-20500 Åbo/Turku, Finland
[2]Perstorp Oy, Technology Centre, P.O. Box 350, FI-06101 Borgå/Porvoo, Finland

Pierre Gallezot and Alexander B. Sorokin
Institut de recherches sur la catalyse et l'environnement de Lyon 2, avenue Albert Einstein, 69626 Villeurbanne, France

Paolo Bondioli[1], Laura Della Bella[1], Nicoletta Ravasio[2] and Federica Zaccheria[2]
[1]Stazione Sperimentale Oli e Grassi, via G. Colombo79, 20133 Milano, Italy
[2]CNR –ISTM and Dip. CIMA, University of Milano, via G.Venezian 21, 20133 Milano, Italy

Rajiv Banavali[1], Robert T. Hanlon[1], Karel Jerabek[2] and Alfred K. Schultz[1]
[1]Rohm and Haas Company, LLC, Spring House, PA 19477
[2]Institute of Chemical Process Fundamentals, Academy of Sciences of the Czech Republic, Rozvojova 135, 165 02, Prague 6, Czech Republic

Anton A. Kiss and Gadi Rothenberg
van 't Hoff Institute for Molecular Sciences, University of Amsterdam, Nieuwe Achtergracht 166, 1018 WV Amsterdam, the Netherlands

Note: The underlined authors signify those who were presenters at the conference.

Board of Editors

Chronology of Organic Reactions
Catalysis Society Conferences

Conf	Year	Chairman	Location	Proceedings Publisher
1	1967	Joseph O'Connor	New York City	NY Acad. Sci.
2	1969	Joseph O'Connor	New York City	NY Acad. Sci.
3	1970	Mel Rubensdorf	New York City	NY Acad. Sci.
4	1973	Paul Rylander	New York City	NY Acad. Sci.
5	1974	Paul Rylander Harold Greenfield	Boston	Academic Press
6	1976	Gerard Smith	Boston	Academic Press
7	1978	William Jones	Chicago	Academic Press
8	1980	William Moser	New Orleans	Marcel Dekker
9	1982	John Kosak	Charleston	Marcel Dekker
10	1984	Robert Augustine	Williamsburg	Marcel Dekker
11	1986	Paul Rylander Harold Greenfield	Savannah	Marcel Dekker
12	1988	Dale Blackburn	San Antonio	Marcel Dekker
13	1990	William Pascoe	Boca Raton	Marcel Dekker
14	1992	Thomas Johnson John Kosak	Albuquerque	Marcel Dekker
15	1994	Mike Scaros Michael Prunier	Phoenix	Marcel Dekker
16	1996	Russell Malz	Atlanta	Marcel Dekker
17	1998	Frank Herkes	New Orleans	Marcel Dekker
18	2000	Michael Ford	Charleston	Marcel Dekker
19	2002	Dennis Morrell	San Antonio	Marcel Dekker
20	2004	John Sowa	Hilton Head	CRC Press
21	2006	Stephen Schmidt	Orlando	CRC Press
22	2008	Michael Prunier	Richmond	CRC Press

Preface

This volume of *Catalysis of Organic Reactions* compiles 64 papers presented at the 22nd conference organized by the Organic Reactions Catalysis Society (ORCS) (www.orcs.org). The conference was held March 30-April 3, 2008 in Richmond, Virginia at The Jefferson Hotel where these papers reported recent developments in catalysis as applied to the production of chemicals. Each of the papers documenting these presentations and published here was edited by ORCS members (drawn from both academia and industry) and was peer-reviewed by experts in related fields of study.

The focus of the 22nd conference was on applied catalysis. In the preface to the proceedings of the 2006 conference Dr. Schmidt noted the *"remarkable diversity of topics, some of which defied our simple attempts at sorting into a small number of categories. This reflects the current degree of specialization within our broad subject area."* This concept of diversity within the field of catalysis was the guiding principle for planning the 22nd conference and the presentations were chosen based on a wide breadth of topics rather than the traditional symposia organized around central topics. The only organized symposium at the 22nd conference was *Catalysis in the Pharmaceutical Industry*. The other presentations were grouped by common themes where possible but no attempts were made to create themed symposia based on predetermined topics.

Key presentations by the Society's Paul N. Rylander and Murray Raney Award winners were given as plenary lectures at the conference. The winners honored at the 22nd conference include the 2007 Paul N. Rylander Award winner Professor Brian R. James of the University of British Columbia, the 2008 Paul N. Rylander Award winner Professor John F. Hartwig of the University of Illinois and the 2008 Murray Raney Award winner Dr. Daniel J. Ostgard of Evonik Degussa Corporation.

The support we received from our sponsors was much appreciated and greatly contributed to the success of the 22nd conference. On behalf of ORCS, I specifically thank these organizations: *Avantium, BASF Catalysts LLC, Eli Lilly and Company, Evonik Degussa Corporation, W.R. Grace (Davison Catalysts), Parr Instrument, Air Products, Amgen, Eastman, Umicore, Bristol-Myers Squibb, DuPont, Headwaters, HEl, Lummus Technology, OMG, Seton Hall University (c/c Dr. John Sowa) and Süd Chemie.*

As chairman I very much appreciate the dedication and hard work of the ORCS board members Bert Chandler, Baoshu Chen, Pierre Gallezot, Kathy Hayes, Steve Jacobson, Steve Quimby, John Super, and Jim White and executive committee members Alan Allgeier, Steve Schmidt and Helene Shea. I also wish to thank the many individuals who contributed their time and efforts as presenters, members of the editorial board, peer reviewers, session chairs, A/V assistants and those who performed the multitude of "invisible" activities that are required to organize and run the conference. A special thanks goes to Dr. Tom Puckette of Eastman Chemicals for his tremendous efforts helping John Super with our sponsorship program and to Dr. Jim White for his wise mentoring and help with many different activities that greatly helped move the conference organizing forward.

I must thank my supervisor John Masters and Lilly management for supporting my term as Chairman and giving me encouragement and support as well as the time and resources I needed to perform my duties and plan and run the conference. I also thank Janet Boggs, my lab associate, who kept my Lilly laboratory functioning smoothly while I focused on my ORCS duties.

I thank my family for their support and efforts including help with folding and sealing the Call for Abstracts and Meeting Announcements that were sent to ORCS' contact list, Web site updates, the on-line registration process, design of the registration bags, proofreading of the newsletters and standardizing the fonts of the manuscripts. Most importantly I thank my wife Judy for her tremendous help, support, understanding and love during this endeavor.

Finally, I thank the members of ORCS for the honor and privilege of being their chairman.

Michael L. Prunier
Chairman of the 22nd Conference of the Organic Reactions Catalysis Society

To Judy
Without your support and understanding, this would not have been important or even possible

1. New Chemistry, Including Pulp Bleaching Processes, of the Golden-Aged, Water-Soluble Compound, Tris(hydroxymethyl)phosphine

Brian R. James

Department of Chemistry, University of British Columbia, 2036 Main Mall, Vancouver, B.C., Canada, V6T 1Z1

brj@chem.ubc.ca

Abstract

Collaborative studies with the Pulp and Paper Institute of Canada (Paprican) have led serendipitously to development of a pulp bleaching process using tris(hydroxymethyl)phosphine, $P(CH_2OH)_3$ (abbreviated THP), and this on-going work is reviewed. Work on hydrogenation of aromatic residues in lignin, using transition metal catalysts under homogeneous, aqueous-phase conditions, led to the discovery that commercially available THP, *in the absence of any metal*, bleached selected pulps. Use of THP or a related diphosphine is competitive with the currently used bleaching agents, hydrosulfite $(S_2O_4^{2-})$ and alkaline peroxide. (Hydroxymethyl)phosphines react as strong nucleophiles and/or reductants toward unsaturated chromophore moieties of lignin, decreasing the degree of conjugation, and hence leading to bleaching. The resulting phosphorus-containing moieties in the lignin also give a more permanent brightening of the pulp. The organo-phosphorus chemistry realized is relevant to other processes catalyzed homogeneously by transition metal-phosphine systems. In situ metal-THP species can catalyze hydrogenation of C=C and/or C=O bonds in lignin chromophores, and investigations on such systems have unearthed new THP-promoted transition metal chemistry.

Introduction

For paper manufacture, wood (in the form of wood chips or sawmill residue) is first converted to chemical or mechanical pulp by a pulping process. Chemical pulp is produced in yields of ~55% through dissolution of wood lignin using, for example, $NaOH/Na_2S$ at temperatures >160°C; mechanical pulp, formed by action of mechanical forces on wood, is obtained with retention of lignin in yields >90% (1). Canada is the largest manufacturer and exporter of mechanical pulp (~11 million tonnes/year, 1/3rd of the total world production) (2). The pale-yellow color of mechanical pulps and natural wood is due to the presence of lignin chromophores such as coniferaldehyde (see Fig. 1.1) and quinones, and their removal or modification (i.e., bleaching of the pulp) is needed for papermaking. Alkaline hydrogen peroxide and dithionite ($Na_2S_2O_4$, named 'hydrosulfite') have been the usual bleaching agents for over 50 years. Neither process involves catalysis and both are 'atom inefficient' and 'non-green.' Peroxide systems require $Na_2(SiO_3)$ and

MgSO$_4$ as stabilizers, and are non-selective in degrading also cellulose and hemicelluloses, while the reductant S$_2$O$_4^{2-}$ is more selective but less effective, and generates various sulfur-containing chemicals including thiosulfate that is corrosive to paper machines (3).

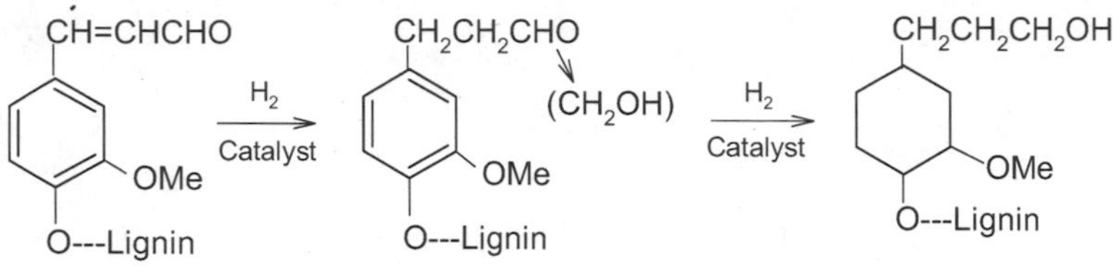

Figure 1.1. Hydrogenation of coniferaldehyde moieties in lignin.

This review describes collaborative research over the last 15 years between my group and the Pulp and Paper Research Institute of Canada (Paprican). The initial goal was to inhibit the yellowing of pulps and paper by reducing formation of quinones and other chromophores, but this evolved into development of a greener bleaching process primarily for mechanical pulps. This has been moderately successful, and the research has led also to significant findings in organo-phosphorus chemistry and transition metal chemistry of water-soluble, tris(hydroxymethyl)-phosphine, P(CH$_2$OH)$_3$ (abbreviated THP).

Catalytic Hydrogenation

The first goal was to catalytically hydrogenate either aromatic residues or conjugated C=C–C=O moieties (Fig. 1.1), including quinone residues, using a water-soluble, recyclable catalyst. Hydrogenation of lignin via heterogeneous and homogeneous catalysts had been reported but using severe conditions of temperature and pressure, with accompanying hydrogenolysis of substituent side-chains (4).

Our studies revealed that in situ Rh- and Ru-based species in 1-phase aq./alcohol and 2-phase aq./organic systems under H$_2$ could reduce under mild conditions organics with a reactivity order: C=C > aromatic C–C ≈ C=O bonds (5). The optimum catalysts were colloidal, zero-valent metal species stabilized by long-chain tetraalkylammonium salts or polyvinylpyrrolidone (for Rh^0), or long-chain trialkylamines (for Ru^0). Within lignin model compounds (LMCs) such as substituted phenols, the reactions gave high chemoselectivity for reduction of the aromatic bonds (*vs.* hydrogenolysis products), and showed high stereoselectivity with hydrogenation to cyclohexyl products occurring on one side of the aromatic ring (5). The Rh species systems were generally ineffective for hydrogenation of isolated, so-called mill-wood lignin, although a catalyst formed from a Rh$_6$(μ_3-O)$_4$ cluster (in retrospect, almost certainly colloidal) did slowly hydrogenate, at ~20°C and 1 atm H$_2$, up to ~70% of the aromatic rings of the lignin as measured by ^{1}H NMR (6). A colloidal species formed from RuCl$_3$/(^{n}C$_8$H$_{17}$)$_3$N was similarly effective at 80°C under 50 atm of H$_2$ (5, 7). The cyclohexyl products from LMCs are white and completely photostable (7), while hydrogenated mill-wood lignin reveals a yellowing inhibition of ~50% (7).

The colloidal catalysts are impractical for pulp bleaching in that they eventually precipitate as inactive black metal, and any such partially hydrogenated pulps always have a greyish tinge! Arene-Ru species, including water-soluble polyhydrido-Ru$_4$ cluster complexes, initially reported to be effective homogeneous hydrogenation catalysts for aromatics (8), were tested but found to be colloidal (5), and later, more detailed studies by other groups (9) have confirmed this finding. With ancillary sulfonated phosphine ligands, arene-Ru species did maintain homogeneity in solution, but the systems showed marginal activity for hydrogenation of substituted phenols and mill-wood lignin (5).

The well-known yellowing (color reversion) of pulp and paper, intimately related to the bleaching, results from irreversibility in bleaching processes. For example, bleaching induced by a reducing agent such as $S_2O_4^{2-}$ can be reversed by atmospheric photo-oxidation processes that proceed via phenoxy radicals, which readily oxidize to *o*-quinones (10). As noted above, hydrogenation of aromatic moieties is effective for inhibition of yellowing, as is conversion of α-carbonyl and α-hydroxyl to CH_2 groups within LMCs. These conversions were accomplished in non-aqueous solutions using, respectively, a colloidal Rh catalyst under H_2 (11) or a Pd-catalyzed hydrogen transfer system (11, 12). Although application of such catalysis for treatment of pulps in aqueous media is impractical, the findings did lead to increased understanding of the chemistry of yellowing and bleaching.

Discovery of New Bleaching Agents via Catalytic Hydrogenation Studies

The homogeneous, sulfonated phosphine-arene-Ru species, which were ineffective hydrogenation catalysts for aromatic residues (see Section 2), were not tested for possible hydrogenation of the C=C–C=O within α,β-unsaturated aldehydes/ketones (*cf.* Fig. 1.1). Attention instead was turned to in situ water-soluble, transition metal systems incorporating low-cost THP (see Section 4). Among many systems studied for hydrogenation of LMCs and pulps, one containing a Ru-THP species revealed the key finding that a thermomechanical pulp (TMP) was visibly bleached. Figure 1.2 shows some data for a spruce TMP (13) [% ISO Brightness values are determined by a standard method (14), and consistency is the weight % of pulp in a pulp/water mixture]. Complementary hydrogenation experiments on LMCs were also carried out

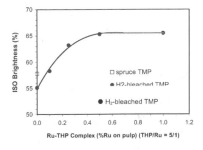

Figure 1.2. Bleaching of a spruce thermomechanical pulp (TMP) in the presence of a Ru-THP species; 100°C, 340 psi H_2, 3 h, 1.5% consistency.

(15). An aqueous solution of the Ru species was prepared by a ligand exchange reaction utilizing THP in water and a Ru precursor in CH_2Cl_2 (exemplified ideally by eq. 1), and could be used in air. Although the composition of the Ru species in solution at room temperature changed with time, even for a fixed THP:Ru ratio, the activity did not at the conditions used (e.g., see Figs. 1.2, and 1.3), implying that the various catalyst precursors are converted to the same active hydride species. The nature of the catalyst remains unknown, and we and others (see Section 6) have found that THP can be a non-innocent ligand when coordinated to Pt-metals. Nevertheless, the air-stable catalyst operates in 2-phase systems, the substrate being essentially insoluble in water, and the systems look promising. Studies on LMCs

$$RuCl_2(PPh_3)_3 + n\,THP \rightarrow n\,PPh_3 + RuCl_2(THP)_n(PPh_3)_{3-n}\,;\, n = 0.5\text{--}6.0 \qquad (1)$$

(substituted -styrenes, -acetophenones, -benzaldehydes, -cinnamaldehydes, -carbinols, and -propenylbenzenes) revealed that the Ru-THP species catalyzed hydrogenation in aqueous or H_2O/EtOH solution of activated olefins, carbonyl groups, and, of more interest, hydrogenolysis of some aromatic alcohols (Fig. 1.3), all these functionalities being found in lignin; aromatic residues were not reduced (15). Hydrogenolysis of an aldehyde or ketone to $>CH_2$ is an important transformation and, although there are procedures for doing this (16), we are unaware of any earlier reported homogeneous systems that catalyze such hydrogenolysis; this is a key process when applied to carbohydrate and fatty acid substrates in terms of renewable resources, and recently other water-soluble Ru catalysts have been found effective (17).

$$Ar\text{--}C(R)\text{=}O \rightarrow Ar\text{--}CH(R)OH \rightarrow Ar\text{--}CH_2R$$

Figure 1.3. Catalytic hydrogenation of LMCs (Ar = 3-MeO-4-OH-C_6H_3; R = H or Me); THP/Ru = 3, 500 psi H_2, 80°C, 22 h, aq. buffer (pH = 10.0), substrate/Ru = 25.

Encouraged by the results with the Ru-THP catalysts, THP systems with the cheaper metals Cu, Co and Ni were studied, but only marginal hydrogenation activity for LMCs was observed. Of interest, however, the zwitterionic complex $Cu^+(THP)_3\{P(CH_2OH)_2(CH_2O^-)\}$ was isolated, and treatment of this with HCl yielded $[Cu(THP)_4]Cl$; both complexes were characterized crystallographically (18), but another group reported simultaneously the former species (19). Although neither complex was effective for hydrogenation activity, the zwitterion was tested with a mechanical pulp at 80°C and 340 psi H_2, and visible bleaching became evident (20). Subsequently, control experiments revealed that neither H_2 nor the Cu complex was needed; i.e., the phosphine itself was the bleaching agent (20)! Section 4 discusses the use of THP in beaching, a topic that led us into new organo-phosphorus chemistry (Section 5). Then, during attempts to effect catalytic homogeneous hydrogenation under mild conditions, cheapness was ignored, and THP complexes of Rh (the hydrogenation metal 'par excellence') were tested, and these studies have revealed interesting inorganic aspects of THP chemistry (Section 6).

Bleaching Activity of Phosphines Containing Hydroxymethyl Groups

The water-soluble $P(CH_2OH)_3$ (THP) was found to have a bleaching power (for spruce TMP) similar to that of the industrially used $S_2O_4^{2-}$ at similar loadings (see Fig. 1.5 below), but has advantages in that it can be used over wider ranges of pH (~4–9), temperature (20–130°C) and consistency (1.5–20%), without the need for removal of O_2 or transition metals (20). Such tolerance is unprecedented in pulp bleaching where it is difficult to control, for example in hydrosulfite bleaching, the optimal pH of 5.5–6.5 because of pH changes caused by hydrolysis or oxidation of $S_2O_4^{2-}$, and because of increasing recycling of mill process water with high contaminant levels. Additionally, THP can be readily formed in situ from the air-stable tetrakis(hydroxymethyl)phosphonium chloride, $[P(CH_2OH)_4]Cl$, or the sulfate, by addition of base to the pulp slurry (eq. 2). These water-soluble salts are available commercially in >3000 tonnes/year, and are used to make flame-retardants, and biocides for oil and waste-water industries (21). The presence of a hydroxyalkyl

$$[P(CH_2OH)_4]Cl + base \rightarrow P(CH_2OH)_3 + HCHO + base \cdot HCl \qquad (2)$$

substituent at the P-atom is essential for bleaching activity (20). A further benefit is that THP-bleached pulps show higher heat-, moisture- and light-stability than those bleached with $S_2O_4^{2-}$ or peroxide; i.e., THP is a brightness stabilizing agent (20), a finding compatible with a report that $[P(CH_2OH)_4]Cl$ prevents yellowing of fluorescently whitened wool in simulated sunlight (22).

Analysis of the phosphorus introduced into the pulp and pulp waste water for different doses of added bleaching agent (23) shows that the bleaching results from organo-phosphorus moieties formed in the pulp, the brightness gain correlating with the amount of covalently bonded-P (Fig. 1.4). Qualitative UV-visible data on LMCs, and mill wood lignin isolated from the bleached pulp, suggested that the phosphorus interacts with coniferaldehyde (see Fig. 1.1) and *p*-quinone chromophores (23). This has been confirmed by more detailed NMR and crystallographic studies on LMCs that include these chromophore units (Section 5).

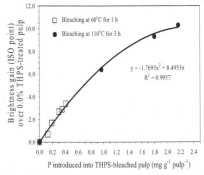

Figure 1.4. Brightness gain over the control (0.0% P-treated spruce TMP) *vs.* P found in the bleached pulp; $[P(CH_2OH)_4]_2SO_4$ (THPS) was used as bleaching agent.

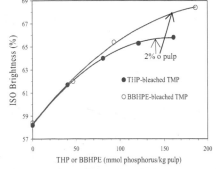

Figure 1.5. Relative bleaching powers of BBHPE and THP for a thermomechanical pulp.

A commercial development using THPS/THP for bleaching pulps has resulted from the finding that selected pulps treated with a mixture $S_2O_4^{2-}$ and THPS give higher brightness than can be achieved with either reagent alone, i.e., there is synergistic bleaching using these reagents (24, 25). A successful application in a mill, which led to a reduced dosage of brightness-enhancing additives, required no capital expenditure: stainless steel tubing was simply welded to the in-use $S_2O_4^{2-}$ addition line, with the THPS solution being added via Nylon tubing attached to the steel tubing at a rate controlled via a small pump (25). Of importance, the generated formaldehyde (see eq. 2), which is a respiratory irritant, reacts with bisulfite (a by-product of hydrosulfite bleaching) and is removed mainly as the sulfonate, $Na(O_3SCH_2OH)$. It should be noted that the chemistries of $S_2O_4^{2-}$ and of THP are complex (depending especially on pH and temperature), and the chemistry of their combination is even more so (26)!

1,2-Bis[bis(hydroxymethyl)phosphino]ethane, $(HOCH_2)_2P(CH_2)_2P(CH_2OH)_2$, abbreviated BBHPE, has a bleaching power greater than that of THP at the same P-loading (Fig. 1.5), and is almost as effective as alkaline peroxide. Testing in a 2-stage process revealed even more promise. Mechanical pulps are often bleached via a so-called *PY* process, the *P* and *Y* identifying, respectively, with successive, individual treatments with peroxide and hydrosulfite; use of BBHPE in place of *Y* was more effective *and* also allowed for a 70% reduction in the amount of *P* needed to reach a required brightness (27). Thus, BBHPE shows promise as a complementary bleaching agent to peroxide. The diphosphine is also a brightness stabilizing agent (28) and, of at least equal interest, BBHPE is significantly more effective than hydrosulfite for bleaching blue-stained pine TMP to acceptable brightness values (27). This is important because the mountain pine beetle epidemic is a serious threat to British Columbia's $2 billion/year mechanical pulp and paper sector (29).

Reported syntheses (30, 31) for the (hydroxymethyl)phosphines and phosphon-ium salts are outlined in Figure 1.6. We have extended the hydroxymethyl-P theme, by synthesizing new oligophosphines **1** and the corresponding tetra- and hexa-phosphines, and the corresponding phosphonium salts such as **2** (see Section 5 for the structures) (32). Treatment of the salts with base (*cf.* eq. 2) should yield 'poly-THP' compounds, which are expected to be powerful metal sequestering agents - THP has been used to remove trace Ru species from polymers synthesized by Grubbs-type Ru-carbene catalysts (33).

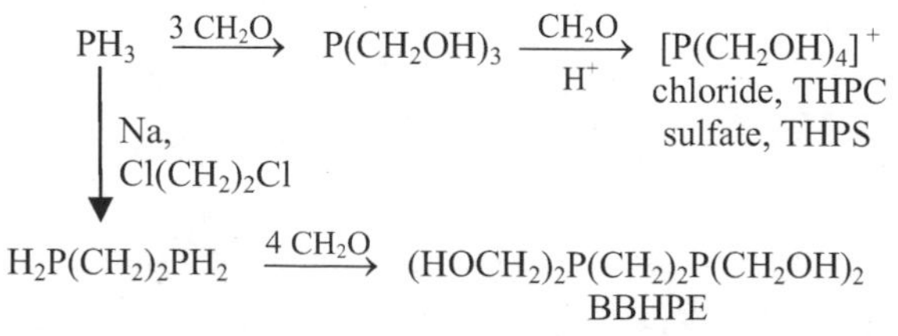

$$PH_3 \xrightarrow{3\ CH_2O} P(CH_2OH)_3 \xrightarrow[H^+]{CH_2O} [P(CH_2OH)_4]^+ \quad \text{chloride, THPC} \quad \text{sulfate, THPS}$$

$$\downarrow \begin{array}{c} Na, \\ Cl(CH_2)_2Cl \end{array}$$

$$H_2P(CH_2)_2PH_2 \xrightarrow{4\ CH_2O} (HOCH_2)_2P(CH_2)_2P(CH_2OH)_2 \quad \text{BBHPE}$$

Figure 1.6. Syntheses of THP, phosphonium salts, and BBHPE.

Organo-phosphorus Chemistry

To obtain a better understanding of the bleaching chemistry, studies were carried out on the interaction (in aqueous media, when possible) of (hydroxymethyl)-phosphines with LMCs, especially substituted aromatic moieties found in lignin. The (hydroxypropyl)phosphine $P[(CH_2)_3OH]_3$, which is also a pulp bleaching agent (20), was mostly used because it gives cleaner reactions than does THP. Although THP is usually described as moderately air-stable in the solid state and in aqueous solution (19, 31), our studies reveal more specifically that there is no oxidation at 25°C over 24 h at pH 7, but at pH 10, THP is oxidized ($t_{1/2} \sim 30$ min) by OH⁻ to the oxide with H_2 being the co-product (26, 34). At lower pH, THP slowly decomposes with loss of CH_2O according to eq. 3 (26, 35).

$$P(CH_2OH)_3 \rightarrow HP(CH_2OH)_2 + HCHO \qquad (3)$$

The relatively basic (hydroxyalkyl)phosphines act toward LMCs as reductants and, compatible with this, also as strong nucleophiles. We have studied such reactions in aqueous and D_2O solutions by ^{31}P-, ^1H-, and ^{13}C-NMR spectroscopies (including 2D correlation methods), product isolation and, when possible, X-ray analysis of isolated compounds or their derivatives. Thus, aromatic aldehyde moieties present in lignin (e.g., **3**) are reduced to the corresponding alcohols (see **4**) with co-production of the phosphine oxide; in D_2O, –CH(D)OD is formed selectively (36). The mechanism proceeds via a phosphonium species formed by initial nucleophilic attack of the P-atom at the carbonyl C-atom, i.e., via ArCH(OH)P⁺R₃, where Ar is the aromatic residue and R is the hydroxyalkyl substituent (36). When the aldehyde contains a 4-OH substituent, the alcohol product

subsequently reacts with the phosphine more readily than the aldehyde precursor to give a phosphobetaine; in D_2O, exchange of both benzyl protons of the phosphobetaine occurs (e.g., to give **5**) via phosphorus ylide intermediates, which can also lead to exchange of the R-CH₂ protons adjacent to the P-atom (36). Reduction of aromatic ketones is more sluggish because the carbonyl C-atom is less electrophilic than that present in an aromatic aldehyde. There is extensive literature on reactivity of phosphines toward carbonyl-containing organics, but the observed redox reaction is new because our reactions are carried out in water where OH⁻ plays a key role in forming the phosphonium intermediate and the phosphine oxide co-product (36). Any such aldehyde reduction or phosphobetaine formation decreases the extent of conjugation, and could presumably contribute to the bleaching of pulp.

Reaction in aqueous solution of $P[(CH_2)_3OH]_3$ with cinnamaldehyde (**6**), which has the phenyl-unsaturated propanoid backbone found in lignin, revealed self-condensation of **6** to the isomers **7** and **8**. The reactions occur via initial nucleophilic attack of the phosphine at the γ-carbon; in D_2O, all but the Ph and -CHO protons in the products are exchanged by D (37). NMR data are consistent with formation of a carbanion and subsequent condensation of two phosphonium-containing aldehydes to generate the products with concomitant elimination of phosphine oxide, in which available OH⁻ provides the O-atom (37). In aqueous DCl, the phosphine now attacks the aldehyde C-atom to form first **9** and then the bis(phosphonium) salt **10**; species **9** and **10**, although having up to 3 chiral carbon centers, are formed with high stereoselectivity just in enantiomeric forms, which is attributed to electrostatic interactions between a phosphonium center and a carbanion lone-pair. When the phenyl group of **6** has a *p*-OH substituent, different chemistry is seen, resulting from deprotonation of the OH-substituent. The substrates studied were the lignin components, coumaraldehyde (**11**, X = Y = H), coniferaldehyde (**11**, X = OMe, Y = H, *cf.* Fig. 1), and sinapaldehyde (**11**, X = Y = OMe) (38). Initial phosphine attack is again at the γ-carbon, but the product is the zwitterionic **12**, and this undergoes self-condensation to give a zwitterionic di(phosphonium) diphenolate that can react with HCl to give the dichloride **13**. All the intermediates were well characterized, and X-ray structures were obtained for the PEt$_3$ and PMe$_3$ derivatives of **13**; in these, the P-atoms bonded to the chiral C-centers are prochiral, containing magnetically inequivalent, diastereotopic Et/Me groups, and the resulting 'exotic' NMR spectra were successfully interpreted (38). Again, there is extensive literature describing reactivity of phosphines with α,β-unsaturated carbonyl compounds, but the aqueous medium leads to novel chemistry seen with $P[(CH_2)_3OH]_3$ (37, 38). A phosphobetaine of type **14** is formed from *p*-quinones with OMe and/or Me substituents (26), the essential chemistry of which has long been known (37). Increased understanding of organo-phosphorus chemistry is also important within metal-phosphine catalyzed hydroformylation processes where aldehydes are formed and reactions, for example, between α,β-unsaturated aldehydes and phosphines have been documented (39).

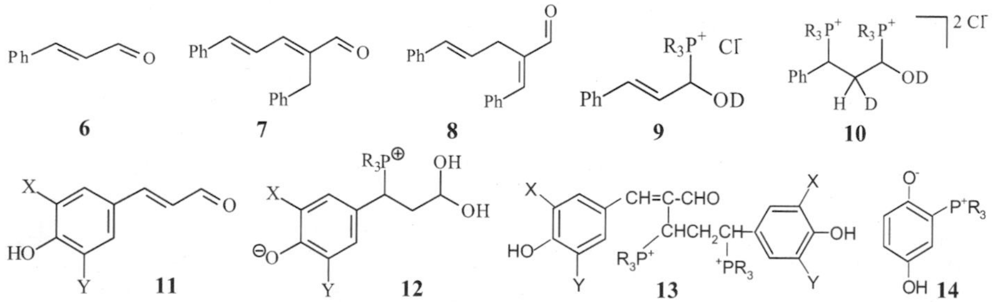

The synergic bleaching effect described in Section 4 using a combination of $S_2O_4^{2-}$ with THPS/THP led us to study the interaction between $S_2O_4^{2-}$ and the P-chemicals at ambient conditions (26). As noted in Section 4, use of THPS generates CH_2O, which is removed as $Na(O_3SCH_2OH)$; this reaction also generates, via eq. 3, bis(hydroxylmethyl)phosphine, $HP(CH_2OH)_2$. This enforced us to investigate reactions of secondary phosphines with LMCs such as cinnamaldehydes. A 1:1 reaction of Ph$_2$PH gives hydrophosphination of the olefinic bond to give new tertiary phosphines of type **15** as a racemic mixture, while a 2:1 reaction shows hydro-phosphination of both the C=C and C=O bonds to give the new diphosphine **16** as a diastereomeric mixture with a diastereomeric ratio of ~2.3 (eq. 4, R = H, Me) (40). Both reactions lead to loss of conjugation in α,β-unsaturated carbonyl compounds, and could play a role in lignin bleaching. The new phosphines **15** and **16** show rich coordination chemistry that can include oxidative addition of the aldehyde moiety at Rh and Ir to give hydrido-carbonyl complexes (41).

$$\text{PhCH=C(R)CHO} \xrightarrow{\text{Ph}_2\text{PH}} \underset{\textbf{15}}{\text{PhCH(PPh}_2\text{)C(H)RCHO}} \xrightarrow{\text{Ph}_2\text{PH}} \underset{\textbf{16}}{\text{PhCH(PPh}_2\text{)C(H)RCH(PPh}_2\text{)OH}} \quad (4)$$

The above organo-phosphorus chemistry exemplifies how bleaching activity likely occurs by loss of conjugation, and the less conclusive data on lignin itself lead to the same conclusion (Section 4). The hydrogenations of C=C and C=O functionalities of the LMCs in the Ru-THP/H$_2$ catalytic systems (Section 3) similarly lead to decreased conjugation; whether the pulp bleaching by these systems (Fig. 2), involves any catalytic hydrogenation process is an interesting question that requires further study, particularly using ^{1}H- and ^{31}P-NMR data from experiments with mill-wood lignin.

There is a possibility that (hydroxymethyl)phosphines might be catalyzing hydration of activated olefinic moieties in lignin. The Michael addition reaction shown in eq. (6a) is catalyzed by 5% THP in water at ambient conditions, with 70% conversion of the acrylonitrile; no such reaction is seen with acrylic acid or the methyl ester, but analogous hydromethoxylation of these compounds is seen in MeOH (42) (eq. (6b), R = H or Me). There is a report on similar catalytic use of trialkylphosphines, which, like THP, are strong nucleophiles (43).

(a) CH$_2$=CHCN → (HO)CH$_2$CH$_2$CN ; (b) CH$_2$=CHCO$_2$R → (MeO)CH$_2$CH$_2$CO$_2$R (6)

Tris(hydroxymethyl)phosphine (THP) Complexes of the Platinum Metals

Although THP has been known for 50 years (44), developments in its coordination chemistry are relatively recent, perhaps because of complicated 'side effects' of the compound (see 1st paragraph of Section 5). Water-soluble phosphine complexes are of interest for potential application in the areas of aqueous or aqueous/organic 2-phase catalysis (*cf.* the Ruhrchemie/Rhône-Poulenc propylene hydroformylation process (45)), and biomedical applications using water-soluble drugs (46). Our interest in catalytic Ru-THP species was discussed in Section 3. We were unable to identify the catalyst precursor species, although others have isolated the structurally characterized Ru(THP)$_2$[PH(CH$_2$OH)$_2$]$_2$Cl$_2$ via the chemistry outlined in eq. 1 (47, 48); note here that two of the THP groups have eliminated formaldehyde (eq. 3), and this complex does effect catalytic hydrogenation of CO$_2$ (48). A structurally characterized dimer, [Ru(μ-PCH$_2$OH){μ-*P,O*-P(CH$_2$O)(CH$_2$OH)$_2$}Cl$_2$]$_2$, isolated from a RuCl$_3$/THP reaction (49), shows extensive THP decomposition, and contains bridged phosphido groups and bridged monoalkoxides Water-soluble Ru complexes with intact THP ligands are known within species containing ancillary arene (50) and Cp ligands (51), and the latter are effective hydrogenation catalyst precursors for olefinic acids (51).

We have elucidated some interesting chemistry of Rh-THP systems during attempts to devise some water-soluble hydrogenation catalysts. As a follow-up to our Rh-colloid catalysis work (Section 2), various in situ RhCl$_3$·3H$_2$O/Rh^I(diene)-THP systems were tested in aqueous or EtOH solution as hydrogenation catalysts for olefinic bonds in LMCs, but Rh metal precipitated when the substrate had been completely hydrogenated. However, addition of 1 mole ethylenediamine (en) to solutions containing a 1:2 ratio of RhCl$_3$:THP gave marked improvement in activity and prevented precipitation of metal. The structurally characterized complex *trans*-[RhCl$_2$(en)(THP)$_2$]Cl was isolated, and when dissolved in aqueous solution regenerated the catalytically active system, presumed to be a Rh^I species (52). In efforts to elucidate the nature of the Rh^{III}/en/H$_2$ chemistry, we synthesized definitely RhCl(cod)(THP), [Rh(cod)(THP)$_2$]Cl, RhCl(THP)$_3$ and *trans*-RhCl(CO)(THP)$_2$, and tentatively RhCl(en)(THP), but the catalysis has not been simulated using the en species (52). Earlier literature has reported isolation of *fac*-RhCl$_3$(THP)$_3$ (53, 54), possibly *trans*-RhCl(CO)(THP)$_2$ (53, 55), and in situ synthesis of *cis*-[Rh(H)$_2$-(THP)$_4$]$^+$, although the source of the hydride ligands is uncertain (56, 57). The *fac*-RhCl$_3$(THP)$_3$ complex and in situ Rh^I/THP species are reported to be ineffective olefin hydrogenation catalysts (53), but [Ir(cod)(THP)$_3$]Cl is reported to catalyze hydrogenation of cinnamaldehyde (**6**) to give mainly the unsaturated alcohol (56).

Attempts to synthesize mixed THP/PR$_3$ complexes of Rh via reaction of RhCl(cod)(THP) with PR$_3$ (R = alkyl/aryl) under Ar revealed instead a THP-promoted P–C bond cleavage of the PR$_3$ to generate *trans*- and *cis*-RhCl(PR$_3$)[*P,P*-R$_2$POCH$_2$P(CH$_2$OH)$_2$] containing a chelated *P,P*-phosphine-phosphinite ligand (58). One R group has been converted to hydrocarbon co-product (RH), the hydrogen coming from a THP-hydroxyl group that becomes the alkoxy group at the residual PR$_2$ moiety (Fig. 1.7). Cleavage of a P–C bond concurrent with formation of a P–O bond is rare (58). A related example to our work is formation of [M{*P,P*-(HOCH$_2$)$_2$POCH$_2$P(CH$_2$OH)$_2$}$_2$]Cl$_2$ from reaction of *cis*-MCl$_2$(THP)$_2$ (M = Pt, Pd) with THP, a non-mixed phosphine system (59). Cleavage of P–C bonds within

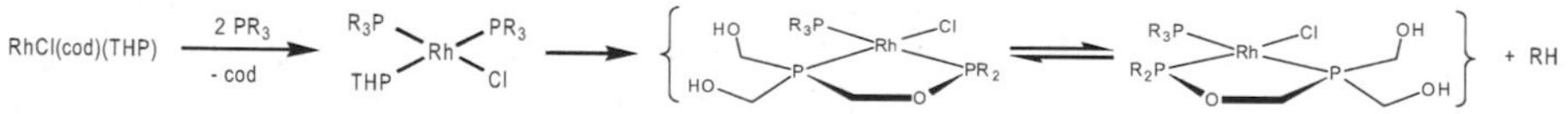

Figure 1.7. Synthesis of *trans*- and *cis*-RhCl(PR$_3$)[*P,P*-R$_2$POCH$_2$P(CH$_2$OH)$_2$].

metal-phosphine systems can be critical in determining catalytic activity in, for example, homogeneous hydroformylation and hydrogenation conditions (60). Reactions of RhCl(cod)(THP) with PRPh$_2$ (R = Me or cyclohexyl) under H$_2$ in acetone/MeOH generate in high yield crystals of *cis,mer*-Rh(H)$_2$Cl(PRPh$_2$)$_3$, while the more obvious route to the dihydride using [RhCl(cod)]$_2$ as precursor under identical conditions does not precipitate the dihydride (61). It is possible that THP promotes oxidative addition of H_2 to a RhCl(PRPh$_2$)$_2$(THP) intermediate (*cf.* Fig. 1.7), with subsequent displacement of the THP by PRPh$_2$. However, free THP is not detected as co-product, and the observation of what appears to be the presence a 2-phase solvent system where the dihydride crystals form suggests other factors are involved. An important, unanswered question in the syntheses is the fate of the THP; as discussed in Section 5, it is a reducing reagent, can lose CH_2O to give HP(CH$_2$OH)$_2$ and, as noted by others (59), can readily form phosphonium salts. Formation of trace amounts of *trans*-RhCl(CO)(PR$_3$)$_2$ via decarbonylation of the CH_2O has been seen in systems outlined in Figure 7 (58), as well as in the systems that generate the dihydrides (61, 62). A phenomenon perhaps related to formation of the dihydrides is the reproducible and remarkable observation of the immediate precipitation of crystals of RhCl(P-N)(THP) from an acetone solution of this complex (where P-N is *P,N*-chelated ligand *o*-PPh$_2$(C$_6$H$_4$NMe$_2$), and THP is trans to the N-donor), when the solution is subjected to an atmosphere of H_2, even though there is no evidence for formation of any hydrido species (63)!

Whether THP-promoted reactions occur generally within other platinum metal systems, besides those of Ru, Rh, and *cis*-MCl$_2$(THP)$_2$ (M = Pt, Pd) mentioned above, remains to be investigated. Within the Ni subgroup metal systems, the complexes M(THP)$_4$ [M = Ni, Pd, or Pt (31)], and *cis*-PtR$_2$(THP)$_2$ [R = alkyl or aryl (59, 64)] are also known.

Conclusions

This review describes on-going research that was initially aimed at decreasing the degree of conjugation in lignin for the purpose of inhibiting the yellowing of mechanical pulps. Catalytic hydrogenation of aromatic residues using water-soluble, homogeneous transition metal species was the chosen approach. A wide range of metal complexes was tested, and this eventually led to the serendipitous finding that water-soluble, (hydroxymethyl)phosphines in the absence of metals or H_2 were effective bleaching agents and brightness stabilizers for pulps. Such phosphines represent a new class of bleaching agents, and P(CH$_2$OH)$_3$ (THP), which can be readily generated from commercially available salts of [P(CH$_2$OH)$_4$]$^+$, is competitive with the currently used bleaching agent 'hydrosulfite' (Na$_2$S$_2$O$_4$). Indeed, THP in conjunction with Na$_2$S$_2$O$_4$ has found industrial use at least in one pulp mill. Investigations into the 'bleaching chemistry' have led to the discovery of new types of organo-phosphorus compounds, while continuing interest in the metal-catalyzed hydrogenations has revealed novel chemistry of the 'non-innocent' THP both in its free and coordinated state.

Acknowledgements

I thank in particular my collaborator at Paprican, Thomas Q. Hu ('the pulp man'), and the contributions of many others (postdoctoral fellows, students, technicians, industrial colleagues, and supervisors of the crystallography, NMR and MS facilities at UBC); their names are in appropriate publications listed in the references. The

Natural Sciences and Engineering Research Council of Canada (NSERC) provided financial support via Mechanical and Chemi-mechanical Pulps Network, Strategic, and Discovery grants, while contributions from Paprican, Cytec of Canada, and Colonial Metals Inc. (for loans of $RuCl_3 \cdot 3H_2O$) are also acknowledged.

References

1. C. J. Biermann, Handbook of Pulping and Papermaking, Academic Press, San Diego, CA, 1996.
2. International Fact and Price Book, Pulp Paper Intern., Miller Freeman, Brussels, 2001, p.278.
3. J. R. Presley and R. T. Hill, *In* Pulp Bleaching–Principles and Practice (eds. C. W. Dence and D. W. Reeve), TAPPI PRESS, 1996, Ch. 1; M. E. Ellis, *In* Pulp Bleaching–Principles and Practice (eds. C. W. Dence and D. W. Reeve), TAPPI PRESS, 1996, Ch. 2.
4. S. W. Eachus and C. W. Dence, *Holzforschung*, **29(2)**, 41 (1975).
5. T. Q. Hu, B. R. James, and C.-L. Lee, *J. Pulp Pap. Sci.*, **23(4)**, J153 (1997); T. Q. Hu, B. R. James, and C.-L. Lee, *J. Pulp Pap. Sci.*, **23(5)**, J200 (1997); T. Q. Hu, B. R. James, S. J. Rettig, and C.-L. Lee, *Can. J. Chem.*, **75**, 1234 (1997); T. Y. H. Wong, R. Pratt, C. G. Leong, B. R. James, and T. Q. Hu, Chemical Industries (Dekker), **82**, (*Catal. Org. React.*), 255 (2001).
6. B. R. James, Y. Wang, and T. Q. Hu, Chemical Industries (Dekker), **68**, (*Catal. Org. React.*), 423 (1996); T. Q. Hu, B. R. James, and Y. Wang, *J. Pulp Pap. Sci.*, **25(9)**, J312 (1999).
7. B. R. James, Y. Wang, C. S. Alexander, and T. Q. Hu, Chemical Industries (Dekker), **75**, (*Catal. Org. React.*), 233 (1998); T. Q. Hu and B. R. James, *J. Pulp Pap. Sci.*, **26(5)**, 173 (2000).
8. M. A. Bennett, T.-N. Huang, and T. W. Turney, *J. Chem. Soc. Chem. Commun.*, 312 (1979); L. Plasseraud and G. Süss-Fink, *J. Organomet. Chem.*, **539**, 163 (1997); G. Süss-Fink, M. Faure, and T. R. Ward, *Angew. Chem. Int. Ed.*, **41**, 99 (2002).
9. J. A. Widegren, M. A. Bennett, and R. G. Finke, *J. Am. Chem. Soc.*, **125**, 10301 (2003); L. Vieille-Petit, G. Süss-Fink, B. Therrien, T. R. Ward, H. Stoeckly-Evans, G. Labat, L. Karmazin-Brelot, A. Neels, T. Bürgi, R. G. Finke, and C. M. Hagen, *Organometallics*, **24**, 6104 (2005).
10. G. J. Leary, *J. Pulp Pap. Sci.*, **20(6)**, J154 (1999).
11. T. Q. Hu, G. R. Cairns, and B. R. James, *Proc. 11th Intern. Symp. Wood/Pulp Chem.*, **1**, 219 (2001).
12. T. Q. Hu, G. R. Cairns, and B. R. James, *Holzforschung*, **54**, 127 (2000).
13. T. Q. Hu and B. R. James, *In* Chemical Modification, Properties and Usage of Lignin (ed. T. Q. Hu), Kluwer Academic/Plenum, New York, 2002, p.247; T. Q. Hu, B. R. James, T. H. Y. Wong, and A. Z. Lu, *Patent Disclosure, C198*, Paprican, Feb. 2000.
14. PAPTAC (Pulp and Paper Technical Association of Canada) Standard Testing Method E.1 (1990).
15. M. B. Ezhova, A. Z. Lu, B. R. James, and T. Q. Hu, Chemical Industries (Dekker), **104**, (*Catal. Org. React.*), 135 (2005).

16. J. March, Advanced Organic Chemistry, 2nd ed., McGraw-Hill, New York, 1977, p.1119; W. F. Maier, K. Bergmann, W. Bleicher, and P. V. R. Schleyer, *Tetrahedron Lett.*, **22**, 4227 (1981); H. Zahalka and H. Alper, *Organometallics*, **5**, 1909 (1986).
17. H. Mehdi, V. Fábos, R. Tuba, L. T. Mike, and I. T. Horváth, *Topics in Catal.*, in press.
18. G. Ma, B. R. James, and T. Q. Hu, unpublished data.
19. D. S. Bharathi, M. A. Sridhar, J. S. Prasad, and A. G. Samuelson, *Inorg. Chem. Commun.*, **4**, 490 (2001).
20. T. Q. Hu, B. R. James, D. Yawalata, and M. B. Ezhova, *J. Pulp Pap. Sci.*, **30(8)**, 233 (2004); *US Patent* 7,285,181 B2, Oct. 2007; *European Patent EP 1 590 525 B1*, Oct. 2006; *PCT/WO 2004/070110 A1*, Aug. 2004.
21. T. K. Haack, B. Downward, and B. Talbot, *Proc. 1997 TAPPI Engin. Papermakers Conf.* p.1115; T. A. Calamari and R. J. Harper, *In* Flame-Retardants for Textiles, Kirk-Othmer Encyclopedia of Chemical Technology, 4th ed., 2000, Vol. 10, p.998.
22. L. A. Holt, B. Milligan, and L. J. Wolfram, *Textile Res. J.*, **44(11)**, 846 (1974).
23. T. Q. Hu, E. Yu, B. R. James, and P. Marcazzan, *Holzforschung*, in press; R. Chandra, T. Q. Hu, B. R. James, M. B. Ezhova, and D. V. Moiseev, *J. Pulp Pap. Sci.*, **33(1)**, 15 (2007).
24. T. Q. Hu, B. R. James, T. Williams, J. A. Schmidt, and D. V. Moiseev, *PCT/WO 2007/016769 A1*, Feb. 2007.
25. T. Q. Hu, T. Williams, J. Schmidt, B. R. James, R. Cavasin, and D. Lewing, *Pulp and Paper Canada*, submitted.
26. D. V. Moiseev, B. R. James, and T. Q. Hu, unpublished data.
27. T. Q. Hu, B. R. James, D. Yawalata, and M. B. Ezhova, *J. Pulp Pap. Sci.,* **31(2)**, 69 (2005).
28. T. Q. Hu, B. R. James, D. Yawalata, and M. B. Ezhova, *J. Pulp Pap. Sci.,* **32(3)**, 131 (2006).
29. P. Watson, *Pulp and Paper Canada*, **107(5)**, 12 (2006).
30. R. I. Wagner, *US Patent* 3,086,053 (1963); A. W. Frank, D. J. Daigle, and S. L. Vail, *Textile Res. J.* **52(11)**, 678 (1982); V. S. Reddy, K. V. Katti, and C. L. Barnes, *Inorg. Chim. Acta* **240**, 367 (1995).
31. J. W. Ellis, K. N. Harrison, P. A. T. Hoye, A. G. Orpen, P. G. Pringle, and M. B. Smith, *Inorg. Chem.*, **31**, 3026 (1992).
32. D. V. Moiseev, B. R. James, B. O. Patrick, and T. Q. Hu, *Inorg. Chem.*, **45**, 2917 (2006).
33. H. D. Maynard and R. H. Grubbs, *Tetrahedron Lett.*, **40**, 4137 (1999).
34. S. M. Bloom, S. A. Buckler, R. F. Lambert, and E. V. Merry, *J. Chem. Soc. D*, 870 (1970).
35. W. J. Vullo, *Ind. Eng. Chem. Prod. Res. Dev.*, **5**, 346 (1966); L. Maier, *Helv. Chim. Acta*, **54**, 1434 (1971).
36. D. V. Moiseev, B. R. James, and T. Q. Hu, *Inorg. Chem.*, **45**, 10338 (2006).
37. D. V. Moiseev, B. R. James, and T. Q. Hu, *Inorg. Chem.*, **46**, 4704 (2007).
38. D. V. Moiseev, B. O. Patrick, B. R. James, and T. Q. Hu, *Inorg. Chem.*, **46**, 9389 (2007).

39. D. R. Bryant, *In* Catalyst Separation, Recovery and Recycling (eds. D. J. Cole-Hamilton, R. Tooze), Springer, Dordrecht, 2006, Ch. 2.
40. D. V. Moiseev, B. O. Patrick, and B. R. James, *Inorg. Chem.*, **46**, 11467 (2007).
41. D. V. Moiseev, F. Lorenzini, and B. R. James, unpublished data.
42. M. B. Ezhova, B. R. James, and T. Q. Hu, unpublished data.
43. I. C. Stewart, R. G. Bergman, and F. D. Toste, *J. Am. Chem. Soc.*, **125**, 8696 (2003).
44. M. Reuter and L. Orthner, *Ger. Patent*, 1035135 (1958).
45. E. Wiebus and B. Cornils, *In* Catalyst Separation, Recovery and Recycling (eds. D. J. Cole-Hamilton, R. Tooze), Springer, Dordrecht, 2006, Ch. 5.
46. K. V. Katti, H. Gali, C. J. Smith, and D. E. Berning, *Acc. Chem. Res.*, **32**, 9 (1999); N. Pillarsetty, K. K. Katti, T. J. Hoffman, W. A. Volkert, K. V. Katti, H. Kamei, and T. Koide, *J. Med. Chem.*, **46**(**7**), 1130 (2003).
47. L. Higham and M. K. Whittlesey, *J. Chem. Soc. Chem. Commun.*, 1107 (1998).
48. Y. Kayaki, T. Suzuki, and T. Ikariya, *Chem. Lett.*, **30**, 1016 (2001).
49. A. Romeroso, T. Campos-Malpartida, M. Serrano-Ruiz, and M. Peruzzini, *Proc. Dalton Discuss. 6, Organomet. Chem. and Catal.*, York, UK, 2003, Poster 15.
50. V. Cadierno, P. Crochet, S. E. García-Garrido, and J. Gimeno, *J. Chem. Soc. Dalton Trans.*, 3635 (2004); J. Čubrilo, I. Hartenbach, T. Schleid, and R. F. Winter, *Z. Anorg. Allg. Chem.*, **632**, 400 (2006).
51. B. Drießen-Hölscher and J. Heinen, *J. Organometal. Chem.*, **570**, 141 (1998)
52. F. Lorenzini, B.O. Patrick, and B.R. James, *J. Chem. Soc. Dalton Trans.*, 3224 (2007).
53. J. Chatt, G. J. Leigh, and R. M. Slade, *J. Chem. Soc. Dalton Trans.*, 2021 (1973).
54. K. Raghuraman, N. Pillarsetty, W. A. Volkert, C. Barnes, S. Jurisson, and K. V. Katti, *J. Am. Chem. Soc.*, **124**, 7276 (2002).
55. F. P. Pruchnik, P. Smolenski, and I. Raksa, *Pol. J. Chem.*, **69**, 5 (1995).
56. A. Fukuoka, W. Kosugi, F. Morishita, M. Hirano, S. Komiya, L. McCaffrey, and W. Henderson, *Chem. Commun.*, 489 (1999).
57. L. J. Higham, M. K. Whittlesey, and P. T. Wood, *J. Chem. Soc., Dalton Trans.*, 4202 (2004).
58. F. Lorenzini, B.O. Patrick, and B.R. James, *Inorg. Chem.*, **46**, 8998 (2007).
59. P. A. T. Hoye, P. G. Pringle, M. B. Smith, and K. Worboys, *J. Chem. Soc., Dalton Trans.*, 269 (1993).
60. P. E. Garrou, *Chem. Rev.*, **85**, 171 (1985).
61. F. Lorenzini, B. O. Patrick, and B. R. James, *Inorg. Chim. Acta,* in press (doi:10.1016/j.ica.2007.10.044).
62. F. Lorenzini, B. O. Patrick, and B. R. James, *Acta Cryst.*, **E64**, m179 (2008); F. Lorenzini, B. O. Patrick, and B. R. James, *Acta Cryst.*, **E64**, m464 (2008).
63. F. Lorenzini, B.O. Patrick, and B.R. James, *Inorg. Chim. Acta,* in press (doi: 10.1016/j.ica.2007.11.017; R. J. Angelici issue).
64. S. Komiya, M. Ikuine, N. Komine, and M. Hirano, *Bull. Chem. Soc. Jpn.*, **76**, 183 (2003).

2. Design and Serendipity in the Discovery and Development of Homogeneous Catalysts for Organic Synthesis

John F. Hartwig

Department of Chemistry, University of Illinois, 600 South Mathews Ave
Urbana, IL 61801

jhartwig@uiuc.edu

Abstract

The origins of several catalysts and catalytic reactions that have become useful for organic synthesis are presented. The progression toward catalysts for selective alkane functionalization, coupling of aryl halides with amines, coupling of aryl halides with enolates, and iridium-catalyzed asymmetric allylic substitution are discussed. The role in identifying catalyst structures and the reactivity of catalytic intermediates to guide the design of improved catalysts is presented.

Introduction

My group has sought to expand the scope of organic transformations catalyzed by organometallic complexes to those that form carbon-heteroatom bonds. During the course of this work, we have also developed related reactions of enolate nucleophiles in which the carbon of the nucleophile shares some similarities in metal-ligand bonding to the nitrogen in amido complexes. In parallel with the discovery and development of these catalytic processes, we have sought to develop a mechanistic understanding of these processes and to use mechanistic data to determine the relationships between metal-ligand bonding and the reactivity of intermediates in these catalytic reactions.

The Rylander Award lecture focused on two classes of catalytic reactions that reveal how the bonding of the heteroatom to the metal influences the reactivity of the catalytic intermediates. One ultimate goal for developing these fundamental principles is to be able to design catalysts for a particular transformation. Considering that the scientific material in this lecture is contained in the primary literature and several recent review articles,[1-6] it seems more appropriate to present a perspective on the work presented, rather than an additional review or selection of original work. In this paper, I have chosen to assess the progress of my group and the field of homogeneous catalysis toward an ability to design homogeneous catalysts for organic synthesis.

In the context of a Rylander Award lecture, it seemed most informative to discuss the discovery phases of the reactions discussed at the meeting and the discovery phases of other reactions developed in my group. In some cases, these reactions were designed from stoichiometric organometallic chemistry. In other cases, new stoichiometric organometallic chemistry was deduced from catalytic chemistry developed from a more classical screening. However, the most synthetically viable chemistry we have developed has in each case stemmed from a mechanistic insight before or after observing the first examples of the catalytic transformation. A brief account of the balance of design, serendipity, and broad screening is described in this paper.

Results and Discussion

Initial serendipity and long-term design leads to regioselective alkane functionalization.

The development of catalytic terminal functionalization of alkanes was not originally a planned target for our work on the reactions of transition metal-boryl complexes, but it quickly became a target when we serendipitously observed the reaction of a simple metal-boryl complex with an arene to form an arylboronate ester.[7] This original observation is shown as equation 1. This equation was originally conducted to homolyze the metal-boron bond to form a two-coordinate boryl radical. Once this reaction was discovered, three sequential steps were taken to develop the catalytic borylation of alkanes. First, a system to functionalize alkanes stoichiometrically was devised by eliminating the aromatic C-H bonds in the cyclopentadienyl ligand and the catecholate group and by using a third-row metal. The result of this design implemented by Karen Waltz is shown in equation 2.[8] To develop this stoichiometric reaction into a catalytic reaction, we used diborane(4) reagents to form the metal-boryl complexes and photochemistry to generate the requisite open coordination sites. This logic led Dr. Huiyan Chen to develop a rhenium catalyst for the borylation of alkanes under photochemical conditions.[9]

$$\text{(1)}$$

$$\text{(2)}$$

The most recent catalysts that operate under thermal conditions were then based on the premise that a Cp*M fragment with ligands that dissociate under thermal conditions could be a catalyst for alkane borylation. After a brief study of Cp*IrH$_4$ and Cp*Ir(ethylene)$_2$, Dr. Chen studied related rhodium complexes. Ultimately, he proposed that the Cp*Rh(η^4-C$_6$Me$_6$) complex would dissociate C$_6$Me$_6$ as an innocent side product, and that Cp*Rh(Bpin)$_2$ from oxidative addition of pinBBpin (pin=pinacolate) would be the active catalyst. The overall catalytic process based on this mechanism is shown in equation 3.[10] Subsequent mechanistic studies have confirmed, at least in a general sense, this hypothesis.[11]

$$2 \quad \diagup\!\diagdown\!\diagup\!\diagdown + B_2pin_2 \xrightarrow[150\ °C]{5\%\ Cp^*Rh(C_6Me_6)} 2 \quad \diagup\!\diagdown\!\diagup\!\diagdown\!\diagup\!\diagdown Bpin + H_2 \tag{3}$$

88%

Thus, the initial result of the stoichiometric borylation of arenes was discovered serendipitously. However, the development of this initial observation into the catalytic borylation of alkane C-H bonds was largely based on the design of complexes for the stoichiometric functionalization of alkanes and then the catalytic functionalization of alkanes.

Catalyst design in the amination of aryl halides.

In contrast to the borylation of alkane C-H bonds, the coupling of aryl halides with amines was based on a literature precedent from another group published about a decade before our initial studies. Kosugi, Kameyama and Migita published the coupling of aryl halides with tin amides.[12] Mechanistic studies we conducted on this process led us to the perhaps obvious realization that the reaction[13] could be conducted with amines and a silylamide base instead of tin amides (equation 4).[14] Surveys of bases with similar pK_a values led Janis Louie to conduct reactions with alkoxide bases. Similar studies were conducted at nearly the same time by Steve Buchwald and coworkers.[15]

$$\text{(4)}$$

Later studies by Michael Driver showed that chelating ligands could improve the scope of this chemistry by inhibiting β-hydrogen elimination.[16] Like the coupling of alkyl groups,[17,18] bidentate ligands discourage the opening of a coordination site necessary for β-hydrogen elimination to occur. Most recently, Qilong Shen studied the coupling of amines with aryl halides using ligands that are more strongly electron donating and sterically hindered than the original bidentate ligands used for C-N coupling.[19,20] These properties improve the rates for the coupling, allow mild couplings of chloroarenes, and lead to reactions that occur with low catalyst loadings (equations 5, 6). Thus, the coupling of amines with haloarenes is a process for which the contributions from my group have largely been based on findings that we revealed during our mechanistic studies of these reactions.

$$\text{(5)}$$

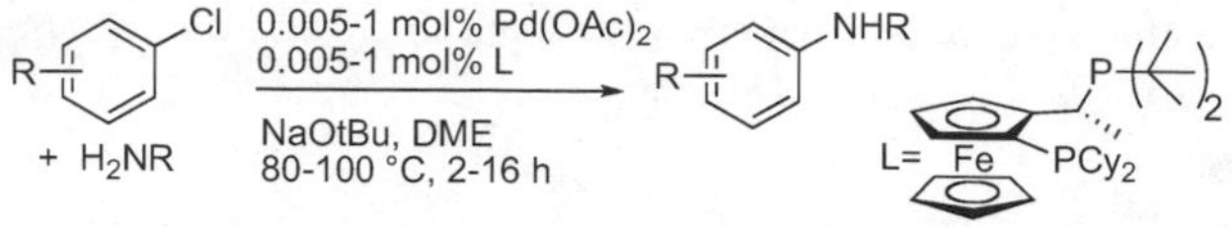

(6)

Serendipity before design in the α-arylation of carbonyl compounds.

However, a related process was discovered by pure accident, though further development was clearly guided by theories of mechanism and reagent development. While conducting the coupling of an aryl halide with an amine and base in acetone solvent, Dr. Blake Hamann observed phenylacetone as product (equation 7). This observation led us to realize that alkali metal enolates generated *in situ* could couple with haloarenes to form α-aryl carbonyl compounds. The original process[21] was not particularly broad in scope. The enolates from aryl ketones reacted, but the enolates of dialkyl ketones reacted in lower yields, and no product was observed from esters or amides. However, the use of sterically hindered alkyl monophosphines by Dr. Motoi Kawatsura led to reactions that occurred with broad scope for the series of carbonyl ligands shown in equation 4.[22] The catalysts were generated from a 1:1 ratio of monodentate ligand to metal. The use of this unconventional ratio of catalyst components was deduced by some preliminary kinetic studies[23,24] on the amination of aryl halides catalyzed by complexes of $PtBu_3$.[25] Thus, an initial chance observation has led to a long-standing project in our laboratory and a synthetically valuable set of processes to form α-aryl carbonyl compounds.[1]

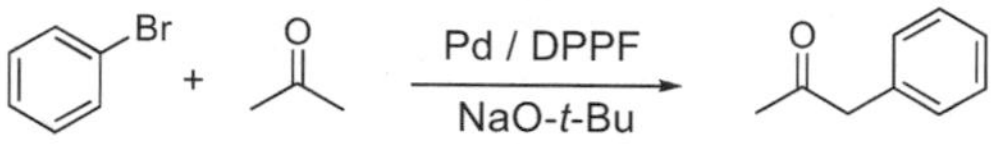

(7)

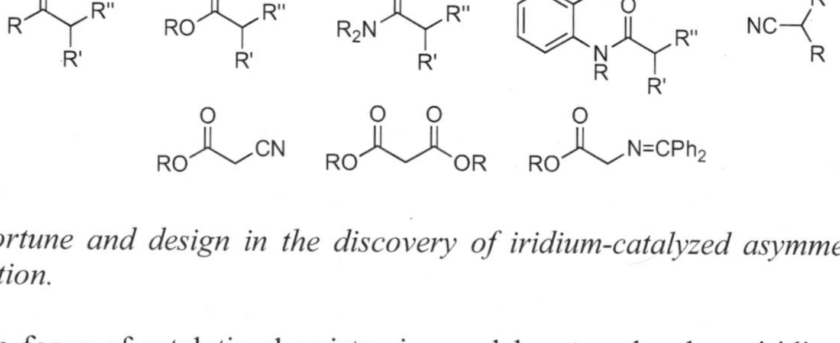

$L = P(t\text{-Bu})_3, Ph_5FcP(t\text{-Bu})_2$

suitable enolate include those from:

(8)

Good fortune and design in the discovery of iridium-catalyzed asymmetric allylic substitution.

A fourth focus of catalytic chemistry in our laboratory has been iridium-catalyzed asymmetric allylic substitution. Dr. Toshimichi Ohmura had been studying additions to rhodium and iridium allyl and benzyl complexes in hopes of developing hydroamination catalysts based on our palladium-catalyzed hydroamination of dienes and vinylarenes. This led to the appreciation of a need to develop allylic substitutions that occur enantioselectively to form branched products. Because of his interest in developing catalysts for hydroamination, he first studied reactions with amine nucleophiles.[26] Takeuchi's work showing that electron-poor phosphorus ligands lead to active catalysts led us to test phosphoramidite ligands,[27] and we investigated for an enantioselective version of this reaction the most hindered ligands developed by Feringa for copper-catalyzed coupling.[28] After just a few experiments, Dr. Ohmura had obtained allylic amine products from reactions of allylic carbonates with high branched to linear selectivity and enantioselectivities of 95% (equation 9).[26] This process would have remained a rather specific reaction that occurs somewhat slowly with high loadings of catalyst if Dr. Christoph Kiener had not identified the active catalyst.[29] The active catalyst is a metalacycle formed by cyclometalation at a ligand methyl group, and this cyclometalation is induced by added amine (equation 10). Once this structure was deduced, we could generate it purposefully[30] and have now generated a family of such metalacyclic catalysts.[31-33] Most striking, Dr. Andreas Leitner was able to trim the stereochemistry of the ligand down to a single stereocenter contained in one phenethylamine used to prepare the phosphoramidite ligand (equation 11).[33]

$$(9)$$

$$(10)$$

$$(11)$$

Thus, our discovery of this process was based on the good fortune that we first tested reactions of amine nucleophiles that were sufficiently basic to generate the active metalacycle. However, once this structure was discovered, we were able to conduct a series of studies that involved the design of simpler ligands and simpler catalysts to use. In addition, we designed methods to generate the active catalyst for reactions of nucleophiles that are not basic enough to generate the active catalyst. Thus, the good fortune has initiated a series of experiments that can be described as a design of improved catalysts and a design of reagents and conditions to improve the scope.   No doubt, these studies have involved extensive evaluation of solvents, additives, and modified ligand structures.

Conclusions

Thus, we are not at a point, and never will be, when one can conduct a single experiment to develop a catalytic homogeneous process.   However, one can use mechanistic data to guide the development of catalysts.   It has been the unraveling of the mechanistic puzzles that has been one fascination of mine and my coworkers. What has been particularly satisfying in recent times is the application of these mechanistic data to the development of new catalytic processes.   Most satisfying has been the development of these reactions to the stage that synthetic chemists routinely use the reactions in their own programs.   These reactions have reached this stage because of the insight and determination of many members of my research group. To those mentioned in this article and the many others who have made similar advances in my laboratory, I express my gratitude and congratulations for being recognized by the Rylander Award.

Acknowledgements

We thank the NSF and NIH for support of these projects, Johnson-Matthey for metal salts, and Frontier Science and Allychem for donations of diboron reagents.

References
1.   Culkin, D. A. and Hartwig, J. F., *Acc. Chem. Res.* 36, 234 (2003).
2.   Hartwig, J. F., *Inorg. Chem.* 46, 1936 (2007).
3.   Hartwig, J.F., in *Handbook of Organopalladium Chemistry for Organic Synthesis*, edited by E.I. Negishi (Wiley-Interscience, New York, 2002), Vol. 1, pp. 1051.
4.   Hartwig, J.F., in *Handbook of Organopalladium Chemistry for Organic Synthesis*, edited by E.I. Negishi (Wiley-Interscience, New York, 2002), Vol. 1, pp. 1097.
5.   Hartwig, J. F., *Nature* submitted (2008).
6.   Hartwig, J. F., *Acc. Chem. Res.* submitted (2008).
7.   Waltz, K. M, He, X., Muhoro, C. N et al., *J. Am. Chem. Soc.* 117, 11357 (1995).
8.   Waltz, K. M. and Hartwig, J. F., *Science* 277, 211 (1997).
9.   Chen, H. and Hartwig, J. F., *Angew. Chem. Int. Ed. Engl.* 38, 3391 (1999).
10. Chen, H., Schlecht, S., Semple, T. C. et al., *Science* 287, 1995 (2000).
11. Hartwig, J. F., Cook, K. S., Hapke, M. et al., *J. Am. Chem. Soc.* 127, 2538 (2005).
12. Kosugi, M., Kameyama, M., and Migita, T., *Chem. Lett.* 927 (1983).
13. Louie, J., Paul, F., and Hartwig, J. F., *Organometallics* 15, 2794 (1996).
14. Louie, J. and Hartwig, J. F., *Tetrahedron Lett.* 36, 3609 (1995).
15. Guram, A. S., Rennels, R. A., and Buchwald, S. L., *Angew. Chem. Int. Ed. Engl.* 34, 1348 (1995).
16. Driver, M. S. and Hartwig, J. F., *J. Am. Chem. Soc.* 118, 7217 (1996).
17. Hayashi, T., Knoishi, M., and Kumada, M., *Tetrahedron Lett.* 21, 1871 (1979).

18. Hayashi, Tamio, Konishi, Mitsuo, Yokota, Kan-ichi et al., *Chem. Lett.* 767 (1980).
19. Shen, Q., Shekhar, S., Stambuli, J. P. et al., *Angew. Chem. Int. Ed.* 44, 1371 (2004).
20. Shen, Q., Ogata, T., and Hartwig, J. F., *J. Am. Chem. Soc.* 130, in press (2008).
21. Hamann, B. C. and Hartwig, J. F., *J. Am. Chem. Soc.* 119, 12382 (1997).
22. Kawatsura, M. and Hartwig, J. F., *J. Am. Chem. Soc.* 121, 1473 (1999).
23. Hartwig, J. F., Kawatsura, M., Hauck, S. I. et al., *J. Org. Chem.* 64, 5575 (1999).
24. Shekhar, S. and Hartwig, J. F., *Organometallics* 26, 340 (2007).
25. Nishiyama, M., Yamamoto, T., and Koie, Y., *Tetrahedron Lett.* 39, 617 (1998).
26. Ohmura, T. and Hartwig, J. F., *J. Am. Chem. Soc.* 124, 15164 (2002).
27. Takeuchi, R., Ue, N., Tanabe, K. et al., *J. Am. Chem. Soc.* 123, 9525 (2001).
28. Feringa, Ben L., *Acc. Chem. Res.* 33, 346 (2000).
29. Kiener, C. A., Shu, C., Incarvito, C. et al., *J. Am. Chem. Soc.* 125, 14272 (2003).
30. Shu, C. T., Leitner, A., and Hartwig, J. F., *Angew. Chem. Int. Ed.* 43, 4797 (2004).
31. Leitner, A., Shu, C. T., and Hartwig, J. F., *Proc. Natl Acad. Sci. USA* 101, 5830 (2004).
32. Leitner, A., Shu, C. T., and Hartwig, J. F., *Org. Lett.* 7, 1093 (2005).
33. Leitner, A., Shekhar, S., Pouy, M.P. et al., *J. Am. Chem. Soc.* 127, 15506 (2005).

3. Asymmetric Hydrogenation of an Amino Acid Intermediate in the Synthesis of Complex Drug Targets: From Kinetic Modeling to Process Development

Daniel S. Hsieh, Dong Lin, Steve S.Y. Wang, Reginald O. Cann, Justin B. Sausker and San Kiang

Bristol-Myers Squibb Company
New Brunswick, NJ 08903

daniel.hsieh@bms.com

Abstract

A highly enantioselective catalytic hydrogenation process was needed for the synthesis of a key amino acid intermediate for a new drug candidate. Et-DuPhos-Rh catalyst produced the chiral intermediate with an enantiomeric purity of over 99%, but with high catalyst loading and low turnover frequency. A kinetic model based on the "Halpern-Brown-Landis" mechanism was established to explore the reaction mechanism. From the model, the coordination of the substrate with the catalyst was proposed as the rate-controlling step. The results indicated that product formation rate is affected by coordination between Rh and non-reacting functionality in the substrate. In addition, the reaction rate is affected by steric hindrance due to a rather bulky protecting group. The presence of an acid can suppress the undesired chelating effect and reduce the induction period. Using a catalyst deactivation rate estimated from the model, we showed that as much as 60% of the initial catalyst is inactivated during the reaction.

We gained a deeper understanding of the process through investigating the kinetics. Ultimately, a commercially viable process was achieved, first by changing the Rh ligand to better accommodate the steric demands of the substrate. Second, the purification of the substrate was required to prevent catalyst poisoning. The selection of a new ligand from screening and purification of the substrate resulted in an increase of the molar ratio of substrate to catalyst from 100 to 1,500 and an increase in the turnover frequency from 10 to 1,500. This process was successfully scaled up to produce kilogram quantities of the chiral intermediate in our facility.

Introduction

The pharmaceutical industry has been giving increased attention to homogeneous asymmetric hydrogenation for the synthesis of chiral molecules due to significant improvements in this technology (1). We recently synthesized a chiral α-amino acid intermediate using Et-DuPhos-Rh catalyst, obtaining enantiomeric purities (EP) of over 99% at 100% conversion. However, the catalyst activity, given as average turnover frequency (TOF), was low, and the catalyst loading was high, making this process costly. Hence, we sought to improve this catalytic process to enable generation of kilogram quantities of the investigational drug for toxicity and pre-clinical studies. Both theoretical and experimental approaches were undertaken to improve this catalytic process.

The theoretical approach involved the derivation of a kinetic model based upon the chiral reaction mechanism proposed by Halpern (3), Brown (4) and Landis (3, 5). Major and minor manifolds were included in this reaction model. The minor manifold produces the desired enantiomer while the major manifold produces the undesired enantiomer. Since the EP in our synthesis was over 99%, the major manifold was neglected to reduce the complexity of the kinetic model. In addition, we made three modifications to the original Halpern-Brown-Landis mechanism. First, precatalyst is used instead of active catalyst in our synthesis. The conversion of precatalyst to the active catalyst is assumed to be irreversible, and a complete conversion of precatalyst to active catalyst is assumed in the kinetic model. Second, the coordination step is considered to be irreversible because the ratio of the forward to the reverse reaction rate constant is high (3). Third, the product release step is assumed to be significantly faster than the solvent insertion step; hence, the product release step is not considered in our model. With these modifications the product formation rate was predicted by using the Bodenstein approximation. Three possible cases for reaction rate control were derived and experimental data were used for verification of the model.

The experimental approach included catalyst screening, reaction rate data collection for kinetic model discrimination and scale-up studies from bench scale to a pilot-scale reactor. The effect of mass transfer on the observed reaction rate was also studied. Criteria for a commercially viable catalytic process in terms of EP, catalyst loading (given as the molar ratio of substrate to catalyst, S/C), and TOF (2) were used to guide the process improvement. We made significant improvements through a better understanding of the kinetics and the search for a more active catalyst and prepared kilogram quantities of the chiral product.

Theoretical Section

Reaction Mechanism. The mechanism for the homogeneous asymmetric hydrogenation of dehydroamino acids to give α-amino acids in high EP, using DuPhos-Rh catalysts and some other catalysts, has been established and documented by several workers (3-6). The general mechanistic scheme (5) shown in Figure 3.1 includes two manifolds with the majority of the total catalyst (>90%) accumulating in the major manifold (which is why it is called the major manifold). The minor manifold produces the desired enantiomeric product (major product), while the major manifold produces the undesirable enantiomeric product (minor product). The reaction sequence starts from the catalyst. The substrate (A in Fig. 3.2) is coordinated to the rhodium atom in the adduct **2** as a chelating ligand through the C=C bond and the protecting group carbonyl. Then the addition of H_2 and insertion of solvent follow in steps 2 and 3, respectively. Finally step 4 releases the product to complete the catalytic cycle. Both the major manifold and the minor manifold have the same reaction sequence, but the majority of the total catalyst resides in the form of $\mathbf{2}^{maj}$. However, by virtue of the much higher reactivity of $\mathbf{2}^{min}$ toward H_2, most of the product emanates from the manifold of the minor diastereomer (5). It should be noted that there is a rapid exchange between $\mathbf{2}^{maj}$ and $\mathbf{2}^{min}$ and they remain at equilibrium (4).

The establishment of a rigorous mathematical kinetic model based on the scheme in Figure 3.1 is very complicated (7). To elucidate the reaction mechanism the complexity of the reaction mechanism can be reduced based upon the following three assumptions:

i.   The major manifold can be neglected at EP over 99%
There are two objectives of setting up a kinetic mathematical model for chiral products. The first is the elucidation of the reaction mechanism with identification of the rate-controlling step. The second is to derive a mathematical expression for the selectivity in terms of the ratio of the major product to the minor product. Then, based upon this expression, the reaction conditions such as pressure or feed ratio are changed to increase the selectivity. However, when the enantiomeric purity is over 99%, the selectivity is extremely high; hence, the reaction mechanism for the major manifold can be neglected to simplify the establishment of the kinetic model.

ii.   Irreversibility
If the forward reaction rate is significantly faster than the reverse reaction rate, the reaction can be considered to be irreversible. For the coordination step in Figure 3.1, the ratio of k_1^{min}/k_{-1}^{min} can be very high (3); hence, the assumption of an irreversible step is reasonable.

iii.   The release of product is fast
The product elimination step at extremely low temperatures ($< -40°C$) was reported as the rate-controlling step (3). However, when the reaction is run at room temperature, this step is assumed to be much faster than the solvent insertion step ($k_4 >> k_3$). Hence this product release step can be neglected. This simplification has been applied for asymmetric hydrogenation and published in the literature (10).

Based upon the above-mentioned assumptions, the reaction scheme in Figure 3.1 is reduced to the scheme shown in Figure 3.2A. It should be noted that active catalyst is used in the reaction scheme in Figure 3.1 while most asymmetric hydrogenation processes use a pre-catalyst (11). Hence, the relationship between the precatalyst and active catalyst needs to be established for the kinetic model. The pre-catalyst used in this study is [Et-Rh(DuPhos)(COD)]BF$_4$ where COD is cyclooctadiene. The active catalyst (X_0) in Figure 3.2A is formed by removal of COD via hydrogenation, which is irreversible. We assume that the precatalyst is completely converted to the active catalyst X_0 before the start of catalytic reaction. Hence, the kinetic model derived here does not include the formation of the active catalyst from precatalyst.

Although the major manifold is neglected due to high enantioselectivity, the intermediate in the major manifold formed from the coordination of the substrate with the active catalyst X_0 should be included in the kinetic model. This is because this intermediate is the resting station for bulk of the catalyst but in a rapid exchange with X_1 in the minor manifold. The intermediates in the kinetic model and the relationship among the intermediates are expressed in Figure 3.2B.

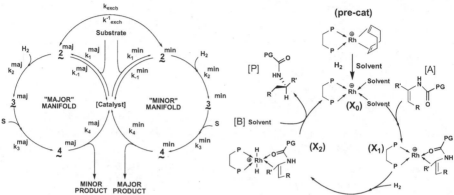

Figure 3.1 Mechanistic Scheme for Asymmetric Hydrogenation.

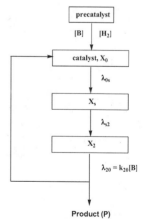

Figure 3.2A Modified Halpen-Brown-Landis Reaction Mechanism.

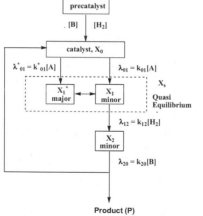

Figure 3.2B Reaction Network for Modified Halpen-Brown-Landis Reaction Mechanism.

Figure 3.2C Simplified Reaction Network.

After the precatalyst is completely converted to the active catalyst X_0, three steps are required to form the desired reduction product. The first step is the coordination of dehydroamino acid (A) to the rhodium atom forming adducts (X_1) and (X_1^*) through C=C as well as the protecting group carbonyl. The next step is the oxidative addition of hydrogen to form the intermediate (X_2). The insertion of solvent (B) is the third step, removing the product (P) from X_2 and regenerating X_0. Hence, the establishment of the kinetic model involves these three irreversible steps.

Rapid exchange between X_1^* and X_1 is reported in reference (3). This means that the forward and reverse reaction rates of this step are much faster than all others, and hence this particular step can be treated as a quasi-equilibrium. The two intermediates in that step are present at all times in concentrations related to one another by a thermodynamic equilibrium constant and can be lumped into one pseudo-intermediate $[X_S]$. This approach is very useful in reducing the number of terms in the denominator of the rate equation, which is equal to the square of the number of intermediates in the cycle (7).

Kinetic Model Establishment. The notation of all the reactions described in Figure 3.2B is summarized below to facilitate the establishment of the mathematical kinetic model:

$$[X_1^*] \xleftrightarrow{K} [X_1] \qquad (1) \qquad [X_0]+[A] \xrightarrow{k_{01}} [X_1] \qquad (2)$$

$$[X_0]+[A] \xrightarrow{k^*_{01}} [X_1^*] \qquad (3) \qquad [X_1]+[H_2] \xrightarrow{k_{12}} [X_2] \qquad (4)$$

$$[X_2]+[B] \xrightarrow{k_{20}} [X_0]+[P] \qquad (5)$$

There are four species containing Rh in the reaction scheme in Figure 3.2B: X_0, X_1, X_1^* and X_2. Since quasi-equilibrium between X_1^* and X_1 is assumed, these two can be lumped into one pseudo-component X_s, thereby reducing the total number of intermediates containing Rh from 4 to 3. This results in a simplified reaction network as shown in Figure 3.2C. However, the mathematical expressions for $[X_s]$, λ_{0s}, and λ_{s2} need to be established; the detailed derivation is described below.

$$[X_1^*] = K[X_1] \qquad (6) \qquad\qquad [X_s] = [X_1^*]+[X_1] = (K+1)[X_1] \qquad (7)$$

where K is the equilibrium constant between $[X_1^*]$ and $[X_1]$. Since $[X_1^*]$ is significantly greater than $[X_1]$ (K>>1), the relationship between $[X_s]$ and $[X_1]$ can be simplified as follows:

$$[X_s] \cong K[X_1] \qquad (8)$$

In addition, the reaction rate to form the intermediate X_2 from X_1 is equal to that from the pseudo-component X_s; hence the following equation can be written:

$$\lambda_{s2}[X_s] = \lambda_{12}[X_1] \qquad (9)$$

in which $\lambda_{12}=k_{12}[H_2]$. Substituting equation (8) into equation (9), the relationship between λ_{s2} and λ_{12} is obtained:

$$\lambda_{s2} = \frac{\lambda_{12}}{K} \qquad (10)$$

The expression for λ_{0s} can be derived from the following two equations:

$$\lambda_{0s}[X_0] = \lambda^*_{01}[X_0] + \lambda_{01}[X_0] \quad (11) \qquad \lambda_{0s} = \lambda^*_{01} + \lambda_{01} = (k^*_{01} + k_{01})[A] \quad (12)$$

Where $\lambda^*_{01} = k^*_{01}[A]$ and $\lambda_{01} = k_{01}[A]$.

The rate expression for each intermediate in Figure 3.2C can be derived based on the Bodenstein approximation of quasi-stationary states of trace-level intermediates (7). These expressions are:

$$\frac{d[X_0]}{dt} = \lambda_{20}[X_2] - \lambda_{0s}[X_0] = 0 \quad (13) \qquad \frac{d[X_s]}{dt} = \lambda_{0s}[X_0] - \lambda_{s2}[X_s] = 0 \quad (14)$$

$$\frac{d[X_2]}{dt} = \lambda_{s2}[X_s] - \lambda_{20}[X_0] = 0 \quad (15) \qquad C_{\Sigma cat} = [X_0] + [X_s] + [X_2] \quad (16)$$

$$r_P = \lambda_{20}[X_2] \quad (17)$$

in which r_p is product formation rate; [A], [B], [P] are the concentrations of the substrate, the solvent and the product, respectively; X_0, X_s, and X_2 are the intermediate concentrations; $C_{\Sigma Cat}$ is the total catalyst concentration; k_{01}, k_{12}, k_{20} are reaction rate coefficients for the reactions from X_0 to X_1, X_1 to X_2 and X_2 to X_0, respectively; $\lambda_{01} = k_{01}[A]$, $\lambda_{12} = k_{12}[H_2]$, and $\lambda_{20} = k_{20}[B]$.

Equations from (13) to (17) are solved to provide the mathematical expression for the concentration of each intermediate and the product formation rate in terms of measurable quantities, including the total catalyst concentration and reactant concentration.

$$[X_0] = \frac{\lambda_{s2}\lambda_{20}}{\lambda_{s2}\lambda_{20} + \lambda_{20}\lambda_{0s} + \lambda_{0s}\lambda_{S2}} * C_{\Sigma cat} \quad (18)$$

$$[X_s] = \frac{\lambda_{20}\lambda_{0s}}{\lambda_{s2}\lambda_{20} + \lambda_{20}\lambda_{0s} + \lambda_{0s}\lambda_{S2}} * C_{\Sigma cat} \quad (19)$$

$$[X_2] = \frac{\lambda_{0s}\lambda_{s2}}{\lambda_{s2}\lambda_{20} + \lambda_{20}\lambda_{0s} + \lambda_{0s}\lambda_{S2}} * C_{\Sigma cat} \quad (20)$$

$$r_P = \frac{(k_{01} + k^*_{01})k_{12}k_{20}[A][H_2][B]C_{\Sigma Cat}}{k_{12}k_{20}[H_2][B] + K(k_{01} + k^*_{01})k_{20}[A][B] + (k_{01} + k^*_{01})k_{12}[A][H_2]} \quad (21)$$

Reduction of Complexity. Equation (21) can be further simplified by applying the concept of the rate-controlling step, resulting in three possible scenarios:

Scenario A: Hydrogenation insertion is the rate-controlling step.

When the hydrogenation insertion is the rate-controlling step, the rate constant k_{12} is much smaller than the two other rate constants k_{01} and k_{20}. Hence, the two terms involving k_{12} in the denominator become negligible, allowing the rate equation (21) to be reduced to the following expression:

$$r_p = \frac{k_{12}[H_2]}{K} C_{\sum Cat} \tag{22}$$

Equation (22) indicates that the production formation rate is proportional to the hydrogen gas pressure. When the hydrogen gas pressure doubles, the product formation rate should be doubled. The product formation rate is inversely proportional to the equilibrium constant K. When the K value decreases, the product formation rate increases.

Scenario B: Coordination of dehydroamino acid (A) to the rhodium atom is the rate-controlling step.

As in the aforementioned approach, when coordination of the dehydroamino acid (A) is the rate-controlling step, the rate equation (21) can be reduced to the following expression:

$$r_p = (k_{01} + k^*_{01})[A]C_{\sum Cat} \tag{23}$$

Equation (23) indicates that the product formation rate is proportional to substrate concentration [A].

Scenario C: Insertion of solvent (B) is the rate-controlling step.

The rate equation for this case can be derived as follows when the rate-controlling step is the insertion of solvent (B) in the step from X_2 to X_0.

$$r_p = k_{20}[B]C_{\sum Cat} \tag{24}$$

Equation (24) indicates that the production formation rate is proportional to the solvent concentration [B].

Experimental Section

Semi-Batch Reactors. We used an Endeavor® catalyst screening system (Biotage, VA) for catalyst screening and kinetic studies. A Buchi autoclave equipped with a Buchi pressflow gas controller bpc® (BuchiGLASuster, Switzerland) was used for scale up. Large pilot plant reactors were used for kilogram-level production. Hydrogen pressure was kept constant, and hydrogen uptake was monitored and recorded.

General Experimental Procedures. Charge 1.0 eq. of substrate, 0.005 eq. or less catalyst, 8 volumes of dichloromethane, 8 volume of methanol. Purge the system three times each with nitrogen and hydrogen. Maintain the temperature at 25°C and hydrogen pressure at 45 psig, until the reaction is deemed complete.

Analytical Methods. A Schimadzu Liquid Chromatograph was used to monitor the reaction conversion and to assign chemical and chiral purity to the final product. Structures were verified by ^{1}HNMR spectra obtained on a Bruker (Model UltraShield 400 spectrometer). Optical rotations were measured on a Perkin Elmer Model 341 Polarimeter.

Solvent, Catalysts and Reagents Used. All solvents were reagent grade and used as received. Reagents were purchased from Aldrich. Catalysts were purchased from Strem Chem and DowPharma.

Results and Discussion

Kinetic Model Discrimination. To discriminate between the kinetic models, semi-batch reactors were set up for the measurement of reaction rates. The semi-batch terminology is used because hydrogen is fed to a batch reactor to maintain a constant hydrogen pressure. This kind of semi-batch reactor can be treated as a batch reactor with a constant hydrogen pressure. The governing equations for a batch reactor, using the product formation rate for three possible scenarios, were derived, as described in reference (12) with the following results:

Case I. Hydrogenation Insertion is the Rate-Controlling Step

$$X = \beta \bullet t \quad (25) \qquad\qquad \beta \equiv \frac{(k_{12}/K)V[H_2]C_{\Sigma cat}}{N_{A0}} \qquad (26)$$

in which t is the reaction time; X is the conversion of the SM (starting Material or substrate); N_{A0} is the number of moles of SM at time t = 0; and V is the volume of the reaction mixture. Equation (25) shows that the conversion is proportional to the reaction time if the insertion of hydrogen is the rate-controlling step.

Case II. Coordination of Dehydroamino Acid (A) to the Rhodium Atom is the Rate-Controlling Step

$$\ln(1-X) = -\alpha \bullet t \quad (27) \qquad\qquad \alpha \equiv (k_{01} + k^*_{01})C_{\Sigma Cat} \qquad (28)$$

Equation (27) shows that the conversion for Case II obeys the first-order reaction kinetics.

Case III. Insertion of Solvent (B) is the Rate-Controlling Step

$$X = \gamma \bullet t \qquad\qquad (29) \qquad\qquad \gamma \equiv \frac{k_{20}V[B]C_{\Sigma Cat}}{N_{A0}} \qquad (30)$$

Equation (29) shows that the conversion is proportional to the reaction time if the insertion of solvent is the rate-controlling step.

The governing equations for all three cases require the determination of reaction conversion. Since the chiral selectivity in this study is over 99%, as one mole of SM is consumed, one mole of hydrogen is also consumed. Hence, the consumption of hydrogen can be used for the calculation of reaction conversion X:

$$X = \frac{Q_t}{Q_\infty} \tag{31}$$

in which Q_∞ is the hydrogen consumption when the reaction conversion is complete and Q_t is hydrogen consumption at reaction time t. HPLC analysis was also used to check the reaction conversion and EP.

The conversion as a function of time for Et-DuPhos-Rh catalyst determined in a Buchi reactor is presented in Figure 3.3. It takes more than 700 minutes to get complete conversion. It should be noted that the relationship between the reaction conversion and time is not linear except at the beginning of the reaction as shown in this figure.

Case I and Case III Evaluation. If either the hydrogen insertion or the solvent insertion is the rate-controlling step, the relationship between the reaction conversion and reaction time should be linear, as shown in equation (25) for Case I and equation (29) for case III. The non-linearity of the experimental conversion vs. time plot in Figure 3.3 suggests that neither hydrogen nor the solvent are rate-controlling steps.

Case II Evaluation. When kinetic data in terms of reaction conversion vs. time are used to fit the kinetic model expressed by equation (28) with the value of α, 0.0102, determined by the best curve fit, the calculated conversion vs. reaction time over the entire reaction period presented in Figure 3.4 is in good agreement with the experimental data.

It should be noted that the reaction using Et-DuPhos-Rh catalyst is not limited by hydrogen mass transfer since the hydrogen mass transfer rate is at least 5 times as fast as the initial reaction rate. Furthermore, the overall reaction time, 700 minutes, remained the same regardless of the size of the reactor.

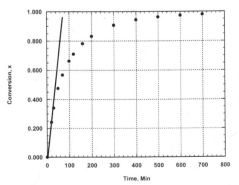

Figure 3.3 Conversion of SM vs. Reaction Time Using Et-DuPhos-Rh Catalyst.

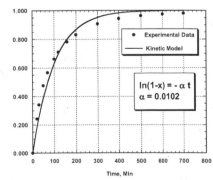

Figure 3.4 Comparison between Experimental Data and Correlation for Case II.

Catalyst Decay. Asymmetric hydrogenation of the SM using the Et-DuPhos-Rh catalyst exhibits a catalyst threshold behavior. When the initial charge of the catalyst is below this threshold value, the reaction is not completed. This indicates that the catalyst may become deactivated.

The catalyst deactivation can be calculated with equation (27). Figure 3.5 shows that the slope of the plot of $\ln(1-x)$ vs. reaction time is 0.0138 at the beginning of the reaction. If there is no catalyst deactivation, the data of $\ln(1-x)$ vs. time should follow a straight line. The deviation from this straight line indicates that the total catalyst concentration decreases as the reaction progresses. Using equation (27), the value for α, proportional to total catalyst concentration, can be determined from the conversion X and reaction time. As shown in this figure, the value for α becomes 0.0054 at the end of the reaction.

If the initial concentration of the catalyst is considered to be 100%, the catalyst concentration becomes 40% at the end of the reaction, as shown in Figure 3.6. This means that 60% of the catalyst becomes non-active near the end of the reaction. The explanation for the catalyst deactivation is not known. Possible reasons include the presence of metallic impurities or organic impurities remaining from synthesis of the starting material.

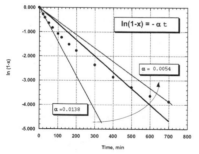

Figure 3.5 Plot of $\ln(1-x)$ vs. Reaction Time.

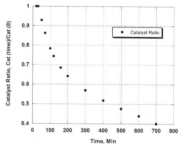

Figure 3.6 Catalyst Decay as a Function of Reaction Time.

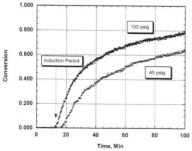

Figure 3.7A Effect of Pressure on Induction Period and Reaction Rate Pressure.

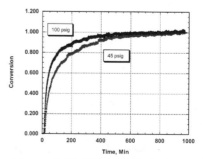

Figure 3.7B Effect of Hydrogen on Reaction Rate and Reaction Time.

Pressure Effect The reaction was conducted in an Endeavor system at 45 psig and 100 psig. The pressure effects on reaction time are presented in Figures 3.7/A and 3.7B. Figure 3.7A indicates that this catalytic system exhibit induction behavior and that the induction period depends upon the reaction pressure. One explanation for the induction period is the time required to form X_0 from pre-catalyst in the presence of hydrogen and solvent as shown in Figure 3.2. As the hydrogen pressure becomes higher, less time is required to form the required concentration of X_0 for the reaction to start. The effect of hydrogen pressure on the reaction rate is presented in Table 3.I. The reaction rate is determined from the slope of the plot of reaction conversion vs. reaction time, as shown in Figures 3.7A and 3.7B. As the reaction nears completion, the reaction rate is very slow, and hence not reported in the table. Table 3.I shows that the ratio of the reaction rate at 114.7 psia to the reaction rate at 59.7 psia ranges from 1.8 to 2.0, indicating that the reaction rate is proportional to the absolute pressure.

Obviously, the effect of hydrogen pressure on the reaction rate cannot be explained using case II in which coordination of the substrate with the catalyst is the rate-controlling step. However, two possible reasons (without experimental support such as measurement of the active catalyst concentration) are offered to explain the effect of pressure on reaction rate. First, the initial active catalyst concentration at the higher pressure is higher. This is related to the observed induction behavior. As the precatalyst is hydrogenated, there are two competing reactions, one for the formation of active catalyst and the other for the inactivation of the precatalyst by poisonous impurities. At a higher pressure, the formation rate of the active catalyst is faster, resulting in a higher initial active catalyst concentration. Second, the reaction rate may be controlled by both the coordination of substrate with the active catalyst and the addition of hydrogen.

Table 3.I Effect of Hydrogen Pressure on Reaction Rate and Time at Absolute Hydrogen Pressure Ratio of 1.92

	Reaction Time, Min		Reaction Rate, dx/dt		
Conversion	45 psig	100 psig	45 psig	100 psig	Reaction Rate Ratio
0.1	5	3	0.022	0.04	1.8
0.2	9	5	0.022	0.04	1.8
0.4	27	15	0.0085	0.017	2.0
0.6	64	34	0.003	0.006	2.0
0.8	183	88	0.001	0.002	2.0
1.0	1000	600	-	-	-

Process Evaluation and Improvement. As homogeneous asymmetric hydrogenation processes are scaled up, one major concern is cost because the catalyst is usually expensive. Hence, several criteria for a commercially viable process (2), including selectively, conversion, catalyst loading (S/C, the molar ratio of substrate to catalyst), reaction time, and TOF (turnover frequency, the ratio of catalyst loading to reaction time), should be considered to evaluate the process and provide a guide for improvement.

The results of the kinetic study and model discrimination show that insertion of SM is rate-controlling. Two reasons may explain why this step is rate-controlling. First, the protection group in our SM is very bulky, making the reaction slow, which is consistent with literature data (8) showing the size effect on reactivity. Second, a free aniline group in the SM could bond with Rh and reduce the catalyst reactivity.

The results from the kinetic study using Et-DuPhos-Rh catalyst lead to the following suggestions for reactivity improvement:

- Add acid to block interference from the aniline group
- Change process condition such as pressure

Addition of a strong acid such as methanesulfonic acid (MSA) to the reaction mixture has a positive impact on the reactivity, as shown in Figure 3.8. The induction time is shortened by 10 minutes and the reaction rate almost doubled. Due to the reaction rate increase from the acid addition, the catalyst loading could be lowered. In addition, the hydrogen pressure could be doubled to reduce the reaction time by half. However, improvements from addition of acid and pressure increase are not sufficient to make this process commercially viable because the catalyst loading and the TOF are significantly lower than the criteria listed in Table 3.II. Therefore, we initiated a search for catalysts more active than Et-DuPhos-Rh catalyst.

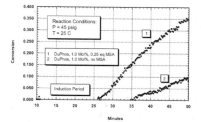   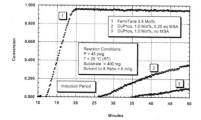

Figure 3.8 Impact of MSA Addition on Induction Time and Reactivity.   **Figure 3.9** Reactivity of Et-FerroTane-Rh Catalyst vs. Et-DuPhos-Rh Catalyst.

Table 3.II Criteria for Commercially Viable Process Et-DuPhos-Rh Catalyst

Catalyst	Ideal	DuPhos-Rh	DuPhos-Rh
Action		Acid Addition	Acid Addition and Pressure Increase
Pressure, psig		45	100
EP, %	> 95	99	99
Conversion, %	99.5	> 99.5	> 99.5
S/C	> 1,000	200	200
Reaction time, Hr	Short	17	10
TOF	> 500	12	20

Search for More Active Catalyst. An extensive screening effort was undertaken to find a catalyst more active than Et-DuPhos-Rh. As a result of this effort, Et-FerroTane-Rh and some other competitive catalysts were found. The reactivity of Et-FerroTane-Rh and Et-DuPhos-Rh, is presented in Figure 3.9. The reaction rate with Et-FerroTane-Rh catalyst is very high with a small induction period, and the total time for reaction completion is drastically less than with Et-DuPhos-Rh.

Hydrogenation using Et-FerroTane-Rh catalyst falls into the mass-transfer controlled regime. The relationship between the reaction conversion and time at various initial substrate charges is linear, as shown in Figure 3.10, indicating that the reaction rate is a zero order with respect to substrate. Also the time required for completion is proportional to the initial charge of substrate as shown in Figure 3.11, characteristic of a mass-transfer limited reaction. Since hydrogen mass transfer is the limiting factor using Et-FerroTane catalyst, the reaction time is dictated by good hydrogen dispersion and mixing performance upon scale-up.

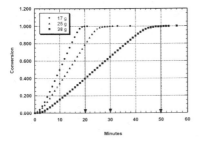 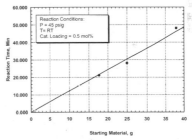

Figure 3.10 Reaction Conversion vs. Time at Various Initial Substrate Charges.   **Figure 3.11** Effect of Initial Substrate Charge on Total Reaction Time.

Scale-up to Pilot Plant Scale. Et-FerroTane-Rh catalyst was selected over Et-DuPhos-Rh catalyst for further scale-up due to the short reaction time using this catalyst. However, a catalyst threshold behavior was also observed for FerroTane-Rh catalysts, probably caused by impurities in the substrate. Hence purification of the substrate was done to remove impurities. As shown in Table 3.III, the ideal S/C target is over 1,000. When the substrate was purified by formation of the MSA salt followed by conversion of the salt back to the free amine, the S/C value was only 200. However, when the substrate was purified further by recrystallization or chromatography (LC), the S/C value increased, meeting the criteria for a commercially viable process. It should be noted that the S/C value is higher for Me- FerroTane-Rh than for Et-FerroTane-Rh, probably because the bulky substrate complements the reduced ligand size of the catalyst, resulting in the higher S/C value. The process using Et-FerroTane-Rh catalyst has been run at pilot-plant scale to produce over 10 kg of chiral product.

Table 3.III Criteria for Commercially Viable Process-FerroTane-Rh Catalysts

Catalyst	Ideal	Et-FerroTane	Et-FerroTane	Et-FerroTane	Me-FerroTane
Substrate purification		MSA	MSA/Re-cryst.	MSA/LC	MSA/LC
EP, %	> 95	99	99	99	99
Conv., %	99.5	> 99.5	> 99.5	> 99.5	> 99.5
S/C	> 1,000	200	500	1,000	1,500
Reaction time, Hr	Short	0.5	0.5	0.5	0.5
TOF	> 500	400	1,000	2,000	3,000
Stage		Glass Plant	Pilot Plant	Endeavor	Endeavor
Kg		4.5	12	0.01	0.01

Conclusions

We conducted a kinetic study of the asymmetric hydrogenation of an enamide intermediate, using Et-DuPhos-Rh catalyst and proposed a mathematical kinetic model to demonstrate that the rate-controlling step is the insertion of the SM into the rhodium complex. The experimental data in terms of reaction conversion as a function of time agree with the model very well. These results are consistent with literature information suggesting that low reactivity is due to steric hindrance from the bulky protecting group in the substrate. Using the kinetic model, catalyst deactivation during the reaction was calculated, with more than 60% of the initial catalyst inactivated, presumably due to the presence of impurities. The model in which hydrogen insertion is the rate-controlling step has not been rigorously excluded, but we feel that the observed effect of hydrogen pressure is due to a higher initial active catalyst concentration generated at the higher pressure. Although reactivity was improved by adding acid and raising the reaction pressure, the reaction using Et-DuPhos-Rh catalyst was not commercially viable due to low reactivity and high catalyst load. Through further catalyst screening, Et-FerroTane-Rh was identified as a superior catalyst. With purification of the substrate, asymmetric hydrogenation using Et-FerroTane-Rh or Me-FerroTane-Rh catalyst meets the criteria for a commercially viable process. We demonstrated the scale-up of this asymmetric hydrogenation process using Et-FerroTane-Rh catalyst and produced kilogram quantities of the chiral product.

Acknowledgements

We thank Dr. James Ramsden and Dr. Ian C. Lennon from Dowpharma for their technical discussion and support. Appreciation is extended to the project team members for their support. Special thanks are expressed toward Dr. R. Brent Nielsen, Dr. William A. Nugent, Dr. Larry Parker, and Dr. Rodney L. Parsons.

References

1. M. Thommen, *Homogeneous asymmetric hydrogenation: Mature and fit for early stage drug development, Specialty Chemicals Magazine*, May (2005).
2. Hans-Ulrich Blaser, Felix Spindler and Marc Thommen, *Industrial Applications* in *Handbook of Homogeneous Hydrogenation*, (ed. J.G. de Vries and C. J. Elsevier), Wiley, (2007).
3. C. R. Landis, Halpern, J.: *J. Am. Chem. Soc.*, **109**, 1746 (1987).
4. S. K. Armstrong, J. M. Brown, and M. J. Burk, *Tetrahedron Letters*, **34(5)**, 879 (1993).
5. C. R. Landis, Hiffenhaus, P., and Steven Feldgus, *J. Am. Chem. Soc.*, **121**, 8741 (1999).
6. X. Zhang and W. Tang, *Chem. Rev.* **103**, 3029 (2003).
7. F. G. Helfferich, *Kinetics of Homogenous Multi-step Reactions*, Elsevier, (2001).
8. H.-J. Kreuzfeld, et al., *Tetrahedron Asymmetry*, **4(9)**, 2047 (1993).
9. W. Wang, et al., *Tetrahedron* **58**,7365 (2002).
10. J. Burk, et al., *J. Org. Chem*, **64**, 3290 (1999).
11. C. Cobley, et al., *Tetrahedron Letters*, **42**, 7481 (2001).
12. O. Levenspiel, *Chemical Reaction Engineering*, 3rd Edition, John Wiley & Sons, (1999).

4. Process Intensification. Continuous Two-Phase Catalytic Reactions in a Table-Top Centrifugal Contact Separator

Gerard N. Kraai[1], Boelo Schuur[1], Floris van Zwol[1], Robert M. Haak[2], Adriaan J. Minnaard[2], Ben L. Feringa[2], Hero J. Heeres[1] and Johannes G. de Vries[2, 3]

[1]Dept. of Chemical Engineering and [2]Stratingh Institute for Chemistry, University of Groningen, Nijenborgh 4, 9747 AG Groningen, the Netherlands
[3]DSM Pharmaceutical Products-Advanced Synthesis, Catalysis & Development, P.O. Box 18, 6160 MD, the Netherlands

Hans-JG.Vries-de@dsm.com

Abstract

Production of fine chemicals is mostly performed in batch reactors. Use of continuous processes has many advantages which may reduce the cost of production. We have developed the use of centrifugal contact separators (CCSs) for continuous two-phase catalytic reactions. This equipment has previously only been used for mixing and separating two liquids. We have converted sunflower oil into biodiesel and glycerol by continuous two-phase reaction with methanolic sodium methoxide in the CCS in 96% conversion. Two-phase catalytic epoxidation was more problematic due to instability of most catalysts in a two-phase system. In a first attempt we managed to convert cyclooctene into the epoxide using hydrogen peroxide as oxidant and a tungsten-based polyoxometalate as catalyst with 20% conversion. Surprisingly, it was also possible to perform a lipase-catalysed esterification of oleic acid in the CCS. In spite of the aqueous environment a very high conversion up to 87% was reached. In this table-top device, production of 100 Kg amounts is possible in just a matter of days. Larger-sized equipment also exists that can be used for ton-scale production.

Introduction

The production of fine chemicals and pharmaceuticals is almost exclusively performed in multi-purpose batch reactors. This allows maximum flexibility. The downside of their use is the relatively large size of the reactors and purification trains and the associated high costs. In addition, there are limited options for process control. Many groups have started to work from different angles on methods to make chemical production processes more efficient by combining unit operations, by debottlenecking the limiting factors (often heat or mass transfer) and by making processes continuous where possible (1). This should result in production units that are much more efficient and hence are strongly reduced in size. This combined effort is called process intensification (2).

It is clear that the use of batch reactors has some serious drawbacks. For the production of larger quantities, multiple batch runs have to be performed and this often leads to batch to batch variation in product quality and performance. Furthermore, the productivity is often lower than for dedicated continuous reactors and fixed costs are higher caused by high operator efforts. Therefore, switching to continuous processes holds great appeal, if we can find a device that is suitable for multiple products.

In view of the above, many researchers are investigating the use of microreactors (3). Microreactors usually consist of one or more interconnected channels in a flat device with diameters in the micrometer range, typically 10-500 µm. In its simplest form, the device is used for the rapid mixing of two solutions containing substrates and reagents and for the very fast transfer of the heat of reaction enabled by the very large surface/volume. Thus, they are very useful for highly exothermic reactions. In addition, they have great merit for reactions that contain highly toxic or explosive components. Although initially many researchers predicted that scale-up could be affected by the parallel use of thousand of these microreactors, this is neither feasible nor cost-effective. In practice, much larger micro-structured reactors in which the channels still roughly have the same diameter as in the laboratory device are used for large scale production although these reactors are still one or two orders of magnitude smaller than conventional batch reactors. DSM has recently announced the use of such a reactor for the performance of a Ritter reaction on ton-scale (4). A Ritter reaction is a typical example of a highly exothermal reaction containing toxic reagents.

A disadvantage of the microreactor is the very short residence time. In fact, a recent analysis shows that only about 6% of reactions are intrinsically suitable for use in a microreactor (5). For this reason, we have focused our attention on devices in which we can perform two-phase catalytic reactions in flow mode. Two-phase catalysis in flow devices has been reported by de Bellefon (6) and by Claus (7), however, without integrated phase separation. Ryu has reported a two-phase Heck reaction in a microreactor in which the phase containing the palladium catalyst was an ionic liquid. In this case the catalyst separation and recycle was fully integrated (8). Ideally, we would want to have a reactor in which a closed loop containing a solution of the catalyst is continuously contacted with the continuous phase containing the substrate. After completion of the reaction, the phase that now contains the product is separated off. The degree of conversion can be determined by the relative flow rates, the size of the mixed phase, the degree of mixing and the amount of catalyst; thus assuring that the majority of catalytic reactions can be performed in this way. What we need for this is a device that combines the functions of mixing and separating in a continuous mode. The device we have chosen is a centrifugal contact separator (CCS) (9) that has been used for oil-water separation (e.g., for cleaning oils spills) (10), for the continuous extraction of fermentation products, such as penicillin (11) and phenylalanine (12) and in the atomic waste industry for the purpose of extraction and purification of radioactive waste (13). In essence, it is a centrifuge (Figure 4.1). The two incoming liquids are introduced on its side through separate openings into the angular space that is formed between the outside of the rotor and the outer housing. Here, very effective intimate mixing takes

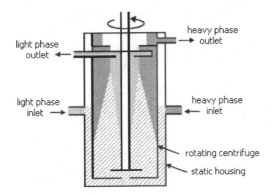

Figure 4.1 Schematic cross-section of a centrifugal contact-separator (light gray = light phase, dark gray = heavy phase, striped = mixed phase).

place at high speed caused by the rotation of the centrifuge. The mixed phase is then sucked inside the rotor through a narrow opening in the bottom plate. Here the liquids slowly move upward and the two phases are gradually separated again to leave the device through separate openings in the top.

Ley and Baxendale have shown that it is possible to complete a total synthesis by using a number of flow devices in series, which contained immobilized reagents (14). Use of a number of CCSs in series would allow cascade catalysis in which the catalytic systems have been separated (Figure 4.2). This is quite attractive in view of the fact that often each catalyst needs highly specific operating conditions, such as temperature and pH.

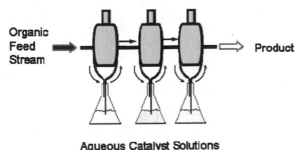

Figure 4.2 Cascade catalysis using centrifugal contact separators in series.

Experimental section

Biodiesel from sunflower oil
The experiments were performed in a CINC V-02 separator also known as the CS-50 (15), equipped with a high-mix bottom plate. Two Watson-Marlow 101U/R

MK2 peristaltic tube pumps were used to feed the CCS. The CCS was connected to a Julabo MV basis temperature controlled water bath (accuracy ± 0.01°C). The water bath was set at 70°C. The CCS was operated in a once-through operation. The CCS was fed with 12.6 ml/min of pure sunflower oil. Subsequently, the centrifuge was started (40 Hz, which corresponds to 2400 rpm). As soon as the oil started flowing out of the heavy phase outlet, the reaction was started by feeding the methanolic NaOMe solution (containing 1% w/w NaOMe with regard to sunflower oil) at 3.15 mL/min.

After about 9 minutes, the oil exiting the heavy phase outlet became clouded. After about 35 minutes the heavy-phase outlet stopped flowing and the light-phase outlet started flowing, giving a FAME/sunflower oil mixture. After about 120 minutes, the heavy-phase outlet started again, consisting of a glycerol/methanol mixture. Samples were taken at regular intervals and were analysed using 1H-NMR.

Experimental procedure for the enzymatic esterification of oleic acid with butanol

The experiments were performed in a CINC V-02 separator also known as the CS-50 (15). Two Verder VL 500 control peristaltic tube pumps equipped with a double pump head (3,2 x 1,6 x 8R) were used to feed the CCS. In case of the enzymatic reaction, the low mix bottom plate was applied. To operate the reactor at a desired temperature, it was equipped with a jacket which was connected to a temperature controlled water bath with an accuracy of ± 0.01°C. The CCS was fed with pure heptane and pure water, both with a flow rate of 6 mL/min. Subsequently, the centrifuge was started (40 Hz, which corresponds to 2400 rpm) and the set-up was allowed to equilibrate for a period of 1 h. At this point, the heptane feed stream was replaced by the organic feed stream (oleic acid (0.6M) and 1-butanol (0.9M) in heptane). After equilibration for 10 minutes, the reaction in the CCS was started by replacing the water stream with the aqueous feed stream (0.1 M phosphate buffer pH 5.6 containing 1 g/l of the lipase form *Rhizomucor miehei*). Samples were taken at regular intervals and analysed by GC.

Results and Discussion

Biodiesel from sunflower oil

In our first experiment we decided to test the conversion of sunflower oil into biodiesel (16). Treatment of sunflower oil (**1**) with NaOMe in MeOH results in formation of a mixture of fatty acid methyl esters (FAME), also known as biodiesel, and glycerol (**2**) (Figure 4.3). The reaction was performed with a six-fold molar excess of methanol with respect to sunflower oil at elevated temperatures (60°C) using a basic catalyst (NaOMe, 1% w/w with respect to sunflower oil). The CCS was equipped with a heating jacket to ensure isothermal conditions. The sunflower oil was preheated to 60°C and was pumped at 12.6 ml/min into one entrance of the CCS. Subsequently, a solution of NaOMe in MeOH was introduced through the other entrance at a flow rate of 3.1 ml per minute. After about 40 minutes, the system reaches steady state and the FAME containing some residual sunflower oil is coming out as the light phase, whereas the heavy phase consists of a solution of glycerol in MeOH.

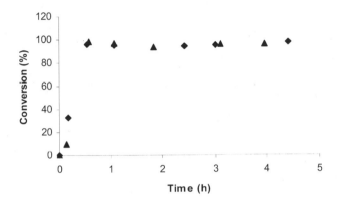

Figure 4.3 Continuous production of biodiesel in CCS (2 duplo runs).

The conversion reaches a maximum at 30 Hz. At a higher rate of rotation the increased separatory power of the centrifuge leads to a reduction of the volume of the mixed phase in which the reaction takes place. At reduced rotational speeds of the centrifuge the mixing process becomes less efficient, resulting in larger average drop sizes in the dispersed phase and thus to reduced mass transfer rates and conversion levels.

Using the optimum settings, biodiesel is produced at a volumetric production rate of 61 kg biodiesel/m^3·min, which compares well with 42 kg/m^3·min reported for typical batch processes (17). In addition, the current process is much more efficient since there is no separate separation step and reactor cleaning in between batches can be omitted (18).

Continuous olefin epoxidation

Having established the feasibility of the method in this relatively simple catalytic conversion, we next attempted a metal catalysed reaction. Since the average residence time in the CCS is about 14 minutes, the reaction has to be relatively fast. Although in case of a somewhat slower reaction the catalyst concentration can be increased, or the temperature or both. As most epoxidation reactions are relatively fast, we decided to develop a continuous two-phase catalytic epoxidation of an olefin. We thus screened a number of catalysts for applicability in a two-phase system. This requires that the catalyst is soluble only in the aqueous or polar phase.

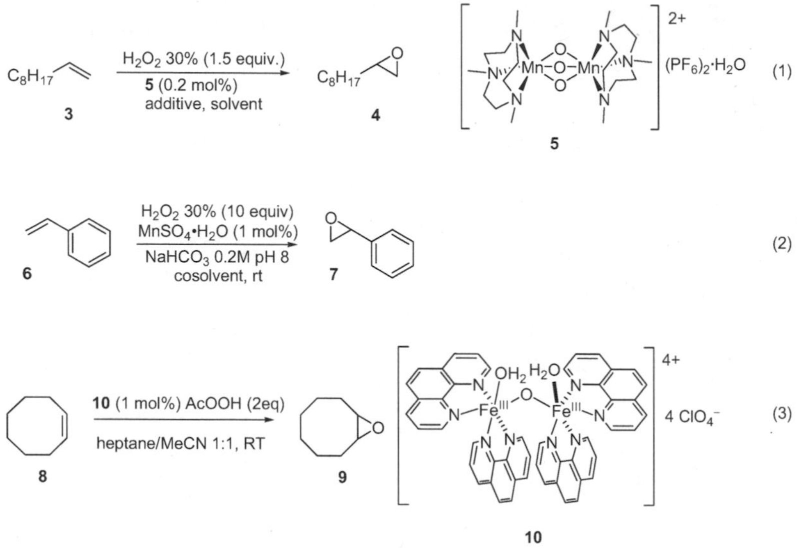

Scheme 4.1 Epoxidation catalysts tested in two-phase solvent systems.

As depicted in Scheme 4.1, the dinuclear manganese catalyst (**5**) developed by Hage et al. seemed to be highly suitable as this catalyst is currently used on a large scale in dishwashing applications; it is an established catalyst for the epoxidation of olefins with hydrogen peroxide and it is sufficiently water-soluble (eq. 1) (19). Nevertheless, when we attempted to perform this reaction in an aqueous two-phase set-up, no reaction ensued. Conversion was only obtained in the presence of acetone or acetonitrile. Similarly, the protocol developed by Burgess et al. (20) based on the use of $MnSO_4 \cdot H_2O$ as catalyst could be transferred to a two-phase aqueous system, but reaction with hydrogen peroxide was exceedingly slow leading to 46% conversion after 5 days (eq. 2). In addition, extensive decomposition of hydrogen peroxide occurred. We did manage to epoxidise an olefin in a two-phase system based on heptane and acetonitrile using a phenanthroline based iron catalyst (eq. 3) and peracetic acid as oxidant (21), but here the exothermicity of the reaction led us to investigate the safety aspects of this reaction more carefully. A number of calorimetric experiments made us decide not to engage in a large-scale operation of this reaction in the CCS. In addition, we observed extensive catalysts degradation in these reactions. This was indeed a common theme in many of these experiments: the presence of catalysts and oxidant in a single phase, with the substrate largely present in a second phase often led to rapid decomposition of the catalyst. Thus we searched for more stable epoxidation catalysts. As much instability of oxidation catalysts is related to the oxidative instability of the ligands we decided to choose a ligand-free catalyst. Tungsten-based catalysts function well without ligand and have been used by many groups in various oxidation reactions. Recently, Alsters et al. made a comparison of 12 different tungsten-based catalysts and concluded that the polyoxometalate $Na_{12}[WZn_3(ZnW_9O_{34})_2]$ (NaZnPOM) in combination with a suitable phase transfer catalyst such as methyltrioctylammonium chloride is the preferred catalyst for olefin epoxidation, particularly viewed from the perspective of industrial application (22). The only slight disadvantage of this catalyst is the somewhat higher temperatures needed to obtain reasonable rates. The catalyst is easily formed by self-assembly according to eq. 4.

$$19\ Na_2WO_4 + 5\ Zn(NO_3)_2 + 16\ HNO_3 \rightarrow Na_{12}[WZn_3(ZnW_9O_{34})_2] + 26\ NaNO_3 + 8\ H_2O \quad (4)$$

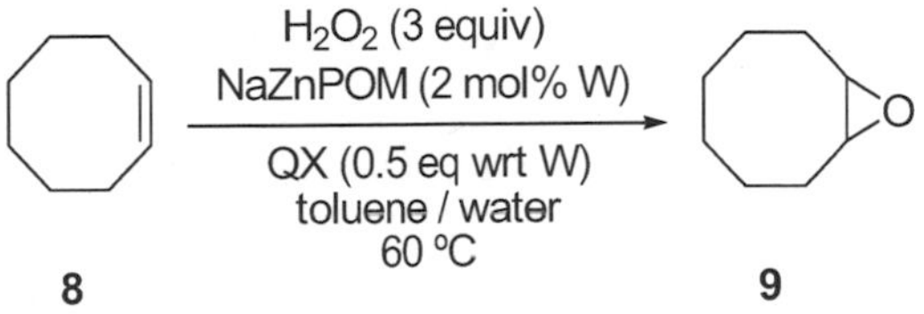

Scheme 4.2 Cyclooctene epoxidation.

In batch, this catalyst system (1 mol%) will convert cyclooctene in a two-phase toluene-water system at 60°C to the epoxide in 87% yield after 90 minutes and in 99% yield after 160 min (Scheme 4.2). The nature of the PTC is extremely important; best results were obtained with $MeNOct_3HSO_4$.

Initial experiments in the CCS led to disappointing results. It became obvious that in this case having the catalyst in the same phase as the hydrogen peroxide at 60°C or above for prolongued periods led to decomposition of the hydrogen peroxide. Thus, hydrogen peroxide was dosed to the aqueous stream, immediately prior to entering the CCS. Although no conversion was observed during the first run, reuse of the aqueous phase and renewed addition of H_2O_2 led to a 20% conversion of cyclooctene. It thus seems that the catalyst needs to be oxidatively activated causing a substantial induction period. This is in contrast to the findings of Alsters et al. who found that this catalyst needs no activation (22). We are currently optimising this system further.

Continuous lipase catalysed esterification of oleic acid

Since enzymes perform best in an aqueous environment, a continuous enzymatic reaction in the CCS, in which the enzyme remains in the aqueous phase and the substrate and product in the organic phase, would seem a feasible option. In view of the high rotational speed we feared that the enzyme might be damaged either by the high shear forces in the centrifuge or through the turbulence of the two-phase system. For this reason we replaced the bottom plate of the CCS, which is indented for optimal mixing with a low-mix bottom plate, which in addition to having a smooth surface, carries an additional set-up which forms a barrier between the inlet and the rotating centrifuge.

In a first experiment, the esterification of oleic acid with 1-butanol catalysed by a *Rhizomucor miehei* lipase was investigated (Scheme 4.3). Lipases usually function at the water/organic interface, which make them extremely suitable for use in the CCS.

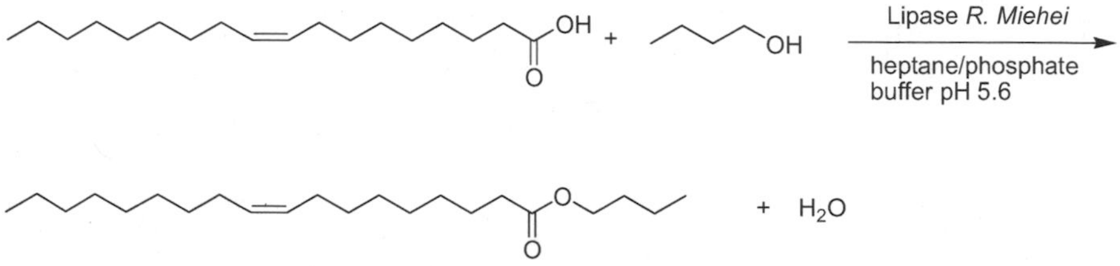

Scheme 4.3 Two-phase lipase catalysed esterification of oleic acid.

The lipase-catalysed reaction between oleic acid and ethanol is known (23), but we found replacement of ethanol with 1-butanol led to much higher conversions; in addition the reaction is faster (24). The esterification of oleic acid with butanol using a crude extract of *penicillium coryophilum* in a micellar system has also been described (25). In batch, this reaction goes to full conversion in spite of the large excess of water present. Presumably, this is driven by the lipophilicity of the reactants. In a first series of experiments an organic phase consisting of a mixture of oleic acid (0.6 mol/L) and 1-butanol (0.9 mol/L) in heptane was used. The aqueous phase consisted of a solution of *R. miehei* lipase (1g/L) in a phosphate buffer pH=5.6. We first examined the effect of the flow rates of both phases and the rotational speed of the centrifuge on the conversion. At the conditions used, the highest steady state conversion (70%) was found at a rotational speed of 40 Hz, and a flow rate of both phases of 6 ml/min. The conversion shows a clear maximum with respect to the flow rate of each phase. At lower flow rates, phase separation in the CCS takes place more efficiently at the expense of the mixed phase, comparable to the effect of high spinning rates. At higher flow rates the residence time in the CCS is too short, leading to lower conversions as well. In this particular case, the optimum flow rate was 6 ml/min for each phase. Similar to the biodiesel case, the rotational speed of the centrifuge has a profound effect on the oleic acid conversion and an optimum was found at 40 Hz.

Using the optimal settings determined above, a lipase-catalysed esterification reaction at a higher enzyme loading (3.0 instead of 1.0 g/L) was performed (Figure 4.4a). After about 1 h, the conversion becomes reasonably steady, fluctuating between 78 and 87% with an average of 82% over the run. Repeat runs showed good reproducibility.

If the aqueous phase containing the enzyme is recycled, the reaction slowed down. Presumably, this is caused by the build-up of an organic impurity in the aqueous phase. Nevertheless, it is still possible to obtain a reasonable conversion of about 78% if the organic phase is also recycled for 90% (Figure 4.4b). This process was run for up to 13 h. During this period the conversion diminishes somewhat. It is clear that this process also needs some optimization with respect to enzyme stability.

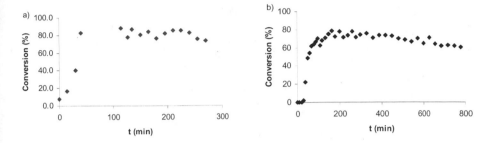

Figure 4.4 Continuous two-phase lipase catalysed esterification of oleic acid in the CCS: (a) Both phases once through. (b) Aqueous phase full recycle; organic phase 90% recycle.

Conclusion

We have shown in this work that it is indeed possible to use a table-top centrifugal contact separator to perform a continuous two-phase catalytic reaction using a homogeneous catalyst with simultaneous separation of the product from the catalyst. Biodiesel was produced continuously with 96% conversion and with an excellent space-time velocity. An unoptimised epoxidation reaction of an olefin with hydrogen peroxide, using a tungsten-based polyoxometalate led to 20% conversion. Surprisingly it was possible to execute a lipase catalysed esterification reaction in the CCS with up to 87% conversion. We are currently refining this new concept, both with respect to hardware development as well as regarding long-term stability of the homogeneous catalysts. The method has been patented by DSM (26).

Acknowledgements

We thank Ir. Gerard Kwant, DSM Research, Chiara Tarabiono and Prof. Dr. Dick B. Janssen, University of Groningen for helpful discussions. This work was funded by the Dutch IBOS program, which contributions from DSM, Organon, the Dutch Ministry of Economic Affairs, the University of Groningen and the Dutch Science Foundation.

References

1.   N. G. Anderson, *Org. Proc Res. Dev.* **5**, 613 (2001).
2.   (a) C. Ramshaw, *Green Chem.* **1**, G15-G17 (1999). (b) A. I. Stankiewicz and J. A. Moulijn, *Chem. Eng. Progress* **96**, 22-34 (2000). (c) A. Stankiewic and J. A. Moulijn, *Re-engineering the chemical processing plant.* Marcel Dekker, New York, (2004).

3. (a) W. Ehrfeld, V. Hessel and H. Löwe, *Microreactors: New technology for modern chemistry*, Wiley-VCH, Weinheim, (2000). (b) V. Hessel, S. Hardt and H. Löwe, *Chemical Micro Process Engineering*, Wiley-VCH, Weinheim, (2004). (c) V. Hessel and H. Löwe, *Chem. Eng. Technol.* **28**, 267-284 (2005). (d) J.-C. Charpentier, *Chem. Eng. Technol.* **28**, 255-258 (2005). (e) J. Yoshida, A. Nagaki, T. Iwasaki and S. Suga, *Chem. Eng. Technol.* **28**, 259-266 (2005). (f) P. Watts and S. J. Haswell, *Chem. Eng. Technol.* **28**, 290-301 (2005). (g) D. M. Roberge, L. Ducry, N. Bieler, P. Cretton and B. Zimmerman, *Chem. Eng. Technol.* **28**, 318-323 (2005). (h) L. Kiwi-Minsker and A. Renken, *Catal. Today.* **110**, 2-14 (2005). (i) J. Kobayashi, Y. Mori and S. Kobayashi, *Chem. Asian J.* **1**, 22-35 (2006). (j) B. P. Mason, K. E. Price, J. L. Steinbacher, A. R. Bogdan and D. Tyler McQuade, *Chem. Rev.* **107**, 2300-2318 (2007).

4. (a) http://isic2.epfl.ch/webdav/site/lcbp/shared/Renken/workshop/presentations/7-Poechlauer.pdf. (b) P. Poechlauer, M. Kotthaus, M. Vorbach, M. Zich and R. Marr, WO2006/125502, 2006 to DSM.

5. (a) D.M. Roberge, *Org. Proc. Res. Dev.* **8**, 1049-1053 (2004). (b) D. M. Roberge, L. Ducry, N. Bieler, P. Cretton and B. Zimmerman, *Chem. Eng. Technol.* **28**, 318-323 (2005).

6. C. de Bellefon, N. Tanchoux, S. Caravieilhes, P. Grenouillet and V. Hessel, *Angew. Chem. Int. Ed.* **39**, 3442-3445 (2000).

7. Y. Onal, M. Lucas and P. Claus, *Chem. Eng. Tech.* **28**, 972-978 (2005).

8. S. Liu, T. Fukuyama, M. Sato and I. Ryu, *Org. Proc. Res. Dev.* **8**, 477-481 (2004).

9. For a preliminary communication about this work see: G. N. Kraai, F. van Zwol, B. Schuur, H. J. Heeres and J. G. de Vries, *Angew. Chem. Int. Ed.* **47**, (2008), accepted for publication.

10. A. Jha, *The Guardian*, July 31, 2003. (http://www.guardian.co.uk/life/thisweek/story/0,,1008930,00.html)

11. (a) E. J. Vandamme, *Biotechnology of industrial antibiotics*. Marcel Dekker, New York, 1984. (b) Z. Likidis and K. Schügerl, *Biotechnol. Bioeng.* **30**, 1032–1040 (1987).

12. Rüffer, U. Heidersdorf, I. Kretzers, G. A. Sprenger, L. Raeven and R. Takors, *Bioprocess Biosyst. Eng.* **26**, 239–248 (2004).

13. (a) R. J. Taylor, I. May, A. L. Wallwork, I. S. Denniss, N. J. Hill, B. Y. Galkin, B. Y. Zilberman and Y. S. Fedorov, *J. Alloys. Comp.* **271-273**, 534-537 (1998). (b) D. H. Meikrantz, L. L. Macaluso, H. W. Sams, III, C. H. Chardin, Jr. and A. G. Federici, US 5,762,800, 1998, to Costner Industries Nevada, Inc.

14. (a) S. Saaby, K. R. Knudsen, M. Ladlow and S. V. Ley, *Chem. Commun.* 2909-2911 (2005). (b) I. R. Baxendale, J. Deeley, C. M. Griffiths-Jones, S. V. Ley, S. Saaby and G. K. Trammer, *Chem. Commun.* 2566-2568 (2006). (c) N. Nikbin, M. Ladlow and S. V. Ley, *Org. Proc. Res. Dev.* **11**, 458-462 (2007).

15. See: www.auxill.nl/uk/cinc.separators.php and www.cincsolutions.com

16. (a) F. Ma and M. A. Hanna, *Bioresour. Technol.*, **70**, 1-15 (1999). (b) J. M. Marchetti, V. U. Michel and A. F. Errazu, *Renew. Sust. Energy Rev.* **11**, 1300-1311 (2007).

17. Gerpen, Van J., Shanks, B., Pruszko, R., Clements D. and Knothe, G., *Biodiesel Production Technology*. NREL/SR-510-36244 (2004).

18. The pharmaceutical industry uses more solvents for reactor cleaning than for the actual chemical production. D. J. C. Constable, P. J. Dunn, J. D. Hayler, G. R. Humphrey, J. L. Leazer, Jr., R. J. Linderman, K. Lorenz, J. Manley, B. A. Pearlman, A. Wells, A. Zaks and T. Y. Zhang, *Green Chem.* **9**, 411–420 (2007).
19. R. Hage, J. E. Iburg, J. Kerschner, J. H. Koek, E. L. M. Lempers, R. J. Martens, U. S. Racherla, S. W. Russell, T. Swarthoff, M. R. P. van Vliet, J. B. Warnaar, L. van der Wolf and B. Krijnen, *Nature* **369**, 637-639 (1994).
20. B.S. Lane and K. Burgess, *J. Am. Chem. Soc.* **123**, 2933-2934 (2001).
21. G. Dubois, A. Murphy and T. D. P. Stack, *Org. Lett.* **5**, 2469-2472 (2003).
22. P. T. Witte, P. L. Alsters, W. Jary, R. Müllner, P. Poechlauer, D. Sloboda-Rozner and R. Neumann, *Org. Proc. Res. Dev.* **8**, 524-531 (2004).
23. G. N. Kraai, J. G. M. Winkelman, J. G. de Vries and H. J. Heeres, *Biochem. Eng. J.*, 2008, accepted for publication.
24. A. C. Oliveira, M. F. Rosa, M. R. Aires-Barros and J. M. S. Cabral *J. Mol. Cat. B: Enz.* **11**, 999-1005 (2001).
25. A. M. Baron, M. I. M. Sarquis, M. Baigori, D. A. Mitchell and N. Krieger, *J. Mol. Catal. B.:Enz.* **34**, 25-32 (2005).
26. J. G. de Vries, G. J. Kwant and H. J. Heeres, WO 2007/031332, to DSM IP Assets, bv.

5. Application of Scavengers for the Removal of Palladium in Small Lot Manufacturing

Alan M. Allgeier, Emilio E. Bunel, Tiffany Correll, Mina Dilmeghani, Margaret Faul, Jacqueline Milne, Jerry Murry, Joseph F. Payack, Christopher Scardino, Bradley J. Shaw and Xiang Wang

Amgen Inc., 1 Amgen Ctr Dr., Thousand Oaks, CA 91320

Allgeier@amgen.com

Abstract

Scavengers are useful for the removal of palladium from organic products originating from palladium catalyzed reactions. Their ease of use and reliable performance with low risk for byproduct formation make them excellent candidates for early-phase process development efforts. A workflow for development of Pd scavenging process is described from screening to implementation. A small volume (1 mL) screening protocol is described and compared to a literature method. Screening experiments employ analysis by ICP/MS and HPLC to identify hits for further evaluation and process development. Some scavengers, including certain polypropylene bound scavengers and trimercaptotriazine, have a marked influence of temperature on Pd removal efficiency, making room temperature screening unreliable. Other scavengers, including some silica bound scavengers were less dependent on temperature under the screening protocol. Process development studies define the reaction progress of removing Pd, the influence of scavenger loading and the loss of product to adsorption. Demonstration of Pd scavenging on multi-kilogram production scale is described.

Introduction

The increasing role of palladium catalyzed coupling reactions (1) in the synthesis of fine chemicals and pharmaceuticals has engendered the need for efficient palladium removal strategies from organic products. For pharmaceutical products the demands for purity are set by regulatory agencies and often, a maximum of 20 ppm Pd in the active pharmaceutical ingredient (API) is targeted. To meet this purity requirement manufacturers have employed a variety of Pd removal strategies, including distillation, crystallization away from the Pd compound(s), extraction and adsorption. (2) Extraction and adsorption have been aided by use of trisodium trithiocyanuric acid, also known as trimercaptotriazine (TMT-Na$_3$) (3) and other molecular agents. (4) Recently a large variety of adsorption agents or scavengers have been introduced to aid in the facile development of Pd removal processes. These scavengers take the form of carbonaceous, polymeric or oxide materials functionalized with strongly binding ligands for Pd and are the subject of the current manuscript.

Scavengers are a particularly attractive option for early-phase process development, where speed to production is of great value. They offer a simple and generally applicable solution to Pd removal (stir and filter) with only slight risk for side reactions and yield loss by product adsorption. A drawback to scavengers is increased costs, comprising both cost of goods for the scavenger and cost of processing for an additional unit operation. Often this cost disadvantage is far overcome by the benefit in the speed of development.

Selection of a scavenger for a given application is guided by: efficiency of palladium removal, feasibility in the process and process cost. The efficiency of palladium removal is influenced by many process parameters including the stability of the palladium complex or particle in solution, temperature, solvent, diffusion limitations in the scavenger and the binding strength of the scavenger. Screening is often employed in the selection of a scavenger and a protocol has been described in the literature. (5) The protocol involves low volume screening (often 100 mg of product in 1.0 mL solution and 50 mg of scavenger) at room temperature to identify lead candidates for further evaluation. We have observed that some resins perform poorly at room temperature and better at elevated temperature both in terms of rate of Pd removal and in the ultimate achievable Pd concentration. With this motivation we describe, herein, an improved methodology for scavenger screening, as well as key follow-up experiments essential for designing an efficient process.

Experimental Section

Typical procedure for scavenger screening: 20 mg of each scavenger were dispensed to individual 2 mL vials using the Powdernium automated solid dispenser. A solution of 100 mg of product, typically in 1 mL total volume and containing >100 ppm Pd, was dispensed to each vial. Afterward sealing vials were stirred overnight at room temperature of 65°C in a vial block. The vial blocks were centrifuged for five minutes and a portion of the supernatant decanted for analysis by HPLC and ICP/MS.

Pd removal was determined as follows. An aliquot of a representative liquid or solid sample was accurately weighed and subsequently digested by refluxing in nitric and/or hydrochloric acid using a closed vessel microwave procedure (CEM MARS5 Xpress or Milestone Ethos EZ). Cooled, digested samples were diluted, matrix matched to standards, and referenced to a linear calibration curve for quantitation; an internal standard was employed to improve quantitation. All samples were analyzed by an Inductively Coupled Plasma Mass Spectrometer or ICP/MS (Perkin Elmer SCIEX Elan DRC II) operated in the standard mode.

Product recovery was determined by chromatographic analyses, conducted on an Agilent 1100 HPLC with Halo C18 column and water/methanol/trifluoroacetic acid mobile phase.

Results and Discussion

For both screening and process development it is critical to evaluate at what stage of a process the scavenging step should be conducted. Selection criteria include minimizing impact to the process cycle time and cost, while providing efficient removal of Pd and not increasing the concentration of impurities or degradants. For truly homogeneous reactions (e.g., homogeneous catalyst hydrogenations) the scavenging step can simply be appended to the reaction step with no extra unit operation, other than filtration. Unfortunately, many reactions catalyzed by homogeneous Pd complexes are still heterogeneous due to the presence of poorly soluble bases, which must be quenched in an aqueous workup. In these cases, scavenger treatments often follow the quench. Still other processes might involve a solvent switch before a crystallization, opening the possibility of conducting the scavenger treatment in either of the two solvents. In these cases efficiency of Pd removal in each solvent is often the decisive factor. The efficiency of Pd removal is also affected by the temperature of the scavenger step. Temperature selection must be done balancing this efficiency with any thermal stability issues associated with the product of interest.

To facilitate the scavenger screening process, a scavenger kit was assembled. Candidate scavengers covered a range, spanning polymer-supported, silica-supported, and carbonaceous scavengers as well as certain organic molecule scavengers known to have low solubility in common organic solvents (trithiocyanuric acid, a.k.a. trimercaptotriazine, TMT, and its tri-sodium salt, TMT-Na$_3$). To focus our efforts, candidates were selected only from materials known to have reliable commercial supply for small lot manufacturing. Manufacturers of such materials include: Reaxa, Johnson-Matthey, Degussa (now Evonik), Silicycle, Sybron Chemicals, Norit, Polymer Laboratories and others and many of the products are distributed through Sigma-Aldrich and Strem. To prepare a scavenger screening kit, a solid-dispensing robot (Flexiweigh or Powdernium) was employed to dispense 20 mg of each scavenger to a 2 mL HPLC vial.

The scavenger screening protocol involved dispensing 1.0 mL of crude product solution containing about 100 mg of product to each vial. The vials were sealed and agitated overnight either with a magnetic stir bar or by shaking, optionally at room temperature or 65°C. The 20% loading of scavenger relative to product, the longer exposure time and the higher temperature screening are more process relevant conditions than those in the literature protocol.[5] After cooling to room temperature, the vials were centrifuged in a plate centrifuge and aliquots are withdrawn for analysis by ICP/MS and HPLC assay.

The protocol was applied to the Pd catalyzed sulfonamide coupling (6) shown in equation 1. This reaction utilized 2.5 mol% Pd and required an acidic quench of the base to generate a toluene solution of the product. Scavenger screening was conducted on this product solution with a library of 31 scavengers at both room temperature and 65°C. Results are shown in Figure 5.1. Control experiments with no scavenger provided Pd concentrations of 423 ppm for room temperature and 384 ppm for 65°C, suggesting the possibility that some Pd may be colloidal and agglomerate on high-temperature treatment.

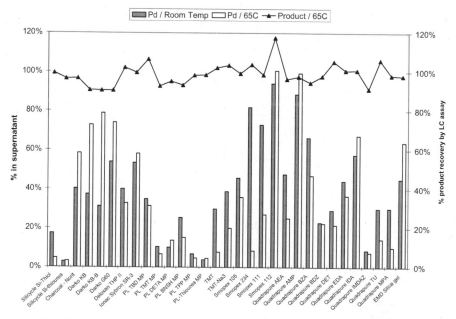

$$Ar\!-\!Br \;+\; \underset{O}{\overset{Ar'}{\underset{\displaystyle \|}{S}}}\!\!-\!NH_2 \xrightarrow[\text{Toluene}]{\substack{Pd_2dba_3CHCl_3 \\ t\text{-BuXphos} \\ \text{base}}} \; Ar'\!-\!\overset{O}{\underset{O}{\overset{\displaystyle \|}{S}}}\!\!-\!\overset{H}{N}\!\!-\!Ar \qquad \text{(eq 1)}$$

It will be noted that there was some variability in the LC assay data, which may reflect small changes in solvent volume due to evaporation or adsorption into the vial septa overnight. Accordingly, LC assays on such small scale heated experiments were not particularly reliable in differentiating performance. Product adsorption was best determined in larger scale experiments (screening validation). The scavenger providing the lowest Pd content after treatment was Silicycle Thiourea leaving only 3% Pd (i.e., <14 ppm) in the supernatant. This scavenger was used extensively in early process development work. For many of the scavengers, higher-temperature treatments provided greater reduction in Pd content, with some notable exceptions, many of which were carbon adsorbents. That counter-intuitive behavior was not well understood. For some of the scavengers the influence of temperature was quite pronounced. Examples included TMT (30% Pd remaining with room temperature treatment vs. 8% for 65°C treatment) and Smopex 234

Figure 5.1. Results for scavenger screen. Percent Pd remaining in the supernatant, as determined by ICP/MS, is shown in the bars for room temperature (dark) and 65°C (light). Percent product remaining in supernatant, as determined by HPLC assay, is shown, for 65°C data set only, as ▲.

(82% Pd remaining with room temperature treatment and 9% for 65°C). These results were notable because they highlighted the need for high-temperature screening methodology, which opens the possibility of lower cost scavengers.

In evaluating process options, the contribution of the scavengers to overall cost was evaluated. Using the data of Figure 5.1, the performance (ppm Pd removed) was normalized to the cost per unit mass for the subset of scavengers, which removed at least 80% of the Pd. Results are shown in Figure 5.2 and helped to focus further process development work. By this analysis the group of 13 potential hits was reduced to 4 options for further evaluation: the incumbent Silicycle Thiourea, TMT, TMT-Na$_3$, and Smopex 234.

Screening validation was conducted on a 1 g product basis (10 mL of product solution) with 20% w/w loading of scavenger on product at 65°C overnight. With this scale of experiment each scavenger experiment was independently filtered and the filter cake washed with toluene. The filtrate was analyzed for residual Pd by ICP/MS and for product recovery by HPLC. Results are shown in Table 5.1. For these analyses it was important to look not only at the ppm Pd in solution but also at the ppm Pd normalized to the concentration of product, which is more reflective of

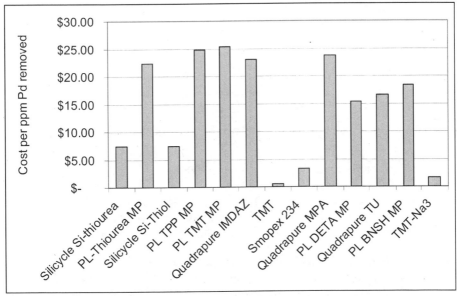

Figure 5.2. A cost comparison for scavengers removing at least 80% of the Pd.

the Pd content on a "solids basis." In evaluating the data of Table 5.1 it is notable that the starting solution was more concentrated than the other samples shown, having not been diluted in a filter/wash step. The screening validation affirms that the best performance was observed for Silicycle Thiourea, also the most expensive scavenger in the screening validation. The exercise also highlighted potential scale-up issues. With TMT the treated mixture was a fine suspension of TMT in product solution and presented filterability concerns on scale. Admixtures of TMT and a filter agent, such as diatomaceous earth, may mitigate filterability problems but our timeline did not allow for development of this option. TMT-Na₃ is a very basic compound and its use on 1 g scale provided a product solution that slowly turned to a gel and gave low product assay. Notably, the product in equation 1 has a base sensitive functional group (not shown). Smopex 234 was of interest due to favorable cost but the product recovery did not meet our criteria. For applications where metal reclamation is desirable, there would be an advantage to carbonaceous scavengers such as Smopex 234, TMT or carbons, because these are amenable to incineration and recovery of Pd from the ash.

With a tight production timeline, we chose to move forward with Silicycle Thiourea, which offered good Pd removal and filterability, albeit at a higher cost. In subsequent process development work the progress of Pd removal versus time was evaluated at both room temperature and 65°C. Results are shown in Figure 5.3 for 3 g scale reactions with 30% w/w Silicycle Thiourea. Pd removal was not only faster at elevated temperature but also more Pd was removed.

Table 5.1. Screening Validation Results.

	ppm Pd in solution - screening	ppm Pd in solution - 1 g scale	Assay product recovery (1 g scale)	ppm Pd on "solids basis"[a] (1 g scale)
Starting Sol'n		381		3577
Smopex 234	36	38	84%	496
TMT	33	15	104%	264
TMT-Na₃	85	81	78%	1555
Silicycle Thiourea	14	11	102%	158

a. "solids basis" refers to dividing solution ppm Pd by the product assay.

A challenge of compressed timelines to production is the difficulty of setting meaningful specifications for Pd content. In the example of equation 1, it was known that excess Pd would complicate downstream chemistry and purification steps but it was not known what level of Pd removal was required to produce a quality product. Accordingly, a conservative approach was applied and excess Silicycle Thiourea was incorporated into the process. In a 70 g experiment, successive charges of approximately 15% w/w Silicycle Thiourea relative to product theoretical yield were utilized with at least 2 h stirring time at 65°C in between charges. The results are shown in Figure 5.4. The asymptotic decrease of Pd concentration suggested an equilibrium limited process or the presence of multiple Pd species, at least one of which was not readily adsorbed. Efforts were not devoted to evaluate potential diffusional limitations.

Using the conditions developed above the Pd scavenging process was conducted on a 1.5 kg product scale giving a toluene product solution with 474 ppm Pd, which on treatment with 15% w/w Silicycle Thiourea gave 69 ppm Pd. An additional 15% w/w Silicycle Thiourea gave 17 ppm Pd in solution. The same solution contained 0.104 g product per mL solution, giving 163.5 ppm Pd on a "solids basis" (17 / 0.104 = 163.5). A crystallization further reduced the Pd content of the product to 6 ppm. The Si content of the product was 87 ppm. Notably, throughout the studies, Si contamination in the product solution was similarly minimal.

Opportunities for further study abound. For the current process we are returning to studies of TMT to define and improve filterability. Additionally, we have observed, in some cases, better scavenger performance in laboratory studies than in kilo-lab or pilot scale equipment, perhaps owing to the extent of oxygen exposure. Identification of the actual Pd species in solution before scavenger application may be effected by NMR studies, examination of model compounds and liquid chromatography with coupled ICP/MS detection. Our observations and those in the literature (5) suggest many parameters influence the efficiency and selectivity of Pd scavenging including the nature of the product, solvent, and ligand. The variety of parameters influencing Pd removal efficiency necessitates multiple scavenger options and a screening protocol. Results must be coupled with sound process development strategies to successfully produce pharmaceutical ingredients with low Pd content.

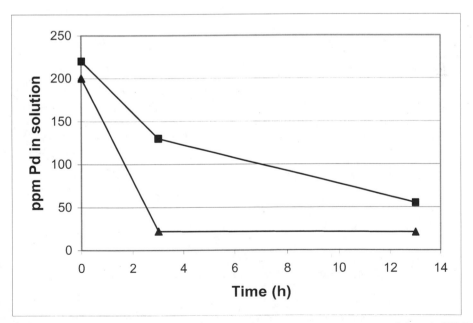

Figure 5.3. Progress of Pd removal versus time at room temperature (■) and 65°C (▲) for Silicycle Thiourea.

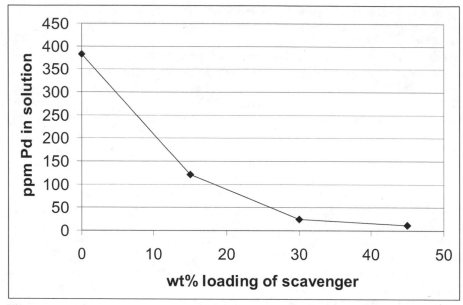

Figure 5.4. Pd removal versus successive additions of Silicycle Thiourea using 65°C treatment and at least 2-hour stirring time between charges.

Conclusions

Scavengers offer a practical and expedient option for removing Pd from process streams to ensure quality of organic products. A scavenger screening and process development workflow has been exemplified for a Pd catalyzed sulfonamide coupling. The screening protocol involves treatment of a candidate process stream with 20% w/w scavenger on product at both room temperature and 65°C followed by analysis of Pd and product adsorption. High-temperature treatment increased the efficiency of Pd removal and some scavengers were notable in that they only showed acceptable performance at higher temperature. Evaluation of process costs is key to identifying Pd removal solutions. While scavengers add cost to a process, their use is often justified by the speed to production in early phase development.

Acknowledgements

Dr. Jessica Tan is acknowledged for development of the HPLC assay.

References
1. King, A. O. and Yasuda, N., *Topics in Organometallic Chemistry*, **6**, 205-245, (2004).
2. Garrett, C. E. and Prasad, K., A*dv. Synth. Catal.*, **346**, 889-900 (2004).
3. Rosso, V. W.; Lust, D. A.; Bernot, P. J.; Grosso, J. A.; Modi, S. P.; Rusowicz, A.; Sedergran, T. C.; Simpson, J. H.; Humora, M. J. and Anderson, N. G., *Org. Proc. Res. Devel.*, **1**, 311-314 (1997).

4. Flahive, E.; Ewanicki, B.; Yu, S.; Higginson, P. D.; Sach, N. W. and Morao, I., *QSAR Comb. Sci.* **26**, 679-685 (2007).
5. Welch, C. J.; Albaneze-Walker, J.; Leonard, W. R.; Biba, M.; DaSilva, J.; Henderson, D.; Laing, B.; Mathre, D. J.; Spencer, S.; Bu, X. and Wang, T., *Org. Proc. Res. Devel.*, **9**, 198-205 (2005).
6. For examples of sulfonamide coupling see: (a) Fujita, K-I; Yamashita, M.; Puschmann, F.; Alvarez-Falcon, M. M.; Incarvito, C. D. and Hartwig, J. F., *J. Am. Chem. Soc.*, **128**, 9044-9045 (2006). (b) Ikawa, T.; Barder, T. E.; Biscoe, M. R. and Buchwald, S. L., *J. Am. Chem. Soc.*, **129**, 13001-13007 (2007).

6. Large-Scale Synthesis of Thienobenzazepine Derivatives Using Two Efficient Metal Catalyzed Processes Telescoped: Nitro Reduction and Intramolecular Aminocarbonylation

William G. Holloway[1], Kumiko Takeuchi[1], Michael L. Prunier[1] and Antonio Navarro[2]

[1]*Lilly Research Laboratories, Lilly Corporate Center, Indianapolis, IN, 46285*
[2]*Lilly Research Laboratories, Lilly Technology Center, Indianapolis, IN, 46360*

navarro_antonio@lilly.com

Abstract

In this chapter we describe a novel, safe and efficient large-scale synthetic approach to tricycle thienobenzazepines. The key steps in the synthesis include a chemoselective hydrogenation of an aryl-nitro functionality in the presence of a 3-bromo thiophene and a subsequent palladium-catalyzed intramolecular aminocarbonylation; telescoped sequentially after simple catalyst and solvent exchange.

Introduction

The tricycle system thieno[2,3-*c*]benza(oxa,aza)zepine (**1**, Figure 6.1) is an interesting molecular construct which is contained within several important drug molecules that exhibit a diversity of biological activities.[1,2] For example, olanzapine contains this tricyclic motif and is one of the most well-known atypical antipsychotic drugs used for the treatment of schizophrenia and bipolar mania.[3,4] Other drug compounds that also contain the thienobenzazepine core are being developed for the treatment of central nervous system (CNS), cardiovascular, and/or gastrointestinal disorders (**2**, Figure 6.1)[5] and have been useful in treating diseases characterized by excess renal reabsorption of water (**3**, Figure 6.1).[6]

Several syntheses of thienobenzazepines have been reported in the literature;[7,8] however, they generally require the use of reagents and/or reaction conditions which present challenges on a preparative scale. In this manuscript we describe a new and efficient synthesis of the thieno[2,3-*c*]benzazepine tricycle that involves a key selective reduction-intramolecular amidation cyclization process and enables large-scale preparation of these important compounds.

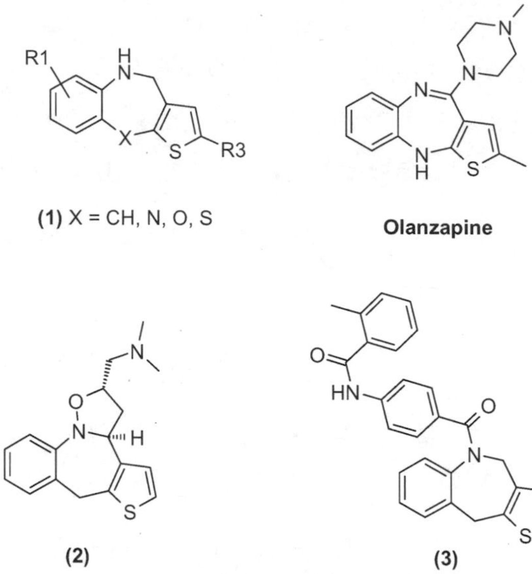

(1) X = CH, N, O, S   **Olanzapine**

(2)   **(3)**

Figure 6.1

Experimental Section

Hydrogenation: To a 2250 ml glass Parr bottle was charged 5% Pt/C (sulfided) (14.56 g) and the reaction vessel was purged with nitrogen (3X). Tetrahydrofuran (THF) (50 ml) was charged to the purged vessel to wet the catalyst and a solution of 3-bromo-2-(4-methoxy-2-nitro-benzyl)-5-methyl-thiophene (30.64 g, 0.0895 mol) dissolved in THF (450 ml) was then added to the pressure vessel. The reaction vessel was sealed, purged with nitrogen (3X), purged with hydrogen (3X), pressurized with hydrogen (60 psi, 400 KPa), and agitated at ambient temperature for 10 hours, while the hydrogen pressure was maintained at 60 psi. The reaction was very exothermic and cooling was needed during the early part of the reaction to control the temperature. After 10 hours the agitation was stopped and the reaction was vented to remove the excess hydrogen and was purged with nitrogen (3X). The reaction mixture was sampled for analysis by LC/MS to confirm that the desired product was the main product. No de-bromination product was observed.

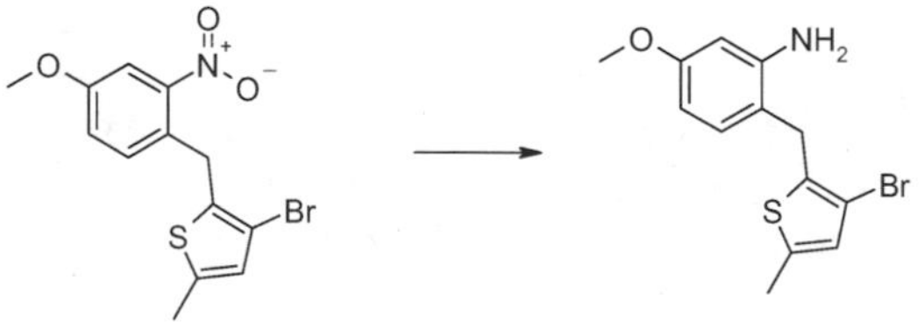

(1)

The reaction mixture was filtered to remove the Pt/C catalyst and was concentrated to tan solids (26.9 g) using a rotary evaporator.

Carbonylation: To a 1 L hastalloy Parr pressure reactor were charged palladium acetate (2.999 g, 0.0134 mol), 1,1'bis-(diphenylphosphino) ferrocene (DPPF) (10.92 g, 0.0197 mol), the tan solids recovered from the hydrogenation above (26.9 g), dry methanol (220 ml, 174 g, 5.43 mol), dry triethylamine (62.0 ml, 45.0 g, 0.445 mol), and dry acetonitrile (330 ml). The reaction vessel was purged with nitrogen (4X), purged with carbon monoxide (4X) and pressurized with carbon monoxide (100 psi, 690 KPa), sealed, and agitated at 90°C for 24 hours while the carbon monoxide pressure was maintained at 100 psi. The reaction was then cooled to ambient temperature and the carbon monoxide was vented from the reaction vessel. The reaction mixture was filtered through a pad of Celite® and crystallized from dichloromethane-hexanes to give 7-methoxy-2-methyl-4,9-dihydro-3-thia-9-aza-benzo[f]azulen-10-one as a yellow solid.

$$+ \quad O\overset{+}{\equiv}C^{-} \ {}^{+}HO- \quad \longrightarrow \qquad\qquad (2)$$

Results and Discussion

The development of new and efficient routes that enable large-scale synthesis and manufacturing of new drugs is an active area of research in the pharmaceutical industry. During the early stages of the drug discovery process, small amounts of structurally related compounds (~10-100 mg) are made using synthetic routes and methodologies that, more often, are not optimized and/or amenable to large-scale preparation. In the later stages of the drug development process, only selected compounds with the best pharmacological profile are tested in assays where larger quantities of material are required (>100 g). In these instances, it is imperative to access the desired molecules in an efficient synthetic manner using protocols that are safe and work well in preparative scale syntheses.

Previous syntheses: An example of this point can be recognized by examination of one known synthesis of thienobenzazepines (Scheme 6.1).[7] This synthetic route involves a key palladium-catalyzed cross-coupling of stannyl intermediate **3**, prepared by method of Gronowitz et al.,[9] with 2-nitrobenzyl bromide. Acetal deprotection and reductive cyclization afforded the desired thienobenzazepine tricycle **4**. In support of structure activity relationship studies, this intermediate was conveniently acylated with various acyl chlorides to yield several biologically active compounds of structure type **5**. While this synthetic approach does access intermediate **4** in relatively few synthetic transformations for structure activity relationship studies, this route is seemingly unattractive for preparative scale requiring stoichiometric amounts of potentially toxic metals that are generally difficult to remove and present costly purification problems at the end of the synthesis.

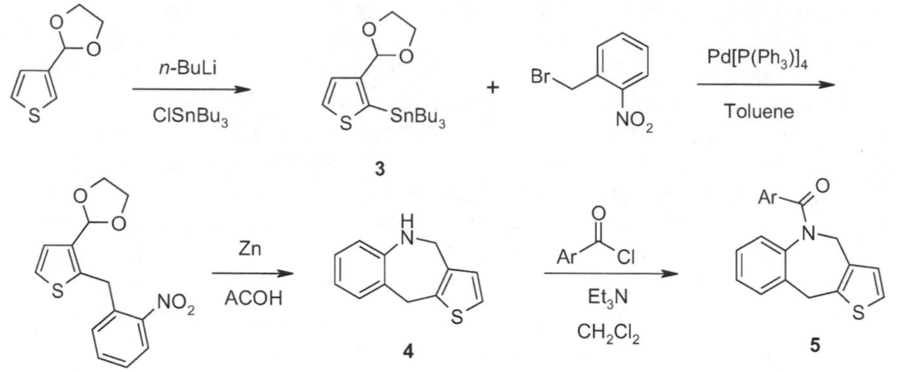

Scheme 6.1

Another examination involves a synthesis of thienobenzazepines based on the formation of key intermediate **6**; prepared according to the method of McDowell and Wisowaty (Scheme 6.2).[10] Selective reduction of this intermediate using zinc dust in 28-30% ammonia solution afforded the benzoic acid **7**, which upon subsequent *Curtius* rearrangement and aluminum trichloride-mediated cyclization furnished the oxo-azepine **8**. While this synthetic approach gave the tricycle in a few synthetic transformations, many of the same concerns as above exist when considering large scale preparation of **8**; the use of large amounts of zinc, sodium azide, and aluminum trichloride.

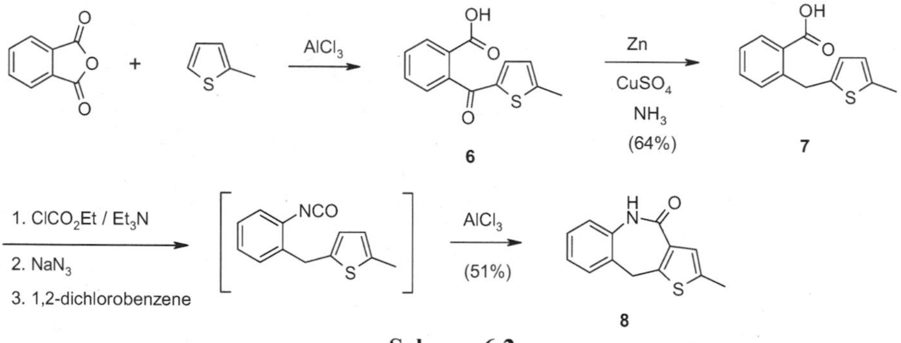

Scheme 6.2

Synthetic routes that access appropriately substituted thienobenzazepines are also quite important for medicinal chemistry structure activity relationship studies, and many involve similar bond connectivity strategies. One notable example employs the use of commercially available 4-methyl-3-nitrophenol (Scheme 6.3). Methylation of the phenol followed by bromination, hydrolysis, and oxidation of the benzylic alcohol afforded aldehyde **9** in quantitative yield. Treatment of this aldehyde with 5-lithio-2-methylthiophene provided, after dehydroxylation, nitro intermediate **A** in good overall yield. Reduction of the nitro functionality and treatment with phosgene presented the corresponding isocyanide; which upon cyclization using aluminum trichloride in a *Friedel-Crafts* fashion afforded the desired tricycle thienobenzazepinone **10**. This process involved 9 synthetic steps and afforded final compound **10** in *c.a.* 29% overall yield; however, this approach also presented issues when considered for a preparative scale; it was long, tedious, and used reagents such as phosgene, the carcinogenic agent dimethyl sulfate, and/or conditions that are generally not large-scale friendly.

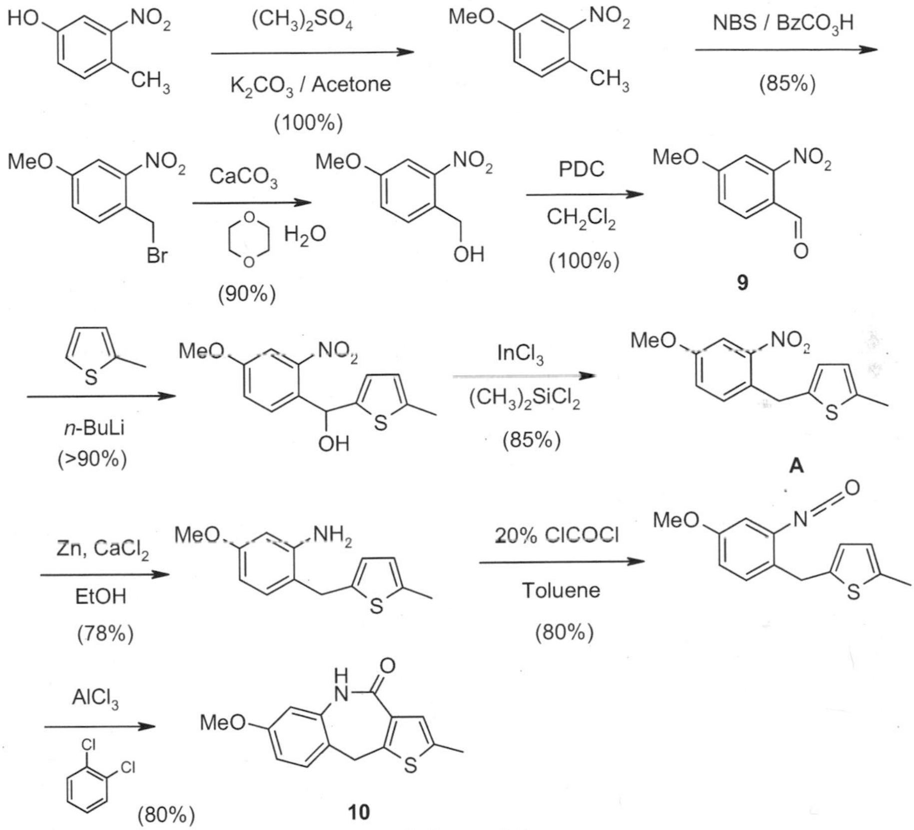

Scheme 6.3

A New Improved Synthesis of Tricycle Thienobenzazepines: Application of chemistry recently developed by Knochel[11] combined with the well-described halogen dance (HD) reaction,[12] allowed preparation of our key intermediate **A'** in only three synthetic transformations (Scheme 6.4). In this respect, treatment of 2-bromo-5-methylthiophene with lithium diisopropylamide followed by dimethylformamide afforded aldehyde **11** in good yield. Iodo-magnesium exchange with commercial 4-iodo-3-nitro anisole followed by reaction with **11** afforded the thiophene carbinol **12**. Dehydroxylation of **12** provided our key intermediate **A'** which presented the requisite functionality to examine our approach to the construction of the seven-member ring system.

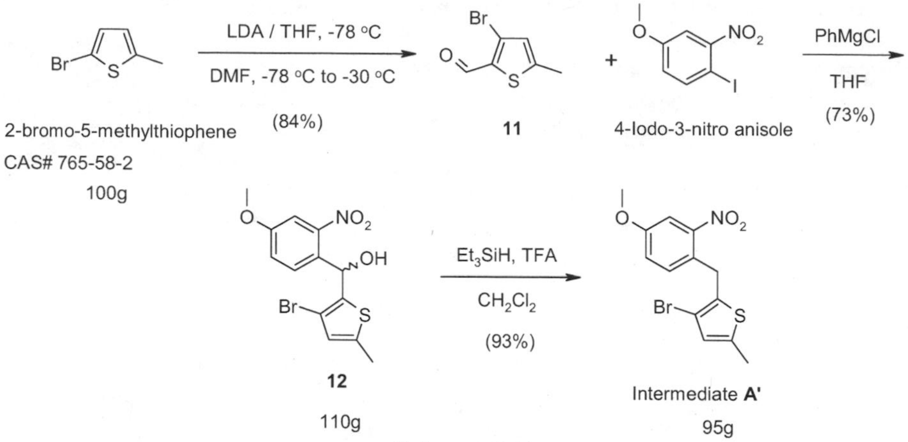

Scheme 6.4

With intermediate **A'** in hand, we envisioned two synthetic approaches to the tricyclic thienobenzazepine that relied upon a palladium catalyzed carbonylation wherein the thieno-bromide functionality served as the electrophilic component (see Scheme 6.5 and 6.6). In the first approach, a palladium-mediated carbonylation reaction followed by reduction of the nitro group, through standard hydrogenation conditions, would spontaneously cyclize to the desired lactam (Scheme 6.5). The methoxy-carbonylation did indeed proceed quite well affording the corresponding methyl ester intermediate. Without isolation, the reaction mixture was then subjected to standard hydrogenation conditions but no appreciable amount of the lactam was detected. This result was likely due to contamination of the heterogeneous catalyst (Pd/C) with residual phosphine ligand from the previous carbonylation reaction, since chromatography isolation of the nitro-ester and subsequent reduction with Pd/C did provide acceptable yields of the desired lactam.

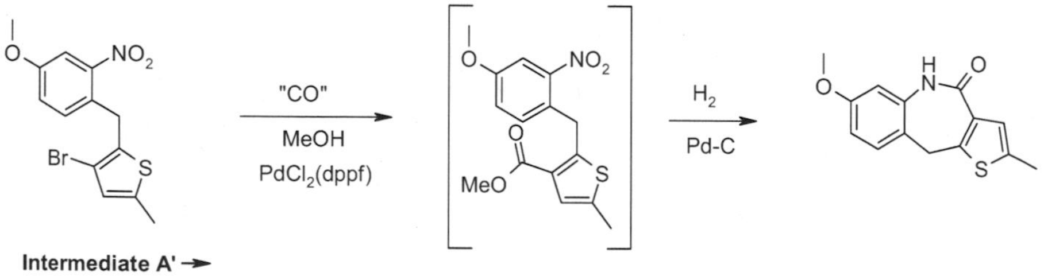

Scheme 6.5

In our second approach, we considered reversal of these functional group transformations, wherein reduction of the nitro group followed by palladium-mediated intramolecular amidation would provide the desired tricylcic lactam (Scheme 6.6). One concern with this approach, however, was to identify reaction conditions that would selectively reduce the nitro functionality and not lead to appreciable amount of de-bromination of the thiophene prior to the intramolecular cyclization. Treatment of intermediate **A'** with 5% Pt-C (sulfided) in THF, with an optimized reaction time of ~10h, cleanly afforded the corresponding aniline and no appreciable de-bromination product(s). Filtration of the catalyst and evaporation of the solvent gave a crude solid that was immediately subjected to treatment with carbon monoxide (100 psi), palladium acetate, and 1,1'bis-(diphenylphosphino) ferrocene (DPPF) in a solution of methanol, triethylamine, and acetonitrile. Upon heating at 90°C, spontaneous intramolecular amidation resulted, affording the desired tricycle thienobenzazepine system in good overall yield. The overall process is considered as an intramolecular aminocarbonylation reaction starting from a masked aniline.

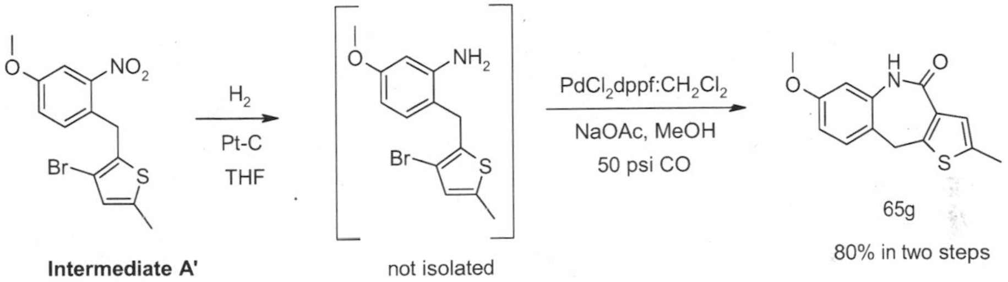

Scheme 6.6

Conclusions

In summary, we have described a novel and efficient synthesis of thienobenzazepine derivatives in which the key transformation includes a telescoped process involving a selective nitro reduction followed by palladium-mediated intramolecular amidation. The process developed is quite amenable for preparative scale (multi-gram) and presents significant advantage to those reported previously with respect to overall yield (e.g., 50% vs. 17% overall yield), total number of synthetic transformations (4 vs. 9), and reagents and/or conditions that are suitable for large-scale synthesis.

Acknowledgements

We would like to thank Dr. Paul Ornstein for his advice and technical direction during this work. Also, we thank Dr. John Masters for his help and discussions during the writing of this manuscript and Dr. James Audia for his revisions.

References

1. Di Cesare, Maria Assunta; Campiani, Giuseppe and Butini, Stefania, PCT Int. Appl. (2005), 32 pp. CODEN: PIXXD2 WO 2005097797.
2. Povlen, U. J.; Noring, U.; Fog, R.; Gerlach, *J. Acta. Psychiatr. Scand.* **1985**, *71*, 176.
3. Beasley, C. M.; Tollefson, G.; Tran, P.; Satterlee, W.; Sanger, T. and Hamilton, S., *Neuropsychopharmacology* **1996**, *14*, 111.

4. Rasmussen, K.; Benvenga, M. J.; Bymaster, F. P.; Calligaro, D. O.; Cohen, I. R.; Falcone, J. F.; Hemrick-Luecke, S, K.; Martin, F. M.; Moore, N. A.; Nisenbaum,L. K.; Schaus, J. M.; Sundquist, S. J.; Tupper, D. E.; Wiernicki, T. R. and Nelson, D. L., *J. Pharm. Exp. Therapeutics* **2005**, *315*(3), 1265.
5. (a) WO1997039001A. (b) Andres-Gil, J. I. et al., *Drug Data Rep* **1998**, *20*(2): 112.
6. (a) WO1997047624. (b) Aranapakam, V.; Albright, J. D.; Grosu, G. T.; Chan, P. S.; Coupet, J.; Saunders, T.; Ru, X. and Mazandarani, H. *Bioor. & Med. Chem. Lett.* **1999**, *9*, 1733.
7. Kohara, T.; Tanaka, H.; Kimura, K.; Fujimoto, T.; Yamamoto, I. and Arita, M. *Synthesis* **2002**, *3*, 355.
8. Aranapakam, V.; Albright, J. D.; Grosu, G. T.; Chan, P. S.; Coupet, J.; Saunders, T.; Ru, X. and Mazandarani, H. *Bioor. & Med. Chem. Lett.* **1999**, *9*, 1733.
9. (a) Gronowitz, S.; Timari, G. *J. Heterocycl. Chem.* **1990**, *27*, 1159. (b) Gronowitz, S. and Timari, G. *J. Heterocycl. Chem.* **1990**, *27*, 1127.
10. MacDowell, D. W. H. and Wisowaty, J. C. *J. Org. Chem.* **1971**, *36*, 3999.
11. Sapountzis, I.; Dube, H.; Lewis, R.; Gommermann, N. and Knochel, P. *J. Org. Chem.* **2005**, *70*, 2445.
12. See for example: Schnurch, M.; Spina, M.; Khan, A. F.; Mihovilovic, M. D. and Stanetty, P. *Chem. Soc. Rev.* **2007**, *36*(7), 1046.

7. Manufacture and Application of Asymmetric Hydrogenation Catalysts

Lee T. Boulton,[1] Christopher J. Cobley, Ian C. Lennon, Graham A. Meek, Paul H. Moran, Céline Praquin[2] and James A. Ramsden

Dowpharma, Chirotech Technology Limited, 162 Cambridge Science Park, Milton Road, Cambridge, CB4 0GH, United Kingdom
[1] Chemical Development, GSK Research and Development Limited, Medicines Research Centre, Gunnels Wood Road, Stevenage, Hertfordshire, SG1 2NY, United Kingdom
[2] Pfizer Ltd, IPC 533, Ramsgate Road, Sandwich, Kent, CT13 9NJ, United Kingdom

gmeek@dow.com

Abstract

This chapter discusses the development of scaleable and robust manufacturing processes for rhodium and ruthenium containing precatalysts that are used for the asymmetric hydrogenation of a diverse range of olefins, ketones and imines. The application of these precatalysts to the preparation of a variety of pharmaceutical intermediates, many of which have been operated on commercial scale, is also discussed.

Introduction

Increasingly stringent regulatory demands by the Food and Drug Administration (FDA) and the continuing drive for economic processes has led to the greater adoption of purely synthetic methods to produce small molecule pharmaceuticals as single enantiomer compounds. Catalytic asymmetric hydrogenation is one such method and has become an accepted addition to commercially relevant technologies for pharmaceutical manufacture and four drugs recently approved by the FDA have reports of this technology being used in their preparation.[1,2] Despite the apparent maturity of catalytic asymmetric hydrogenation, with over 3000 ligands being known for this transformation,[3] only a handful of catalyst systems are truly available on a commercially relevant scale within a reasonable lead time.

Chirotech, now part of Dowpharma, obtained the exclusive license to the use of DuPhos, BPE and 5-Fc based catalysts for the production of pharmaceuticals and intermediates from DuPont in 1995. It has been demonstrated that the DuPhos technology has a broad substrate scope and Chirotech has reported, in collaboration with our pharmaceutical customers, asymmetric hydrogenation processes for Tipranavir,[1] Candoxatril,[1] and Pregabalin.[1] While certain catalyst systems can be prepared *in situ*, isolation and use of the defined ligand/metal complex allows for greater quantification, which is an important consideration for cGMP manufacture. Additionally, in general and in our experience, when compared to *in situ* prepared catalyst systems pre-formed catalyst systems provide more robust and reliable asymmetric hydrogenation processes.   To further develop, and potentially commercialise, such hydrogenation processes required the diphosphine rhodium precatalysts to be produced on multi-kilogram scale and it was clear that the initial synthetic methods were not practical in terms of yield, purity and economics.

Ruthenium catalysis is complementary to rhodium catalysis as it is effective with different substrate classes.   Dowpharma has in-licensed world-class technologies in this important area from the Japan Science and Technology Corporation (JST), including [diphosphine-RuCl$_2$-diamine] systems for asymmetric ketone hydrogenation, developed by professors Noyori and Ikariya.[4]   As we found more commercial applications for the ketone hydrogenation technology, it became obvious that the original procedure was not ideal in terms of productivity and yield for the preparation of precatalysts on a kilogram scale.

Results and Discussion

Manufacture of rhodium precatalysts for asymmetric hydrogenation. Established literature methods used to make the Rh-DuPhos complexes consisted of converting (1,5-cyclooctadiene) acetylacetonato Rh(I) into the sparingly soluble bis(1,5-cyclooctadiene) Rh(I) tetrafluoroborate complex which then reacts with the diphosphine ligand to provide the precatalyst complex in solution.[5] Addition of an anti-solvent results in precipitation of the desired product. Although this method worked well with a variety of diphosphines, yields were modest and more importantly the product form was variable. The different physical forms performed equally as well in hydrogenation reactions but had different shelf-life and air stability.

A scaleable process was developed that delivers crystalline rhodium precatalysts in very high yields with excellent chemical purity and substantially improved chemical and physical stability (Scheme 7.1).[6] The new process involves taking (1,5-cyclooctadiene) acetylacetonato Rh(I) in an ethereal solvent, treating it with an alcohol solution of strong acid, such as tetrafluoroboric acid, to give a soluble bis-solvato species, which is then reacted with an ethereal solution of the diphosphine ligand. Shortly after the addition of the ligand crystallisation of the precatalyst complex is observed. The rate of nucleation is controlled at higher temperatures through rate of ligand addition furnishing granular, free-flowing precatalyst in exceptionally high yields. The crystalline form with each diphosphine ligand is subtly different and the crystallisation protocol requires tailored optimisation for each product although the basic process is the same for each precatalyst. Once crystallisation has been carried out, the precatalyst can be readily weighed in air, as long as it is stored under nitrogen, and has a long shelf-life (>12 months). With this process working exceptionally well for Me-DuPhos rhodium complexes, it was extended to the manufacture of a variety of diphosphine containing precatalysts (Scheme 7.1).

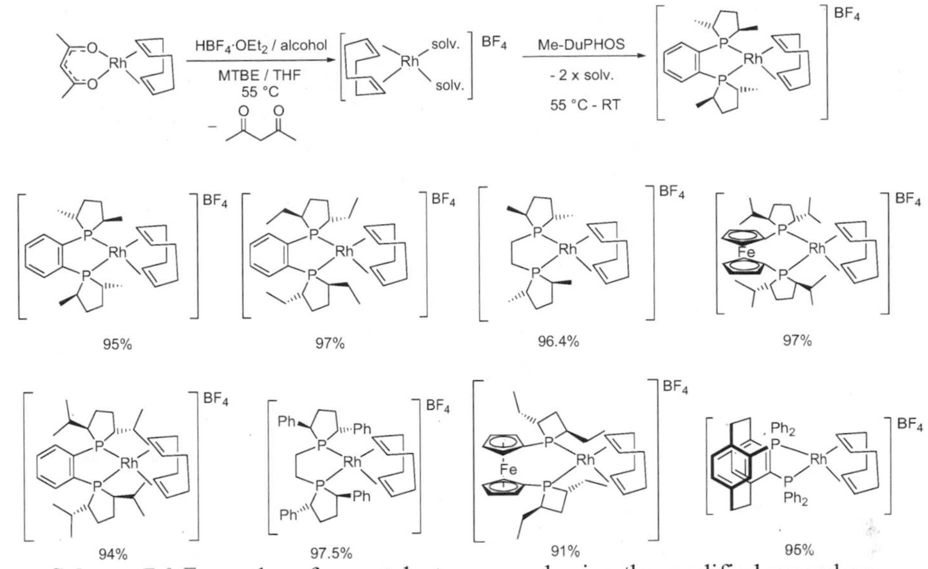

Scheme 7.1 Examples of precatalysts prepared using the modified procedure.

Asymmetric hydrogenation applications of rhodium catalysts. Rhodium DuPhos and BPE catalysts have been used for the asymmetric hydrogenation of a wide range of enamide substrates. Working with AstraZeneca an asymmetric hydrogenation process was developed for the synthesis of ZD6126, a water soluble phosphate pro-drug of *N*-acetylcolchinol.[7] This was a particularly challenging substrate owing to the presence of the seven-membered ring system in the hydrogenation substrate. While the asymmetric hydrogenation of five- and six-membered cyclic enamides has been reported, there are relatively few examples of seven-membered ring systems being reduced with high enantioselectivity.[8] Extensive screening at a molar substrate to catalyst ratio (S/C) of 100 revealed that rhodium precatalysts containing DuPhos and BPE ligands provided poor enantioselectivity and variable conversion for this substrate. However, [(*R*,*R*)-^{t}Bu-FerroTANE Rh(COD)]BF_4 afforded good enantioselectivity and the activity of this

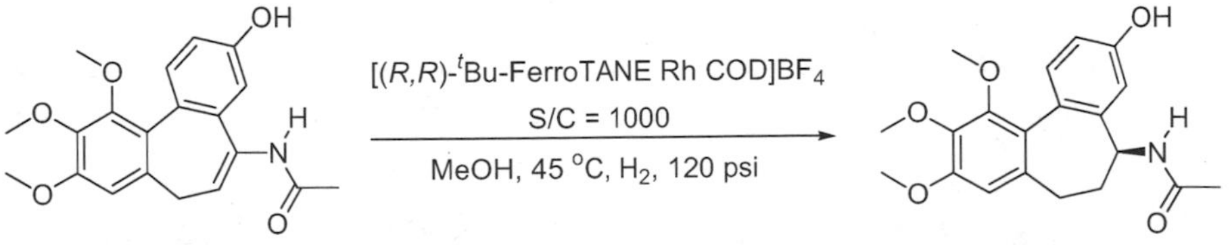

ZD6126 Enamide

(*S*)-ZD6126 Phenol, 91% ee

Scheme 7.2 Synthesis of (S)-ZD6126 by asymmetric hydrogenation.

precatalyst was sufficiently high that at a S/C of 1000 the product was obtained with 91-94% ee (Scheme 7.2). This particular enantiomer of the catalyst provided (*S*)-ZD6126, although the ready availability of both enantiomers of the ligand means that a process based on asymmetric hydrogenation to provide the desired (R)-ZD6126 could be readily obtained.

Unnatural α-amino acids are an important class of building blocks within the pharmaceutical industry and are commonly incorporated into peptide isosteres. Owing to the imparted increased metabolic stability, drug substances containing unnatural α-amino acids frequently have a greater bioavailability and hence more potency.[9] One such group of unnatural α-amino acids are 3,3-diarylalanines, in which the two aryl substituents are typically the same, and such fragments have been incorporated into a thrombin inhibitor,[10] a factor IIa antagonist[11] and an HIV aspartyl protease inhibitor.[12] In spite of their clear importance, there is currently no efficient, scaleable, economic and general synthetic route to 3,3-diarylalanines that provides an individual enantiomer in high selectivity, with routes based on the use of a chiral auxiliary[13] or classical resolution[14] or asymmetric phase transfer catalysis[15] being the state of the art. We therefore investigated an asymmetric hydrogenation approach to this class of unnatural α-amino acid and initially focused on 3,3-diphenylalanine.[16]

The use of rhodium catalysts for the synthesis of α-amino acids by asymmetric hydrogenation of N-acyl dehydro amino acids, frequently in combination with the use of a biocatalyst to upgrade the enantioselectivity and cleave the acyl group which acts as a secondary binding site for the catalyst, has been well-documented.[17] While DuPhos and BPE derived catalysts are suitable for a broad array of dehydroamino acid substrates, a particular challenge posed by a hydrogenation approach to 3,3-diphenylalanine is that the olefin substrate is tetra-substituted and therefore would be expected to have a much lower activity compared to substrates which have been previously examined.

The hydrogenation substrate was initially prepared from the commercially available dehydroamino acid derived from cinnamic acid, which was readily transformed into the corresponding methyl ester (Scheme 7.3). Bromination followed by treatment with triethylamine provided the β-bromoenamide ester as an approximately 1.3:1 mixture of stereoisomers in excellent yield.[18] Despite the steric encumbrance, the second phenyl group was introduced under relatively mild conditions by performing a Suzuki cross-coupling reaction on the mixture of stereoisomers catalysed by ligand-free palladium(II) acetate.[19] This step produced α-acetamidoacrylate as a by-product *via* de-bromination, which was readily removed by performing a solvent slurry, affording the 3,3-diphenyl substituted dehydroamino acid in good yield.

As expected initial examination of the hydrogenation of this substrate revealed its relatively low activity compared to dehydroamino acids that provide 3-aryl-α-amino acids. By carrying out the hydrogenation at an elevated temperature, however, the inherent low activity could be overcome. A screen of the Dowpharma catalyst collection at S/C 100 revealed that several rhodium catalysts provided good conversion and enantioselectivity while low activity and selectivity was observed with several ruthenium and iridium catalysts. Examination of rate data identified [(R)-PhanePhos Rh (COD)]BF₄ as the most active catalyst with a rate approximately

360 times faster than the best catalyst derived from the DuPhos/BPE family of ligands. Reducing the catalyst loading to S/C 250 using [(R)-PhanePhos Rh (COD)]BF$_4$ afforded (R)-3,3-diphenylalanine with 88% ee with a t$_{1/2}$ for the hydrogenation of 3 minutes. Although this rate of reaction suggests the catalyst system is highly active, further efforts to develop this into an economic process were hindered by a significant reduction in the rate of reaction upon decreasing the catalyst loading. This behaviour suggested that the substrate may have contained an impurity which poisoned the active hydrogenation catalyst.

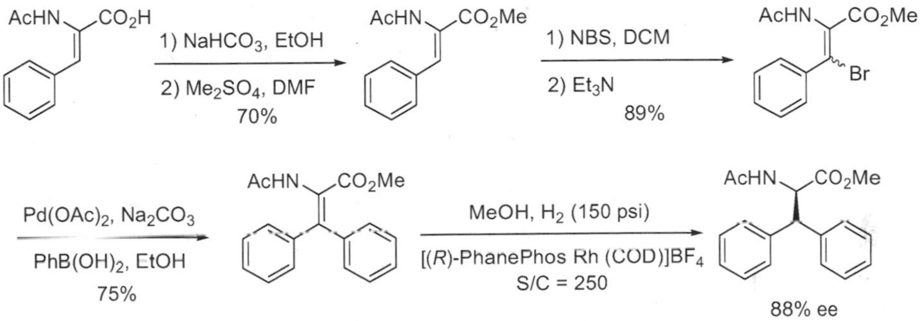

Scheme 7.3 Asymmetric hydrogenation approach to (R)-3,3-diphenylalanine.

The deleterious effects of catalyst poisoning when carrying out asymmetric hydrogenations at low catalyst loading cannot be overemphasised. In order to eliminate the possibility that the substrate synthesis introduced inhibitory impurities, an alternative synthetic protocol was examined (Scheme 7.4). The use of a brominating agent and an expensive palladium catalysed step in the initial route could limit the development of this as an economically favourable process and this was further motivation to examine alternative routes to the hydrogenation substrate.

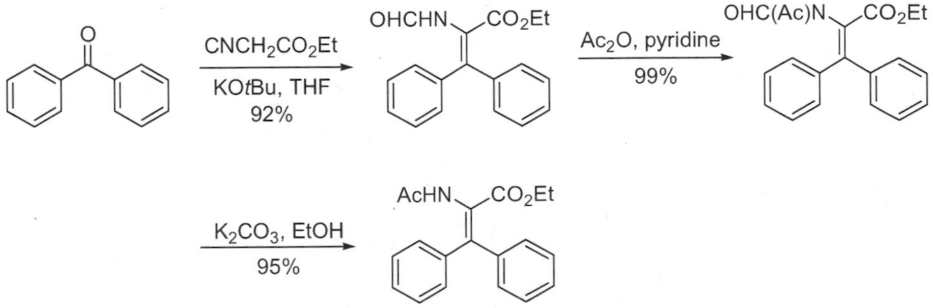

Scheme 7.4 Modified synthesis of the hydrogenation substrate.

The commercially available biphenyl ketone and ethyl isocyanoacetate underwent a smooth condensation reaction and following protecting group manipulations the hydrogenation substrate was produced in excellent yield.[20] Examination of the asymmetric hydrogenation of material produced using this protocol revealed that the intermediate *N*-formyl species was a catalyst poison and in order for this route to be adopted high purity hydrogenation substrate had to be obtained.

Although the asymmetric hydrogenation route to 3,3-diphenylalanine *via* this modified substrate preparation was not developed further, Dowpharma had a requirement to rapidly develop and scale up the manufacture of a related 3,3-diarylalanine product. The work to 3,3-diphenylalanine centred around substrate preparation and removal of impurities leading to high activity associated with the PhanePhos catalyst system allowed for a facile transfer from laboratory scale experiments to the commercial manufacture of the related diphenylalanine derivative by a robust, reproducible and scaleable procedure.

Manufacture of ruthenium precatalysts for asymmetric hydrogenation. The technology in-licensed from the JST for the asymmetric reduction of ketones originally employed BINAP as the diphosphine and an expensive diamine, DAIPEN.[4] Owing to the presence of several patents surrounding ruthenium complexes of BINAP and Xylyl-BINAP, [HexaPHEMP-RuCl$_2$-diamine][21] and [PhanePHOS-RuCl$_2$-diamine][22] were introduced as alternative catalyst systems in which a cheaper diamine is used. Compared to the BINAP-based systems both of these can offer superior performance in terms of activity and selectivity and have been used in commercial manufacture of chiral alcohols on multi-100 Kg scales.

These complexes were originally prepared using the procedures of Noyori,[4] whereby the diphosphine ligand was reacted with an [(arene)RuCl$_2$]$_2$ species in DMF at 100°C followed by treatment with a suitable diamine, typically DPEN, DACH or DAIPEN, to provide the desired product. Significant by-product formation and uneconomic yields (~70%) were observed using this protocol. However, it was found that an isolated [diphosphine(arene)RuCl]Cl complex could be reacted with a diamine at moderate temperatures in ethereal solvents, routinely leading to >95% yield in excellent purity and this process has been applied to a variety of catalysts (Figure 7.1).[23]

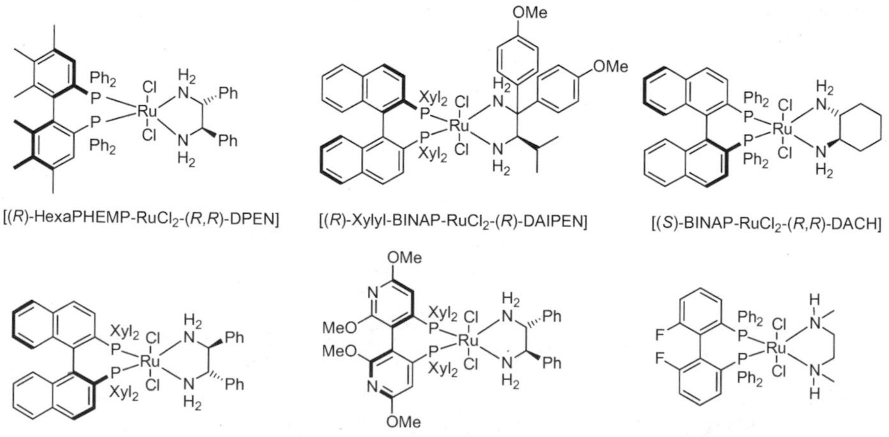

[(*R*)-HexaPHEMP-RuCl$_2$-(*R*,*R*)-DPEN]   [(*R*)-Xylyl-BINAP-RuCl$_2$-(*R*)-DAIPEN]   [(*S*)-BINAP-RuCl$_2$-(*R*,*R*)-DACH]

[(*S*)-Xylyl-BINAP-RuCl$_2$-(*S*,*S*)-DPEN]   [CTH-(*R*)-Xylyl-P-Phos-RuCl$_2$-(*R*,*R*)-DPEN]   [F-BIPHEP-RuCl$_2$-DMEDA]

Figure 7.1 [Diphosphine-RuCl$_2$-diamine] precatalysts produced in >95% yield.

Asymmetric hydrogenation applications of ruthenium catalysts. Compared to the asymmetric hydrogenation of ketones, the metal catalysed asymmetric hydrogenation of imines is a significant challenge. The presence of *syn/anti* imine isomers, enamine tautomers and the hydrolytic instability of imines, as well as the catalyst poisoning effect of amines, account for this difference in the behaviour of ketones and imines. Given that [diphosphine-RuCl$_2$-diamine] complexes mediate hydrogenation by an outer-sphere mechanism,[24] we postulated that such inhibitory behaviour of amines may not be prevalent with these precatalysts and we were amongst the first to report their use for the catalytic asymmetric hydrogenation of imines.[25]

Working in collaboration with Oril Industrie we examined asymmetric hydrogenation approaches to S 18986,[26] which has potential application for the treatment of memory and learning disorders, central nervous system (CNS) trauma and neurodegenerative disease. This had previously been prepared in a modestly enantioselective manner by use of lithium aluminium hydride in the presence of a chiral modifier, which provided S 18986 in up to 76% ee. This route has several disadvantages in addition to the hazards associated with using lithium aluminium hydride and the modest selectivity. A stoichiometric quantity of the chiral modifier was required and the reaction was incomplete. The unreacted starting material had to be removed prior to carrying out two recrystallisations in order to enhance the enantioselectivity to >99% ee. Transfer hydrogenation or pressure hydrogenation using a variety of rhodium, ruthenium and iridium complexes reported for related imines were unsuccessful when applied to producing S 18986 with either high conversion or enantioselectivity.

Screening the sulfoximine substrate against over 40 [diphosphine-RuCl$_2$-diamine] precatalysts revealed that only biaryl diphosphine containing complexes were useful and based on these results and the ready availability of the ligand, [(R)-BINAP-RuCl$_2$-(R,R)-DPEN] was identified as the preferred precatalyst. Active hydrogenation catalysts capable of carrying out asymmetric hydrogenation of ketones are generated from [diphosphine-RuCl$_2$-diamine] complexes by treatment with a quantity of a base which is typically in the order of 1-5 mol% relative to the substrate. For this substrate, however, the proportion of base proved critical with between 0.8 and 1.5 equivalents being required to effect full conversion. Examination of the experimental parameters by factorial design revealed that conversion correlated positively with temperature while hydrogen pressure has a small effect. Moreover, the enantioselectivity correlated negatively with hydrogen pressure while increasing temperature had a marginally deleterious effect. The addition of toluene as a co-solvent led to a higher enantioselectivity albeit at a reduced conversion owing to limited solubility of the substrate. By taking these factors into account, the asymmetric hydrogenation could be performed using [(R)-BINAP-RuCl$_2$-(R,R)-DPEN] at an economically attractive loading of S/C of 2,500 by using toluene as co-solvent and carrying out the reaction at a higher temperature than used during the screening process (Scheme 5). In this manner, S 18986 was produced in 87% ee and excellent yield. A single recrystallisation from acetonitrile afforded S 18986 with >99% ee and this also provided material with an industrially acceptable low level of ruthenium.

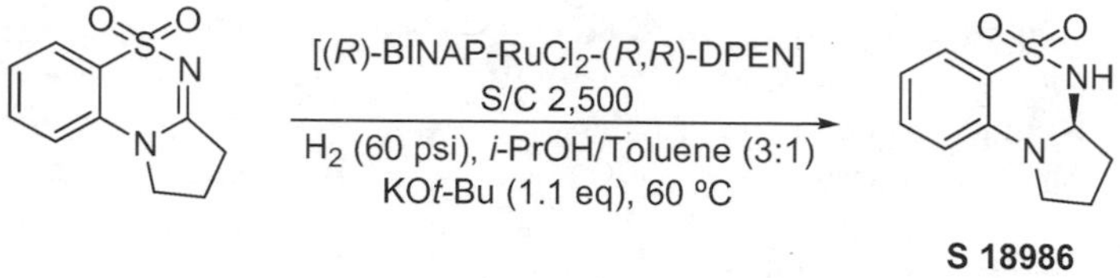

Scheme 7.5 Asymmetric hydrogenation approach to S 18986.

Conclusions

The development of efficient, scalable and economic manufacturing protocols for rhodium and ruthenium precatalysts on a commercial scale is pivotal if asymmetric hydrogenation is going to be further embraced by the pharmaceutical industry. In this chapter we have reported a protocol which provides [diphosphine Rh (COD)]BF$_4$ precatalysts in excellent yield and with improved physical form, providing longer shelf-life and greater ease of use. Furthermore, the preparation of [diphosphine-RuCl$_2$-diamine] precatalysts has been improved. The application of these two types of complexes for asymmetric hydrogenation with commercially relevant examples illustrates their value and underlines the point that the improved catalyst preparation will further advance the adoption of catalytic asymmetric hydrogenation for the production of pharmaceutical intermediates.

Acknowledgements

The work into the asymmetric hydrogenation approaches to ZD6126 and S 18986 was carried out in collaboration with AstraZeneca and Oril Industrie, respectively.

References

1. C. J. Pilkington and I. C. Lennon, *Synthesis*, 1639 (2003).
2. K. Fukatsu, O. Uchikawa, M. Kawada, T. Yamano, M. Yamashita, K. Kato, K. Hirai, S. Hinuma, M. Miyamoto and S. Ohkawa, *J. Med. Chem.*, **45**, 4212 (2002); Y. Hsiao, N. R. Rivera, T. Rosner, S. W. Krska, E. Njolito, F. Wang, Y. Sun, J. D. Armstrong, III, E. J. J. Grabowski, R. D. Tillyer, F. Spindler and C. Malan, *J. Amer. Chem. Soc.*, **126**, 9918 (2004); J. A. F. Boogers, U. Felfer, M. Kotthaus, L. Lefort, G. Steinbauer, A. H. M. De Vries and J. G. De Vries, *Org. Process. Res. Dev.*, **11**, 585 (2007).
3. W. Tang and X. Zhang, *Chem. Rev.*, **103**, 3029 (2003).
4. T. Ohkuma, H. Ooka, S. Hashiguchi, T. Ikariya and R. Noyori, *J. Amer. Chem. Soc.*, **117**, 2675 (1995); R. Noyori and T. Ohkuma, *Angew. Chem. Int. Ed. Eng.*, **40**, 40 (2001).
5. M. Green, T. A. Kuc and S. H. J. Taylor, *J. Chem. Soc (D), Chem. Comm.*, **22**, 1553 (1970).
6. J. A. Ramsden and P. H. Moran, WO 2005032712 (2005).

7. I. C. Lennon, J. A. Ramsden, C. J. Brear, S. D. Broady and J. C. Muir, *Tetrahedron Lett.*, **48**, 4623 (2007).
8. J. D. Armstrong, K. K. Eng, J. L. Keller, R. M. Purick, F. W. Hartner, W. B. Choi, D. Askin and R. P. Volante, *Tetrahedron Lett.*, **35**, 3239 (1994).
9. A. Glannis and T. Kolter, *Angew. Chem. Int. Ed. Eng.*, **32**, 1244 (1993).
10. W. L. Cody, C. Cai, A. M. Doherty, J. J. Edmunds, J. X. He, L. S. Narasimhan, J. S. Plummer, S. T. Rapundalo, J. R. Rubin, C. A. Van Huis, Y. St-Denis, P. D. Winocour and M. Arshad Siddiqui, *Bioorg. Med. Chem. Lett.*, **9**, 2497 (1999).
11. A. -R. Kim, K. -H. Cho, J. -H. Lee, S. -H. Lee, H. -J. Kim, H. -J. Maeng, H. -D. Park, T. -H. Kim, B. -C. Kim, S. -J. Kim and K. -S. Choo, WO 2004073747 (2004).
12. R. B. Stranix, J. -F. Lavallee, N. Leberre and V. Perron, WO 2004056764 (2004).
13. H. Josien, A. Martin and G. Chassaing, *Tetrahedron Lett.*, **32**, 6547 (1991).
14. V. Beylin, H. G. Chen, O. P. Goel and J. G. Topliss, US5,198,548 (1993).
15. D. E. Patterson, S. Xie, L. A. Jones, M. H. Osterhout, C. G. Henry and T. D. Roper, *Org. Proc. Res. Dev.*, **11**, 624 (2007).
16. I. T. Boulton, WO 2006127273 (2006).
17. C. J. Cobley, N. B. Johnson, I. C. Lennon, R. McCague, J. Ramsden and A. Zanotti-Gerosa, in *Asymmetric Catalysis on Industrial Scale* (ed. H. U. Blaser and E. Schmidt), Wiley-VCH, Weinhein, 2004, p.269.
18. M. J. Burk, J. G. Allen, W. F. Kiesman and K. M. Stoffan, *Tetrahedron Lett.*, **38**, 1309 (1997).
19. Y. Deng, L. Gong, A. Mi, H. Liu and Y. Jiang, *Synthesis*, 337 (2003).
20. S. Y. Sit, R. A. Parker and J. J. Wright, *Bioorg. Med. Chem. Lett.*, **2**, 1085 (1992).
21. J. P. Henschke, M. J. Burk, C. G. Malan, D. Herzberg, J. A. Peterson, A. J. Wildsmith, C. J. Cobley and G. Casy, *Adv. Synth. Catal.*, **345**, 300 (2003).
22. M. J. Burk, W. Hems, D. Herzberg, C. Malan and A. Zanotti-Gerossa, *Org. Lett.*, **2**, 4173 (2000).
23. P. H. Moran, WO 2007005550.
24. C. A. Sandoval, T. Ohkuma, K. Muniz and R. Noyori, *J. Amer. Chem. Soc.*, **125**, 13490 (2003).
25. C. J. Cobley and J. P. Henschke, *Adv. Synth. Catal.*, **345**, 195 (2003).
26. C. J. Cobley, E. Foucher, J. P. Lecouve, I. C. Lennon, J. A. Ramsden and G. Thominot, *Tetrahedron Asymm.*, **14**, 3431 (2003).

8.   Hydrogenation of Nitro-Substituted Acetophenones

S. David Jackson, Rebecca McEwan and Ron R. Spence

WestCHEM, Dept. of Chemistry, The University, Glasgow G12 8QQ, Scotland, UK

sdj@chem.gla.ac.uk

Abstract

We have examined the hydrogenation of ortho-, meta- and para-nitroacetophenone to the respective aminoacetophenone in the liquid phase over a Pd/Carbon catalyst. The meta- and para-substituted nitroacetophenones gave high selectivity to the aminoacetophenones; however, the ortho-nitroacetophenone, which was the least reactive of the three isomers, produced 1-indolinone as well as ortho-aminoacetophenone. The selectivity to 1-indolinone was typically ~10%. The formation of this product revealed an internal cyclisation reaction. Formation of such a cyclic product would indicate nitrene insertion into a C-H bond, hence suggesting Ar-N as a surface intermediate in the hydrogenation of the nitro-group. This species is rarely considered in mechanistic interpretations of nitro-group hydrogenation over metal surfaces. However, in this system we have, in effect, a chemical trap that allows us to identify the intermediate giving a far clearer insight into the reaction mechanism.

Introduction

Aminoacetophenones are important as food flavourings (1), photopolymerization initiators (2), and pharmaceutical intermediates (3). The hydrogenation of nitroacetophenones to the respective aminoacetophenones has, to the best of our knowledge, not been subject to a systematic study, although there is an early patent by Rylander (4) describing a process for selective hydrogenation of the nitro group in nitroacetophenones using Pd catalysts. Even earlier studies going back to 1946 (5) reveal differences between the isomers but due to the limitations of the time the analysis of the products is incomplete. Many of the early studies and patents used Raney Ni and platinum oxide but often found over-hydrogenation (6). In this study we have examined the hydrogenation of the three nitroacetophenone isomers over a palladium catalyst in a 3-phase batch reactor.

Experimental Section

The catalyst used throughout this study was a 1% w/w palladium on graphite powder (S.A. ~10 m^2g^{-1}) supplied by Johnson Matthey. 2-nitroacetophenone, 3-nitroacetophenone and 4-nitroacetophenone (all Aldrich >99 %) were used without further purification. No significant impurities were detected by GC. Gases (BOC, >99.99 %) were used as received.

The reaction was carried out in a 0.5 l Buchi stirred autoclave equipped with an oil jacket and a hydrogen-on-demand delivery system. 0.05 g of 1% Pd/Al$_2$O$_3$ was added to 280 ml of degassed solvent, propan-2-ol. Reduction of the catalyst was performed *in situ* by sparging the system with H$_2$ (300 cm^{3}min^{-1}) for 30 minutes at 323 K while stirring the contents of the autoclave at 300 rpm. After reduction, the autoclave was adjusted to the appropriate reaction temperature of between 298 and 323 K under a nitrogen atmosphere. Nitroacetophenone (typically 5.6 mmole) was added into the reaction vessel and flushed through with 20 ml of de-gassed propan-2-ol. The autoclave was then mixed briefly at a stirrer speed of 800 rpm and pressurised to 1 barg with nitrogen and a sample was taken. The vessel was then depressurised and then pressurised with hydrogen to the appropriate pressure. Following this the stirrer was set to a speed of 1000 rpm and samples were taken via a sample valve at defined time intervals. The liquid samples were analysed by gas chromatography using a Varian 3400 fitted with a CPsil-19CB column. Standard checks were undertaken to confirm that the system was not under mass transport control.

Results and Discussion

The hydrogenation of the three nitroacetophenone isomers was compared at 3 barg and 323 K. The results are shown in Table 8.1, where it can be seen that the 3-isomer is by far the most reactive with little difference between the 2- and 4-isomers.

Table 8.1. % Conversion of Nitroacetophenone (NAP) Isomers.

Time (min)	2-NAP	3-NAP	4-NAP
15	9	62	11
30	33	82	34
50	50	91	44

The 3- and 4-isomers are selectively hydrogenated to the respective aminoacetophenone; however, the 2-nitroacetophenone is hydrogenated to not only 2-aminoacetophenone but also 1-indolinone (Figure 8.1).

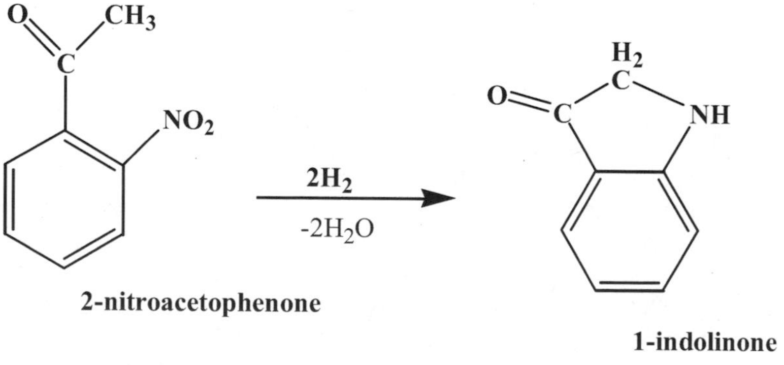

Figure 8.1. Conversion of 2-NAP to 1-indolinone.

A typical reaction profile is shown in Figure 8.2.

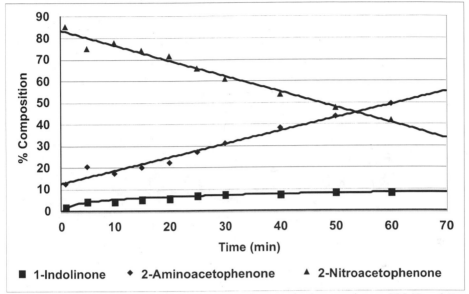

Figure 8.2. Reaction profile for 2-nitroacetophenone hydrogenation at 323 K and 5 barg hydrogen pressure.

It is not immediately obvious why the 3-isomer should be so reactive. The functional groups are meta-directing but are slightly deactivating; however, the adsorption characteristics may play a large part in determining the reaction behaviour and these are reflected in the order of reaction.

The reactions were also performed with deuterium instead of hydrogen. All isomers showed a kinetic isotope effect. Figure 8.3 shows the loss of 3-NAP for both hydrogen and deuterium with a clear kinetic isotope effect calculated to be 1.7. This value compares well with that reported for nitrobenzene hydrogenation of 1.9 (7). 4-NAP reacted similarly. With 2-NAP the loss showed a clear kinetic isotope effect and, as can be seen in Figure 8.4, this is reflected in the production of both products. NMR analysis of the 2-NAP reaction mix showed that the ring protons Hb and Hd (Figure 8.5) are exchanged for deuterium both in 2-NAP and 2-AAP suggesting a flat bonding geometry. Note that the protons exchanged are in a meta position to the nitro-group and that the nitro-group is meta directing.

The activation energies determined for the conversion of 2-, 3- and 4-NAP to 2-, 3- and 4-aminoacetophenone (AAP) are reported in Table 8.2, as is the activation energy for the formation of 1-indoline. As might have been expected the activation energies for the aminoacetophenone isomers are indistinguishable. However, the activation energy for 1-indoline is significantly different.

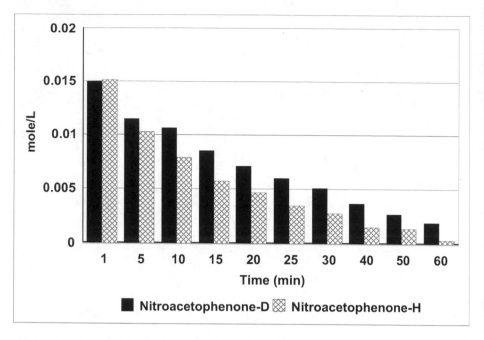

Figure 8.3. Reaction of 3-nitroacetophenone with hydrogen and deuterium at 3 bar and 323 K.

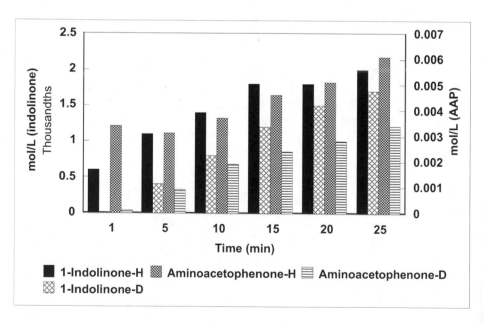

Figure 8.4. Production of 1-indolinone and 2-AAP by reaction of 2-nitroacetophenone with hydrogen and deuterium at 3 bar and 323 K.

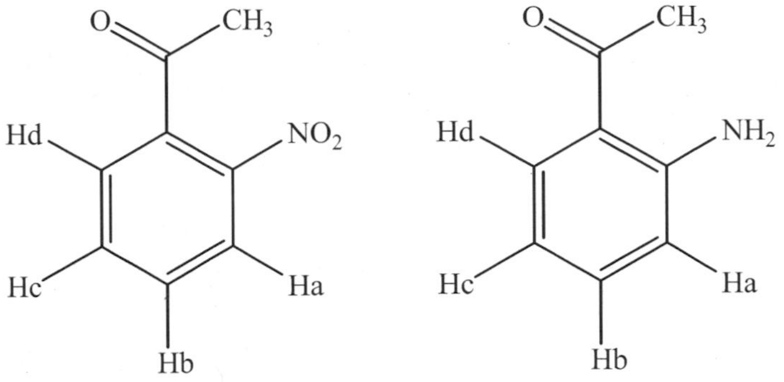

Figure 8.5. 2-NAP and 2-AAP showing the protons Hb and Hd that are exchanged in the presence of D_2.

Table 8.2. Kinetic Parameters for the Hydrogenation of 2-, 3- and 4-Nitroacetophenone.

Formation of	Order of reaction in		Activation Energy (kJ.mol^{-1})
	Hydrogen	2-, 3- and 4-NAP	
2-AAP	0.4±0.2	-0.2±0.4	34±2
3-AAP	-0.1±0.3	-1.1±0.2	33±3
4-AAP	0.8±0.1	-0.7±0.0	36±11
1-indolinone	0.7±0.1	0.8±0.2	48±3

The orders of reaction that were found vary significantly depending upon the isomer being hydrogenated. This variation is in keeping with the literature on nitrobenzene hydrogenation where although a first order dependence on hydrogen pressure is widely reported (8-10), there are also a number of studies reporting different values dependent upon pressure regime and temperature (11-13). For example, Aramendia and coworkers (11), using pressures of 4-7 bar, found a partial reaction order with respect to hydrogen that increased towards first order as the reaction temperature was increased. Similarly, wide variations in the reaction order with respect to the nitrobenzene concentration have been described in the scientific literature. Zero order (13), first order (14), negative order (12) and partial order (15) have all been reported.

The hydrogen order for aminoacetophenone formation goes from zero order for 3-AAP, through half order for 2-AAP, to first order for 4-AAP. The nitroacetophenone orders go from negative first order for 3-AAP, negative half order for 4-AAP, to around zero order for 2-AAP. As the catalyst is the same we would expect that the mode and strength of hydrogen adsorption, in the absence of a co-adsorbate, would be the same. Therefore the changes in hydrogen order reflect the impact of the nitroacetophenone isomer on the hydrogen adsorption. In the case of 3- and 4-NAP there is strong adsorption to the surface with increasing concentration resulting in lower activity. Both these isomers have the potential to bind to the surface through both functional groups and the benzene ring. With 2-NAP it is much more difficult to achieve the correct geometry to allow such adsorption but the deuterium study confirms that the rings and the nitro group are bonded to the surface. Given the large negative effect of 3-NAP it is surprising that the hydrogen order is zero. Zero order would typically imply that the surface was saturated with the reactant (hydrogen) and indeed this would fit in with the high reactivity of the 3-NAP.

        Mechanistically the formation of 1-indolinone is interesting. Formation of ring structures in this way usually occurs through insertion of the nitrogen into the C-H bond of the CH_3 via a nitrene intermediate (16, 17). A general mechanistic description is shown in Scheme 8.A.

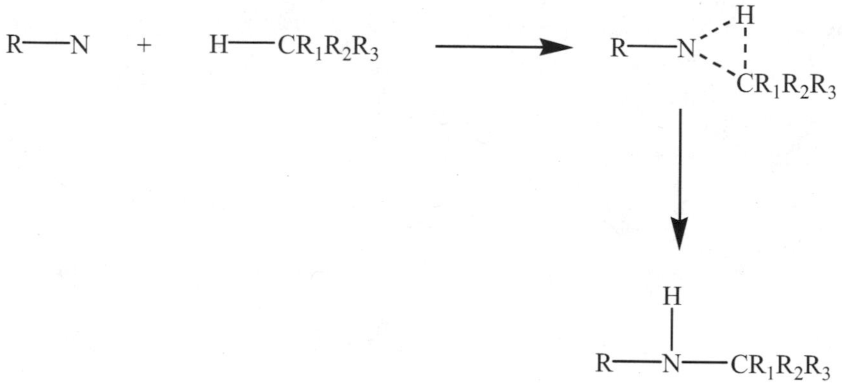

Scheme 8.A.

This implies the presence of Species A adsorbed on the surface. Note that this geometry, with the CH_3 pointing towards the nitrogen, is what would be expected in the starting 2-NAP.

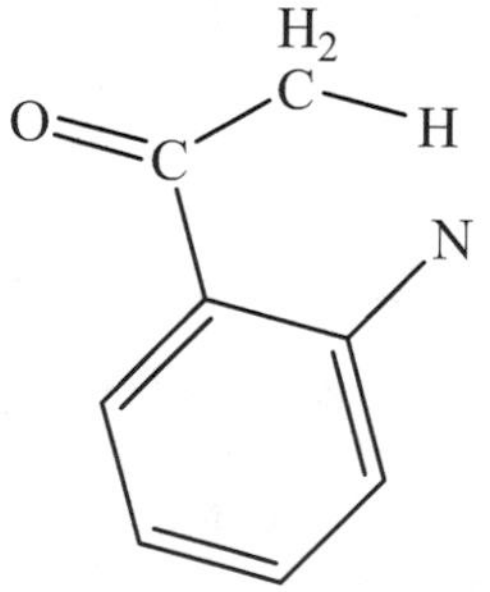

Species A

        In the hydrogenation of nitrobenzene the presence of a nitrene is rarely postulated (7, 8). In a previous study of nitrobenzene hydrogenation (7) we postulated, as part of the overall mechanism, the following step:

2 Ph-N(OH)H(a) → Ph-N=N-Ph(a) + $2H_2O$ → azobenzene.

However, in light of our current results it may be more correct to split this step into the following:

Ph-N(OH)H(a) → Ph-N(a) + H_2O
2 Ph-N(a) → azobenzene

Hence the question arises as to whether $CH_3CO.Ph.N$(ads) is a major intermediate. In the hydrogenation of nitrobenzene the yield of azobenzene was < 1%, and phenyl hydroxylamine can be a significant product which would mitigate against the nitrene as the primary intermediate in the route to aniline. However, as the formation of azobenzene is competing with hydrogenation through to aniline, so although it is unlikely to be the major route to aniline, it may still be significant.

A similar nitrene intermediate can also be postulated in the mechanism of nitrosobenzene hydrogenation. Indeed a standard way of producing Ph-N is from the reaction of Ph-NO with PPh_3 (17). In the hydrogenation of nitrosobenzene the principal product in the early stages is azoxybenzene [7, 18]. It was suggested that azoxybenzene was formed by the following sequence:

Ph-NO → Ph-NO(a)
Ph-NO(a) + H(a) → Ph-N(OH)(a)
2Ph-N(OH)(a) → Ph-N=N(O)-Ph(a) + H_2O

However, if a nitrene is present another route is available, namely:

Ph-N(a) + Ph-NO(a) → Ph-N=N(O)-Ph(a)

Note that the nitrene now allows a common intermediate that can be hydrogenated to aniline from both nitrobenzene and nitrosobenzene. It is also interesting to speculate on another route to form a nitrene on the surface. Phenyl hydroxylamine is a known intermediate whose concentration in solution is highly dependent upon reaction conditions. On adsorption of phenyl hydroxylamine one can write the following:

Ph-N(H)OH(a) ↔ Ph-N(a) + H_2O

This will clearly be an equilibrium with its position determined by a large number of system variables.

Conclusions

The hydrogenation of nitroacetophenones has been studied and considerable kinetic and mechanistic information obtained. Differences in reaction rate, bonding and selectivity have been observed. The formation of 1-indolinone from 2-NAP was unexpected and revealed the presence of a surface nitrene. This intermediate has not been postulated in nitroaromatic hydrogenation previously. Hydrogenation in the presence of deuterium revealed, as well as a kinetic isotope effect, that it is likely that the NAP bonds to the surface in a flat geometry allowing the protons meta to the nitro-group to exchange.

Acknowledgements

The authors would like to acknowledge the help of Prof. R. Hill in the interpretation of the NMR spectra and the support of Engineering and Physical Sciences Research Council and Johnson Matthey plc under the ATHENA grant. One of the authors (R.McE.) also acknowledges the receipt of a Nuffield school bursary.

References

1. B. French, *Bats*, **21(2)**, 12 (2003).
2. Japanese Patent Publication Kokai No. 9-169957.
3. Canadian Patent No. CA 2495386.
4. US Patent No. 3423462.
5. N. J. Leonard and S. N. Boyd Jr., *J. Org. Chem.*, **11**, 405 (1946).
6. US Patent No. 2797244.
7. E. A. Gelder, S. D. Jackson and M. Lok, Chemical Industries (CRC), **115**, (*Catal. Org. React.*), 167 (2007).
8. V. Holler, D. Wegricht, I. Yuranov, L. Kiwi-Minsker and A. Renken, *Chem. Eng. Technol.*, **23**(3), 251 (2000).
9. H.-C. Yao and P.H. Emmett, *J. Am. Chem. Soc,* **83**, 796 (1961).
10. C. Li, Y.-W. Chen and W.-J. Wang, *Appl. Catal. A,* **119**, 185 (1994).
11. M.A. Aramendia, V. Borau, J. Gomez, C. Jimenez and J.M. Marinas, *Appl. Catal.*, **10**, 347 (1984).
12. A. Metcalfe and M.W. Rowden, *J. Catal.*, **22**, 30 (1971).
13. D.J. Collins, A.D. Smith and B.H. Davis, *Ind. Eng. Chem. Prod. Res. Dev.*, **21**, 279 (1982).
14. P.B. Kalantri and S.B. Chandalla, *Ind. Eng. Chem. Prod. Res. Dev.*, **21**, 186 (1982).
15. H.-C. Yao and P.H. Emmett, *J. Am. Chem. Soc.*, **84**, 1086 (1961).
16. P. Muller and C. Fruit, *Chem. Rev.* **103,** 2905 (2003).
17. W. Lwowski in *Nitrenes* (Ed. W. Lwowski), Interscience, New York, 1970, p. 1.
18. G. V. Smith, R. Song, M. Gasior and R. E. Malz Jr., Chemical Industries (Dekker), **53**, (*Catal. Org. React.*), 137 (1994).

9. Tuning Selectivity through the Support in the Hydrogenation of Citral over Copper Catalysts

I.V. Deliy[1], I.G. Danilova[1], I.L. Simakova[1], F. Zaccheria[2], Nicolletta Ravasio[2] and R. Psaro[2]

[1]*Boreskov Institute of Catalysis, Pr. Akademika Lavrentieva 5, 630090, Novosibirsk, Russia*
[2] *CNR –ISTM and Dip. CIMA, University of Milano, via G.Venezian 21, 20133 Milano, Italy*

n.ravasio@istm.cnr.it

Abstract

IR spectroscopy of two supports was used for the determination of their surface acidity. The presence of Lewis acid sites on the surface of sepiolite allowed the preparation of a catalyst able to transform citral into menthol in fairly good yield under very mild conditions (90°C, 1 bar H_2).

Introduction

In the last few years much attention has been devoted to the set up of bi- and poly-functional processes exploiting different active sites on a catalyst surface or, in a few cases, the existence of multifunctional sites [1]. Most popular among this kind of reactions is the cyclization-hydrogenation of citronellal to menthol. Some of us already reported that citronellal can be converted into menthol in a one pot-one step reaction over a Cu/SiO_2 catalyst at 90°C and 1 atm H_2 [2]. More recently some group reported interesting results in the one pot transformation of citral into menthol [3,4]. Although citral is somewhat more expensive than citronellal this transformation represents a valid model reaction for the set-up of domino reactions. Thus the catalyst should be very selective towards the hydrogenation of the conjugated double bond in citral but inactive in the hydrogenation of the carbonyl bond both in citral and citronellal. It should also show some acidity in order to cyclise citronellal while keeping some activities to hydrogenate isopulegol to menthol, possibly with some stereoselectivity to the desired (-)menthol.

Here we wish to report that the support acidity, investigated through IR spectroscopy of adsorbed CO, allows one to tune the selectivity towards different products in the hydrogenation of citral over Cu catalysts.

Experimental Section

Catalysts Cu/SiO_2 and Cu/Sepiolite, with a metal loading ~8% were prepared as already reported [5] by using silica (BET=723 m^2/g, PV=0.66 ml/g, Nippon Gosei, Japan), sepiolite (natural Mg silicate, BET=240 m^2/g, PV=0.4 ml/g, Tolsa, Spain).

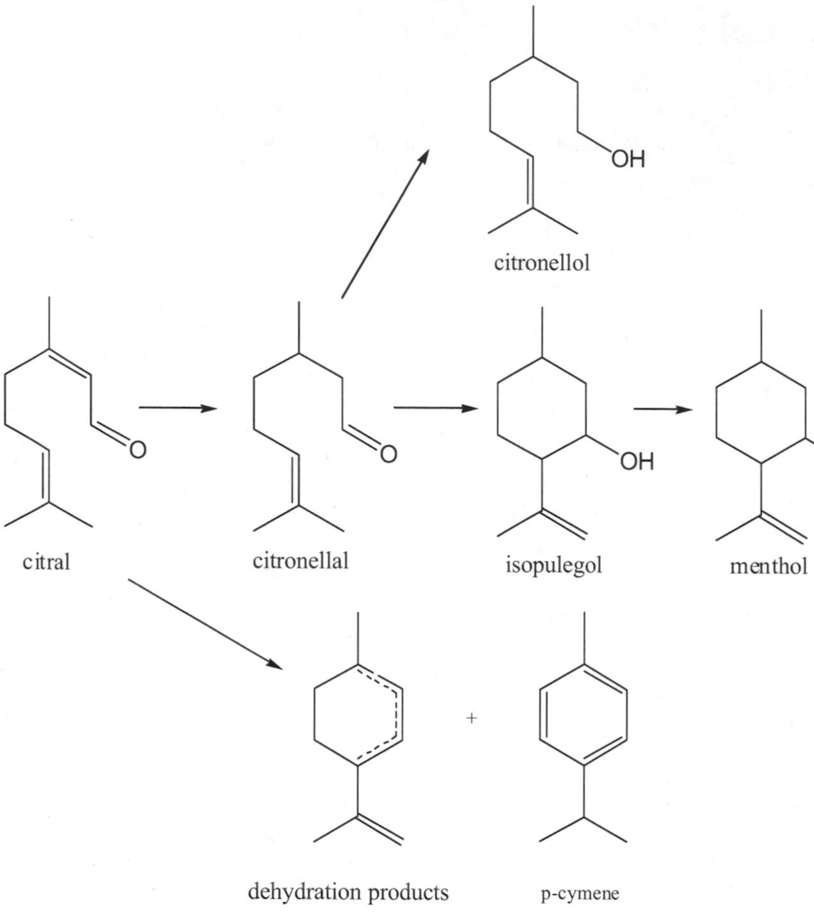

citronellol

citral   citronellal   isopulegol   menthol

+

dehydration products   p-cymene

The acidic properties of the bare supports were studied by IRS method using CO adsorption at 77 K. The IR spectra were measured on a Shimadzu FTIR-8300 spectrometer over a range of 700-6000 cm^{-1} with a resolution of 4 cm^{-1}. Before spectra registration, sample of the supports powder was pressed in wafer ($\rho = 0.007$-0.016 g/cm^{2}) and treated in vacuum (450°C, 1 hr., $< 10^{-3}$ Torr).

The concentration of surface sites was measured by the integral intensity of the CO absorbance band using the formula: C [µmol/g] = A/A$_0$, where A is the integral intensity of the CO stretching band and A$_0$ is the coefficient of the integral absorbance [6]. Concentrations are measured to within ± 20%.

Catalytic Reactions. The copper catalysts were pre-reduced at 270°C with H$_2$ before the catalytic test. Citral (0.1 g) was dissolved in toluene or heptane dehydrated over zeolites (8 ml) or not and the solution transferred under N$_2$ into a glass reaction vessel where the catalyst (0.1 g) had been previously treated. Reactions were carried out at 90°C with magnetic stirring under N_2 or H_2.

Product analysis: Reaction mixtures were analyzed by GC using a crosslinked 5% phenyl methyl silicone (HP5, 30 m) or a nonbonded, poly(80% biscyanopropyl/20% cyanopropylphenyl siloxane; SP2330, 60 m) capillary column. Reaction products were identified through their MS (HP 5971 series) and ^{1}H NMR spectra (Bruker 300 MHz).

Results and Discussion

Results obtained in the acid sites determination are summed up in Figure 9.1 and Figure 9.2. CO adsorption on SiO_2 leads to the shift towards lower frequencies of OH-groups absorption bands with ν (OH) 3747 cm^{-1} (Δν~ 87 cm^{-1}) showing formation of a very weak hydrogen bond of CO with hydroxyl groups. Simultaneously in the area of CO stretching vibrations the appearance of the absorption bands with ν (CO) 2156 cm^{-1} is observed.

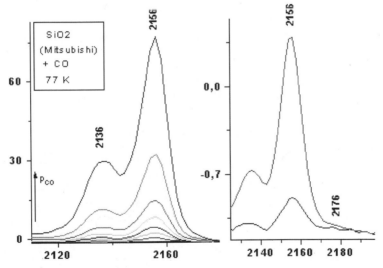

Figure 9.1: IR spectra of CO adsorbed on silica.

Rising of CO partial pressure resulted in the appearance of a series of absorption bands with ν (CO) 2176, 2156 and 2136 cm^{-1} in the area of CO stretching vibrations. The band at ν (CO) 2176 cm^{-1} can be attributed to the interaction of CO with weak Bronsted acid sites while the band at ν (CO) 2156 cm^{-1} can be attributed to the complex of CO with non-acid hydroxyl groups. The band with wavenumber ν (CO) 2136 cm^{-1} can be related to physical-adsorbed CO over Silica.

A very different picture comes from the IR spectra of sepiolite. The different spectra of adsorbed CO over sepiolite are represented in Figure 9.2. Introduction of CO resulted in the appearance of a series of absorption bands with ν

(CO) 2200, 2190, 2214, 2154 and 2136 cm^{-1}. CO absorption at ν (CO) 2200, 2190 and 2214 cm^{-1} are ascribed to the complex of CO with Mg^{2+} cations. The formation of the CO complex with the strongest adsorption sites (ν (CO) 2214 cm^{-1}) is observed at higher pressure with respect to the formation of the complexes with the weakest Lewis acidic sites (ν (CO) 2200 and 2190 cm^{-1}). The concentration of sites with ν (CO) 2200 cm^{-1} (Q(CO)=39 kJ/mol) is about 53 μmol/g, the concentration of sites with ν (CO) 2190 cm^{-1} (Q(CO)=34 kJ/mol) is about 310 μmol/g, the concentration of sites with ν (CO) 2214 cm^{-1} (Q(CO)=46.5 kJ/mol) is about 15 μmol/g.

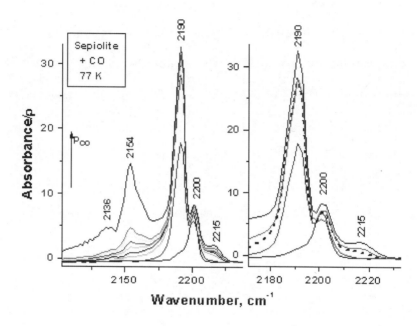

Figure 9.2: IR spectra of CO adsorbed on sepiolite.

The intense absorption band at ν (CO) 2154 cm^{-1} that appears at high CO pressure can be attributed to the hydrogen-bonded complex of CO with silanol OH-groups. Its appearance corresponds to a slight shift to lower frequencies of the absorption band at ν (OH) 3742 cm^{-1} corresponding to the stretching of Si-OH groups. The absorption band at ν (CO) 2136 cm^{-1} can be ascribed to physical-adsorbed CO over sepiolite. From these results we can conclude that on this particular kind of silica no acid sites are present, whereas sepiolite shows both weak and medium Lewis acid sites.

Results obtained in the hydrogenation of citral by using Cu catalysts on these two supports are reported in Figure 9.3 and Figure 9.4.

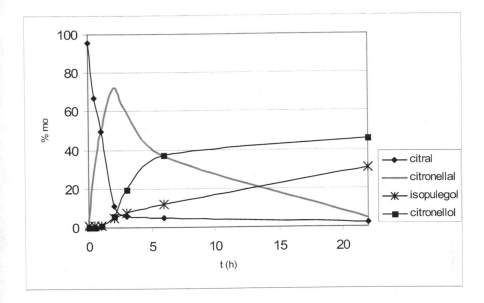

Figure 9.3: Hydrogenation of citral over Cu/SiO₂.

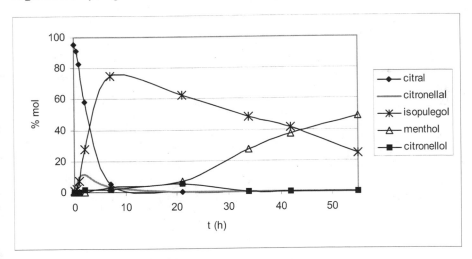

Figure 9.4: Hydrogenation of citral over Cu/sepiolite (90°C, 1 bar H₂.)

    A striking difference in selectivity was observed. According to the non-acidic character of the support Cu/SiO₂ showed to be an excellent hydrogenation catalyst. The conjugated double bond was selectively reduced giving citronellal with fairly good selectivity and then citronellol.

On the contrary, the Lewis acid sites present on the surface of sepiolite make the Cu/sepiolite catalyst extremely active in promoting the *ene* reaction of citronellal. Thus, citronellal never accumulates in the reaction mixture but it is converted into isopulegol as soon as it forms. Hydrogenation of isopulegol is very slow under these reaction conditions, but this simple catalyst is able to produce menthol in a one-pot-one-step reaction under very mild experimental conditions. Notably dehydration products, which give account of 40% of the reaction mixture obtained over Ni-H-MCM-41 [4], are kept under 20% over both Cu catalysts.

Conclusions

Determination of the acidic sites through IR spectroscopy of adsorbed CO is a valuable tool for the choice of the support when selective or multifunctional processes are to be set up. This technique allowed to identify a particular kind of silica as the support of choice for the selective hydrogenation of citral to citronellal and sepiolite as a Lewis acid support able to promote the one-step transformation of citral into menthol.

Acknowledgements

We thank the CNR-RAS cooperation program 2005-2007 for financial support.

References

1. F. Zaccheria, N. Ravasio, A. Fusi, M. Rodondi and R. Psaro, *Advanced Synthesis & Catalysis*, **347**, 1267 (2005).
2. N. Ravasio, N. Poli, R. Psaro, M. Saba and F. Zaccheria, *Top. Catal.*, **213**, 195 (2000).
3. A.F. Trasarti, A.J. Marchi and C.R. Apesteguia, *J. Catalysis* **247**, 155 (2007).
4. P. Mäki-Arvela, N. Kumar, D. Kubicka, A. Nasir, T. Heikkilä, V.-P. Lehto, R. Sjöholm, T. Salmi and D.Yu. Murzin, *J. Mol. Catal. A* **240**, 72 (2005).
5. N.Ravasio, R.Psaro, F.Zaccheria and S.Recchia, Chemical Industries (Dekker), **89**, (*Catal. Org. React.*), 263 (2003).
6. E.A. Paukshtis, Infra-red spectroscopy in heterogeneous acid-base catalysis, Novosibirsk, Nauka, 1992.

10. Improved Activity of Catalysts for the Production of Hydroxylamine

Jaime R. Blanton[1], Takuma Hara[2], Konrad Moebus[3] and Baoshu Chen[3]

[1]*Evonik Degussa Corporation, 5150 Gilbertsville Highway, Calvert City, KY 42029*
[2]*Evonik Degussa Japan Co., Ltd., Tsukuba Minami Daiichi Kogyo Danchi, 21 Kasuminosato Ami-machi, Inashiki-gun, Ibaraki-ken, Japan 300-0315*
[3]*Evonik Degussa GmBH, Rodenbacher Chaussee 4, Hanau (Wolfgang), Germany 63457*

jaime.blanton@evonik.com

Abstract

Hydroxylamine (hyam) is used in the production of caprolactam, a key raw material for the manufacture of Nylon-6. Several technologies exist for the production of caprolactam with a key difference being the amount of byproduct ammonium sulfate, a low cost fertilizer, formed. The hyam used in the HPO^{TM} process is produced by selective hydrogenation of nitric acid over a Pd/C catalyst (Equation 1).

$$\overset{Pd/C}{NO_3^- + 3H_2 + 2H^+ \longrightarrow NH_3OH^+ + 2H_2O}$$

Equation 1

Numerous parameters affect a catalyst's performance in a particular reaction. In general, the precious metal(s), support and preparation method are three of the most important variables. In this work, a range of carbon supports and preparation methods were screened with the goal of identifying the optimal catalyst for the production of hyam. The optimal catalyst would have an activity > 25 g hyam/g Pd, selectivity > 90% and fast filtration rate. The effects of varying the Pd loading and the addition of modifiers were also investigated.

Introduction

Production of Nylon-6 from caprolactam is an important global industrial process. Of the several billions of pounds of caprolactam produced annually, most is polymerized to Nylon-6 [1]. Nylon-6 polymer is used in the manufacture of carpets, automotive parts and sporting goods as well as in films and packaging.

At least nine manufacturing technologies are available for the production of caprolactam and, in most, hydroxylamine (hyam) is one of the important raw materials. In particular, in the HPO^{TM} process the hydroxylamine is made by using a precious metal powdered catalyst to selectively hydrogenate nitric acid. Evonik

Degussa has done extensive work to develop highly active catalysts for the production of hydroxylamine in this process.

Experimental Section

Six catalysts available from Evonik Degussa were tested initially (Catalysts A-F in Table 10.1).

Catalyst Designation	Evonik Degussa Catalyst Name
A	E 1529 BB/W 10% Pd
B	E 1527 BB/W 10% Pd
C	E 1002 BB/W 10% Pd
D	E 1533 BB/W 10% Pd
E	E 101 BB/W 9% Pd
F	E 1097 BB/W 10% Pd
A'	E 1529 BBF/W 10% Pd
C'	E 1002 BBF/W 10% Pd
F'	E 1097 BBF/W 10% Pd

Table 10.1. Evonik Degussa catalysts corresponding to catalysts A-F.

These catalysts were prepared by the same proprietary procedure and each contains 10% Pd except D which contains 9% Pd. Samples A', C' and F' were prepared using a stronger base for fixing of Pd onto the carbon support.

In a typical test 750 mg of catalyst was added to a continuous stirred tank reactor containing the nitrate ions in 1 L of phosphate buffer solution. This suspension contained 85% H_3PO_4 (331 g), $NaNO_3$ (198 g), NaOH (84g), and GeO_2 dissolved in water and was stirred under a H_2 flow of 150 L/h. The amount of hyam formed and selectivity after 90 min at 30°C were measured by titration [2-3]. Catalysts A and C were also chosen for studying the effect of Pd loading and Pt addition.

Catalyst filtration rates were measured by passing a slurry of buffer solution containing 5 g (dry) catalyst through a filter cloth (BHS Werk Sonthofen PAN 2326 multiful Luftdl. 20 L / m2s bei 20 mm WS) at 0.1 MPa N_2 pressure.

Results and Discussion

Several catalysts were prepared and tested for their hyam activity. Numerous preparation methods were investigated. Catalysts prepared using the method that provided the most active catalysts were used in this study. The aim was to see how varying the carbon support, Pd loading and modifier addition would affect the activity, selectivity and filterability of the catalyst in hopes of identifying the optimal catalyst for the hyam reaction (activity greater than 25 g hyam/g Pd, selectivity > 90% and fast filtration rate).

Of the catalysts prepared on different activated carbon supports, Catalyst A was found to be the most active and most selective catalyst for the hyam reaction. However, this catalyst had the poorest filterability of the catalysts evaluated. Catalyst F had good activity, but the lowest in this study, and had excellent filtration properties. A direct correlation between hyam activity and filtration rate was found (Figure 10.1).

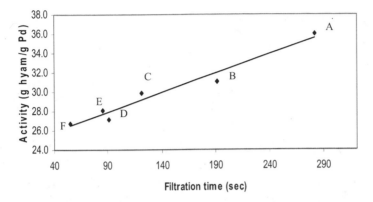

Figure 10.1. Hyam activity vs. filtration of catalysts.

To probe the details of this correlation, the hyam activity was plotted against the d_{50} and % fines of each catalyst (Figures 10.2 and 10.3). In general it was found that increasing d_{50} lead to lower activity catalysts while increasing fines (% particles < 3 μm) yielded more active catalysts. Therefore, particle size, particularly the amount of fines, seems to be important to the catalyst activity.

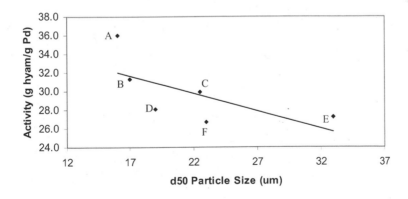

Figure 10.2. Particle size (d_{50}) vs. hyam activity.

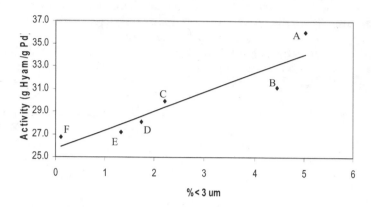

Figure 10.3. Particle size (% < 3 μm) vs. hyam activity.

Increased Pd loading and Addition of Modifiers

Catalyst metal loadings as well as bimetallic catalysts were also investigated. In one study the Pd content was varied from 9% to 15%. It was found that the catalyst efficiency increased with decreasing Pd loading (Table 10.2). One reason for this could be the fact that recipes giving eggshell metal distribution yielded the most active catalysts. This has been reported in the literature and was confirmed in our own studies [3]. Therefore, the exposed Pd located on the surface will be decreased with increasing metal loading.

Cat. Type	% Pd	% Pt	Activity (g hyam/g Pd)	Selectivity
A	9	0	37.2	100
A	10	0	36.0	95.7
A	12	0	32.7	94.6
A	15	0	28.9	98.0
C	12	0	27.7	93.2
C	12	0.1	30.7	92.3
A	12	0.1	35.0	93.3
A	8	2	57.3	86.0

Table 10.2. Activity (g hyam/g Pd) and selectivity for bimetallic catalysts and catalysts of varying Pd loadings.

It is known that the addition of Pt enhances the activity of hyam catalysts while sacrificing selectivity [4-5]. Bimetallic catalysts were prepared and tested to see if one could improve the activity of these catalysts while maintaining the good selectivity. As expected, Catalyst A with an 8% Pd + 2% Pt loading showed an increased activity of 57.3 compared to 32.7 for Catalyst A with a 10% Pd loading (Table 10.2). On the other hand, the selectivity of the bimetallic catalyst was reduced from 94.6% to 86%. By fine tuning the Pd and Pt loadings, however, bimetallic catalysts were developed that matched the selectivity of Pd only catalysts with a boost in activity. Only a minimal amount (0.1%) of the more expensive Pt was required to give a noticeable improvement in activity.

The effect of other modifiers has also been studied by Evonik Degussa for Pd/C and Pd + Pt/C systems [6]. Of the numerous metals investigated, only iron was found to increase the hyam activity while maintaining high selectivity. To further optimize this catalyst, a range of 10% Pd catalysts were prepared with Fe loadings varying from 0.5 – 5.0% Fe (Table 10.3). An increase in yield was observed up to a maximum Fe loading of 2%.

Promoter metals	Hydroxylamine yield	Selectivity
None	22.7	99.2
Fe (0.5 wt%)	23.3	100.0
Fe (1.0 wt%)	24.5	99.3
Fe (2.0 wt%)	24.8	98.8
Fe (5.0 wt%)	23.4	100.0

Table 10.3. Effect of Fe loading on standard 10% Pd/C catalyst.

Improved Catalyst Preparation Method

In addition to carbon support properties and metallic modifiers, recipe variations were also investigated. It was found that an adjustment to the recipe also yielded more active hydroxylamine catalysts. By using a stronger base to hydrolyze Pd more active catalysts could be developed. On average, a 9% increase in activity was observed (Table 10.4). One explanation could be that by using a stronger base Pd crystallites are smaller leading to higher dispersion and a more egg-shell type metal distribution, both of which appear to have a positive effect on hyam activity.

Cat.	Activity (g hyam/g Pd)	% increase
A	36.6	
A'	41.0	8.8
C	29.9	
C'	32.3	9.3
F	26.7	
F'	29.6	9.0

Table 10.4. Activity of catalysts prepared by modified recipe compared to original catalysts.

Conclusions

Several highly active hyam catalysts have been developed for the HPO^{TM} process. Depending on which performance indicator is most important (activity, selectivity or filterability) one can select the best Evonik Degussa catalyst for their application.

Not surprisingly, PSD of the support had a significant impact on the hyam activity of the catalyst. The most active catalyst, A, had the slowest filtration rate. Also, by adding a small amount of Pt an improvement in activity with minimal effect on selectivity could be achieved. Lastly, by a slight modification in the preparation method, and/or by the addition of Fe the already very active catalysts could be even further improved.

References

1. S. van der Linde and G. Fischer, *Fibres and Textiles in Eastern Europe* **12**, 1 (2004).
2. P. Mars, C.J. Duyverman and M.J. Gorgels, Pat. NL 6,717,085, to Stamicarbon (1969).
3. A. van Montfoort and J. J. F. Scholten, Pat. US 4,052,336, to Stamicarbon (1977).
4. C. J. Duyverman, Pat. NL 7,109,068, to Stamicarbon (1973).
5. C. G. M. van de Moesdijk, Chemical Industries (Marcel Dekker), **18**, *Catal. of Organic Reactions*, 379-407 (1984).
6. T. Hara, Y. Nakamura and J. Nishimura, *Applied Catalysis* A, **320**, 144 (2007).

11.   A *"Trans*-Effect" in Carbon Double-Bond Hydrogenation

Arran S. Canning, S. David Jackson, Andrew Monaghan, Ron R. Spence and Tristan Wright

WestCHEM, Dept. of Chemistry, The University, Glasgow G12 8QQ, Scotland, UK

sdj@chem.gla.ac.uk

Abstract

The *trans*-effect in co-ordination chemistry is well known; however, what has not been examined in any detail is the effect of *trans*-molecules in heterogeneous catalytic hydrogenation. In this paper we will show that "trans" molecules hydrogenate more slowly than other isomers and can poison reactions of species that would be expected to be more strongly bound. However, if a *cis/trans*-mixture is used this strong adsorption can be disrupted.

Introduction

Hydrogenation of olefins has been known and practiced for almost a century (1). In some early reports (2-4) careful reading reveals that the *trans*-isomer was less reactive than the other isomers. In this paper we will confirm that *trans*-isomers do react more slowly and that they can have a negative effect on the hydrogenation of other, notionally more strongly bound, molecules.

Experimental Section

The catalyst used throughout this study was a 1% w/w palladium on alumina supplied by Johnson Matthey. The support consisted of θ-alumina trilobes (S.A. ~100 m^2g^{-1}) and the catalyst was sized to < 250 μ for all catalytic studies. The alkenes and alkynes (all Aldrich >99%) were used without further purification. No significant impurities were detected by GC. Gases (BOC, >99.99%) were used as received.

The reaction was carried out in a 0.5 l Buchi stirred autoclave equipped with an oil jacket and a hydrogen-on-demand delivery system. 0.05 g of 1% Pd/Al$_2$O$_3$ was added to 280 ml of degassed solvent, propan-2-ol for the cinnamonitrile or 2,2,4-trimethylpentane for the alkenes. Reduction of the catalyst was performed *in situ* by sparging the system with H$_2$ (300 cm^3min^{-1}) for 30 minutes at 323 K while stirring the contents of the autoclave at 300 rpm. After reduction, the autoclave was adjusted to the appropriate reaction temperature of between 298 and 323 K under a nitrogen atmosphere. Reactant (typically 1 ml) was added into the reaction vessel and flushed through with 20 ml of degassed solvent. The autoclave was then mixed briefly at a stirrer speed of 800 rpm and pressurised to 1 barg with nitrogen and a sample was taken. The vessel was then depressurised and then pressurised with hydrogen to the appropriate pressure. Following this, the stirrer was set to a speed of 1000 rpm and samples were taken via a sample valve at defined time intervals. The liquid samples were analyzed by gas chromatography. Standard checks were undertaken to confirm that the system was not under mass transport control.

Results and Discussion

When the pentene isomers are hydrogenated (Table 11.1) the results show that the order in the rates of hydrogenation derived from rate constants over a 1% Pd/Al$_2$O$_3$ catalyst is *cis*-2-pentene > 1-pentene >> *trans*-2-pentene.

Table 11.1. First-order rate constants for hydrogenation of pentenes.

Reactant	1-pentene	*Cis*-2-pentene	*Trans*-2-pentene
Rate Constant (min^{-1})	0.014	0.035	0.003

Bond and Winterbottom (4) observed a similar trend for n-butene hydrogenation over alumina-supported palladium, where the hydrogenation activity of the *trans*-isomer was significantly slower than *cis*-2-butene and 1-butene (4). The reason for the sluggish hydrogenation activity of *trans*-2-pentene may result from it being the thermodynamically more stable isomer (thermodynamic mixture: 81%, 17.5%, and 1.5% for *trans*-2, *cis*-2 and 1-pentene, respectively) (2) and thus more reluctant to isomerise or hydrogenate when it reaches this preferred state. To investigate this further, we examined the effect of the pentene isomers on 1-pentyne hydrogenation. When *trans*-2-pentene is hydrogenated over a catalyst for one hour, then 1-pentyne is added to the solution, the alkyne reaction rate is reduced by 90% compared to the reaction rate in the absence of the *trans*-2-pentene(Figures 11.1 and

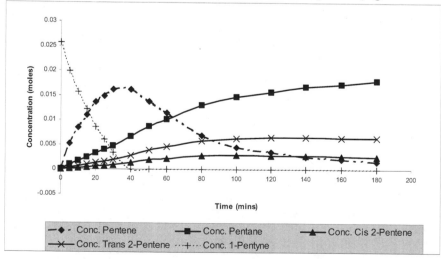

Figure 11.1. Hydrogenation of 1-pentyne at 313 K and 2 barg.

11.2). However if the *cis*-2-pentene isomer or the 1-pentene isomer is added first, the rate of alkyne hydrogenation is unaffected. The *trans*-molecule acts as a poison for the alkyne hydrogenation whereas the *cis*- and 1-pentene isomers have no effect.

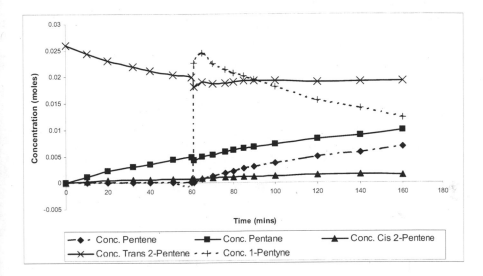

Figure 11.2. Hydrogenation of 1-pentyne following hydrogenation of *trans*-2-pentene for 1h.

This suggests that the *trans*-2-pentene is more strongly bound than the pentyne and that the slow rate of hydrogenation may be due to strong adsorption rather than the thermodynamic stability of the *trans*-isomer.

A similar effect is found with other "*trans*"-molecules. When *trans*-cinnamonitrile is hydrogenated at 343 K to 3-propionitrile the reaction is slow (k = 3.4×10^{-3} min^{-1}); however, when a *cis/trans*-mixture is hydrogenated under identical conditions the rate of *trans*-hydrogenation is significantly increased (k = 7.1×10^{-3} min^{-1}). This may be due to a reduction in the poisoning effect of the *trans*-molecule. This was observed in the hydrogenation of *trans*-1,3-pentadiene where the reaction rapidly self-poisons (5); however, when a *cis/trans*-mixture is used this poisoning effect is dramatically reduced (5). With the *trans*-cinnamaldehyde there is significant mass laydown in the early stages of reaction, which is not seen with the *cis/trans*-mixture.

Conclusions

Trans-isomers have been shown to be much less reactive than the respective *cis*-isomer. This is due to strong adsorption of the *trans*-isomer. The strength of adsorption is such that in many systems the *trans*-isomer acts as a poison. Co-adsorption/reaction of a *cis/trans*-mixture disrupts this strong *trans*-isomer adsorption, removing the poisoning effect and enhancing rate.

Acknowledgements

The authors would like to acknowledge the support of Engineering and Physical Sciences Research Council and Johnson Matthey plc under the ATHENA grant.

References

1. Nobel Lectures, Chemistry 1901-1921, P. Sabatier 1912, Elsevier Publishing Company, Amsterdam, 1966.
2. G. C. Bond and J. S. Rank in W. H. M. Sachtler, G. C. A. Schuit and P. Zwietering (Eds), Proceedings of the Third International Congress on Catalysis Vol.2, North Holland, Amsterdam, 1965, p. 1125.
3. G. C. Bond, G. Webb and P. B. Wells, Trans. Faraday Soc., 64 (1968) 3077.
4. G. C. Bond and J. M. Winterbottom, Trans. Faraday Soc., 65 (1969) 2794.
5. S. D. Jackson and A. Monaghan, Catalysis Today, 128 (2007) 47.

12. Detailed Kinetic Analysis Reveals the True Reaction Path: Catalytic Hydrogenation, Hydrolysis and Isomerization of Lactose

Tapio Salmi, Jyrki Kuusisto, Johan Wärnå and Jyri-Pekka Mikkola

Åbo Akademi, Process Chemistry Centre, Laboratory of Industrial Chemistry, FI-20500 Åbo/Turku, Finland

tsalmi@abo.fi

Abstract

Lactose, the milk sugar is a very attractive and inexpensive starting material for the production of lactitol, which has a huge potential as well-tasting, harmless and health-promoting sweetener. Experimental kinetic investigations and kinetic modelling of lactose hydrogenation are important not only from the viewpoint of fundamental understanding of the reaction mechanism, but also for the design and optimization of lactitol production units. Since it concerns alimentary production the amounts of by-products should be suppressed by selection of catalyst and operation conditions.

Very precise kinetic experiments were performed with sponge Ni and Ru/C catalysts in a laboratory-scale pressurized slurry reactor (autoclave) by using small catalyst particles to suppress internal mass transfer resistance. The temperature and pressure domains of the experiments were 20-70 bar and 110-130°C, respectively. Lactitol was the absolutely dominating main product in all of the experiments, but minor amounts of lactulose, lactulitol, lactobionic acid, sorbitol and galactitol were observed as by-products on both Ni and Ru catalysts. The selectivity of the main product, lactitol typically exceeded 96%.

Rival stoichiometric schemes and kinetic equations were investigated for the formation of lactitol and the by-products. Pseudo-first order kinetics was applicable for the organic components involved and a detailed analysis in the stoichiometric space became possible. A comparison of the proposed kinetic schemes revealed that the best description of the process is provided by a kinetic scheme, in which galactitol and sorbitol are formed through cleavage of lactitol molecule followed by subsequent hydrogenolysis and hydrogenation steps rather than hydrolysis of lactose and followed by hydrogenation steps. The parameters of the kinetic models were determined by non-linear regression analysis by using a hybrid simplex-Levenberg-Marquardt model. The fit of the model to the experiments was very satisfactory. The model can be used for optimization of the reaction conditions for lactitol production.

Introduction

Lactose, the milk sugar, is a reducing disaccharide consisting of glucose and galactose moieties. The estimated annual worldwide availability of lactose as a by-product from cheese manufacture is several million tons [1,2], but only about 400 000 t/a lactose is processed further from cheese whey [3]. Non-processed whey is an environmental problem due to its high biochemical and chemical oxygen demand [2]. The use of lactose as such is limited by two main factors: relatively low solubility of lactose in most solvents and lactose intolerance in human body [1].

Lactose, can however be transformed to value-added products, such as lactitol (by hydrogenation), lactulose (by isomerisation) and lactobionic acid (by oxidation). In addition, the hydrolysis and hydrogenation products, D-galactose, D-glucose, galactitol and sorbitol are valuable molecules for pharmaceutical and alimentary industry [4,5]. The hydrogenation product of lactose, lactitol, is an ancariogenic sugar alcohol tolerated by people suffering from lactose intolerance and diabetes. Lactitol is a widely used ingredient for sugar-free chocolate, baked goods and ice cream.

The classical heterogeneous catalyst used for hydrogenation of sugars to sugar alcohols is sponge nickel (Raney nickel), but recently, supported ruthenium catalyst have got more attention, because of the high stability of ruthenium catalyst. Kinetics of D-glucose and D-xylose hydrogenation has been reported in several publications [6-16], but few studies about D-lactose hydrogenation have been published so far [17]. A particular challenge is to reveal the true reaction path, the adequate reaction scheme for the hydrogenation, since the amounts of by-products are small, but significant, since high purity is required in pharmaceutical and alimentary products. To suppress the separation costs, the catalyst and the reaction conditions should be carefully selected in such a way that the amounts of the by-products are minimized. In this paper we demonstrate how detailed kinetic analysis can discriminate between rival stoichiometric schemes and contribute to the optimization of the reaction conditions. The kinetic analysis is based on experimental lactose hydrogenation data obtained from a slurry reactor, which was operated in kinetic regime.

Experimental section

D-lactose (40 wt% in water, 1.31 mol/litre) hydrogenation experiments were carried out batchwise in a three-phase laboratory scale reactor (Parr Co.) operating at 20-70 bar and between 110 and 130°C. The reactor was equipped with a heating jacket, a cooling coil, a filter (0.5 μm metal sinter) in a sampling line and a bubbling chamber (for removing dissolved air from the liquid phase prior to the hydrogenation experiments). The effective liquid volume of the reactor was about 125 ml (total volume 300 ml) and it was equipped with a hollow shaft concave blade impeller to ensure efficient mixing and gas dispersion into the liquid phase. The impeller rate was fixed at 1800 rpm in all of the kinetic experiments to ensure operation at the kinetically controlled regime [15-16]. A Parr 4843 controller was used for the temperature control and for monitoring the impeller speed and the reactor pressure. The temperature and pressure profiles were stored on a computer. Lactose solutions were prepared by dissolving D-lactose monohydrate (Valio, purity > 99.5% of dry substance and dry substance content 95%) in deionized water. Too high lactose dissolution temperatures were avoided to suppress lactose hydrolysis prior to the hydrogenation. The amount of 5% Ru/C (Johnson Matthey) varied between 1.5 and 2.5 wt% (dry weight) of the lactose weight throughout the kinetic hydrogenation series. The amount of Mo-promoted sponge nickel catalyst (Acticat) varied between 2.5 and 10 wt% (dry weight).

Prior to the first hydrogenation batches, the supported ruthenium catalysts were reduced in the autoclave under hydrogen flow at 200°C for 2 hours (10 bar H_2, heating/cooling rate 5°C/min). As the catalyst had been reduced, a lactose solution saturated with hydrogen was fed into the reactor rapidly and the hydrogen pressure and reactor temperature were immediately adjusted to the experimental conditions. Simultaneously, the impeller was switched on. This moment was considered as the initial starting point of the experiment. No notable lactose conversion was observed before the impeller was switched on.

The withdrawn liquid-phase samples were analyzed with an HPLC (Biorad Aminex HPX-87C carbohydrate column. 1.2 mM $CaSO_4$ in deionized water was used as a mobile phase, since calcium ions improve the resolution of lactobionic acid [17]). Dissolved metals were analysed by Direct Current Plasma (DCP). The catalysts were characterized by (nitrogen adsorption BET, XPS surface analysis, SEM-EDXA, hydrogen TPD and particle size analysis).

Catalyst characterization results

Ruthenium leaching determined at the end of different hydrogenation batches, both in kinetic experiments and deactivation series, remained quite constant (around 13 ppm). The specific surface areas, pore volumes and pore size distributions of fresh and recycled catalyst samples were determined by nitrogen physisorption. The specific surface areas of the sponge Ni and Ru catalysts were $90 m^2/g$ and $800 m^2/g$, respectively. The comparison of pore size distributions of fresh and recycled sponge nickel and Ru/C catalysts revealed that Ru/C has relatively more large pores (5-100 nm) and sponge nickel pores are mainly in the range of 2-5 nm. Smaller catalyst pores around 2 nm may get easily blocked by lactose molecule and the hydrogenation products. SEM-EDXA analysis (analysis area ~10×10 μm) revealed that ruthenium was not homogeneously distributed on carbon support. The recycled ruthenium catalyst contained 6.1 wt% Ru. The catalyst particle size distributions were measured by Malvern 2600. The median particle sizes of the fresh Ru/C and sponge Ni catalysts were 20.8 μm and 32.4 μm, respectively. The reduced Ru/C catalysts` ability to adsorb hydrogen was determined by hydrogen TPD. First catalyst sample was reduced with a gas mixture (H_2:Ar=5:1) at 200°C for two hours. After completed reduction, the temperature was cooled down to a typical lactose hydrogenation temperature 120°C. Hydrogen adsorption was continued at this temperature with the same gas mixture for 50 min. Before hydrogen TPD, physisorbed hydrogen was removed with an argon flow. The hydrogen desorption temperature was increased 10°C/min from adsorption temperature until 600°C, at which temperature desorption was continued for 40 min. According to the measurement, the reduced catalyst was able to adsorb 0.290 mmol H_2/g catalyst at 120°C.

Qualitative kinetics

A typical lactose hydrogenation experiment is displayed in Figure 12.1. The main hydrogenation product was lactitol and the selectivity varied between 96.5 and 98.5% for both catalysts studied. Increased hydrogenation temperature improved the reaction rate at the experiments carried out between 110 and 130°C. Elevated reaction temperatures increased to some extent the formation of the by-products, thus impairing the lactitol selectivity. An increased hydrogen pressure had a positive effect on the reaction rate and lactitol selectivity, especially at lower temperatures. The lactitol selectivity improved, as the hydrogen pressure increased and the reaction temperature decreased within the experimental range. These qualitative observations were valid both for sponge nickel and ruthenium catalysts. Related to the amount of active metal on the catalyst, Ru/C gave a substantially higher reaction rate compared to the sponge nickel catalyst.

Small amounts of lactulose, lactulitol, sorbitol, galactitol and lactobionic acid were detected as by-products. Lactulose is formed through isomerization and hydrogenated further to lactulitol. Lactobionic acid was detected as a by-product depending on the reaction conditions - typically high temperatures and low hydrogen pressures favour lactobionic acid formation. In long run, some part of lactobionic acid was hydrogenated away and the concentration curves of lactobionic acid exhibited a soft, but still visible maximum. Sorbitol and galactitol mainly appeared in equal amounts (within the accuracy of chemical analysis) at the late stage of the reaction.

The crucial issue is, however, whether sorbitol and galactitol are formed through hydrolysis of lactose to glucose and galactose and hydrogenation of them to sorbitol and galactitol or through a complex surface reaction mechanism involving hydrolysis of lactitol with a subsequent hydrogenation step. Alternative reaction schemes are displayed below (1-3+4a and 1-3+4b). The latter reaction scheme is illustrated on a molecular level in Figure 12.2.

Lactose → lactitol	(1)
Lactose → lactulose → lactulitol	(2)
Lactose ↔ lactobionic acid	(3)
Lactose → glucose + galactose →sorbitol + galactitol	(4a)
Lactitol →sorbitol + galactitol	(4b)

Glucose and galactose are assumed to be rapidly reacting intermediates. A detailed kinetic analysis and comparison of schemes (1-3+4a) and (1-3+4b) is provided in the next chapter.

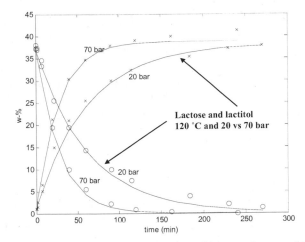

Figure 12.1. Hydrogenation kinetics of lactose (sponge Ni, 120°C, 20 and 70 bar).

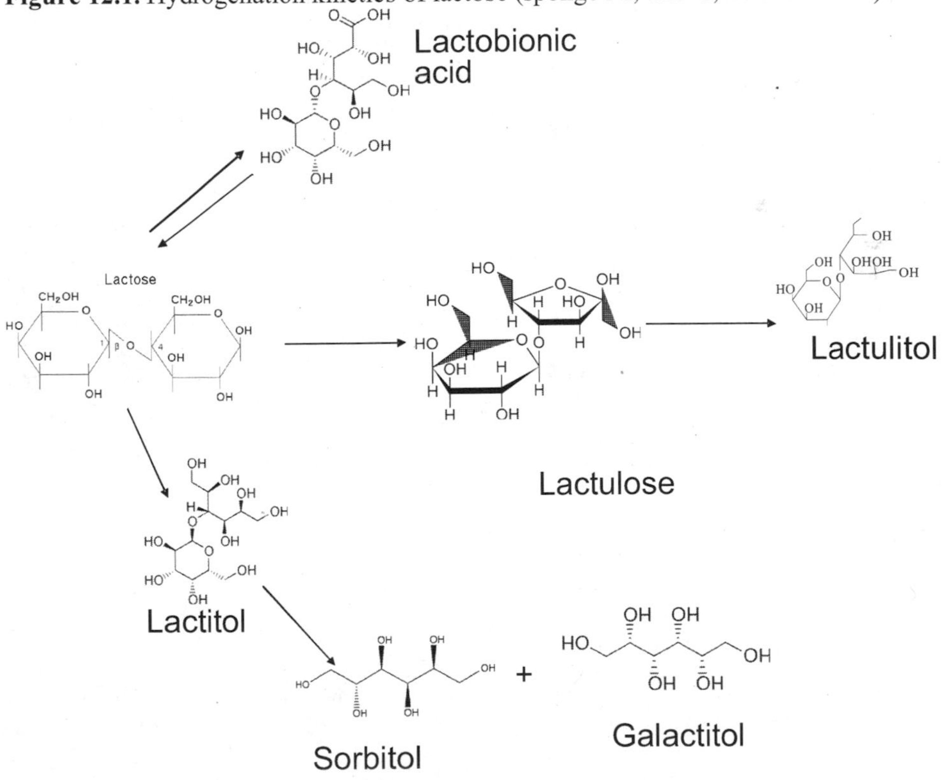

Figure 12.2. Reaction scheme (1-3, 4b) for formation of lactitol and by-products.

Detailed kinetic analysis

The mass balance equations of all the organic components involved can be expressed in a very simple way. The volatilities of the sugars and sugar alcohols are extremely low, thus they can be treated as pure liquid-phase components. In addition, stirring in the batch reactor was very vigorous removing the external mass transfer limitations. Small catalyst particles were used (Experimental section), so internal mass transfer resistance was negligible. Gas-phase hydrogen pressure was maintained constant during the course of the reaction. Since hydrogen mass transfer from gas to liquid phase did not affect the overall hydrogenation rate, the concentration of dissolved hydrogen in liquid phase was constant throughout the reaction. The mass balance of hydrogen is thus not needed. For the liquid-phase components in the reactor autoclave the mass balances can be written as follows:

$$\frac{dc_i}{dt} = \sum v_{ij} r_j \; \rho_B \tag{5}$$

where ρ_B is the catalyst bulk density, and v_{ij} and R_j denote the stoichiometric coefficients and the reaction rates, respectively.

Previous experience has shown that the hydrogenation kinetics of sugar molecules can very well be described with a non-competitive adsorption model for hydrogen and the organics. This is due to the fact that the molecule sizes are very different for hydrogen and sugar/sugar alcohol molecules. Consequently, some interstitial sites remain accessible for the adsorption of hydrogen even on a 'fully-covered' surface of organics. This leads to a semi-competitive adsorption model, which is described in detail in an overview paper of us [18]. The limiting case of semi-competitive adsorption is, however, non-competitive adsorption as the differences between the molecule sizes increases. Therefore, the simpler non-competitive adsorption model was used here. Adsorption steps are assumed to be rapid and reversible, while the hydrogenation step is presumed to be irreversible and rate-determining. After applying adsorption quasi-equilibria separately for hydrogen and organic molecules, the surface concentrations can be solved explicitly from the model and inserted in the rate equations for hydrogenation, hydrogenolysis and isomerisation steps. The final form of the rate equation of a hydrogenation step becomes

$$R_j = \frac{k_j c_A c_{H2}^{nH2}}{\left(1 + K_{H_2} c_{H2}^{nH2}\right)\left(1 + \sum K_l c_l\right)} \tag{6}$$

where $n_{H2} = 1/\alpha$ ($\alpha=1$ in case of molecular hydrogen and $\alpha=2$ in case of dissociative hydrogen adsorption). Subscript 'l' refers to the organic component taking part in reaction j. Previous studies of our group have also indicated that the solubility of hydrogen in a sugar is practically equal to that in the corresponding hydrogenation product, sugar alcohol. This implies that the solubility remains constant during the experiment. Henry's law can thus be used and the hydrogen concentration can be replaced by its partial pressure in the above equation. During an isothermal experiment the factor $k_j c_{H2}^{n2}/(1+ K_{H2} c_{H2}^{n2})$ is constant and denoted by k'_j in further treatment.

The roles of the various organic components in the kinetics is interesting - to which extent the different components influence the denominator of the rate equation (1)? First of all, lactose and lactitol are very dominant components during the experiment, by-products appearing only in minor quantities (typically less than 1 wt%). If lactose adsorption is strong, the reaction order with respect to lactose is suppressed; for weaker lactose adsorption the reaction order one is approached. This limit value is attained also for simultaneous strong adsorption of lactose and lactitol if their adsorption strengths are similar (since the amounts of by-products are small). Again we are close to pseudo-first order kinetics.

Pseudo-first order kinetics with respect to lactose implies (at constant hydrogen pressure) that the mass balance of lactose becomes

$$\frac{dc_A}{dt} = (\sum k'_j) c_A \rho_B \tag{7}$$

which upon integration and insertion of the limits yields

$$- \ln(c_A / c_{0A}) = \rho_B k'' t \tag{8}$$

Where $k'' = \Sigma k'_j$. The concentration ratio expressed with the aid of lactose conversion (X) is $c_{0A}/c_A = 1-X$. A series of first-order test plots were prepared and they revealed that lactose obeys pseudo-first order kinetics very well.

Analogously, by assuming first order kinetics with respect to the other organic components, a product distribution analysis can be performed. The reaction rates of steps (1)-(5) in scheme (Figure 12.2) are expressed as

$$R_1 = k'_1 c_A$$
$$R_2 = k'_2 c_A$$
$$R_3 = k'_3 c_A - k'_{-3} c_D \tag{9a-e}$$
$$R_4 = k'_4 c_C$$
$$R_5 = k'_5 c_B$$

Furthermore, the reaction stoichiometry gives the generation rates according to eq. (5),

$$dc_A / dt = -\rho_B(R_1 + R_2 + R_3)$$
$$dc_B / dt = \rho_B(R_1 - R_5)$$
$$dc_C / dt = \rho_B(R_2 - R_4) \tag{10a-g}$$

$$dc_D / dt = \rho_B R_3$$
$$dc_E / dt = \rho_B R_4$$
$$dc_F / dt = \rho_B R_5$$
$$dc_G / dt = \rho_B R_5$$

By inserting the expressions for R_1, R_2 and R_3 in eq. (10a) and integrating we obtain

$$c_A / c_{0A} = \exp(-(k'_1 + k'_2 + k'_3)\rho_B t)) \qquad (11)$$

It should be observed that k_{-3} was approximated to zero in the above treatment. The differential equations describing B and C concentrations are linear ones with respect to the participating concentrations. The expression for the A-concentration is inserted in eqs. (10b) and (10c) and the first order differential equations are solved with the initial conditions $c_B=0$ and $c_C=0$ at $t=0$. The solutions become

$$c_B / c_{0A} = (k'_1 /(k'_5 - k''))(\exp(-k''\rho_B t) - \exp(-k'_5 \rho_B t)) \qquad (12)$$

$$c_C / c_{0A} = (k'_2 /(k'_4 - k''))(\exp(-k''\rho_B t) - \exp(-k'_4 \rho_B t)) \qquad (13)$$

where $k''=k'_1+k'_2+k'_3$.

The concentration of D is obtained directly by integration of eq. (10d) with the initial condition $c_D=0$ at $t=0$:

$$c_D / c_{0A} = (k'_3 / k'')(1 - \exp(-k''\rho_B t)) \qquad (14)$$

The analysis continues in a straightforward manner. Components E, G and F are in equivalent positions in the stoichiometric scheme. The concentration expressions of C and B are inserted in eqs. (10e) and (10f), respectively, and integrated with the initial conditions $c_C=0$ and $c_B=0$ at $t=0$. The concentrations of G and F are equal. We obtain finally

$$c_E / c_{0A} = (k'_1 k'_4) /(k'_4 - k'')(1 / k'')(1 - \exp(-k''\rho_B t)) - (1 / k'_4)(1 - \exp(-k'_4 \rho_B t))$$

$$c_F / c_{0A} = (k'_1 k'_5) /(k'_5 - k'')(1 / k'')(1 - \exp(-k''\rho_B t)) - (1 / k'_5)(1 - \exp(-k'_5 \rho_B t))$$

$$(15\text{-}16)$$

The product distribution can be analyzed with respect to lactose conversion $X=1-c_A/c_{0A}$. From eq. (11) we get $\exp(-k''\rho_B t)=1-X$, which is inserted in the expressions for B, C, D, E and F. The equations are listed below:

$$c_B/c_{0A} = \alpha_1/(\alpha_5-1)((1-X)-(1-X)^{**}\alpha_5) \tag{17}$$
$$c_C/c_{0A} = \alpha_2/(\alpha_4-1)((1-X)-(1-X)^{**}\alpha_4) \tag{18}$$

$$c_D/c_{0A} = \alpha_3 X \tag{19}$$

$$c_E/c_{0A} = \alpha_2/(\alpha_4-1)(\alpha_4 X + (1-X)^{**}\alpha_4 - 1) \tag{20}$$

$$c_F/c_{0A} = \alpha_1/(\alpha_5-1)(\alpha_5 X + (1-X)^{**}\alpha_5 - 1) \tag{21}$$

where $c_F = c_G$ according Figure 12.2. From the above equations, a simple but interesting relationship can be deduced between the C- (lactulose) and E- (lactulitol) concentrations:

$$c_C/c_{0A} + c_E/c_{0A} = \alpha_2 X \tag{22}$$

The concentration versus lactose conversion (X) plots have maxima for B and C, while the concentrations of D, E, F and G increase monotonously with conversion; E, F and G showing a steep decrease at high lactose conversions.

Corresponding analytical pseudo-first order rate expressions can be derived for the rival reaction scheme (eqs. 1-3, 4a), but for the sake of brevity, they are not presented here. The rival models were compared by data fitting, i.e., estimation of parameters for both models. The fit of the experimental data to the kinetic model was carried out by Modest software [19] by using a combined Simplex-Levenberg-Marquardt method for the objective function minimization. The underlying ordinary differential equations (5) describing the component mass balances were solved by backward difference method suitable for stiff differential equations. In order to describe precisely the experimental features, the hydrogenation step of lactobionic acid (rate parameter k-3) was included in the models. The following objective function was used in data fitting:

$$Q = \sum \left(c_{i,exp} - c_{i,calc}\right)^2 w_i \tag{23}$$

where the weight factors (w_i) were selected as follows: w = 1 for lactose and lactitol and w = 0.05 for by-products. Using higher weight factors for low-concentration components improved the estimation of them. The parameter estimation was performed for all experiments at different temperatures and pressures merged together.

Some data fitting results are displayed in Figures 12.1 and 12.3. The general conclusion is that both models describe the behaviours of the main components, lactose and lactitol very well, both for sponge nickel and ruthenium catalysts. In this respect, no real model discrimination is possible. Both models also describe equally well the behaviour of lactobionic acid (D), including its concentration maximum when the reversible step is included (k_{-3}) (Figure 12.3).

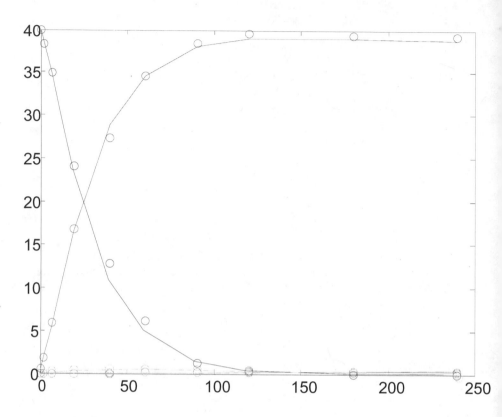

Figure 12.3. Lactose hydrogenation at 120°C and 50 bar on Ru/C (main product: lactitol, by-product with maximum:lactulose). Fit of model (1-3, 4b).

     The picture is completely changed as the behaviours of sorbitol (F) and galactitol (G) are examined more closely. Both models describe the monotonic increase of sorbitol and galactitol with lactose conversion, but the consecutive model (1-3, 4b) displayed in Figure 12.2 follows much better the concentration trends of both sugar alcohols (F and G) than the parallel model (1-3, 4a). The data fitting of the rival models is compared in Figure 12.4, which can be taken as a clear evidence for a mechanism, where the main part of sorbitol and galactitol is formed from lactitol but not from lactose. This conclusion is valid for sponge nickel and ruthenium/C catalysts, suggesting thus similarities for the reaction mechanisms. It is clear that the parallel-type reaction route (1-3, 4a) cannot be completely discarded by this kind of model comparison, but more detailed experimental studies incorporating separate hydrogenation/hydrogenolysis experiments of lactitol, glucose and galactose are needed in the future.

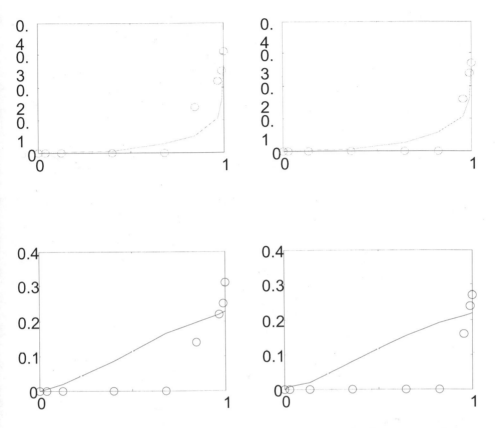

Figure 12.4. Galactitol (left) and sorbitol (right) concentrations versus lactose conversion: comparison of consecutive model (1-3, 4b; upper figures) and parallel model (1-3, 4a, lower figures). Reaction conditions: 120°C, 50 bar (left) and 60 bar (right), Ru/C catalyst.

Conclusions

Hydrogenation of lactose to lactitol on sponge nickel and ruthenium catalysts was studied experimentally in a laboratory-scale slurry reactor to reveal the true reaction paths. Parameter estimation was carried out with rival and the final results suggest that sorbitol and galactitol are primarily formed from lactitol. The conversion of the reactant (lactose), as well as the yields of the main (lactitol) and by-products were described very well by the kinetic model developed. The model includes the effects of concentrations, hydrogen pressure and temperature on reaction rates and product distribution. The model can be used for optimization of the process conditions to obtain highest possible yields of lactitol and suppressing the amounts of by-products.

Acknowledgements

This work is part of the activities at the Åbo Akademi Process Chemistry Centre within the Finnish Centre of Excellence Programme (2000-2011) by the Academy of Finland. Financial support from the Finnish Funding Agency for Technology and Innovation (TEKES), Danisco Sweeteners and Waldemar von Frenckells stiftelse is gratefully acknowledged.

Notation

c	concentration
K	adsorption constant
k	rate constant
k', k''	merged rate parameters
n	molar amount
p	pressure
Q	objective function in parameter estimation
R	reaction rate
r	component generation rate
t	time
w	weight factor in parameter estimation
X	conversion
α	rate parameter, in general $\alpha_j = k'_j/k''$
ρ_B	catalyst bulk density (=mass of catalyst/liquid volume), kg/dm^3
ν	stoichiometric coefficient

Subscripts and superscripts

i	component index
j	reaction index

References

1. P. Linko, Lactose and Lactitol, Nutritive Sweeteners, eds. Birch, G.; Parker, K., Applied Science: NJ, 1982, p.109.
2. M. Hu, M.J. Kurth, Y.-L. Hsieh and J.M. Krochta, HPLC and NMR study of the reduction of sweet whey permeate, J. Agric. Food Chem. 44 (1996) 3757.
3. K.-G. Gerling, Large-scale production of lactobionic acid – used and new applications, International Dairy Federation, Whey proceedings of the second international whey conference (1998) 251.
4. A. Abbadi, K.F. Gotlieb, J.B.M. Meiberg and H. van Bekkum, New food antioxidant additive based on hydrolysis products of lactose, Green Chemistry 5 (2003) 47.
5. H. Berthelsen, K. Eriknauer, K. Bottcher, H.J.S. Christensen, P. Stougaard, O.C. Hansen and F. Jorgensen, Process for manufacturing of tagatose, WO 2003008617, 2003.

6. P. Gallezot, P.J. Cerino, B. Blanc, G. Flèche and P. Fuertes, Glucose Hydrogenation on Promoted Raney-Nickel Catalysts, Journal of Catalysis 146 (1994) 93.
7. P. Gallezot, N. Nicolaus, G. Flèche, P. Fuertes and A. Perrard, Glucose Hydrogenation on Ruthenium Catalysts in a Trickle-Bed Reactor, Journal of Catalysis 180 (1998) 51.
8. B. Kusserow, S. Schimpf and P. Claus, Hydrogenation of Glucose to Sorbitol over Nickel and Ruthenium Catalysts, Adv. Synth. Catal. 345 (2003) 289.
9. K. van Gorp, E. Boerman, C.V. Cavenaghi and P.H. Berben, Catalytic hydrogenation of fine chemicals: sorbitol production, Catalysis Today 52 (1999) 349.
10. B.W. Hoffer, E. Crezee, P.R.M. Mooijman, A.D. van Langeveld, F. Kapteijn and J.A. Moulijn, Carbon supported Ru catalysts as promising alternative for Raney-type Ni in the selective hydrogenation of D-glucose, Catalysis Today 79-80 (2003) 35.
11. P.H. Brahme and L.K. Doraiswamy, Modelling of a Slurry Reaction. Hydrogenation of Glucose on Raney Nickel, Ind. Eng. Chem. Process Des. Dev. 15 (1976) 130.
12. E. Crezee, B.W. Hoffer, R.J. Berger, M. Makkee, F. Kapteijn and J.A. Moulijn, Three-phase hydrogenation of D-glucose over a carbon supported ruthenium catalyst – mass transfer and kinetics, Applied Catalysis A: General 251 (2003) 1.
13. J. Wisniak and R. Simon, Hydrogenation of Glucose, Fructose and Their mixtures, Ind. Eng. Chem. Prod. Res. Dev. 18 (1979) 50.
14. N. Déchamp, A. Gamez, A. Perrard and P. Gallezot, Kinetics of glucose hydrogenation in a trickle-bed reactor, Catalysis Today 24 (1995) 29.
15. J.-P. Mikkola, T. Salmi and R. Sjöholm, Modelling of kinetics and mass transfer in the hydrogenation of xylose over Raney nickel catalyst, J. Chem. Technol. Biotechnol. 74 (1999) 655.
16. J.-P. Mikkola, R. Sjöholm, T. Salmi and P. Mäki-Arvela, Xylose hydrogenation: kinetic and NMR studies of the reaction mechanisms, Catalysis Today 48 (1999) 73.
17. J. Kuusisto, J.-P. Mikkola, J. Wärnå, M. Sparv, H. Heikkilä, R. Perälä, J. Väyrynen and T. Salmi, Hydrogenation of lactose over sponge nickel catalyst – kinetics and modelling, Ind. Eng. Chem. Res. 45 (2006) 5900.
18. T. Salmi, D. Murzin, J.-P. Mikkola, J. Wärnå, P. Mäki-Arvela, E. Toukoniitty and S. Toppinen, Advanced Kinetic Concepts and Experimental Methods for Catalytic Three-Phase Processes, Ind. Eng. Chem. Res. 43 (16) (2004) 4540.
19. H. Haario, MODEST User's Manual, Profmath Oy, Helsinki, 1994.

13. Hydrogenation of α,β-Unsaturated Aldehydes over Ir Catalysts Supported on Monolayers of Ti-Si and Nb-Si

J.J. Martínez[1], H. Rojas[1], G. Borda[1] and P. Reyes[2]

[1]*Grupo de Catálisis (GC-UPTC) Universidad Pedagógica y Tecnológica de Colombia, Escuela de Química, Facultad de Ciencias. Av. Norte - Tunja, Colombia*
[2]*Facultad de Química, Universidad de Concepción, Chile. Concepción, Casilla 160-C, Concepción, Chile*

jjmartinezz@unal.edu.co

Abstract

Results of hydrogenation of various α,β-unsaturated aldehydes over Ir catalysts supported on monolayers of mixed oxides (Ti/Si, Nb/Si) are compiled. The catalysts present an amorphous structure with a high surface area ($S_{BET} > 280$ m^2g^{-1}); the strong surface interaction of silica with TiO_2 or Nb_2O_5 as Ti-O-Si and Nb-O-Si bonding was observed using FTIR and XPS analysis. TEM results revealed that monolayers of Ti/Si and Nb/Si allow a high metal distribution with a small Ir particle size. On the other hand, H_2 chemisorption results showed low H/Ir ratios for HT series, as a consequence of surface decoration of Ir crystals by partially reduced of TiO_{2-x} or Nb_2O_{5-x} species and consequently a decrease in the H_2 uptake. SMSI effect affects the activity and selectivity, being favored the production of unsaturated alcohol when the catalysts are reduced at high temperature (773 K). The structure of the substrate also affects the selectivity. Thus, for molecules with a ring plane, such as furfural and cinnamaldehyde, the adsorption mode can influence the products obtained.

Introduction

Binary oxides of Ti-Si and Nb-Si are very promising model systems for metal catalyst preparation and research of different aspects of strong metal support interaction (SMSI) effect, because at low concentration Nb and Ti ions may be considered as isolated and immobilized species on the SiO_2 matrix (1). SMSI effect describes the suppression of the extent of chemisorption observed in metals supported over reducible supports (2); however, this suppression is only observed when the catalyst is reduced at high temperature reduction (HTR: 773 K). The grade of SMSI effect in binary oxides of Ti-Si depends naturally on the amount of reducible oxide addition and the interaction of metal as preparation conditions (3-5); thus, the catalytic activity of partially reducible oxides is completely modified by the interaction with the silica support, which is associated with changes in the molecular structure and coordination environments (6). Since the supported oxides of Nb_2O_5/SiO_2 and TiO_2/SiO_2 have been used as catalysts and supports for different reactions (7,8), fewer results have been reported for hydrogenation, especially regarding preferential hydrogenation of unsaturated aldehydes and as support for noble metal catalysts. We had shown recently with use of Ir catalyst supported over binary oxides of Ti-Si ($Ir/TiO_2/SiO_2$) a high selectivity in the preferential hydrogenation of carbonyl group of citral, due to SMSI effect (3). The principal aim of this work is to determine the maximum amount of surface oxide of Nb_2O_5 and TiO_2 and the interactions with noble metal (Ir) to favoring selectivity improvements in the hydrogenation of various α,β-unsaturated aldehydes.

Experimental Section

For preparation of mixed oxides of Ti/Si and Nb/Si, a commercial Ti or Nb alkoxide was impregnated in an organic solvent over Aerosil silica (Syloid-266-Grace Davidson) previously activated under vacuum at 423 K. The proportions added were 1 $mmolg^{-1}$ of silica and 2 $mmolg^{-1}$ of silica of Ti or Nb alkoxide (titanium isopropoxide (Aldrich) and niobium ethoxide (Merck). The resultant mixture was stirred for 12 h under inert atmosphere at the solvent reflux temperature. The obtained solid was dried for 5 h under vacuum at 423 K. The mixed oxides were impregnated with 1 wt.% of metallic component using a H_2IrCl_6 aqueous solution. The impregnated solid was dried at 343 K for 6 h, calcined in air at 673 K for 4 h and reduced at 473 K (LT) and 773 K (HT) for 2 h.

Nitrogen physisorption at 77 K and hydrogen chemisorption at 298 K were carried out in a Micromeritic ASAP 2010 apparatus; TEM micrographs were obtained in JEOL Model JEM-1200 EXII System and XRD in a Rigaku apparatus. FTIR spectra of the samples were recorded on Nicolet 550 FTIR spectrometer at room temperature with 64 scans and a resolution of 4 cm^{-1} on sample wafers consisting of 150 mg dry KBr and about 1.5 mg sample. Photoelectron spectra (XPS) were recorded using an Escalab 200 R spectrometer provided with a hemispherical analyzer, and using non-monochromatic Mg Kα X-ray radiation (*hv =1253.6 eV*) source. The surfaces Ti/Si, Nb/Si, Ir/Ti/Si and Ir/Nb/Si atomic ratios were estimated from integrated intensities of Ir $4f_{7/2}$, Ti $2p_{3/2}$ Nb $3d_{5/2}$ and Si $2p_{3/2}$ lines after background subtraction and corrected by the atomic sensitivity factors. The spectra were fitted to a combination of Gaussian-Lorentzian lines of variable proportion.

Catalytic reactions were conducted in a batch reactor at a constant stirring rate (1000 rpm). For all reactions, hydrogen partial pressure was of 0.62 MPa, catalyst weight of 200 mg, 25 ml of a 0.10 M solution of aldehyde α,ß-unsaturated (furfural, citral, cinnamaldehyde) in n-heptane and reaction temperature of 363 K. The absence of oxygen was assured by flowing He through the solution, as well as when the reactor was loaded with the catalyst and reactants at atmospheric pressure during 30 min. Prior to the experiment, all catalysts were reduced *in situ* under hydrogen flow of 20 cm^3min^{-1} at atmospheric pressure and temperature of 363 K. Samples were withdrawn at various time intervals during hydrogenation under intense speed of agitation. Samples were analyzed using a HP-5 capillary column (30 m × 0.25 μm i.d.) in a gas chromatograph-mass spectrometer (Varian 3800-

Saturn 2000). The initial oven temperature was set at 90°C, then increased to 120°C with 10°C/min. The injector ports and the detector were held at 250°C.

Results and Discussion

Characterization results. The XRD patterns of Ir supported on mixed oxides of Ti-Si and Nb-Si showed that SiO_2 remains unchanged after grafting with low content of TiO_2 or Nb_2O_5. An increase in Ti content produces a significant signal of anatase, while increasing Nb produced more poorly defined XRD diffraction peaks. This difference could be the result of the partial reaction of niobia with silica to form an amorphous silicate-like (8). No peaks corresponding to iridium could be observed in the XRD patterns of Ir/Ti-Si and Ir/Nb-Si, which is indicative of the presence of very small metal particles, only for Ir supported over single oxides the signal of Ir becomes visible. XRD results revealed that the TiO_2 and Nb_2O_5 are highly dispersed on the support suggesting that silica is partially covered by fraction of TiO_2 or Nb_2O_5 monolayer at low contents. This phenomenon is verified by decreased IR bands characteristic of the sylanols groups to 3400 cm^{-1}, indicating a strong interaction of the reducible oxides over the silica surface.

Ir catalysts supported over mixed oxides Ti-Si and Nb-Si exhibit a high surface area (Table 13.1) that decreases with the increase in TiO_2 and Nb_2O_5 content, which is explained as a gradual coverage of the SiO_2 by the oxide partially reducible, similar results were found by Asakura et al. and Castillo et al. (8,9). The H/Ir ratio is relatively low for the catalysts supported on single oxide and it is considerably superior for Ir supported over mixed oxides (Table 13.1), this is explained due to the possible creation of other anchorage sites for the metallic precursor. In the series of HTR catalysts, the titania or niobia presence leads to a decrease in the H/Ir ratio. This behavior SMSI is frequently reported when partially reduced species of the support that migrates on the metallic particle covering part of the metal sites generating a surface decoration of the Ir crystals. SMSI effect cannot be attributed to an increase in the metal particle size, since the Ir crystals exhibit similar sizes as TEM experiments have demonstrated. Only the catalysts supported on pure oxides showed larger particle size, close to 4.0, 4.5 and 8.5 for Ir/TiO₂, Ir/SiO₂, Ir/Nb₂O₅, respectively, being the result of the lower interaction of iridium precursor with the SiO_2 support and the low surface area of TiO_2 and Nb_2O_5, respectively. Supported catalysts in mixed oxide of Ti-Si showed a narrow distribution of size of Ir; however, in those supported on Nb-Si the distribution is wider, indicating that the creation of TiO_2-SiO_2 or Nb_2O_5-SiO_2 interface generates sites which can interact with iridium species promoting a higher dispersion of iridium crystals.

As it can be observed in Table 13.1, Ir supported over pure oxides exhibited low acidity, but Ir supported on mixed Nb_2O_5-SiO_2 displayed an important enhancement in the surface acidity with surface coverage by niobia increases. Binding energies (BE) of core-level electrons and metal surface composition were obtained from XP spectra. The BE values of Si 2p, Ti $2p_{3/2}$, Nb $3d_{5/2}$ were 103.4, 458.5 and 123 eV respectively, which are exactly the expected values considering the presence of oxides of Si (IV), Ti (IV) and Nb (V). With regard to Ir $4f_{7/2}$ core level, a value of 60.2 eV may be observed for $Ir^{°}$ species, which appear only Ir/SiO_2 catalyst. However, a slight increase in the BE of the pure oxides from 60.3 to 60.6 eV was observed for Ir/TiO_2 and Ir/Nb_2O_5, suggesting the presence of $Ir^{\delta+}$ species due to electron transfer between the metal and the partially reducible species of TiO_2 or Nb_2O_5. For mixed oxides the BE does not increase drastically for Ir/Ti/Si as for Ir/Nb/Si where was observed a similar increase that Ir/Nb_2O_5. It can be attributed to the better dispersion of the monolayer of Nb_2O_5 than for TiO_2 in these types of mixed oxides that facilitate the migration of some TiO_x or Nb_2O_x species over the Ir crystals (Figure 13.1).

Table 13.1. TiO_2 Content (wt.%), Specific Surface Area, H/Ir Ratios and Metal Particle Size Obtained by TEM of Supported Ir Catalysts.

Catalyst	S_{BET},m²g⁻¹		H/Ir		$d_{TEM,nm}$		NH_3, mmol g⁻¹	
	LT	HT	LT	HT	LT	HT	LT	HT
Ir/SiO_2	290	290	0.200	0.186	3.1	3.2	71	72
Ir/Ti-Si1	283	280	0.026	0.390	1.2	1.3	89	89
Ir/Ti-Si2	271	261	0.260	0.086	1.3	1.3	95	95
Ir/TiO_2	39	39	0.090	0.030	4.0	4.0	4.4	2.5
Ir/Nb_2O_5	8.1	8.2	0.077	0.030	8.5	8.7	40	40
Ir/Nb-Si1	262	262	0.085	0.034	3.0	3.0	109	103
Ir/Nb-Si2	216	222	0.015	0.060	2.9	2.9	340	334

Table 13.2. Binding Energies (eV) of Supported Iridium Catalysts Reduced at High Temperature.

Catalyst	BE Ir4f$_{7/2}$	Bulk	Surface	Bulk	Surface
Ir/SiO_2	60.2	-	-	1.650	0.00315
Ir/TiO_2	60.6	-	-	0.0242	0.0271
Ir/Ti-Si1	60.3	0.057	0.021	0.00322	0.0026
Ir/Ti-Si2	60.3	0.119	0.050	0.00327	0.0023
Ir/Nb_2O_5	60.6	-	-	0.0741	0.0045
Ir/Nb-Si1	60.5	0.005	0.008	0.0031	0.0020
Ir/Nb-Si2	60.6	0.0015	0.0023	0.0032	0.0021

In fact, Table 13.2 summarizes the surface and bulk Ti/Si and Nb/Si, although a partial segregation of TiO_2 or Nb_2O_5 crystals takes place, possibly in aggregate forms (clusters) such as has been described in the literature (3,10) is more evident for Ir/Ti/Si than for Ir/Nb/Si catalysts due to the facility of isomorphic substitution of this system. The increases in the Ti/Si and Nb/Si surface ratio is associated with slight increases in the thickness of the TiO_2 or Nb_2O_5 monolayers. In other studies it has been demonstrated that no significant differences comparing the Nb/Si ratios of LT and HT catalysts; this indicates that almost no dispersion neither sintering of the niobia occurs during high temperature reduction. With regard to Ir/Si+Ti ratio, there are no significant differences due to the metal particle size and the surface similar area. The same behavior is encountered for Ir/Si+Nb surface atomic ratio, which does not change significantly with Nb loading suggesting similar metal dispersion or Ir particle size in line with TEM results.

Activity Test: Table 13.3 summarizes the initial activity and TOF for Ir catalysts. It can be observed that the LT series displays lower activity and TOF values, with the exception of the Ir/TiO$_2$ and Ir/Nb$_2$O$_5$ catalysts. The HT series poses different behavior, depending on the reducible oxide and molecule studied, an increase in Ti or Nb loading leads to an enhancement or decrease in the catalytic activity. In general pure Ir/TiO$_2$ and Ir/Nb$_2$O$_5$ catalysts are more active than Ir supported over mixed oxide catalysts. The results indicate that the catalytic activity of the catalysts may be attributed, at least as one of the factors, to the SMSI effect, at acidity increased in grafted samples and preferential adsorption mode of molecule. The addition of TiO$_2$ or Nb$_2$O$_5$ over silica support may be attributed to the extent of titania silica interaction (Ti-O-Si bonds) and niobia silica interaction (Nb-O-Si) through the generation of new properties which could favour the reduction and/or the dispersion of iridium, thus the formation of monolayer of Ti to the SiO$_2$ support increases the surface coverage making easier the interaction with the metallic component which is deposited on the mixed oxides by impregnation, occasioning a narrow distribution of size that assure the necessary proximity between acid and hydrogenation sites. These new sites are more active for citral, but not for furfural and cinnamaldehyde that present similar structure and similar behavior catalytic.

Table 13.3. Hydrogenation of Aldehydes α,β-Unsaturated at 363 K and at H$_2$ Pressure of 0.62 MPa Over Ir Catalysts Supported on Mixed Oxides. Initial Activity and TOF at 10 % of Conversion.

Catalysts	Furfural		Citral		Cinnamaldehyde	
	In. act. (μmols^{-1}g^{-1})	TOF (s^{-1})	In. act. (μmols^{-1}g^{-1})	TOF (s^{-1})	In. act. (μmols^{-1}g^{-1})	TOF (s^{-1})
Ir/SiO$_2$ LT	0.021	0.006	0.031	0.003	0.070	0.007
Ir/SiO$_2$ HT	0.019	0.005	0.021	0.005	0.068	0.007
Ir/Ti-Si1 LT	0.087	0.004	0.365	0.018	0.087	0.004
Ir/Ti-Si1 HT	0.139	0.103	0.463	0.344	0.139	0.103
Ir/Ti-Si2 LT	0.130	0.010	0.141	0.018	0.130	0.010
Ir/Ti-Si2 HT	0.091	0.020	0.772	0.347	0.091	0.020
Ir/TiO$_2$ LT	0.069	0.015	0.344	0.074	0.069	0.015
Ir/TiO$_2$ HT	0.278	0.179	0.463	0.298	0.278	0.179
Ir/Nb$_2$O$_5$ LT	0.694	0.174	0.164	0.096	0.694	0.174
Ir/Nb$_2$O$_5$ HT	0.651	0.419	0.077	0.039	0.651	0.419
Ir/Nb-Si1 LT	0.029	0.007	0.064	0.037	0.029	0.007
Ir/Nb-Si1 HT	0.019	0.011	0.110	0.025	0.019	0.011
Ir/Nb-Si2 LT	0.047	0.061	0.077	0.248	0.047	0.061
Ir/Nb-Si2 HT	0.058	0.019	0.086	0.110	0.058	0.019

For furfural and cinnamaldehyde a similar behavior was observed; thus, the order of activity for this molecules in HT series corresponds to $Nb_2O_5 > TiO_2 > Ti–Si2 > Ti–Si1 > SiO_2, > Nb–Si1 > Nb–Si2$. As in the furfural (FAL) hydrogenation, for all catalysts studied the main product is furfuryl alcohol (FOL) (Figure 13.1), in the hydrogenation of cinnamaldehyde (CALD) the production of cinnamyl alcohol (COL) is favored: however, during the reaction the concentration of COL is observed to decrease. This decrease is due to a consecutive reaction to produce phenylpropane (PHP) by hydrogenolysis over acidic sites of the support (Figure 13.1). This fact can be explained considering the uniform distribution of Ir particles over these supports that facilitate the proximity between acid and hydrogenation sites; thus, the unsaturated alcohol formed at the beginning of the reaction remains adsorbed in large amounts on the metal surface and reacts over site acids. Hydrocinnamaldehyde (HCALD) and phenylethane (PHE) also were detected in minor proportions, especially in LT series.

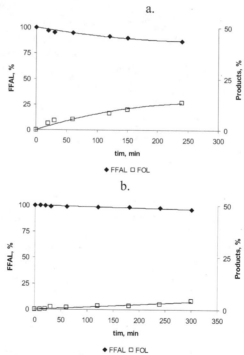

Figure 13.1. Hydrogenation of FFAL over Ir catalysts supported over mixed oxides. (a.) Ir/Ti–Si, (b.) Ir/Nb–Si.

Furfural posseses a furan ring, while cinnamaldehyde has a phenyl group (Figure 13.2). It has been observed that the hindered adsorption of the olefinic bond of the α,β-unsaturated aldehyde by a steric repulsion between the flat metal surface in large particles and the aromatic ring (furan ring or phenyl group) can be responsible for the high selectivity toward C=O group hydrogenation (11). The average particle size of iridium near 8 nm for Ir/Nb_2O_5 could favor the formation of unsaturated alcohol (cinnamyl alcohol and furfuryl alcohol) and reduce the influence of the support. However, small iridium particles of Ir catalysts supported over mixed oxide do make the parallel arrangement of CALD molecules on the metal surface to be less probable. But, for the citral molecule that does not present a flat ring (Figure 13.3c)

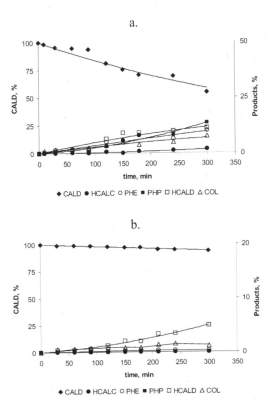

Figure 13.2. Hydrogenation of CALD over Ir catalysts supported over mixed oxides. (a.) Ir/Ti-Si, (b.) Ir/Nb-Si.

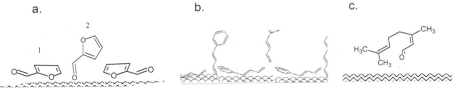

Figure 13.3. Hydrogenation of CAL over Ir catalysts supported over mixed oxides: (a.) furfural, (b.) cinnamaldehyde, (c.) citral.

the larger particles are not necessary, thus the selectivity is independent of the Ir particle size. Similar results have been encountered for other authors in Ru catalysts (12). It can be observed that for the hydrogenation of citral over Ir/Ti-Si catalysts (Figure 13.4a), the main products were unsaturated alcohols (UOL: geraniol and nerol), citronellal (CNAL) and citronellol (CNOL) in minor extension. For Ir/Nb-Si catalyst (Figure 13.4b), several products such as CNOL, UOL as well as acetals and hemiacetals were observed. The formation of acetals is associated with the acidic properties of the support, which confirms that the number of acid sites can be considered as a factor for the activity and selectivity of Ir catalysts in both series (HT and LT). As indicated earlier, Ti-grafting or Nb-grafting modifies the amount of acidic sites. Catalysts supported over binary oxides are more acid than silica and their acidity is lower compared with pure titania or niobia as it was observed for NH_3-TPD analysis. This behavior is more evident with Nb-grafting and thus can explain the lower activities and selectivities towards UOL observed for Ir/Nb-Si catalysts.

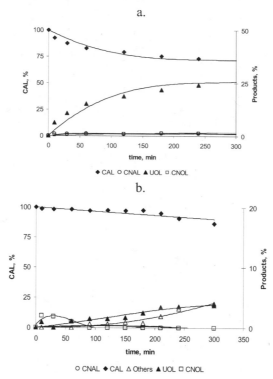

Figure 13.4. Hydrogenation of CAL over Ir catalysts supported over mixed oxides: (a.) Ir/Ti-Si, (b.) Ir/Nb-Si.

Conclusions

Ir catalysts supported on binary oxides of Ti/Si and Nb/Si were prepared and essayed for the hydrogenation of α,β-unsaturated aldehydes reactions. The results of characterization revealed that monolayers of Ti/Si and Nb/Si allow a high metal distribution with a small size crystallite of Ir. The activity test indicates that the catalytic activity of these solids is dependent on the dispersion obtained and acidity of the solids. For molecules with a ring plane such as furfural and cinnamaldehyde, the adsorption mode can influence the obtained products. SMSI effect (evidenced for H_2 chemisorption) favors the formation of unsaturated alcohol.

Acknowledgements

We thank to COLCIENCIAS for their financial support under the project No. 11090517865.

References

1. A. Y. Stahkeev and E.S. Shpiro, J. Apijok, *J. Phys. Chem.* **97**, 5668 (1993).
2. S.J. Tauster, S.C. Fung and R.L. Garten. *J. Am. Chem. Soc.* **100** (1) 170 (1978).
3. H. Rojas, G. Borda, P. Reyes, J.J. Martínez and J. Valencia, J.L.G. Fierro, *Catal. Today*, (2008).
4. P.S. Kumbhar, *Appl. Catal. A.* **96**, 241 (1993).
5. H. Hoffman, P. Staudt, T. Costa, C. Moro and E. Benvenutti, *Surf. Interface Anal.* **33**, 631 (2002).
6. M.C. Capel-Sanchez, J.M. Campos-Martin and J.L.G. Fierro, *J. Catal*, **234**, 488 (2005).
7. X. Gao and X.E. Wachs, *Catal. Today.* **51**, 235 (1999).
8. K. Asakura and Y. Iwasaka, *J. Phys. Chem.* **95**, 1711 (1991).
9. R. Castillo, B. Koch, P. Ruiz and B. Delmon, *J. Mater. Chem.* **4**, 903 (1994).
10. G. Borda, H. Rojas, J. Murcia, J. L. G. Fierro, P. Reyes and M. Oportus. *React. Kinet. Lett* **9**, 369 (2007).
11. Giroir-Fendler, P. Gallezot and D. Richard. *Catal. Lett.* **5**, 169 (1990).
12. S. Galvagno, G. Capannelli, G. Neri, A. Donato and R. Pietropaolo, *Catal. Lett.* **18**, 349 (1993).

14.   4,6-Diamino Resorcinol by Hydrogenation, a Case Study

Antal Tungler[1], Zeno Trocsanyi[2] and Laszlo Vida[1]

[1]Budapest University of Technology and Economics, Department of Chemical and Environmental Process Engineering Budapest, 1111 Hungary, Budafoki ut 8[1]
[2]Repét Company, Petfurdo, Hungary 8105

atungler@mail.bme.hu

Abstract

A hydrogenation process for the preparation of 4,6-diamino resorcinol from 4,6-bisphenylazo resorcinol was studied and developed up to pilot scale. The high reaction rate together with the protection of the product against oxidation could ensure providing the addition of HCl was made after the fast hydrogenation period. Activity and lifetime of commercial catalysts were evaluated also.

Introduction

Zylon® is the registered trademark of Toyobo Corporation for a special polymer of extraordinary strength and thermal stability. Chemically it is polybenzoxazole, consisting of diamino resorcinol and dicarboxylic acids, as terephthalic acid (Scheme 14.1).

Scheme (14.1)

Our task was to develop a feasible synthesis for diamino resorcinol. From several possibilities [1-4] the resorcinol and aniline were chosen as starting materials. The diazotated aniline was coupled with resorcinol among basic conditions giving 4,6-bisphenylazo resorcinol. Hydrogenation of the latter resulted in diamino resorcinol and aniline, which could be recycled (scheme 14.2). This chemistry is well known [5-13]; therefore, the research work focused on finding the optimal parameters and catalyst and finally on elaborating a process for scale-up.

Scheme (14.2)

Experimental Section

The chemicals used, resorcinol, aniline, $NaNO_2$, HCl, NaOH and solvents were all technical grade materials. The palladium catalysts were activated carbon supported ones, all commercial products (see Table 14.1).

Table 14.1. Catalyst Types.

Number	Catalyst Pd/C	Pd content % for dry material	Water content %
1	SQ-6	10	0
2	SQ-3	5	0
3	Heraeus	10	50
4	Heraeus	5	50
5	AMCPMC	5	63

4,6-bisphenylazo resorcinol was prepared by coupling the diazotated aniline (diazotation carried out in acidic water solution on 0°C) with resorcinol in basic solution on r.t.. The precipitated dark reddish material was washed with hot xylene. The recovered substance contained ~90% bis-azo compound beside tris-azo and aminoazobenzene.

Catalytic hydrogenations were performed generally under 2-8 bar of H_2 at 25 to 40°C in methanol or methanol/water (HCl) media, with a total volume of 80-

150 mL; the substrate was essentially insoluble in methanol and water, but dissolved readily during the reaction.

The reaction rate was calculated from hydrogen consumption, from pressure drop. As the hydrogen consumption rate was dependent on amount and purity of substrate, catalyst weight and metal content, conversion rate was given (%/min*mg Pd), the measured total hydrogen consumption was taken equivalent with 100% conversion.

The substrate, catalyst and solvent were placed into a glass liner equipped with a magnetic stir bar. The liner was placed in a 250 mL steel autoclave that was then charged at r.t. first with N_2, then with H_2. The reaction mixture was stirred at r.t. until the hydrogen consumption ceased, after which the H_2 was vented, the autoclave flushed with N_2 and the reaction mixture filtered off on glass filter. The solution was evaporated and the products crystallized and filtered. The crystalline mixture contained the hydrochloride salt of diamino resorcinol and aniline. In order to separate completely diamino resorcinol and aniline, the product mixture has to be recrystallized from hydrochloric acid solution.

Hydrogenated products were identified by GC-MS using the commercially available and fully characterized materials.

Results and Discussion

Already after the first preliminary hydrogenation experiments, it became evident that diamino resorcinol cannot be manipulated in its base form, as contacting with traces of air it is oxidized, which appears in changing its pink color during seconds to dark purple.

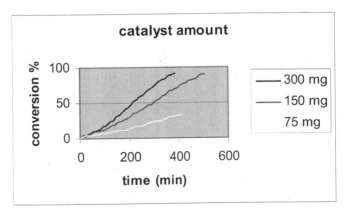

Figure 14.1.
Avoiding the fast oxidation acidic medium was applied, the hydrogenation was carried out in methanol-water mixture containing a stoichiometric amount of hydrochloric acid, 4 mole with respect to 1 mole 4,6-bisphenylazo resorcinol. Under

these conditions the product salt was more stable, but the hydrogenation rate was low (Figure 14.1).

Even at high catalyst loading, 10% with respect to the substrate (substrate 3,2 g, methanol 20 cm³, Pd/C catalyst SQ-6, 10wt% HCl solution 14 cm³, hydrogen pressure 5-6 bar, temperature 25°C, rpm 850 min⁻¹) the completing of the reaction took more than 6 hours. Increasing the pressure within reasonable range didn't help much (Figure 14.2). The preliminary saturation of the catalyst with hydrogen didn't change significantly the reaction rate as well.

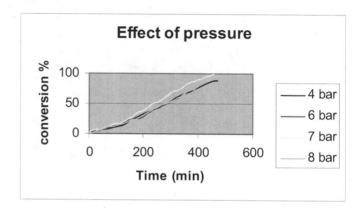

Figure 14.2.

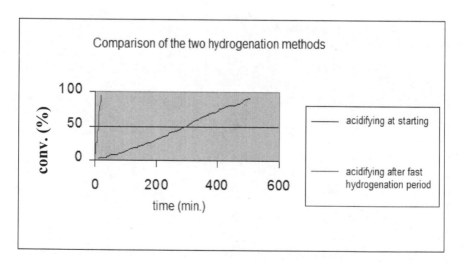

Figure 14.3.

Finally the hydrogenation in methanol and a subsequent acidification under hydrogen atmosphere turned out to be most successful (Figure 14.3.). After ceasing of the hydrogen consumption, hydrochloric acid solution was added to the reaction mixture, which was cooled, removing the neutralization heat, and finally the hydrogenation was continued. The hydrogen consumption was negligible in this last phase, but the operation was important from the aspect of the catalyst lifetime. Without this post-hydrogenation, the second use of the catalysts, the activity decreased to 1/10 of the original value.

Starting with methanol solvent, with less than 1% catalyst (60 mg for 6.4 g substrate) the reaction was completed within 80 minutes (Figure 14.4).

Alternative solvents were also tested. Isopropanol and methyl tertiary-butyl ether (MTBE) with water (HCl), the latter a two-phase system, the displayed reaction rates were much slower.

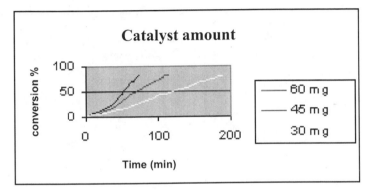

Figure 14.4.

Finally different catalysts had to be compared among the optimized reaction conditions, the results are in Table 14.2 and in Figure 14.5.

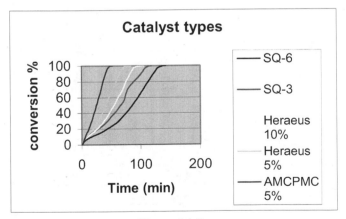

Figure 14.5.

Substrate 6,4 g, methanol 40 cm^3, HCl solution 14 cm^3, hydrogen pressure 5-6 bar, temperature 25°C, rpm 850 min^{-1}.

Table 14.2. Catalyst Activities.

Entry	Catalysts	Catalyst amount mg	Specific rate I. conv%/min*mg Pd	Specific rate II conv%/min*mg Pd
1	SQ-6	50	0,08	0,22
2	SQ-3	100	0,1	0,23
3	Heraeus-10	100	0,12	0,28
4	Heraeus-5	200	0,19	0,35
5	AMCPMC	270,3	0,31	0,52

However the catalyst loadings were calculated for same amount of Pd, the activities were different in both period of the hydrogenation. The first, slower part can be attributed to the dissolution of the substrate. In this period there is a dense slurry in the reactor, which makes mixing difficult in small and in large scale as well. In the second phase a homogeneous solution is present; therefore, the mixing is easier; hydrogen concentration and reaction rate are higher.

Table 14.3. Relative rate = (rate in actual entry/rate in first entry)*100

Catalyst repeated use		Reaction time (min)	Relative rate (%)
AMCPMC	1.	70	100
	2.	140	50
	3.	270	26
Heraeus-5	1.	128	100
	2.	238	54
	3.	458	28
Heraeus-10	1.	100	100
	2.	102	98
	3.	114	88
	4.	167	60
SQ-6	1.	116	100
	2.	151	77
	3.	170	68
	4.	195	59
	5.	196	59

Finally the catalyst lifetime was tested, as together with activity it influences the costs of catalyst. The catalyst after completing the hydrogenation was filtered, washed with methanol and water and used again under similar conditions. This was continued until the activity decreased significantly.

In Table 14.3 the reaction times (until complete conversion) and relative rate values for the repeated use of four catalysts are collected. It was obvious that the two 10% Pd/C catalysts (Heraeus-10, SQ-6) were more stable.

Conclusions

A feasible hydrogenation process was developed for the production of 4,6-diamino resorcinol from 4,6-bisphenylazo resorcinol. In the final process 0.8% catalyst (SQ-6, 10% Pd/C) was used in five subsequent reactions, this means that the specific catalyst consumption was less than 1:500 with respect to the substrate. The yield of the raw product (diamino resorcinol.2HCl) was quantitative, but its aniline content was 3-6 %, after purification the yield was 74-76%, the purity >99%. Even the color of the product was white. The aniline could be recycled with more than 80% yield. The key to the improved results is that the hydrogenation was performed in the absence of HCl. After the reaction, the product was treated with HCl under hydrogen, and this procedure improved catalyst life.

After the laboratory experiments the hydrogenation process was scaled up first in 50 then in 1000 liter volume autoclaves, 200 kg diamino resorcinol.2HCl was produced in proper quality.

Acknowledgements

The authors gratefully acknowledge the support of the Hungarian NKFP3-35/2002 project and the cooperation of Nitrokemia 2000 Co.

References

1. H.U. Blank and U. Heinz, US Patent 5072053, 1991.
2. H. Behre, H. Fiege, H.U. Blank and W. Eymann, US Patent 6093852, 2000.
3. T.A. Morgane, B.S. Nader, P. Vosejpka, W. Wu and A.S. Kende, US Patent 5453542, 1995.
4. Z. Lysenko, R.G. Pews and P. Vosejpka, US Patent 5414130, 1995, Dow Chemical Co.
5. K. Tokunaga, M. Shiratori, K. Akimoto, H. Suzuki and I. Hashiba, US Patent 6222071, 2001.
6. Z. Lysenko and R.G. Pews, US Patent 5399768, 1995, Dow Chemical Co.
7. S. Komatsu, K. Minamisaka, H. Ueda, K. Murata, T. Takahata and I. Yoshida, Japanese Patent 07242604, 1995, Sumitomo Chemical Co. Ltd., Japan.
8. H. Behre, H. Fiege, H-U. Blank, U. Heinz and W. Eymann, Eur. Pat. Appl. European Patent 710644, 1996, Bayer A.-G., Germany.
9. H. Ueda and S. Komatsu, Japanese Patent 09194443, 1996, Sumitomo Chemical Co., Ltd., Japan.

10. H. Behre, H. Fiege, H-U. Blank and W. Eymann, German Patent 19620589, 1997, Bayer A.-G., Germany.
11. J. Kawachi, H. Matsubara, Y. Nakahara and Y. Watanabe, German Patent 19744362, 1998, Daiwa Kasei Industry Co., Ltd., Japan.
12. K. Tokunaga, M. Siratori, K. Akimoto, H. Suzuki and I. Hashiba, PCT Int. Appl. WO 9937601, 1998, Nissan Chemical Industries, Ltd., Japan.
13. Xia, Enjiang; Xia, Ensheng; Yue, Yan; Wu, Yanqiang; Xia, Niancheng and Zhang, Jing. Faming Zhuanli Shenqing Gongkai Shuomingshu, CN 1569813, 2005, Shandong Taishan Dyes Co., Ltd., Peop. Rep. China.

15. Novel Hydrogenolysis Route to Perfluoroalkyl Ethane Thiols

Stephen E. Jacobson[1], Lloyd Abrams[1], Glover Jones[1], Lei Zhang[1] and James F. White[2]

[1]DuPont Company, Wilmington, DE 19880, Contribution #8847
[2]Pacific Northwest National Laboratory, Richland, WA 99352

stephen.e.jacobson@usa.dupont.com

Abstract

We have studied the hydrogenolysis of 2-(perfluorohexyl)ethane thiocyanate to 2-(perfluorohexyl)ethane thiol. It was discovered that perfluoroalkyl thiocyanates can be reduced to thiols and co-product hydrogen cyanide with molecular hydrogen in the presence of a carbon-supported palladium-tin catalyst. This result is surprising since it is known that palladium and other groups 8 to 10 metal catalysts are poisoned by the product thiol, traces of hydrogen sulfide byproduct, and the hydrogen cyanide co-product. For that reason, we characterized the catalyst to understand why it was so robust under conditions that would normally poison such a catalyst.

The effects of tin/palladium ratio, temperature, pressure, and recycling were studied and correlated with catalyst characterization. The catalysts were characterized by chemisorption titrations, *in situ* X-Ray Diffraction (XRD), and Electron Spectroscopy for Chemical Analysis (ESCA). Chemisorption studies with hydrogen sulfide show lack of adsorption at higher Sn/Pd ratios. Carbon monoxide chemisorption indicates an increase in adsorption with increasing palladium concentration. One form of palladium is transformed to a new phase at $140°C$ by measurement of *in situ* variable temperature XRD. ESCA studies of the catalysts show that the presence of tin concentration increases the surface palladium concentration. ESCA data also indicates that recycled catalysts show no palladium sulfide formation at the surface but palladium cyanide is present.

Introduction

2-(Perfluoroalkyl)ethane thiols have been used as precursors to fluorinated surfactants and products for hydro- and oligophobic finishing of substrates such as textiles and leather (1). The synthesis of 2-(perfluoroalkyl)ethane thiol and a byproduct bis-(-2-perfluoroalkylethane)-disulfide (5-10%) has been practiced via the reaction of 2-(perfluoroalkyl)ethane iodide with thiourea to form an isothiuronium salt which is cleaved with alkali or high molecular weight amine as shown in Equation 1 for 2-(perfluorohexyl)ethane iodide (1).

$$\text{Base}$$
$$C_6F_{13}CH_2CH_2I + NH_2C(S)NH_2 \rightarrow C_6F_{13}CH_2CH_2SH + (C_6F_{13}CH_2CH_2S)_2 \qquad (1)$$

Thiourea is a carcinogen and will be eliminated in future manufacturing facilities (2). Its use requires special equipment and expense, such as incinerators, for safety.

Nucleophilic substitution of sodium thiolate on 2-(perfluorohexyl)ethane iodide leads to extensive elimination (10-20%) and formation of (perfluorohexyl)ethylene (3), Equation 2.

$$C_6F_{13}CH_2CH_2I + NaSH \rightarrow C_6F_{13}CH_2CH_2SH + C_6F_{13}CH=CH_2 \qquad (2)$$

The olefin yields are higher with 2-(perfluoroalkyl) ethane iodides than their hydrocarbon analogues apparently because of the strongly electron withdrawing fluorine making the hydrogens more acidic and prone to elimination.

It has been known for many years that 2-(perfluorohexyl)ethane thiocyanate can be synthesized in quantitative yields from 2-(perfluorohexyl)ethane iodide (4), as shown in Equation 3.

$$C_6F_{13}CH_2CH_2I + NaSCN \rightarrow C_6F_{13}CH_2CH_2SCN + NaI \qquad (3)$$

However, lithium aluminum hydride or zinc metal and HCl (5) are required as reducing agents to reduce the thiocyanate to the thiol. These reducing agents are stoichiometric reagents and aren't environmentally acceptable at this time because of their hazardous properties and waste disposal problems on a large manufacturing scale.

It is apparent that a new synthetic methodology, preferably catalytic, is needed for the synthesis of this important class of 2-(perfluoroalkyl)ethane thiols. In this context, a variety of catalysts was examined to determine if they would catalyze the hydrogenolysis of 2-(perfluorohexyl)ethane thiocyanate. In the course of this study, much to our surprise, it was discovered that a carbon supported Pd-Sn would catalyze the reaction. It is known that palladium and other group VIII metal catalysts are poisoned by the product thiol, traces of hydrogen sulfide byproduct, and the hydrogen cyanide co-product (6), but our observations are that this catalyst is surprisingly robust in the reaction medium.

The subjects of this chapter are the exploration of the scope and limitations of the new Pd-Sn catalyzed hydrogenolysis route for the synthesis of thiols via 2-(perfluoroalkyl)ethane thiocyanate, the characterization of the surprisingly active and robust Pd-Sn catalysts, and the attempted correlation of the characterization of the catalysts with observed onset of hydrogenolysis reactivity and surprisingly long lifetime in the presence of known catalyst 'poisons.'

Experimental Section

2-(Perfluorohexyl)ethane iodide was obtained from the Surface Protection Solutions business at DuPont. Sodium thiocyanate was purchased from Aldrich and used without further purification. The 2-(perfluorohexyl)ethane iodide was converted to the 2-(perfluorohexyl)ethane thiocyanate by reaction with sodium thiocyanate using phase transfer catalysis described in a prior patent (7).

The palladium-tin catalysts were prepared by Engelhard on a commercial wood based carbon powder with a BET surface area of approximately 800 m^2/g and a median particle size (D50) of 19 microns. The preferred carbon was chosen mainly for having good filtration properties. Catalysts with essentially equivalent activities for selectivity and conversion could also be made on two other similar carbons. The preparation process is proprietary but is based on the well-known adsorption-deposition technique (8). Reduction during the preparation process was accomplished via an Engelhard proprietary method. A series of catalysts containing from 1 to 7.5 wt% palladium and from 0 to 1 wt% tin were prepared by the same technique and provided for the experimental program.

The hydrogenation reactor consisted of a 1-L "Hastelloy" C autoclave (Model 4641M, Parr instrument company, Moline Illinois) equipped with a belt-driven, magnetic stirrer (1000 rpm).

In a typical hydrogenolysis run, 2-(Perfluorohexyl)ethane thiocyanate (133.7 g, 0.33 mole), ethyl acetate(160g), 5% Pd-0.5% Sn on carbon catalyst (0.5g, 2.4 x 10^{-4} mole Pd, 1400: 1 substrate: Pd molar ratio), and biphenyl (internal standard, 3.0g) were charged to a 1-L autoclave, the solution purged with nitrogen three times and then hydrogen three times, the agitation was set at 1100 RPM, the hydrogen reactor pressure set at 700 psig and the continuous flow of hydrogen set at 1.5 SLPM with a brine condenser. The autoclave was then heated to 175°C for 2 hours. (The hydrogenolysis runs were generally carried out at 140-180°C) Gas chromatography showed 97.9% conversion of thiocyanate, 99.5% 2-(perfluorohexyl)ethane thiol product and 0.3% byproduct bis-(-2-perfluorohexylethane)-disulfide. All %conversion and %selectivity data reported in this chapter are in mole%. All substrate: Pd ratios reported here are molar ratios.

All gas chromatography analyses were done with a Hewlett-Packard 5890A instrument coupled with a 3393A integrator. A 30 m x 0.25 mm RTX-20 capillary column was used with a flame ionization detector.

The catalysts were characterized by chemisorption titrations, *in situ* XRD and ESCA.

For the chemisorption experiments a weighed catalyst sample (wet) was put in a cell and mounted on the Micromeritics 2010 (static) chemisorption instrument. The sample was heated under vacuum to 150°C where it was exposed to hydrogen (0.7 atm) for 0.5 hour. The sample was then evacuated at room temperature, re-exposed to hydrogen for 0.5 hour, then evacuated, and cooled to 30°C under vacuum.

Titrations of carbon monoxide and hydrogen sulfide up to 800 torr were performed at 30°C; each volumetric titration was composed of two adsorption isotherms: the first isotherm was a combination of chemisorption and physisorption, while the second isotherm was assumed to be physisorption only. The sample was evacuated to 10^{-5} torr after each isotherm. Subtraction of the second isotherm from the first resulted in the chemisorption isotherm.

In situ XRD studies of thermally and chemically induced transformations were monitored with a totally automated diffractometer-micrometer system (9). Samples were thermally ramped in 5°C increments from 30 to 160°C, maintained at 160°C for 0.5 hr, and then cooled in 5°C increments to 30°C under 500 torr hydrogen. 2θ scans were obtained after each temperature increment. The individual scans could be stacked to show the temperature evolution of the structure.

ESCA analyses were performed using a Physical Electronics PHI 5600 ci spectrometer equipped with an aluminum monochromatic x-ray source. Hydrogen reduction at 75°C and 150°C was performed *in situ* in a reaction chamber attached to the main ESCA analytical chamber. Gas mixture of 80/20 N_2/H_2 was used for the hydrogen reduction. Charge correction was performed shifting the C-(C,H) peak in C 1s spectra to 284.8 eV. PHI MultiPak® software version 6.0A was used for data analysis.

Results and Discussion

I. Thiocyanate Hydrogenolysis Reaction

We report the discovery of a new Pd-Sn/C catalyzed hydrogenolysis reaction to produce thiols in high yield (10), Equation 3.

$$\text{Pd-Sn/C}$$
$$C_6F_{13}CH_2CH_2SCN + H_2 \rightarrow C_6F_{13}CH_2CH_2SH + (C_6F_{13}CH_2CH_2S\text{-}) + HCN \qquad (4)$$
$$99\% \qquad\qquad 0.2\%$$

The thiol was obtained in >98% yield with trace amounts of the disulfide at 175°C and 700 psig H_2 reactor pressure in 1.5 hours at a 900:1 substrate: catalyst molar ratio. As discussed above, it is known that palladium and other groups 8 to 10 metal catalysts are poisoned by the product thiol, traces of hydrogen sulfide byproduct, and hydrogen cyanide coproduct (6), but it is surprising that this catalyst is so robust. The effects of solvents, temperature, pressure, catalyst, and recycle will be discussed. The characterization of the catalyst by various techniques will help to explain some of these observations.

A variety of solvents was investigated for this reaction, as shown in Table 15.1. As inferred from Table 15.1, the hydrogenolysis performance is best in more polar solvents such as acetonitrile, acetone, ethyl acetate, and acetic acid. Only in *o*-dichlorobenzene is the rate of reaction much lower than predicted by the dielectric constant. The presence of nonpolar solvents such as hexane and the thiol product resulted in large amounts of the disulfide intermediate. It has been shown that the disulfide is the intermediate in stoichiometric reductions such as samarium diiodide reduction of alkyl thiocyanates to thiols (11) so it is reasonable to expect it as the intermediate in this case. This intermediate formation might have been caused by more rapid deactivation of the catalyst in the presence of less polar solvents.

Table 15.1. Effect of Solvents on Thiocyanate Hydrogenolysis to Thiol.

Solvent	Dielectric	Conv(%)	Thiol(%)	Disulfide(%)
Pfhethiol	na**	59	27	72
Hexane	1.9	20	20	79
Bdmpe	2.3*	73	84	0
Ethyl acetate	6.0	99	93	0
Acetic acid	6.2	98	93	0
Tetrahydrofuran	7.6	100	98	0
o-Dichlorobenzene	9.9	52	97	3
Acetone	20.7	100	99	0.1
Acetonitrile	37.5	100	98	0.5

All runs at 140°C with 5% Pd-0.5% Sn catalyst (1000:1 substrate:catalyst molar ratio), 700 psig H_2 for 2 hr.
Abbreviations: Bdmpe[1,1-Bis(3,4-dimethylphenyl)ethane]; Pfhethiol (2-perfluorohexyl)ethane thiol).
*The dielectric constant = 2.3 for 1,3,5-trimethylbenzene, of similar structure to Bdmpe.
**Not available.

The disulfide intermediate was also hydrogenolyzed very efficiently at about 2-3 times the rate of the thiocyanate in selected solvents such as tetrahydrofuran at a lower temperature and pressure, as illustrated in Table 15.2. Of course, in the disulfide hydrogenolysis, no hydrogen cyanide was available as a co-product to poison the palladium catalyst.

Table 15.2. Comparison of Rates of Disulfide and Thiocyanate Hydrogenolysis.

Reactant	% Conv	% Thiol	% Disulfide
2-perfluorohexyl-ethanethiocyanate	30	30	60
Bis-perfluorohexylethane-disulfide	97	99	N/A

Operating conditions: 125°C, 400 psig H_2, THF solvent, 500:1 substrate:Pd ratio molar ratio with 5%Pd-0.5%Sn/C for 2 hr.

It is well known that palladium on carbon catalysts are poisoned by hydrogen cyanide and thiol products or hydrogen sulfide (6). Therefore, it was of interest to investigate the reduction of perfluoroalkyl thiocyanates as a function of tin concentration, keeping the concentration of palladium and reaction conditions constant. Figure 15.1 delineates the % conversion vs. Sn/Pd ratio, under the same reaction conditions of 175°C, 700 psig H_2 for 2 hours with 5% Pd on carbon catalysts in ethyl acetate solvent at a 1000:1 substrate:catalyst molar ratio. The increase in catalyst activity at constant Pd concentration illustrates the dramatic positive effect of Sn on the initial activity of the catalyst and also on retardation of catalyst deactivation.

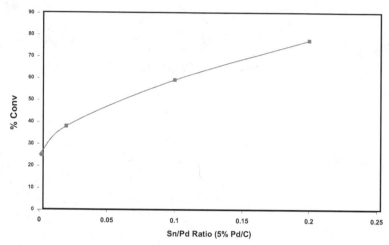

Figure 15.1. Effect of Sn/Pd Ratio on Thiocyanate Conversion.

In order to study the catalyst deactivation, the recycle of the carbon supported 5%Pd-0.3%Sn catalyst was attempted. The catalyst was filtered after reaction under nitrogen, dried under vacuum, and again charged with fresh solvent and substrate under the same reaction conditions as the first cycle. The results are shown in Table 15.3.

Table 15.3. Recycle of 5%Pd-0.3%Sn/C Catalyst.

Run	% Conv	% Thiol	% Disulfide
First cycle	100	99	0.2
Second cycle	68.2	4.2	83.5

Operating conditions: 140°C, 700 psig, THF solvent, 1000:1 substrate:catalyst molar ratio for 2 hours.

These results demonstrate that the catalyst is partially deactivated during recycle. The high disulfide yield on recycle strongly suggests that the disulfide is an intermediate to the thiol final product. Both H_2S and HCN are possible sources of catalyst deactivation.

The effects of temperature and hydrogen pressure on catalyst activity were quite dramatic. No significant hydrogenolysis was observed below 120°C and high activity was only found at temperatures above 140°C. A study of the effect of pressure on conversion, as shown in Figure 15.2, demonstrated the dramatic effect of hydrogen pressure on activity. These runs were carried out at 170°C using a 5%Pd-0.5% Sn/C catalyst in ethyl acetate solvent at a 1000:1 substrate:catalyst molar ratio for 1.5 hour. One hypothesis is that the high hydrogen partial pressure helps to lower interaction with H_2S and/or HCN at the catalyst surface. It was also helpful to have a continuous subsurface hydrogen flow with a condenser (Experimental) to remove HCN and traces of H_2S, as they were formed.

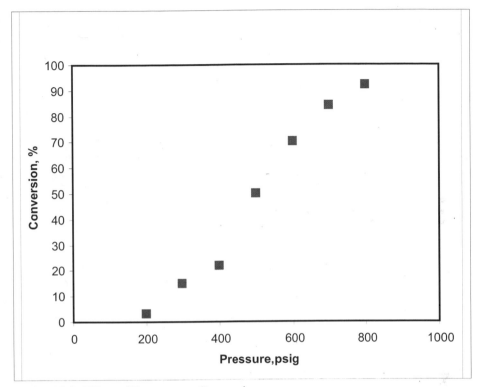

Figure 15.2. Effect of Pressure on Conversion.

II. Characterization of the Pd-Sn Catalyst

a. Chemisorption Titrations

The Pd-Sn/C catalysts (1 to 7.5% Pd containing 0 to 1% Sn) were heated under vacuum at 150°C and then exposed to hydrogen. These preactivated samples were then titrated with carbon monoxide, a very specific ligand for Pd, up to 800 Torr at 30°C. A general linear trend of carbon monoxide concentration with % Pd in Figure 15.3 indicates that the carbon monoxide adsorption is directly correlated to Pd concentration, as expected. The trend is independent of Sn content. This linear Pd-CO trend indicates that the particle size distribution is similar for the different catalysts. However, Figure 15.3 also indicates no relationship between % H_2S irreversibly adsorbed and % Pd.

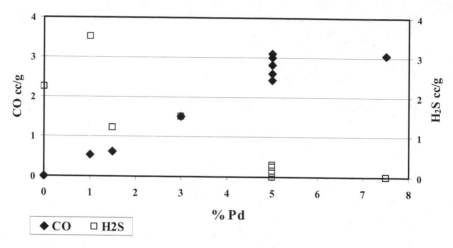

Figure 15.3. CO, H_2S Adsorption vs. % Pd (with Differing Amounts of Sn).

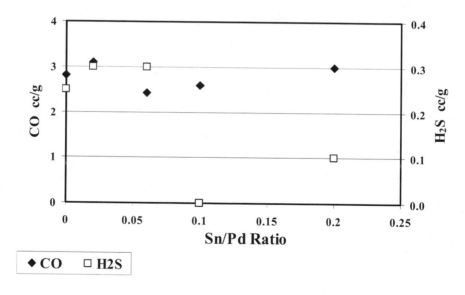

Figure 15.4. CO, H_2S Adsorption vs. Sn/Pd Ratio at 5% Pd.

The 5% Pd/C catalysts, with an increasing amount of tin (from 0 to 1%), were activated in the same way described above and titrated with both carbon monoxide and hydrogen sulfide, up to 800 Torr, at 30°C. Figure 15.4 indicates a constant volume of carbon monoxide adsorbed, as expected from the above relationship for a fixed amount of %Pd on the catalyst. However, there does appear to be a relationship between H_2S adsorption and the Sn/Pd ratio at constant 5 wt% Pd concentration on the catalyst. When there is little or no tin associated with the Pd-catalyst, the H_2S is irreversibly adsorbed, resulting in high volumetric uptake of hydrogen sulfide. At higher tin concentrations, > 0.05 Sn/Pd ratios, much less H_2S is adsorbed.

b. XRD Characterization

Shown in Figure 15.5 are the temperature dependent XRD data for the 5% Pd-1% Sn catalyst. As noted above, the scans were offset in the order that they were obtained (the "Time" axis, as shown, is the scan sequence number and not the actual temperature). The inset of Figure 15.5 illustrates the temperature profile for the scan sequence. The first scan was obtained at room temperature, at which time hydrogen was introduced into the chamber at 500 Torr. The temperature was then ramped in 10°C increments to 160°C and XRD scans were taken after each increment. The sample was held at 160°C for ½ hour, and then cooled to room temperature. After ½ hour at room temperature, the sample was purged with dry nitrogen.

The first scan at room temperature, in the front of Figure 15.5, is typical of a Pd-supported catalyst. The catalyst, heated under H_2, retains this structure until a temperature of ~140°C at which point the Pd phase is transformed into a new phase

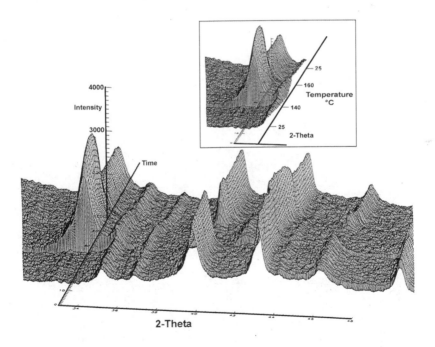

Figure 15.5. Temperature Dependence of Heating the 5% Pd-1% Sn Catalyst under Hydrogen.

whose composition we could not determine. The transformation into a new phase corresponds to the temperature effect of catalyst reactivity data discussed above and clearly demonstrates the need to heat the catalyst above 140° to produce this phase (12). It is this new palladium phase which is resistant to "sulfur poisoning" and/or HCN poisoning (6). Cooling the catalyst to room temperature under hydrogen caused some of the new phase to convert back to Pd while some remained in the new structure. Because of the pressure limitation to 1 atm of H_2 (safety), the effect of using high pressure H_2 could not be studied. It seems plausible that the high H_2 pressures used in the catalytic reaction may further stabilize the new phase from converting back.

These results, coupled with the chemisorption titrations, provide evidence that a new phase of palladium is formed with tin under hydrogen that is more resistant to "sulfur poisoning."

c. ESCA Analyses

The ESCA data indicated that the addition of tin caused the surface concentration of palladium to increase. The 5%Pd-0.5%Sn had the highest concentration of palladium when compared to 5% Pd, indicating the possibility that tin promoted more palladium on the catalyst surface (Table 15.4).

Table 15.4. Surface Concentrations Detected from the Two Catalyst Samples.

Sample	C	N	O	Pd	Sn
5% Pd(no Sn)	67.5	<0.1	22.7	8.4	nd
5% Pd After 30 Min @ 75°C in 80/20 N_2/H_2	70.1	0.2	20.7	7.7	nd
5% Pd After 30 Min @ 150°C in 80/20 N_2/H_2	71.4	nd	19.4	7.5	nd
5% Pd -0.5% Sn	64.0	0.2	15.2	18.9	1.8
5% Pd -0.5% Sn, Exposed to 80/20 N_2/H_2 @ 75°C, 0.5 h	66.6	0.1	13.1	18.6	1.6
5% Pd -0.5% Sn, Exposed to 80/20 N_2/H_2 @ 150°C, 0.5 h	68.6	0.3	11.8	17.8	1.4

The concentrations are normalized to 100%. H and He cannot be detected by XPS.
nd: not detected with detection limit at ~0.1 atom%.

The "as received" 5%Pd-0.5%Sn/C catalyst was dominated with palladium assigned as Pd^{+2} (the binding energy of the major Pd $3d_{5/2}$ peak at ~337.1 eV) and tin as $Sn^{+2/+4}$ on the surface. After 75°C hydrogen reduction, small amounts of oxide species were still observed for both Pd and Sn but with a predominance of reduced species (the binding energy of the Pd $3d_{5/2}$ peak at binding energy of ~335.5 eV). After 150°C reduction, both Pd and Sn were fully reduced to metallic/possibly alloy species. The Pd spectra before and after hydrogen reduction are shown in Figure 15.6.

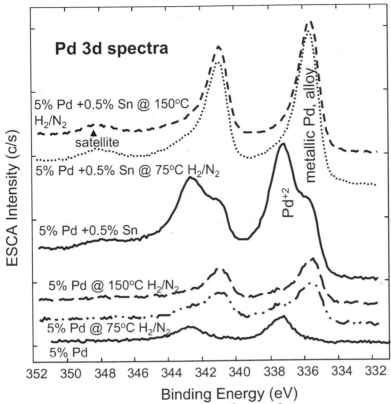

Figure 15.6. Surface Concentration of Pd^{+2} and Pd^{0}.

It is also possible that Pd^{+2} is reduced to a second PdH_x phase. When the metallic Pd chemical shift was compared to PdH_x as reported in the literature (13), the core level Pd $3d_{5/2}$ binding energy shift was only 0.2 eV. The article also found that the asymmetry of the 3d peak was slightly reduced, and a shakeup satellite peak (indicated by the arrow in Figure 15.6) disappeared. However, in the presence of metallic Pd, we cannot determine whether a second PdH_x phase is present.

The ESCA data does not allow one to distinguish between a Pd^{0}/Sn^{0} mixture and a true Pd-Sn alloy with these bimetallic catalysts at low Sn loadings, so alloy formation is possible but this possibility cannot be verified with ESCA.

ESCA studies were also performed on the partially deactivated recycled carbon supported 5%Pd-0.3%Sn catalyst presented in Table 15.3. The sulfur detection limit for this study was ~0.5%. Therefore these data implied that there was no obvious Pd-S formation on the used catalyst, but a large amount of nitrogen was present as shown in Table 15.5. We assign this as a possible Pd-CN species. These data suggested that the presence of HCN rather than H_2S might be the predominant factor for the deactivation of the catalyst.

Table 15.5. Surface Concentrations Detected on the Fresh and Used 5%Pd-0.3%Sn Catalyst Samples.

Sample	C	N	O	Pd	Sn	Na	F	S*	Cl
5% Pd +0.3% Sn Fresh	58.7	0.4	6.1	30.4	2.0	0.8	nd	0.5	1.6
5% Pd +0.3% Sn Used	66.4	7.3	4.5	3.9	1.7	0.2	15.7	0.5	0.3

* The detection limit of sulfur was at ~0.5% for both samples.

Summary

We report the discovery of a new Pd-Sn catalyzed hydrogenolysis reaction to produce thiol product in high yields. The relationship between catalyst activity and surface characterization (chemisorption, ESCA, and *in situ* temperature-dependent XRD,) has aided our understanding of the reasons why these catalysts are sulfur resistant, extremely active, and activated at certain temperatures and pressures. The predominant mode of deactivation appears to be the formation of Pd-CN species rather than the formation of Pd-S species on the surface of the catalyst.

Acknowledgements

The authors would like to thank Steve Blades who did much of the experimental laboratory work described here, and DuPont for the permission to publish this work. The catalysts for this study were generously supplied by Engelhard Corporation (Iselin, NJ), now part of BASF. The authors also thank Michael Duch (ESCA), David Rosenfeld (XRD), and Sourav Sengupta for their analyses and helpful discussions.

References

1. A. Lantz (Atochem), DE 3800392(1988).
2. K. Hooper, J. LaDou, J. Rosenblum, and S. Book, *Am. J. of Ind. Med.*, **22**(6), 793(1992).
3. C. A. Thayer, internal Dupont report(1974).
4. N.A. Rao and B.E. Baker, *Organofluorine Chemistry* (Plenum Press) New York(1994), p. 327.
5. J.T. Barr (Pennsalt Chemicals Corp), US Patent 2,894,991(1959).
6. (a) N. Tamura, *Studies in Surface Science and Catalysis*, **6**, 439(1980). (b) N. Kulishkin and A. V. Mashkina, *Reaction Kinetics and Catalysis Letters*, **45** (1), 41(1991).
7. S.E. Jacobson, US Patent 5,726,337(1998).
8. (a) G.J.K. Acres, et al., The Design and Preparation of Supported Catalysts in Catalysis, eds. C. Kemball and D. A. Dowden, Royal Society of Chemistry, **4**, 1-30(1981). (b) C. H. Bartholomew and M. Boudart, *J. Catal.*, **25**, 173(1972). (c) P. Ehrburger, *J. Catal.*, **43**, 61(1976) (d) M. K. van der lee, *J. Amer. Chem. So.c*, **127,** 13573(2005).

9. D. R. Corbin, L. Abrams, G. Jones, M. M. Eddy, W. T. A. Harrison, G. D. Stucky, and D, E. Cox, *J. Am. Chem. Soc*, **112**, 4822(1990).

10. S.E. Jacobson, US Patent, 5,728,887(1998).

11. Z. Zhan, K. Lang, F. Liu, and L. Hu, *Syn. Comm.*, **34**, 2303(2004)

12. Subsequent to this study, one of the authors, S.E.J., was asked to use the same Pd-Sn catalyst for a different hydrogenolysis reaction (a sulfide-containing reactant) at 80°C. Heating the catalyst with the reactants to 80°C did not effect any reaction. However, by activating the catalyst, under hydrogen, at 140-150°C, and cooling to 80°C, and then adding the reactant mixture, the catalytic reaction proceeded smoothly.

13. P.A. Bennett and J.C. Fuggle, *Phys. Rev. B*, **26**, 6030-6039(1982).

16. New Insight into the Electrocatalytic Hydrogenation of Model Unsaturated Organic Compounds

Jean Lessard[1], Maja Obradović[2] and Gregory Jerkiewicz[2]

[1]*Laboratoire de chimie et électrochimie organiques, Département de chimie, Université de Sherbrooke, Sherbrooke, QC, Canada J1K 2R1*
[2]*Department of Chemistry, Queen's University, Kingston, ON, Canada K7L 3N6*

jean.lessard@usherbrooke.ca

Abstract

We present new results on the influence of $C_6H_{6\,ads}$ on the H_{UPD} and anion adsorption on the Pt(111) electrode in aq. $HClO_4$. The Pt(111) electrode is modified by an overlayer of $C_6H_{6\,ads}$ by its cycling in the 0.05-0.80 V potential range in aq. $HClO_4$ + 1 mM C_6H_6. The $C_6H_{6\,ads}$ overlayer significantly changes the H_{UPD} and anion adsorption, and CV profiles show a sharp cathodic peak and an asymmetric anodic one in the 0.05-0.80 V potential range. The $C_6H_{6\,ads}$ layer blocks the perchlorate adsorption but facilitates the adsorption of one ML of H_{UPD} which is embedded in the Pt(111) surface. Cycling of Pt(111) in the 0.40-1.25 V potential range in C_6H_6-free aq. $HClO_4$ results in complete oxidation of $C_6H_{6\,ads}$ to CO_2, which supports the notion that the overlayer of $C_6H_{6\,ads}$ lies on the Pt(111) surface. Repetitive cycling of the C_6H_6-modified Pt(111) in C_6H_6-free aq. $HClO_4$ from 0.05 to 0.80 V results in a partial desorption of C_6H_6 and in a partial recovery of the CV profile characteristic of an unmodified Pt(111). The presence of an overlayer of $C_6H_{6\,ads}$ modifies the Pt(111)-H_{UPD} attractive interactions. The behavior described above in aq. $HClO_4$ is the same as in aq. H_2SO_4. The behaviour of the Pt(110) and Pt(100) electrodes is somewhat similar to that of the Pt(111) electrode in H_2SO_4 but the strength of the bond between $C_6H_{6\,ads}$ and the Pt surface differs with the Miller index, increasing in the order Pt(111), Pt(110), Pt(100). H_{UPD} does not hydrogenate benzene on Pt(hkl) electrodes in aqueous acid.

Introduction

Electrocatalytic hydrogenation (ECH) involves *in situ* generation of electro-adsorbed hydrogen, H_{ads}, by water reduction in basic electrolytes or by proton discharge in acidic electrolytes [eq. 1 where M represents the active catalyst surface]. Following the generation of H_{ads}, the reaction with an adsorbed unsaturated organic molecule, $Y=Z_{ads}$, takes place [eqs. 2-4]. These three steps are the same as in the classical catalytic hydrogenation. The advantages resulting from the *in situ* generation of adsorbed hydrogen are: (i) mild hydrogenation conditions (low temperatures, from 20 to 100°C, and low pressure, atmospheric or up to a few atm) favoring higher chemoselectivity; (ii) elimination of the utilization of pressurized H_2 gas and associated complex manipulations; and (iii) possibility of using catalysts that have a low activity in classic catalytic hydrogenation (due to inefficient molecular hydrogen dissociation). References (1) to (5) contain useful introductory material and reviews on ECH of organic compounds.

$$2H_2O\ (H_3O^+) + 2e^- \xrightarrow{\ M\ } 2H_{ads} + 2HO^-(H_2O) \tag{1}$$

$$Y=Z \xrightleftharpoons{\ M\ } Y=Z_{ads} \tag{2}$$

$$Y=Z_{ads} + 2H_{ads} \rightleftharpoons YH\text{-}ZH_{ads} \tag{3}$$

$$YH\text{-}ZH_{ads} \xrightleftharpoons{\ -M\ } YH\text{-}ZH \tag{4}$$

$$H_{ads} + H_2O\ (H_3O^+) + e^- \xrightarrow{\ -M\ } H_2 + HO^-(H_2O) \tag{5}$$

$$2H_{ads} \xrightleftharpoons{\ -M\ } H_2 \tag{6}$$

In order to develop highly chemoselective, efficient electrohydrogenation reactions as well as new and active electro-catalysts, it is necessary to: (i) comprehend the mechanism of ECH at the molecular level and identify the mechanism of reaction between H_{ads} and the adsorbed organic substrate, $Y=Z_{ads}$; (ii) identify the nature of the electro-adsorbed H species active in the process (uderpotential deposited hydrogen, H_{UPD}, versus overpotential deposited hydrogen, H_{OPD}); (iii) evaluate the strength and nature of the adsorption of $Y=Z_{ads}$ at the electrode; (iv) identify and understand various factors (electrode potential, current density, temperature, pH, solvent nature, supporting electrolyte) which affect the adsorption of H and $Y=Z$ and their mutual interactions at the electrode's surface; and (v) determine the influence of the electronic nature (Pt *vs.* Ni *vs.* Cu) and surface structure of the electrode on the direction, kinetics, and thermodynamics of the ECH process. In relation to point (iv), it is important to recognize that the lateral interactions between H_{ads} and $Y=Z$ influence the chemoselectivity and the competition between hydrogenation (eqs. 2 to 4) and H_2 desorption (eq. 6) [rate of hydrogenation (V_{Hydrog})/rate of hydrogen desorption ($V_{H\text{-}desorp}$) = current efficiency].

In this contribution, we present new results of a cyclic voltammetry study of the influence of $(C_6H_6)_{ads}$ on H_{UPD} and on anion adsorption at the Pt(111) electrode in aqueous $HClO_4$ and H_2SO_4, and relate them to analogous studies at Pt(110) and Pt(100) electrodes in aqueous H_2SO_4. Perchlorate anions are known to be less strongly adsorbed on Pt than bisulfate or sulfate anions; thus, the cyclic voltammetry behavior in the two electrolytes is expected to reveal important differences.

Experimental Section

Preparation of the Pt(111) Single-Crystal Electrode. The Pt single-crystal electrode was prepared according to the procedure developed by Clavilier (7) and oriented using the methodology of Hamelin (8). It was subsequently polished with alumina to a mirror-like finish. The quality of the Pt(111) surface was verified by recording CV profiles in 0.05 or 0.5 M aqueous acid ($HClO_4$ or H_2SO_4) solutions in the potential range corresponding to H_{UPD} and anion adsorption. Agreement between our results and those reported in the literature (7) indicates that the Pt(111) electrodes were of good quality and precisely oriented, and that the surface was well ordered. The surface area (the roughness factor equals unity), A, was 0.0475 ± 0.0001 cm^2.

Solutions, Electrochemical Cells, and Reference Electrode. The 0.05 and 0.5 M aqueous acidic solutions were prepared from BDH Aristar grade acid and Nanopure water (18 MΩ cm). The experimental work was done in two identical cells, one containing only the aqueous electrolyte and the other containing the electrolyte $+ X$ mM C_6H_6 (HPLC grade, Sigma-Aldrich), where $X = 1$, 5, 10, or 20. The electrochemical cells and all glassware were precleaned according to a well-established procedure (9). The reference electrode was a reversible hydrogen electrode, RHE. The counter electrode, CE, was a Pt wire (99.998% purity, Aesar). High-purity N_2 gas, presaturated with water vapor, was passed through the working electrode, WE, compartment.

Electrochemical Measurements. The electrochemical instrumentation included (a) a PAR model 263A potentiostat, (b) a PAR PowerCV software for data acquisition and analysis, and (c) a Dell Pentium IV computer.

Results and Discussion

Cyclic Voltammetry of Pt(111) in aqueous $HClO_4$ + 1 mM C_6H_6 at 298 K. Figure 16.1 shows three CV profiles for the Pt(111) electrode in the 0.05-0.80 V potential range recorded at a scan rate, s, of 50 mV s^{-1} and a temperature, T, of 298 K. The CV profile in the inset refers to the cell containing 0.05 M aq. $HClO_4$ and reveals features corresponding to the H_{UPD} and perchlorate adsorption; it shows that the Pt(111) surface was atomically well ordered and the system was clean (7, 10-12). The main CV profiles were recorded in the 0.02-0.80 V (*vs.* RHE) potential range at $s = 50$ mV s^{-1} and $T = 298$ K in the cell containing 0.05 M aq. $HClO_4$ + 1 mM C_6H_6 (solid line) or 0.5 M aq $HClO_4$ + 1 mM C_6H_6 (dotted line). These CV transients differ from the one in the inset through the following: (i) there is a sharp, symmetric cathodic peak (0.02-0.15 V) and a well-defined and asymmetric anodic peak (0.04-0.20 V); (ii) there are no features in the 0.37-0.55 V range that correspond to the anion adsorption in a benzene-free aq. $HClO_4$ electrolyte; (iii) the double-layer charging current is very small and there are no other features. These CV profiles are stable and do not change upon prolonged cycling between 0.02 and 0.80 V as long as benzene does not evaporate and its concentration is maintained constant. The cathodic charge-density value q_c (corresponding to H_{UPD} adsorption) and the anodic charge-density value q_a (corresponding to H_{UPD} desorption) are the same ($q_c = q_a = q_{HUPD}$); they are 225 ± 5 and 250 ± 5 μC cm^{-2} in 0.05 and 0.50 M $HClO_4$, respectively. This behavior in aq. $HClO_4$ is very similar to that in aq. H_2SO_4 (13) except for a small shift of the peak potential towards lower values in $HClO_4$. The q_{HUPD} values agree to within $\pm 5\%$ in both media. The conclusions are the same as those drawn from the analysis of CV profiles in aq. H_2SO_4 (13): (i) the fact that $q_c = q_a$ indicates that H_{UPD} does not hydrogenate C_6H_6; (ii) the higher value of q_{HUPD} for the C_6H_6-modified Pt(111) electrode than for the unmodified Pt(111) electrode is indicative that C_6H_6 $_{ads}$ facilitates the adsorption of 1 ML of H_{UPD} (attractive interactions between Pt(111) and H_{UPD}) (10, 11, 14); (iii) the difference in symmetry of CV profiles for H_{UPD} adsorption and H_{UPD} desorption is due to differences in thermodynamics of the two processes (13): the coexisting overlayer of C_6H_6 $_{ads}$ adopts different structures upon H_{UPD} adsorption and H_{UPD} desorption (e.g., difference in Gibbs energy of surface hydration and C_6H_6 ordering).

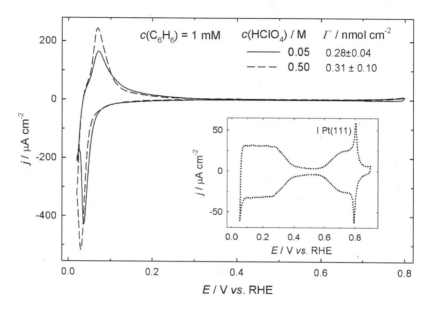

Figure 16.1. CV profiles for the Pt(111) electrode in 0.05 M aq. $HClO_4$ + 1 mM benzene (solid line, main graph), in 0.5 M aq. $HClO_4$ + 1 mM benzene (dashed line, main graph), and in 0.05 M aq. $HClO_4$ (inset); s = 50 mV s^{-1}, T = 298 K.

Determination of the C_6H_6 surface coverage on Pt(111). Figure 16.2 shows three CV profiles in the 0.4-1.25 V potential range recorded at s = 20 mV s^{-1} and T = 298 K in 0.05 M aq. $HClO_4$. The anodic peaks at 0.80 V and 1.08 V in the CV for the unmodified Pt(111) (dashed line labeled "I Pt(111)") correspond to the adsorption of anions and the formation of Pt surface oxide, respectively. In the CV for the C_6H_6-modified Pt(111) (solid line), the wave at 0.95 V corresponds to the oxidation of C_6H_6 $_{ads}$ to benzoquinone and the peak at 1.20 V corresponds to the complete oxidation of the adsorbed species to CO_2 (15, 16). After the fifth scan (dotted line labeled "V C_6H_6 mod. Pt(111)"), we observe the emergence of CV features characteristic of the unmodified Pt(111) electrode (there are slight differences due to some modification of the surface because of the Pt oxide formation and reduction). The sum of the charge density values q_{od} (q_{od} refers to oxidative desorption) of the anodic peaks at 0.95 V and at 1.20 V corresponds to the complete oxidation of C_6H_6 to CO_2, which requires 30 F (15, 16); q_{od} is corrected for the Pt oxidation. Using eq. 7, the surface coverage of C_6H_6 Γ(C_6H_6 1mM) was then calculated to be 0.28 ± 0.04 nmol cm^{-2} in 0.05 M aq. $HClO_4$. In 0.5 M aq. $HClO_4$, Γ

(C_6H_6 1mM) is equal to 0.31 ± 0.10 nmol cm^{-2}. These Γ values are of the same magnitude as the value of 0.42 nmol cm^{-2} calculated for a monolayer of C_6H_6 $_{ads}$ on Pt(111) with all the molecules being parallel to the surface ($\theta = 0.14$ (17)). However, the latter value of $\Gamma(C_6H_6)$ was not determined in the presence of H_2SO_4, thus it does not take into account the presence of the electrochemical environment (13).

$$\Gamma(C_6H_6) = \frac{q_{od}}{30F} = \frac{\Sigma(Q_{o,i} - Q_{r,i})}{30F} \tag{7}$$

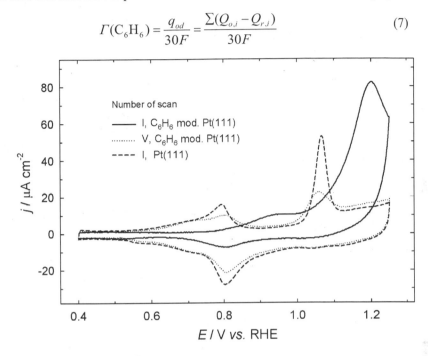

Figure 16.2. CV profiles for the oxidative desorption of C_6H_6 by cycling the C_6H_6-modified Pt(111) electrode in 0.05 M aq. $HClO_4$ (0.4-1.25 V range) after its transfer to the benzene-free cell (solid line: 1st scan; dotted line: 5th scan; dashed line: unmodified Pt(111)); $s = 20$ mV s^{-1} and $T = 298$ K.

These results clearly show that the Pt(111) surface is modified by a saturation layer of C_6H_6 $_{ads}$ with the C_6H_6 $_{ads}$ molecules residing parallel to the Pt(111) surface. The presence of this overlayer of C_6H_6 $_{ads}$ induces an attractive interaction between H_{UPD} and the Pt(111) surface and lower energy of later repulsions between H_{UPD} adatoms, as already pointed out, which leads to the formation of a monolayer of H_{UPD}. This H_{UPD} adatoms are embedded in the Pt(111) surface lattice as discussed in ref. 13 that reports results of analogous investigation in aq. H_2SO_4. This observation has a great significance for the electrocatalytic and gas phase hydrogenation: (i) a complete layer of H_{UPD}, energetically equivalent to H_{chem} (gas phase), develops on the benzene modified Pt(111) prior to the onset of hydrogen evolution; (ii) the partial electrocatalytic hydrogenation of preadsorbed C_6H_6 molecules on Pt(111) to cyclohexane (1 molecule hydrogenated out of 30 desorbed molecules) reported by Baltruschat and coworkers (18) must then take place on H_{UPD}-modified Pt(111) and not on a bare Pt(111) surface. Importantly, the gas-phase catalytic hydrogenation of unsaturated organic molecules might also require a presence of H_{chem} that is embedded in the catalytic material (13).

CV profiles for the C_6H_6-modified Pt(111) in benzene-free aq. $HClO_4$. Figure 16.3 shows CV profiles recorded at $s = 50$ mV s^{-1} and $T = 298$ K for the Pt(111) electrode after its transfer from the benzene-containing to the benzene-free 0.05 M aq HClO$_4$. The data show that the CV profiles undergo significant changes (indicated by arrows) as the Pt(111) electrode is cycled between 0.05 and 0.80 V. The cathodic (H_{UPD} adsorption) and anodic (H_{UPD} desorption) peaks broaden and their intensity decreases. At the same time, there is an emergence of the characteristic CV features in the 0.37-0.55 V potential range that correspond to the anion adsorption-desorption as well as those in the 0.05-0.37 V potential range that represent H_{UPD} adsorption-desorption. After 25 scans (35 scans in 0.50 M HClO$_4$), there is complete recovery of the characteristic features of unmodified Pt(111). Thus, the C_6H_6 adsorption-desorption does not disorder Pt(111), which maintains its long-range order as has been observed also in 0.05 M and 0.50 M aq. H$_2$SO$_4$ (13) and by the group of Itaya (17) in 0.01 M aq. HF.

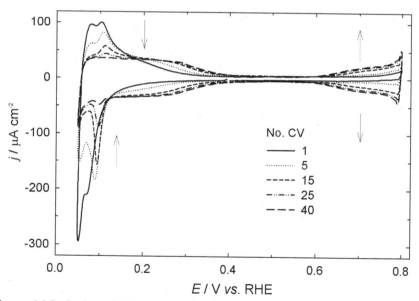

Figure 16.3. Series of CV profiles for the Pt(111) electrode in 0.05 M aq. HClO$_4$ after its transfer from the benzene-containing electrolyte to the benzene-free one; the arrows show changes brought about by electrode cycling in the 0.05-0.80 V range; $s = 50$ mV s^{-1} and $T = 298$ K.

Figure 16.4 shows the evolution of q_{C1} and of $q_{C1}+q_{C2}$ vs. the scan number for repetitive scanning between 0.05 and 0.80 V in 0.05 M (circles) and 0.5 M (triangles) aq. HClO$_4$. The qualitative behavior in these two electrolytes is similar: (i) q_{C1} (155 µC cm^{-2} and 60 µC cm^{-2}, respectively, in 0.05 M and 0.5 M aq. HClO$_4$) is much smaller than in C$_6$H$_6$-containing electrolyte (225-250 µC cm^{-2}); (ii) q_{C1}

increases at first, reaches a maximum, then decreases to level off in both electrolytes to the values equal to those for unmodified Pt(111) (155 and 140 µC cm^{-2}, respectively, in 0.05 M and 0.5 M aq. HClO$_4$); (iii) $q_{C1}+q_{C2}$ increases in both electrolytes then levels off to the values observed for unmodified Pt(111). However, the effect is more pronounced in 0.50 M aq. HClO$_4$ since q_{C1} (60 µC cm^{-2}) is much smaller than in 0.05 M aq. HClO$_4$ (155 µC cm^{-2}). The increase of q_{C1} (adsorption of H$_{UPD}$) to a value higher than that for the unmodified Pt(111) again points to attractive interactions between H$_{UPD}$ and the Pt(111) surface as was concluded above.

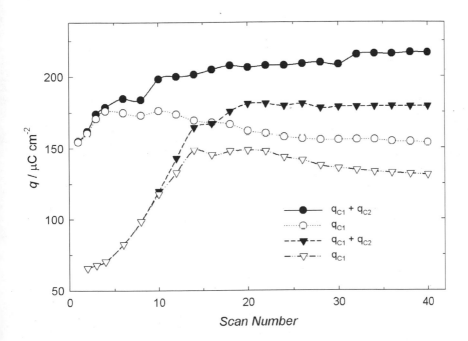

Figure 16.4. Evolution of q_{C1} (H$_{UPD}$ adsorption) and $q_{C1}+q_{C2}$ (q_{C2} corresponds to anions desorption) *vs.* the number of CV transients between 0.05 and 0.80 V in 0.05 M aqueous HClO$_4$ (circles) and in 0.5 M aqueous HClO$_4$ (triangles).

The behavior in aq. HClO$_4$ is quite different from that in aq. H$_2$SO$_4$. In aq. H$_2$SO$_4$, q_{C1} and $q_{C1}+q_{C2}$ are the same as those in the C$_6$H$_6$-containing solution (240 ± 5 µC cm^{-2}) and decrease to the values equal to those for unmodified Pt(111) (13): q_{C1} = 80 µC cm^{-2} and 127 µC cm^{-2} in 0.05 and 0.5 M H$_2$SO$_4$, respectively; $q_{C1}+q_{C2}$ = 117 µC cm^{-2} and 183 µC cm^{-2} in 0.05 and 0.5 M H$_2$SO$_4$, respectively. We have as yet no satisfactory explanation of this difference in behavior in the two aqueous media.

Influence of C$_6$H$_6$ concentration. As seen in Figure 16.5, an increase of the C$_6$H$_6$ concentration from 1 to 20 mM (the benzene saturation concentration is 22 mM) leads to: (i) a shift of the H$_{UPD}$ adsorption and desorption peaks toward less-positive potentials; (ii) a decrease in H$_{UPD}$ charge density; and (iii) the complete disappearance of H_{UPD} adsorption at a C_6H_6 concentration of 20 mM. The inset shows that surface coverage of $C_6H_{6\ ads}$ is close to a monolayer ($\Gamma = 0.28\text{-}0.35$ nmol cm^{-2}) at concentrations of 1 and 5 mM with all molecules being parallel to the surface (for a full monolayer: $\Gamma = 0.42$ nmol cm^{-2} (17)) and then increases to a value of 0.66 ± 0.10 nmol cm^{-2}. The latter value can be most likely explained by the formation of a multilayer of C_6H_6 due to π-stacking of the benzene molecules in a parallel orientation to the surface; such a multilayer can completely block the adsorption of H_{UPD}.

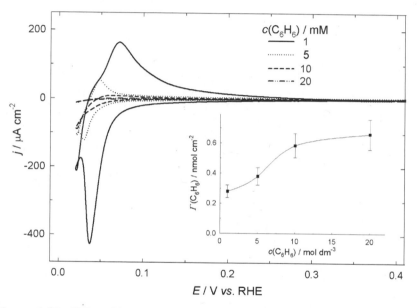

Figure 16.5. CV profiles for the Pt(111) electrode in 0.05 M HClO$_4$ (main graph) and evolution of the C_6H_6 surface coverage (inset) as a function of C_6H_6 concentration; electrode cycling in the 0.05-0.42 V range; $s = 50$ mV s^{-1}; $T = 298$ K.

The behavior in aq. HClO$_4$ is drastically different from that in aq. H$_2$SO$_4$. In aq. H$_2$SO$_4$, the cathodic peak and anodic peak are shifted to less-positive potential values as the concentration of C_6H_6 is increased and their peak current increases. In the case of aq. HClO$_4$, the peaks also shift towards less-positive values but their current density decreases. However, in aq. H$_2$SO$_4$, not only an increase of C_6H_6 concentration does not suppress the adsorption of H_{UPD} as is the case in HClO$_4$, but it even causes an increase of the charge density of H_{UPD} adsorption (q_{HUPD}) from 240 μC cm^{-2} at C_6H_6 concentrations ≤ 5 mM to 280 μC cm^{-2} at a C_6H_6 concentration of 20 mM (13). In addition, $q_c = q_a$ ($= q_{HUPD}$) to within 2-3 μC cm^{-2} over the whole range of concentrations (13). Elsewhere (13), it was clearly stated that such a behavior was unexpected and difficult to justify. The tentative explanation proposed for the increase of q_{HUPD} was the formation of a multilayer of $C_6H_{6\ ads}$ molecules, which would result in an enhanced total interfacial capacity. At present, we have no satisfactory hypothesis to explain the different qualitative behavior in the two electrolytes.

Influence of benzene on the H_{UPD} *and anion adsorption on Pt(110) and Pt(100) electrodes in aqueous* H_2SO_4. A comparative analysis of the influence of C_6H_6 $_{ads}$ on the CV response of the Pt(110), Pt(100), and Pt(111) electrodes in aqueous H_2SO_4 shows some similarities (19): the C_6H_6 $_{ads}$ molecules inhibit the H_{UPD} and anion adsorption on the three single-crystal electrodes and H_{UPD} *does not hydrogenate* C_6H_6 *on any of them*. However, the strength of C_6H_6 adsorption on these low Miller index Pt electrodes strongly depends on the surface geometry (19): it increases in the order Pt(111), Pt(110), Pt(100) as shown by CV profiles in benzene-free aq. H_2SO_4. Scanning the Pt(110) electrode in the 0.05-0.50V range or the Pt(111) electrode in the 0.05-0.80V range leads to a partial desorption C_6H_6, while scanning the Pt(100) electrode in the 0.05-0.75V range does not lead to any desorption of C_6H_6 $_{ads}$ (19). Therefore, the surface bond between Pt(100) and C_6H_6 $_{ads}$ is stronger than that between Pt(110) or Pt(111). In addition, cathodic polarization of the single-crystal Pt electrodes leads to C_6H_6 $_{ads}$ reductive desorption and almost a complete recovery of the CV profiles characteristic of the unmodified electrodes but the time required to desorb reductively C_6H_6 $_{ads}$ is longer in the case of Pt(100) than in the case of Pt(110) (19).

Conclusions

The Pt(111) electrode is modified by an overlayer of C_6H_6 $_{ads}$ in aqueous HX (X = ClO_4, HSO_4) + 1-5 mM C_6H_6 by cycling in the 0.02-0.80 V range with all the C_6H_6 molecules being parallel to the electrode surface. The C_6H_6 $_{ads}$ blocks the anion adsorption but facilitates the adsorption of *ca.* a monolayer of H_{UPD} for $c(C_6H_6) = 1$-5 mM. A monolayer of H_{UPD} adatoms and a monolayer of C_6H_6 $_{ads}$ molecules coexist on the Pt(111) surface with H_{UPD} being embedded in the Pt(111) lattice and C_6H_6 $_{ads}$ residing parallel on its surface. The thermodynamics of H_{UPD} adsorption and desorption on C_6H_6-modified Pt(111) are different due to differences in Gibbs energy of surface hydration and C_6H_6 ordering, thus giving rise to the asymmetry of the CV profiles. The increase of the charge density for H_{UPD} adsorption with $\Gamma(C_6H_6)$ (0.05 *vs.* 0.50 M $HClO_4$) and its value higher than that on unmodified Pt(111) points to attractive interaction between H_{UPD} and the electrode surface in aqueous $HClO_4$, as was also concluded from a thermodynamic study of H_{UPD} adsorption and desorption in aqueous H_2SO_4 (13). At $c(C_6H_6) = 20$ mM in $HClO_4$, a multilayer of C_6H_6 $_{ads}$ is formed on the Pt(111) surface by π-stacking with most molecules oriented parallel to the surface. This multilayer completely blocks the adsorption of H_{UPD}. In H_2SO_4, at $c(C_6H_6) = 20$ mM, the adsorption of H_{UPD} is not blocked and q_{HUPD} is even larger than at $c(C_6H_6) = 1$-5 mM (13). The C6H6 ads molecules also inhibit the H_{UPD} and anion adsorption on the Pt(110) and Pt(100) electrodes but the bond between Pt(100) and C_6H_6 $_{ads}$ is stronger than the bond between Pt(110) and C_6H_6 $_{ads}$, the latter bond being stronger than that between Pt(111) and C_6H_6 $_{ads}$ (19). *Finally,* H_{UPD} *does not hydrogenate benzene on any of the* Pt(hkl) *electrodes studied*.

Acknowledgements

We thank NSERC of Canada, FQRNT du Québec, Queen's University, and Université de Sherbrooke for financial support of this work. M. Obradović (on leave of absence from the Institute of Chemistry, Technology and Metallurgy, University of Belgrade, Belgrade, Serbia) is thankful to the Ministry of Science and Environment Protection of the Republic of Serbia for a postdoctoral fellowship.

References

1. S. Robitaille, G. Clément, J. M. Chapuzet, and J. Lessard, Chemical Industries (CRC Press), **115**, (*Cat. Org. React.*), 281-288 (2007).
2. J. Lessard, Chemical Industries (CRC Press), **104**, (*Cat. Org. React.*), 3-17 (2005), and refs. therein.
3. J. Lessard. In *Proceedings of the 7ᵗʰ International Symposium on Advances in Electrochemical Science & Technology* (ed. Society for Advancement of Electrochemical Technology, India), 2002, Vol. I, p. A100 and refs. therein.
4. J.M. Chapuzet, A. Lasia, and J. Lessard, in *Electrocatalysis, Frontiers of Electrochemistry Series* (eds. J. Lipkowski and P.R. Ross,), Wiley-VCH, Inc., New York, 1998, p. 155, and refs. therein.
5. D. Robin, M. Comtois, A. Kam Cheong, R. Lemieux, G. Belot, and J. Lessard, *Can. J. Chem.*, **68**, 1218-1227 (1990), and refs. therein.
6. B. Mahdavi, J.M. Chapuzet and J. Lessard, *Electrochim. Acta*, **38**, 1377-1380 (1993).
7. J. Clavilier, in *Interfacial Electrochemistry* (ed. A. Wieckowski), Marcel Dekker, New York, 1999, chap. 14 and refs. therein.
8. A. Hamelin, in *Modern Aspects of Electrochemistry* (eds. J.O'M. Bockris, B.E. Conway, R.E. White), Plenum Press, New York, 1985, Vol. 9, Ch. 1 and refs. therein.
9. H. Angerstein-Kozlowska, in *Comprehensive Treatise of Electrochemistry* (eds. E. Yeager, J.O'M. Bockris, B.E. Conway), Plenum, New York, 1984, Vol. 9, Ch. 1.
10. A. Zolfaghari and G. Jerkiewicz, in *Electrochemical Surface Science of Hydrogen Adsorption and Absorption* (eds. G. Jerkiewicz, P. Marcus), The Electrochemical Society, Pennington, NJ, 1997, PV 97-16.
11. A. Zolfaghari and G. Jerkiewicz, *J. Electroanal. Chem.*, **467**, 177-185 (1999).
12. A. Zolfaghari and G. Jerkiewicz, *J. Electroanal. Chem.*, **422**, 1-6 (1997).
13. G. Jerkiewicz, M. DeBlois, Z. Radovic-Haprovic, J.-P. Tessier, F. Perreault, and J. Lessard, *Langmuir*, **21**, 3511-3523 (2005).
14. G. Jerkiewicz, *Prog. Surf. Sci.*, **57**, 137-186 (1998).
15. F. Montilla, F. Huerta, E. Morallon, and J.L. Vazquez, *Electrochim. Acta*, **45**, 4271-4277 (2000).
16. U. Schmiemann and H. Baltruschat, *J. Electroanal. Chem.*, **347**, 93-109 (1993).
17. S.L.Yau, Y.G. Kim, and K. Itaya, *J. Am. Chem. Soc.* **118**, 7795-7803 (1996).
18. T. Hartung, U. Schmiemann, and H. Baltruschat, *Anal. Chem.*, **63**, 44-48 (1991).
19. M. DeBlois, J. Lessard, and G. Jerkiewicz, *Electrochim. Acta*, **50**, 3511-3520 (2005).

17.   Catalytic Reductive Alkylation of Aromatic and Alkyl Amines and Diamines over Sulfided and Unsulfided Platinum Group Metals

Venu Arunajatesan[1], Jaime L. Morrow[1] and Baoshu Chen[2]

[1]*Evonik Degussa Corporation, 5150 Gilbertsville Hwy, Calvert City, KY 42029*
[2] *Evonik Degussa GmbH, Rodenbacher Chaussee 4,63457 Hanau-Wolfgang, Germany*

venu.a@evonik.com

Abstract

The coupling of C-N bond via the reductive alkylation of an amine with a ketone or aldehyde is an industrially important class of reaction to manufacture fine chemicals and intermediates. In this paper we examine the generality of reductively alkylating primary aromatic and aliphatic amines and diamines such as aniline, 4-aminodiphenyl amine, ethylenediamine, and propylenediamine with cyclic and aliphatic ketones (cyclohexanone, acetone, MIBK). The catalysts investigated were sulfided and unsulfided Pd and Pt supported on activated carbon. In general, we found that Pt was more active than Pd as a catalyst, especially when the amine is an aromatic amine. The selectivity to the alkylated product improves with sulfidation; additionally, the reduction of ketones to the corresponding alcohols was more pronounced over Pd catalyst than on Pt catalysts. The reaction between diamine and ketone leads to heterocyclic amines such as imidazolidines, diazepanes, and pyrimidines if the diamine and ketone are brought in contact well before the reductive alkylation could occur. Hence, for these reactions, a continuous reactor configuration with least prior contact between the two reactants is recommended. Based on this study, in general, we recommend sulfided Pt catalysts as the most suitable catalyst for reductive alkylation of primary amines with ketones. In addition, unsulfided Pd or Pt catalysts could be used for longer chain amines such as 1,3-diaminopropane.

Introduction

Secondary amines synthesized by catalytic reductive alkylation of primary amines are used in a variety of fine and specialty chemical industries. For example, derivatives of cyclohexylamine are used as corrosion inhibitors, N-(1,3-dimethylbutyl)-N'-phenyl-p-phenylenediamine (6-PPD) is used as an anti-oxidant in rubber industry, several dialkylated diamines are used in the coatings industry, while they are used in the pharmaceutical industry as pharmacophores (1-7). Harold Greenfield and co-workers have examined the ability of platinum group metals (PGM), base metals, and their sulfides to catalyze reductive alkylation of primary and secondary amines (8-11). They found that different catalysts are optimal for the reaction based on the nature of the amine, ketone, the steric hindrance, and the reaction condition.

The reductive alkylation of a primary amine with ketone leads to the formation of a stable imine. In the presence of hydrogen and a hydrogenation catalyst, the imine is reduced to a secondary amine. Similarly, a diamine reacts step-wise to form dialkylated secondary amines. However, several side reactions are possible for these reactions as outlined by Greenfield (12). The general scheme depicting the reaction between primary amine or diamine to yield secondary amine through a Schiff base is shown in Figure 17.1.

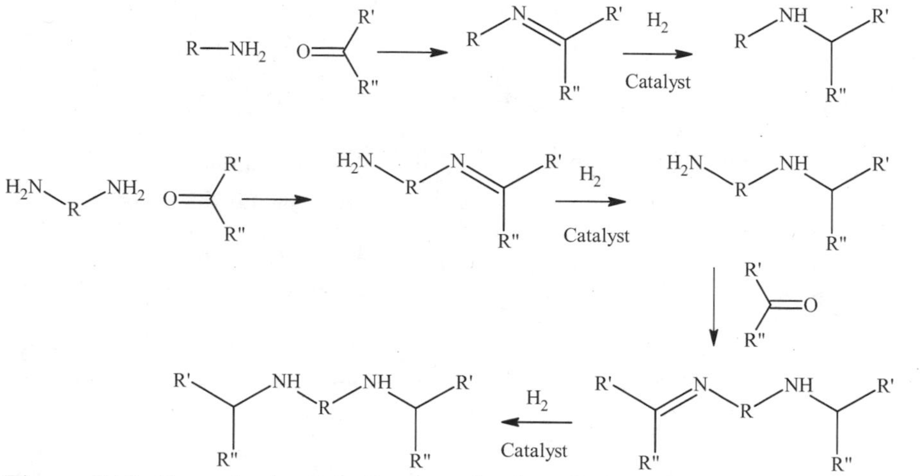

Figure 17.1. General schematic for the reductive alkylation of primary amines and diamines

In this study we examine the generalities in reductive alkylation; however, since the subject is vast, we limited ourselves to the interaction of aromatic and aliphatic primary amines and diamines with ketones. The ketones examined include the cyclic ketone, cyclohexanone, and aliphatic ketones such as acetone, and methyl isobutyl ketone (MIBK). We limited our study to sulfided and unsulfided Pt and Pd catalysts supported on activated carbon that were commercially available from Evonik Degussa Corporation.

Experimental Section

Reactions were carried out in liquid phase in a well-stirred (1000 rpm) high-pressure reactor (Parr Instruments, 300 mL) at 30 bar and 150°C with 370 mg catalyst for two hours, unless otherwise specified. The feed consisted of the amine with slight excess of ketone at ketone/amine-group molar ratio of 1.6 while maintaining a reaction volume of about 150 mL. In a typical experiment, 576 mmol of aniline is mixed with 920 mmol of cyclohexanone and 370 mg of BS2 catalyst in the 300 mL reactor. The reactor is closed and then pressure-tested to 50 psi above operating pressure to ensure that the system is leak proof. Once pressure-tested, the headspace is replaced

H_2 from the high-pressure burette and pressurized to the operating pressure using a manual pressure regulator. The overhead stirrer is then set at 1000 rpm and the temperature of the reactor is ramped up to 150°C. As the H_2 in the headspace is consumed, more flows in from the burette to maintain the reactor pressure. Hence, the reactor is operated in a semi-batch mode at a constant pressure, with the liquid phase in batch mode and the H_2 as a continuous feed. The reaction was monitored by hydrogen uptake while the conversion, and selectivity were determined from gas chromatographic (Agilent Technologies, 6890N) analysis. Conversion was measured as the amount of amine consumed over the reaction time, while selectivity was ascribed to the monoalkylated product in case of monoamines and the sum of mono- and dialkylated products in the case of diamines. The reaction time for more active catalysts was determined by measuring the time taken to reach 95% conversion while for catalysts that were significantly lower in activity, the end time was chosen as the point at which rate hydrogen consumption was negligible (<2 mmol/min).

The amines and ketones used in this study are shown in Table 17.1. Among the seven catalysts tested, one was a 5% Pd/C while two were 3% Pt/C prepared on different carbon supports. Same catalysts were then sulfided to various degrees as shown in Table 17.2 along with their designation.

Table 17.1. Amines and ketones used in the study

Amine	Ketone
Aniline	Cyclohexanone
4-aminodiphenylamine	Acetone
1,3-diaminopropane (DAP)	Methyl isobutyl ketone
1,2-diaminoethane (DAE)	

Table 17.2. Catalysts used in the study and their nomenclature

Catalyst	Catalyst Code
5% Pd/C	A1
3% Pt/C	B1
3% Pt/C	B2
0.1% S-5%Pd/C	AS1
0.3% S-5% Pd/C	AS2
0.3%S-3% Pt/C	BS1
0.3%S-3% Pt/C	BS2

Results and Discussion

Aromatic amine-Ketone. The interaction between aniline, the simplest of the aromatic amines with acetone in the presence of various catalysts to yield N-isopropylaniline was examined (Table 17.3). Among the catalysts tested, sulfided platinum catalysts were found to be the most active catalysts for this reaction.

Unsulfided palladium is more active but less selective than sulfided Pd catalysts. Since Pd catalyzes both reductive alkylation and ring hydrogenation, the desired product could undergo ring hydrogenation to form enamine followed by hydrolysis of the C-N bond to form cyclohexanone and cyclohexanol (12), both of which were observed in the product analysis by GC. Also, the product cyclohexylaniline was observed, potentially from the reductive alkylation of aniline with cyclohexanone (13). Increasing the S loading leads to a decrease in the formation of cyclohexanone, cyclohexanol, and cyclohexylaniline, and is completely eliminated at 0.3% S loading. However, the activity of the catalyst also tends lower with increased S loading. The major side product on AS2 catalyst was cyclohexylamine.

Table 17.3. Activity and selectivity during reductive alkylation of aromatic amines

Catalyst	Reaction Time (min)	Conversion (%)	Selectivity (%)
Aniline-acetone			
A1	78	88	58
AS1	94	75	62
AS2	120	48	87
B1	120	59	96
BS1	93	92	99
BS2	37	100	99
Aniline-MIBK			
A1	99	71	43
BS1	120	87	99
BS2	120	88	100
Aniline-Cyclohexanone			
AS2	120	83	91
B1	63	91	81
BS1	79	100	99
BS2	32	100	99
ADPA-MIBK			
BS1	77	92	99
BS2	75	93	100
ADPA-Acetone			
BS1	56	93*	-

*based on hydrogen consumption

Similar to observations by Greenfield, we found that Pt/C was less active but more selective than Pd/C catalyst, presumably due to poisoning by amines (12). However, as can be seen in Table 17.3, sulfidation of the catalyst led to an increase in activity and selectivity. As shown by Greenfield and others, sulfides of transition metals are excellent catalysts for this reaction and are more resistant to poisoning or the formation of undesired side products. BS2, a sulfided Pt catalyst prepared on a different activated carbon support than BS1, showed significantly improved activity without compromising the selectivity at the same Pt and S loading. This reiterates the importance that catalyst support and other preparation parameters play in the performance of the catalyst.

The activity of both Pd and sulfided Pt catalysts for the synthesis of N-(1,3-dimethylbutyl)-N-phenylamine from aniline and MIBK was lower than that with acetone. This could be attributed to the steric hindrance effect (14). The selectivity over sulfided Pt catalysts was higher compared to unsulfided Pd catalysts due to formation of byproducts as with aniline-acetone reaction.

Both sulfided and unsulfided Pt catalysts were more active for the reaction between cyclohexanone and aniline to yield N-cyclohexylaniline compared to acetone as the ketone. Again, BS2 was significantly more active but as selective as BS1 catalyst. No cyclohexanol was observed with sulfided Pt catalysts, while only a trace amount was found with the B1 catalyst. The sulfided Pd catalyst, AS2 had activity and selectivity similar to that of unsulfided Pt catalyst. This suggests that if catalyst cost is of higher importance than productivity in the commercial process, an optimally sulfided Pd catalyst may be an acceptable alternative to a sulfided Pt catalyst.

The reductive alkylation of 4-aminodiphenylamine with MIBK to yield 6-PPD have been studied previously (2,12,15). Greenfield showed that Pd, Pt and sulfided Pt are active for this reaction and Arunajatesan et al. showed that the optimal mole ratio of S/Pt is around 1.7. Again, we found that the sulfided Pt/C is an effective catalyst for the synthesis of 6-PPD.

The reaction of ADPA with acetone led to a solid product, presumably 4-(isopropylamino)diphenylamine (mp: 72-76°C), with the hydrogen uptake leveling at 56 minutes. Since the product could not be analyzed by GC, the conversion was calculated from the amount of hydrogen consumed during the reaction.

Diamine-Ketone. Dialkylated diamines are often used in coatings as curatives and modifiers, hence we examined the ability of various catalysts to effect the alkylation of diamines with ketones. It is well known that the diamines undergo step-wise reductive alkylation with ketones. The reductive alkylation of DAE with acetone led to a significant formation of side products apart from the monoalkylated N-(isopropyl)ethylenediamine and the dialkylated N, N'-(diisopropyl)ethylenediamine. The side products were attributed to heterocyclic compounds such as imidazolidine and diazepane (16,17). The A1 catalyst led to a higher selectivity than AS1 since it is a more active hydrogenation catalyst and hence was able to effect the C=N hydrogenation thereby lowering the selectivity to the undesired heterocyclic compounds. The reaction between diamines and ketones are very facile and can lead to the formation of imines even at room temperature and in the absence of a catalyst

(17). The BS2 catalyst was more selective toward the formation of the dialkylated product than the Pd catalysts tested. The activity of BS2 for DAE-MIBK reaction was slower than that with acetone due to steric effects posed by the larger ketone. Here again, the imine tends to rapidly cyclize to form imidazolidines or pyrimidines. Figure 17.2 shows the stepwise formation of various side products observed during the reductive alkylation of DAE with acetone.

The presence of hydrogen and a highly active hydrogenation catalyst, both cyclization and hydrogenation of imine will compete to form the heterocycles and the monoalkylated amine. As seen in Figure 17.2, the monoalkylated amine could again undergo the competing reaction of cyclization to yield diazepanes in the case of DAE, or diazocanes in the case of DAP. Such formation of heterocycles was observed when a small amount of diamine or monoalkylated amine was added into acetone or MIBK in a vial allowed to sit at room temperature for 30 minutes and analyzed by GC. The more active the hydrogenation catalyst is, the more the product spectrum is toward the formation of the desired mono- and dialkylated amino compounds. Hanahata et al. (17) have shown that using excess ketone will minimize the formation of heterocycles, while the presence of water tends to favor the formation of heterocycles over the diketimine.

Table 17.4. Activity and selectivity during reductive alkylation of diamines

Catalyst	Reaction Time (min)	Conversion (%)	Selectivity (%)
DAE-Acetone			
A1	95	100	40
AS1	92	100	19
BS2	71	100	84
DAE-MIBK			
B2	93	100	41
BS2	120	100	50
DAP-Acetone			
A1	84	100	98
AS1	120	100	84
B2	42	100	99
BS2	56	100	98
DAP-MIBK			
A1	120	100	69
BS1	120	100	65
BS2	120	100	69

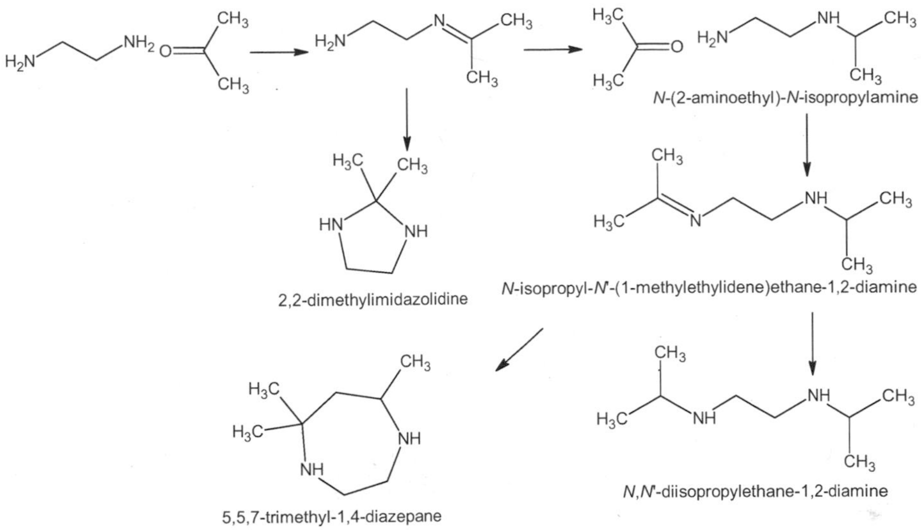

Figure 17.2. Side products formed during the reductive alkylation of ethylenediamine with acetone

The reductive alkylation of DAP with acetone led to high conversions and selectivity to the dialkylated product over A1, B1, and BS2 catalysts. The AS1 catalyst, which typically has lower activity than the A1 or Pt-based catalysts showed greater formation of heterocycles. These results indicate that a more active catalyst, a shorter reaction time, a higher operating temperature, or sterically hindered amines/ketones will help minimize the formation of the heterocycles. Similar high selectivities were obtained with DAP-MIBK reaction over BS1 and BS2 catalysts with no heterocycles being formed. However, over A1, the undesired heterocyclic compound was over 15%. This indicates that the reaction between diamines and ketones has a significant potential to form heterocyclic compounds unless the interaction between these is kept to a minimum by the use of a continuous flow reactor as proposed by Speranza et al. (16) or by other methods.

Conclusions

The current work indicates that sulfided platinum catalysts are, in general, more active and selective than Pt, Pd, or sulfided Pd catalysts for reductive alkylation of primary amines with ketones. The choice of the catalyst preparation parameters, especially the support, plays a major role in determining the performance of the catalyst. Diamines, especially of lower molecular weight, tend to react with ketones even at room temperature to form heterocycles such as imidazolidine, diazepanes, and pyrimidines. Hence, a continuous reactor configuration that minimizes the contact between the amine and the ketone, along with a highly active catalyst is desired to obtain the dialkylated product. In general, sulfided Pt appears to be more suited for the reductive alkylation of ethylenediamine while unsulfided Pd or Pt may also be used if 1,3-diaminopropane is the amine.

References

1. K.S. Hayes, *Appl. Catal. A* **221**, 187 (2001).
2. V. Arunajatesan, M. Cruz, K.M. Moebus and B. Chen, Chemical Industries (CRC Press), **115**, (*Catal. Org. React.*), 481 (2007).
3. J.C. Saukaitis, US Patent 5,644,056 to Hoechst Celanese Corporation (1997).
4. V.L. Mylroie, US Patent 5,861,535 to Eastman Kodak Company (1999).
5. H.E. Petree and J.B. Nabors, US Patent 4,272,238 to Ciba-Geigy Corporation (1980).
6. K.R. Lassila, K.E. Minnich and R.V. Court Carr US Patent 6 103 799 to Air Products and Chemicals (2000).
7. R. N. Salvatore, C. H. Yoon and K. W. Jung, *Tetrahedron*, **57**, 7785 (2001).
8. F. S. Dovell and H. Greenfield, *J. Org. Chem.*, 1265 (1964).
9. F. S. Dovell and H. Greenfield, US Patent 3,336,386 to Uniroyal Inc (1963).
10. R. E. Malz, Jr. and H Greenfield, US Patent 4,607,104 to Uniroyal Chemical Co. (1986).
11. R. E. Malz, Jr. and H Greenfield, *Heterogeneous Catalysis and Fine Chemicals II*, (ed. M. Guisnet et al.) Elsevier Science, Amsterdam, 1991, p. 357.
12. H. Greenfield, Chemical Industries (Dekker), **53**, (*Catal. Org. React.*), 265 (1994).
13. D. Roy, R. Jaganathan and R.V. Chaudhari, *Ind. Eng. Chem. Res.*, **44**, 5388 (2005).
14. S. Gomez, J.A. Peters and T. Maschmeyer, *Adv. Synth. Catal.*, **344(10)**, 1037 (2002).
15. F.S. Dovell and H. Greenfield, *J. Am. Chem. Soc.*, 87, 2767 (1965).
16. G.P. Speranza, J.–J. Lin, J. H. Templeton and W.–Y. Su, US Patent 5,001,267 to Texaco Chemical Company (1991).
17. H. Hanahata, E. Yamazaki and Y. Kitahama, *Polymer J.*, **29(10)**, 811 (1997).

18. Catalytic Synthesis of N,N-Dialkylglucamines

I. K. Meier, Mike E. Ford and R. J. Goddard

Air Products and Chemicals, Inc., 7201 Hamilton Boulevard, Allentown, PA 18195

fordme@airproducts.com

Abstract

Reductive alkylation of N-alkylglucamines with a variety of aldehydes or ketones in the presence of Pd/carbon provides the corresponding N,N-dialkylglucamines in high yield. Only minor amounts of aldol byproducts are formed.

Introduction

N,N-Dialkylglucamines (Figure 18.1) are potentially useful in a spectrum of surfactant

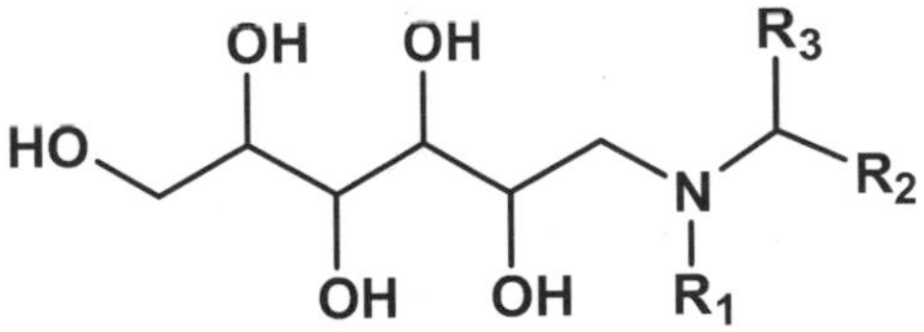

Figure 18.1. N,N-Dialkylglucamines **1**; R_1, R_2, R_3 = alkyl, functionalized alkyl.

applications, including coatings, inks, adhesives, cleaning compositions, personal care products, and formulations for textile processing and oil field applications. Remarkably, few preparations of **1** have been reported. An early patent (1) describes reductive amination of monosaccharides with primary and secondary amines. The products were characterized only as "viscous non-crystalline syrups," an indication of low purity. Reductive amination of hydroxyalkyl compounds with secondary amines is among the claims of a more recent patent (2). A single preparation of N,N-dimethylglucamine (**1**; R_1 = CH_3; R_2 = R_3 = H) is provided; use of higher secondary amines is not exemplified. Although reductive alkylation of N-alkylglucamines with aldehydes has been disclosed (3), only methylation with aqueous formaldehyde is shown. Somewhat surprisingly, water is the requisite solvent. As a result, the range of available products is limited to more water-soluble glucamines; N-methyl-N-hexylglucamine (**1**; R_1 = C_6H_{13}; R_2 = H; R_3 = H) is the highest molecular weight example. Dialkylglucamines with larger substituents have been prepared, albeit by a non-catalytic route. Stoichiometric alkylation of N-methylglucamine with sodium lauryl sulfate was used to make the corresponding **1** (R_1 = lauryl, R_2 = H; R_3 = H) (4). Neutralization with sodium hydroxide was required to obtain the free base.

Results and Discussion

We now report that N,N-dialkylglucamines are readily obtained in high yield and purity by reductive alkylation of an N-alkylglucamine with either an aldehyde (Table 18.1, runs 1, 3 - 14) or ketone (Table 18.1, run 2) (5). In all but one instance (run 7), conversions were greater than 90%. Since minimal quantities of byproducts were formed in these reactions, longer reaction times would be expected to provide higher conversions. Owing to our focus on straight chain alkyl functionality in **1**, acetone was the only ketone evaluated (run 2). Similar results would be anticipated for other unhindered ketones such as methyl ethyl ketone or methyl iso-butyl ketone. Symmetric and non-symmetric products are equally accessible (Table 18.1, cf. run 3 with, e.g., runs 5, 10, 14). Owing to the flexibility in choice of alkyl groups and the ability to incorporate ether functionality, products with moderate to very high hydrophobicity can be obtained (cf. runs 2 and 3 with, e.g., runs 9, 12, and 14). In all instances, the concentration of aldol-derived byproducts was low, and no other byproducts were detected.

In conclusion, N,N-dialkylglucamines are prepared in high yield and purity by palladium-catalyzed reductive alkylation of an N-alkylglucamine with an aldehyde or ketone. A wide variety of **1** with moderate to very high hydro-phobicity are readily accessible.

Experimental Section

A 100 mL Parr stainless steel reactor was charged with 0.05 mole of the monoalkylglucamine, 0.052 mole (1.04 equivalents) of the aldehyde or ketone, 2 wt% (based on starting glucamine and carbonyl compound) of 5% palladium on carbon, and 30 gm of methanol. The reactor was closed, purged with nitrogen and hydrogen, and pressurized to ca 600 psig with hydrogen. The mixture was heated with stirring (1000 rpm) to 100°C and pressurized with hydrogen to 1000 psig. The reaction was maintained at this temperature; pressure was maintained at 1000 psig via regulated hydrogen feed. After 10 hr, the mixture was cooled to 50°C, and the product removed from the reactor by filtration through an internal 0.5 μ sintered metal element. Upon cooling, the product precipitated as a white-cream colored solid. A sample was derivatized with trimethylsilylimidazole/ pyridine. This reagent is available in ready-to-use form as Sylon™ TP from Supelco®. Owing to the requirement for complete derivatization of multiple hydroxyl groups, reaction times of 0.75-1.0 hr were used. After trimethylsilylation, the product was analyzed by GC and GC-MS. Yield and purity data are included in Table 18.1.

Table 1. Preparation of N,N-dialkylglucamines.

Run	R_1	R_2	R_3	Conv(%)[b]	Selectivity[a] Dialkylglucamine	Selectivity[a] Aldol[c]
1	$n\text{-}C_4H_9$	H	$n\text{-}C_5H_{11}$	99	99	1
2	$n\text{-}C_4H_9$	CH_3	CH_3	99	99	1
3	$n\text{-}C_4H_9$	H	$n\text{-}C_3H_7$	98	97	3
4	$n\text{-}C_4H_9$	H	$n\text{-}C_4H_9$	98	>99	ND[d]
5	$n\text{-}C_6H_{13}$	H	CH_3	>99	99	1
6	$n\text{-}C_6H_{13}$	H	$n\text{-}C_3H_7$	>99	97	3
7	$n\text{-}C_6H_{13}$	H	$n\text{-}C_5H_{11}$	82	99	1
8	$n\text{-}C_8H_{17}$	H	CH_3	99	95	5
9	$n\text{-}C_8H_{17}$	H	$n\text{-}C_3H_7$	98	97	3
10	$(\text{2-ethylhexyl})\text{-O-}(CH_2)_3\text{-}$	H	CH_3	93	98	2
11	$C_8H_{17}\text{-O-}(CH_2)_3\text{-}$ and $C_{10}H_{21}\text{-O-}(CH_2)_3\text{-}$ (1:1)	H	CH_3	97	98	2
12	$i\text{-}C_{10}H_{21}\text{-O-}(CH_2)_3\text{-}$	H	CH_3	93	99	1
13	$n\text{-}C_{12}H_{25}$	H	CH_3	99	99	1
14	Cocoalkyl[e]	H	CH_3	>99	99	1

[a] Analyses are GC (A%) of trimethylsilylated product.

[b] Based on monoalkylglucamine.

[c] Aldol reaction of the aldehyde with itself. In some instances, dehydration of the aldol and subsequent hydrogenation of the double bond was also observed, as was reductive alkylation of the monoalkylglucamine by aldol-derived aldehydes.

[d] Not detected.

[e] Mixture of $n\text{-}C_8$, $n\text{-}C_{10}$, $n\text{-}C_{12}$, $n\text{-}C_{14}$, $n\text{-}C_{16}$ and $n\text{-}C_{18}$ species, centered on $n\text{-}C_{12}$.

Acknowledgements

We thank J. Cunningham and S. D. Furdyna for technical assistance, and Air Products and Chemicals, Inc. for permission to publish this work.

References

1. R. B. Flint and P. L. Salzberg, US Pat. 2,016,962, to E. I. DuPont deNemours & Company (1935).
2. H.-J. Weyer, H. J. Mercker, and R. Becker, German Pat. 4,400,591, to BASF AG (1995).
3. F. Weinelt, German Pat. 4,307,163, to Hoechst AG (1994).
4. B. M. Phillips, A. Kumar, A. Smithson, European Pat. 569,904 B1, to Albright & Wilson Ltd (1997).
5. K. Meier, M. E. Ford, and R. J. Goddard, US Pat. 2006 100127 A1, to Air Products and Chemicals (2006).

19. Simple Catalytic Synthesis of N,N'-Dialkyl-N,N'-di(1-deoxyglucityl)ethylenediamines, Sugar-Based Gemini Surfactants

Mike E. Ford, C. P. Kretz, K. R. Lassila, R. P. Underwood and I. K. Meier

Air Products and Chemicals, Inc., 7201 Hamilton Boulevard, Allentown, PA
18195

fordme@airproducts.com

Abstract

Reductive coupling of unhindered N-alkylglucamines with glyoxal in the presence of Pd/carbon provides the corresponding N,N'-dialkyl-N,N'-di(1-deoxyglucityl)-ethylenediamines in good yield. Only minor amounts of byproducts are formed.

Introduction

N,N'-Dialkyl-N,N'-di(1-deoxyglucityl)alkylenediamines   (Figure   19.1)   are potentially

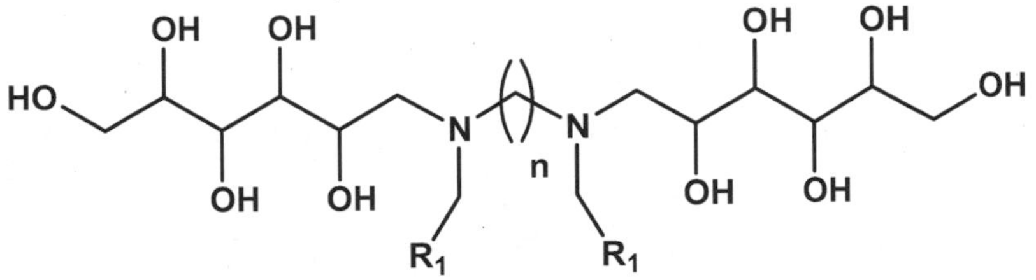

Figure 19.1. N,N'-Dialkyl-N,N'-di(1-deoxyglucityl)alkylenediamines **1**; n = 2 - 6; R_1CH_2 = (cyclo)alkyl.

useful in a spectrum of surfactant applications, including coatings, inks, adhesives, cleaning compositions, personal care products, and formulations for textile processing and oil field applications. Owing to their linked, bifunctional (Gemini) architecture, **1** may be more efficient than simple sugar-based surfactants (1). Remarkably, preparations of only two structural variants of **1** have been reported. Vesicular tertiary amino gemni surfactants have been reported as agents for gene transfer (2). These surfactants were prepared by reductive alkylation of an N,N'-di(1-deoxyglucityl)alkylenediamine with a long chain aldehyde (e.g., eqn. 1). In turn, the latter were prepared in all but one example by stoichiometric (Swern) oxidation of the corresponding alcohols. Although a patent on detergent compositions (3) discloses a broad range of sugar-based gemini surfactants, very few examples are provided. Reductive methylation of N,N'-di(1-deoxyglucityl) propanediamine

$$(1)$$

1; $R_1CH_2 = C_{12} - C_{18}$ alkyl
n = 4, 6

with formaldehyde to form the corresponding N,N'-dimethyl-derivative is the most relevant.

Experimental Section

A 100 mL Parr stainless steel reactor was charged with 0.01 mole of the desired monoalkylglucamine, 4.83×10^{-3} mole (0.966 equivalent) of glyoxal, 0.44 wt% (based on starting glucamine and weight of contained glyoxal) of 5% palladium on carbon, and 30 gm of methanol. The reactor was closed, purged with nitrogen and hydrogen, and pressurized to ca 600 psig with hydrogen. The mixture was heated with stirring (1000 rpm) to 125°C and pressurized with hydrogen to 1000 psig. The reaction was maintained at this temperature; pressure was maintained at 1000 psig via regulated hydrogen feed. After 12 hr, the mixture was cooled to ambient, and the product removed from the reactor by filtration through an internal 0.5 μ sintered metal element. Approximately 1 mL of the homogeneous product was dried in a vac oven, and derivatized with trimethylsilylimidazole/pyridine. This reagent is available in ready-to-use form as Sylon™ TP from Supelco®. Owing to the requirement for complete derivatization of multiple hydroxyl groups, reaction times of 0.75-1.0 hr were used. After trimethylsilylation, the product was analyzed by GC and GC-MS. Yield and purity data are included in Table 19.1.

Results and Discussion

We now report a catalytic route to N,N'-dialkyl-N,N'-di(1-deoxyglucityl)-ethylenediamines (2) which relies on reductive coupling of the corresponding N-alkyl-(1-deoxyglucityl)amines with glyoxal (eqn. 2; see Table 19.1) (4). To drive conversion toward 2, a slight excess (typically, 3-5 eq%) of the N-alkyl-(1-deoxyglucityl)amine was used. In all but one instance (run 7), conversions were greater than 90% and selectivities were greater than 85%. Minor quantities of 3, the adduct of glyoxal with starting N-alkyl-(1-deoxy-glucityl)amine, and of **4**, the product of overhydrogenation of **3**, were typically formed. Owing to our focus on

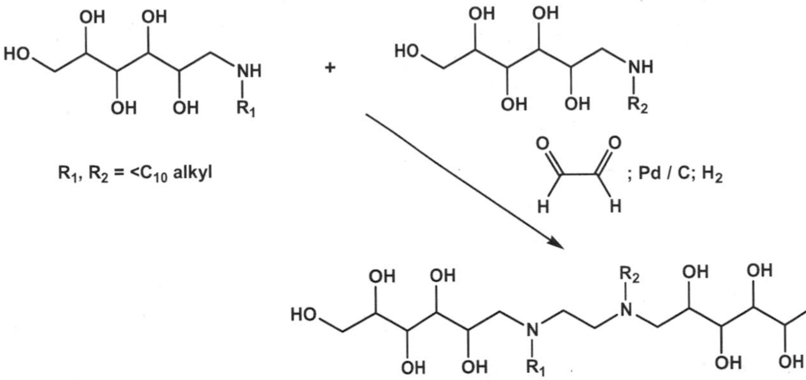

(2)

straight chain functionality in **1**, N-cyclohexyl-(1-deoxyglucityl)amine was the only branched amine evaluated (run 7). Lower conversion and significantly lower selectivity were obtained, presumably as a result of steric hindrance of the

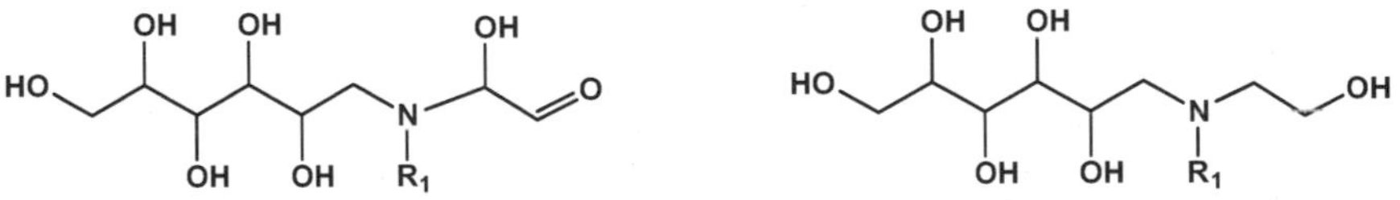

Figure 19.2. Intermediate (**3**) and byproduct (**4**) of coupling reaction.

Table 19.1. Preparation of N,N'-Dialkyl-N,N'-di(1-deoxyglucityl)ethylenediamines

Run	R_1	R_2	Conv (%)[b]	Selectivity[a]		
				2	**3**	**4**
1	n-C_8H_{17}	n-C_8H_{17}	99	88	3	6
2	n-C_6H_{13}	n-C_6H_{13}	95	89	4	4
3	n-C_4H_9	n-C_4H_9	96	91	2	4
4[c]	n-C_8H_{17}	n-C_4H_9	99	90	4	6
5	CH_3-O-$(CH_2)_3$-	CH_3-O-$(CH_2)_3$-	92	90	4	5
6	n-C_4H_9-O-$(CH_2)_3$-	n-C_4H_9-O-$(CH_2)_3$-	99	86	5	5
7	cyclo-C_6H_{11}	cyclo-C_6H_{11}	81	57	17	6

[a] Analyses are GC (A%) of trimethylsilylated product. Selectivities are based on converted amine; balance of product consists of unidentified components.
[b] Based on monoalkylglucamine.
[c] An equimolar mixture of N-butyl- and N-octyl-(1-deoxyglucityl)amine was used as the amine component.

hydrogenation process. This interpretation is supported by observation of a higher concentration of **3** in comparison with runs 1-6. Not surprisingly, reaction of an equimolar mixture of N-butyl-(1-deoxy-glucityl)amine and N-octyl-(1-deoxyglucityl)amine provided a 1:2:1 mixture of N,N'-dibutyl-N,N'di(1-deoxyglucityl)ethylenediamine, N-butyl-N-octyl- N,N'-di(1-deoxyglucityl)ethylene-diamine, and N,N'-dioctyl-N,N'-di(1-deoxy-glucityl)ethylenediamine (run 4). Owing to the flexibility in choice of alkyl groups and the ability to incorporate ether functionality, products with moderate to high hydrophobicity can be obtained (cf. runs 2 and 3 with runs 1 and 6). For unhindered substrates, the concentrations of unreacted intermediates and over-hydrogenation byproducts were low.

In conclusion, N,N'-dialkyl-N,N'-di(1-deoxyglucityl)alkylenediamines (**2**) are prepared in good yield and purity by palladium-catalyzed reductive coupling of straight chain N-alkyl-(1-deoxyglucityl)amines with glyoxal. A wide variety of **2** with moderate to high hydrophobicity is readily accessible.

Acknowledgements

We thank J. Cunningham and S. D. Furdyna for technical assistance, and Air Products and Chemicals, Inc. for permission to publish this work.

References

1. M. J. Rosen, Special Publication - Royal Society of Chemistry (1999), 230 (Industrial Applications of Surfactants IV), 151-161.
2. M. L. Fielden, C. Perrin, A. Kremer, M. Bergsma, M. C. Stuart, P. Camilleri and J. B. F. N. Engberts, Eur. J. Biochem., 268, 1269-1279 (2001); see also P. Camilleri, J. B. F. N. Engberts, M. L. Fielden, and A. Kremer, WO Pat. 00/76954 A1, to SmithKline Beecham and The University of Groningen (2000).
3. J. J. Scheibel, D. S. Connor, Y.-C. Fu, J.-F. Bodet, L. A. Brown, P. K. Vinson and R. T. Reilman, US Pat. 5,669,984, to The Proctor and Gamble Company (1997).
4. M. E. Ford, C. P. Kretz, K. R. Lassila, R. P. Underwood and I. K. Meier, US Pat. 2006 013780 A1, to Air Products and Chemicals (2006).

20. Anchored Wilkinson's Catalyst: Comparison with the Homogeneous Catalyst and Supported Rhodium with Respect to Reaction Selectivity

Setrak K. Tanielyan, Robert L. Augustine, Norman Marin and Gabriela Alvez

Center for Applied Catalysis, Department of Chemistry and Biochemistry, Seton Hall University, South Orange, NJ 07079

augustro@shu.edu

Abstract

A series of anchored Wilkinson's catalysts were prepared by reacting the homogeneous Wilkinson catalyst with several alumina/heteropoly acid support materials. These catalysts were used to promote the hydrogenation of 1-hexene. The results were compared with those obtained using the homogeneous Wilkinson and a 1%Rh/Al$_2$O$_3$ catalyst with respect to catalyst activity and stability as well as the reaction selectivity as measured by the amount of double bond isomerization observed. The effect which the nature of the heteropoly acid exerted on the reaction was also examined.

The results of these hydrogenations will be discussed along with data relating to the composition of the anchored catalyst before and after hydrogenation.

Introduction

The first organometallic homogeneous hydrogenation catalyst, (Ph$_3$P)$_3$RhCl (**1**), was developed by Wilkinson in 1966 (1). Within a few years, attempts at heterogenizing this species began appearing in the literature (2-4). We have shown that anchoring a catalytically active complex onto a support using a heteropoly acid as the anchoring agent gives a stable catalytic material having an activity and selectivity similar to that of the corresponding homogeneous catalyst. While most of the publications describing these 'anchored homogeneous catalysts' have been concerned with the use of chiral catalysts for the hydrogenation of prochiral substrates (5-10), a recent preliminary report described the use of this anchoring procedure to prepare the anchored Wilkinson catalyst (11).

When used in high substrate to catalyst ratios (high turnover numbers, TONs, <5,000) this anchored catalyst was shown to be more stable than the homogeneous material, **1**, which was usually deactivated somewhere between 50-70% conversion when used to promote high TON hydrogenations. One possible reason for this deactivation can be found on examination of the mechanism of alkene hydrogenation over **1** (Scheme 20.1) (1). The first step is the loss of one of the triphenylphosphines to give the bisphosphine, **2**, followed by the oxidative addition of hydrogen and the double bond of the alkene. Subsequent hydrogen transfer produces the alkane product and regenerates the catalytically active **2**. Since **2** is in solution, it can also dimerize to the inactive bischloro species, **3**, a process which leads to the removal of the active entity from the catalytic cycle. With the anchored catalyst the catalytically active species is attached to the support and, thus, is unable to dimerize.

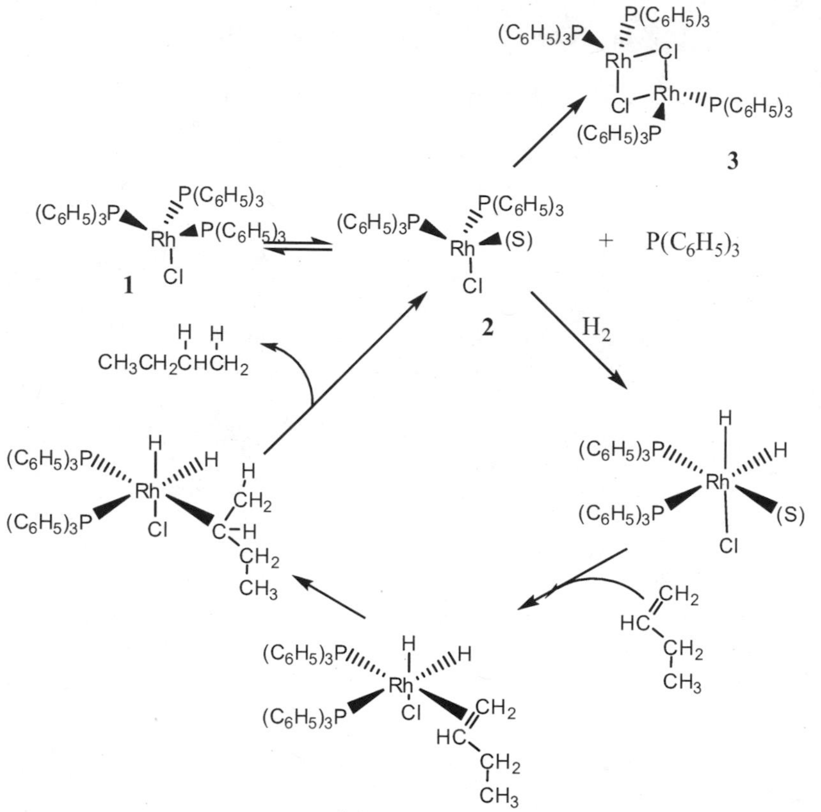

Scheme 20.1

There are, though, some questions which need to be addressed concerning the anchored Wilkinson catalyst. The initial concern is about the nature of the active entity present in this catalyst. Is it analogous to the bisphosphine, **2**, or does the size of the heteropoly acid play a role in the formation of a different active species? Another area of concern is how similar in activity and selectivity are the homogeneous catalyst, **1**, and the anchored material. Finally, what is the role of the heteropoly acid in these catalysts? Does it simply serve as an anchoring agent or does it influence the activity and selectivity of the catalytic moiety? We have examined this later aspect under low TON (<500) conditions and did find that the HPA does play a role in determining the activity and selectivity of a number of reactions (12) but, we have also seen over the years that some of our initial conclusions, which were based on low TON reaction data, needed modification when the catalysts were used to promote high TON reactions. This chapter will attempt to address these concerns.

Experimental Section

The alumina supported anchored Wilkinson catalysts using phosphotungstic acid (PTA), silicotungstic acid (STA), phosphomolybdic acid (PMA) or silicomolybdic acid (SMA) as the anchoring agent, were prepared using the procedures described previously (5,6,12). The amount of the heteropoly acid (HPA) adsorbed onto the alumina was determined by quantitative UV analyses of the reaction solution by which the concentrations of both the starting solution and that present after saturation of the alumina was complete. The Rh complex was added to the HPA/alumina in a 1:1 molar ratio with the amount of HPA present. These values are given in Table 20.1.

The hydrogenations were run using the low pressure apparatus previously described (13). In the standard reaction an amount of catalyst equivalent to 20 μmole of Rh was placed in 15 mL of absolute ethanol in the reactor vessel. After closing the system and venting several times to replace the air with hydrogen, the catalyst was pre-hydrogenated in an ethanol suspension for 2 hr at 36°C under 50 psig of hydrogen. After the pre-hydrogenation, 25 mL of 1-hexene (200 mmoles) were added to the reactor through a septum, the pressure was set at 50 psig and the stirring begun with the hydrogen uptake recorded. In some runs the reaction was stopped periodically with samples taken using a gas tight syringe. These samples were quantitatively analyzed by GC using an internal standard.

When the catalyst was to be used several times the stirring was stopped after complete hydrogen uptake was observed and the catalyst allowed to settle. The reaction liquor was removed under a positive hydrogen pressure using a cannula run through a septum port on the reactor. The reaction vessel was then re-charged through the septum with a further portion of the substrate and solvent using a gas tight syringe and the hydrogenation cycle repeated.

The microanalyses were run by the Schwarzkopf Microanalytical Laboratory, Woodside, NY. The 1%Rh/Al$_2$O$_3$ catalyst was obtained from Johnson Matthey.

Results and Discussion

Nature of the Active Species. It was considered that microanalyses of the starting anchored catalyst and that recovered from a hydrogenation would provide the information needed to define the nature of the active species on the anchored catalyst. For this purpose, STA was used as the agent instead of PTA to simplify the phosphorous analysis. As described later, there is very little difference in the catalytic activity and selectivity between these two anchored catalysts.

Table 20.1. Catalyst loading on HPA modified alumina

Catalyst	Loading [a]
Wilk/PTA	46.8
Wilk/STA	46.5
Wilk/PMA	49.1
Wilk/SMA	49.7

[a] μmoles **1**/g catalyst

Based on the data listed in Table 20.1, a value of 0.42% P was calculated for an anchored catalyst having three triphenylphosphine ligands, 0.28% P with two phosphine groups and 0.14% with one triphenylphosphene. An analytical value of 0.37% P was found which indicates that all three triphenyl-phosphines (TPP) are present in the catalyst as depicted by **4** in Scheme 20.2. However, only 0.11% P was found in the catalyst sample taken after catalyst pre-hydrogenation indicating that only one TPP is present on the active entity. Because of steric constraints between the bulky TPP and the HPA, it would appear that the TPP should be in the axial position as in **5**. A proposed reaction mechanism for the anchored Wilkinson based on that shown in Scheme 20.1 is shown in Scheme 20.2.

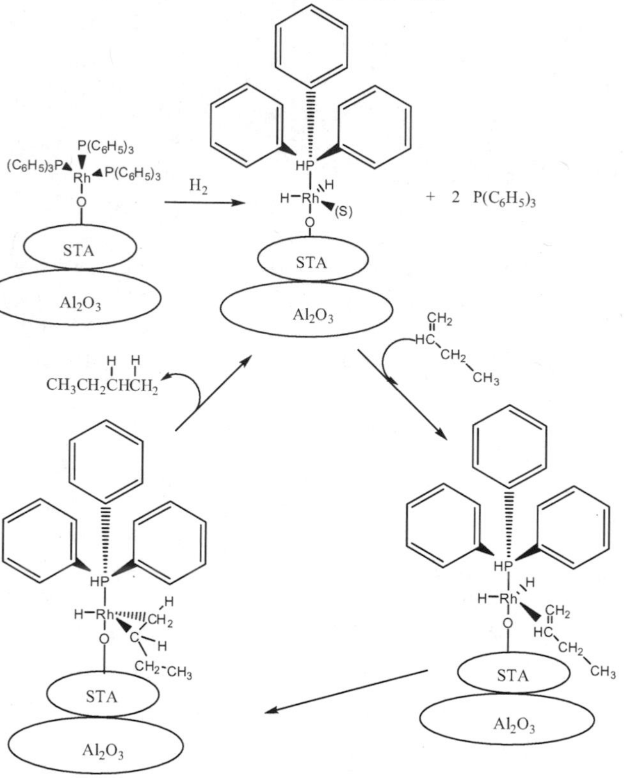

Scheme 20.2

The analytical data for %Cl were inconclusive because of traces of residual Cl in the alumina support.

Anchored Wilkinson Catalysts. To make certain that the data obtained in this study were not influenced by mass transfer factors, it was necessary to define reaction conditions under which this would not be the case. Figure 20.1 illustrates the effect of the stirring rate on the rate of hydrogen uptake at hydrogen pressures of 20 psig and 50 psig. These data indicate that the reaction is apparently run with no mass transport limitations at stirring rates above about 1600 rpm. All of the following reactions were run using stirrers set at 1800 rpm.

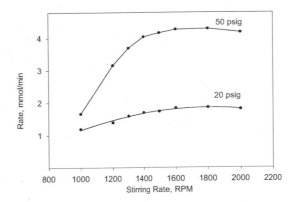

Figure 20.1. Relationship between stirring rate and the initial rate of 1-hexene hydrogenation over Wilk/STA at 36°C and two hydrogen pressures.

From the temperature/rate data plot depicted in Figure 20.2 for the hydrogenation of 1-hexene over Wilk/STA an activation energy of 36.4 kJ/Mol (8.7 kcal/mol) was calculated. While this value is somewhat lower than that commonly accepted for a reaction run in the kinetic regime, it is still sufficiently close that, when taken with the stirring rate data (Figure 20.1) one can be reasonably certain that mass transport is not significant in these reactions. It was also determined that the reaction is first order in hydrogen (Figure 20.3) and at higher substrate concentrations, the reaction approaches pseudo zero order in alkene (Figure 20.4).

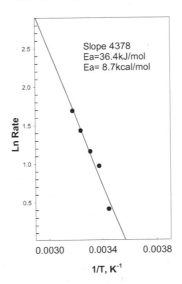

Figure 20.2. Reciprocal temperature-rate plot for the determination of the energy of activation for the hydrogenation of 1-hexene over Wilk/STA.

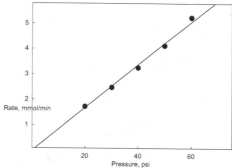

Figure 20.3. Effect of hydrogen pressure on the initial rates of hydrogenation of 1-hexenes over Wilk/STA

A series of alumina supported anchored Wilkinson catalysts were prepared with phosphotungstic acid (PTA), silicotungstic acid (STA), phosphomolybdic acid (PMA), and silicomolybdic acid (SMA) as the anchoring agents. The resulting catalysts are referred to as Wilk/PTA, Wilk/STA, etc. The Wilkinson loading data are given in Table 20.1. These catalysts, along with the homogeneous **1** and a 1% Rh/Al$_2$O$_3$ comparison catalyst, were used to promote the hydrogenation of 1-hexene in ethanol at 36°C under 50 psig of hydrogen. The hydrogen uptake curves are shown in Figure 20.5 with the initial rates and turnover frequencies (TOF hr-1) listed in Table 20.2.

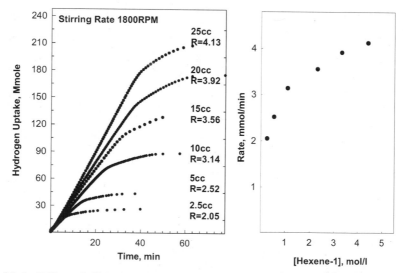

Figure 20.4. Effect of alkene concentration in the hydrogenation of 1-hexene over Wilk/STA at 36°C and 50 psig of hydrogen.

From these data it is obvious that the supported Rh metal is much more active than the homogeneous or anchored Wilkinson catalysts and that the initial rate

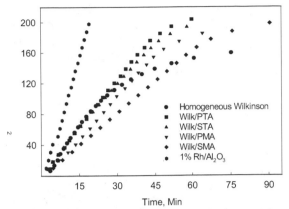

Figure 20.5. Rates of hydrogen uptake in the hydrogenation of 1-hexene at 36°C and 50 psig of hydrogen over different catalysts.

for the homogeneous catalyst is comparable to those of the anchored species regardless of the nature of the anchoring agent. The hydrogenation over the homogeneous catalyst, however, slows down considerably after absorption of about 60% of the theoretical amount of hydrogen. While it might be considered that some of this deactivation may be due to the formation of the dimer, **3**, since the reaction will eventually go to near completion, it may also be due to the slow hydrogenation of the isomeric internal olefins which are being formed by isomerization of the 1-hexene (Equation 1).

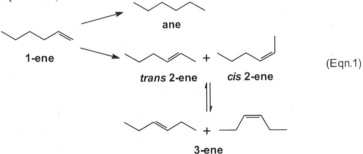

Table 20.2. Initial rates and turnover frequencies observed in the hydrogenation of 1-hexene over anchored Wilkinson catalysts.

Catalyst	Initial Rate[a]	TOF (hr-1)[b]
Homogeneous 1	5.28	4,250
Wilk/PTA	4.77	10,900
Wilk/STA	4.67	9,800
Wilk/PMA	3.37	6,600
Wilk/SMA	3.14	7,750
1% Rh/Al$_2$O$_3$	11.1	30,000

[a]mmole H$_2$/min
[b]mmole H$_2$/mmole RH/hour

A portion of the Wilk/STA catalyst was re-used for the hydrogenation of successive 10,000 TON batches of 1-hexene with the TOF remaining constant for each run. A separate portion of this catalyst was also used for the 100,000 TON hydrogenation of 1-hexene with a TOF of 4800 hr^{-1}. In each of these reactions the amount of Rh in the reaction mixture was below the level of detection, less than 1 ppm.

Table 20.3 lists the product composition for each of these hydrogenations at 70% completion. It can be seen from these data that only a small amount of the isomeric 3-hexenes are formed, regardless of the catalyst. The extent of isomerization to the *trans* and *cis* 2-hexenes, however, is dependent on the nature of the catalyst, going from about 10% over the homogeneous catalyst to about 25% over the supported metal with the anchored catalysts giving values very close to the homogeneous results. Interestingly, the two tungstate (PTA and STA) anchored species have similar initial rate, TOF (Table 20.2) and isomerization rates as do the two molybdate (PMA and SMA) anchored catalysts. Considering that in the PTA and STA anchored species, the active Rh entity is attached to a tungstate fragment of the HPA while in the other two it is connected to a molybdate fragment. It appears that the electronic difference between these two 'ligands' could be sufficient to explain the difference in hydrogenation and isomerization rates.

Table 20.3. Percent product composition at 70% conversion in the hydrogenation of 1-hexene over anchored Wilkinson catalysts.[a]

Catalyst	ane	*trans* 2-ene	*cis* 2-ene	*trans/cis*	3-ene
Homogeneous 1	58.6	5.6	5.2	1.08	1.2
Wilk/PTA	57.2	6.9	5.1	1.35	1.0
Wilk/STA	58.0	6.5	4.8	1.34	1.0
Wilk/PMA	55.1	8.2	5.6	1.47	1.3
Wilk/SMA	52.9	9.6	6.5	1.49	1.2
1% Rh/Al$_2$O$_3$	43.6	15.6	9.5	1.64	1.2

[a] See Equation 1 for product identification

Conclusions

The anchored Wilkinson catalysts are true single site heterogeneous catalysts which have an active entity akin to that proposed in the homogeneous Wilkinson catalytic cycle. This similarity is evident by the fact that the reaction selectivities observed using the anchored species and the homogeneous catalyst are alike but quite different from that found using a supported Rh metal catalyst. These anchored catalysts, though, do have some distinct advantages over the homogeneous Wilkinson species in that the anchored catalysts can be used in reactions having a high substrate to catalyst ratio without loss of activity throughout the reaction. In contrast, in such hydrogenations using the homogeneous catalyst the rate of reaction usually drops off significantly at 50-70% conversion. Another potential advantage of these anchored catalysts stems from the influence which the heteropoly acid anchoring agent has on the outcome of the reaction. In the present study it was found that the tungstate HPAs had different activities and selectivities from the molybdate HPAs. Given the fact that a large number of HPAs are known, it would seem that one could be able to tailor-make anchored catalysts to optimize specific reactions.

References

1. J.A. Osborn, F.H. Jardine, J.F. Young and G. Wilkinson, *J. Chem. Soc., A*, 1711 (1966).
2. J.P. Collman, L.S. Hegedus, M.P. Cooke, J.R. Norton, G. Dolcetti and D.N. Marquardt, *J. Amer. Chem. Soc.*, **94**, 1789 (1972).
3. W. Dumont, J.C. Poulin, T.P. Daud and H.B. Kagan, *J. Amer. Chem. Soc.*, **95**, 8295 (1973).
4. E. Arstad, A.G.M. Barrett and L. Tedeschi, *Tetrahedron Letters,* **44**, 2703 (2003).
5. S.K. Tanielyan and R.L. Augustine US Patents 6,005,148 (1999) and 6,025,295 (2000), to Seton Hall University.
6. R. Augustine, S. Tanielyan, S. Anderson and H. Yang, *Chem. Commun. (Cambridge),* 1257 (1999).
7. R.L. Augustine, S.K. Tanielyan, S. Anderson, H. Yang and Y. Gao, *Chem. Ind.* (Dekker), **82** *(Catalysis of Organic Reactions),* 497 (2000).
8. P. Goel, S. Anderson, J. Nair, C. Reyes, G. Gao, S. Tanielyan and R. Augustine, *Chem. Ind.* (Dekker), **89** *(Catalysis of Organic Reactions),* 523 (2003).
9. C. Reyes, Y. Gao, A. Zsigmond, P. Goel, N. Mahata, S.K. Tanielyan and R.L. Augustine, *Chem. Ind.* (Dekker), **82** *(Catalysis of Organic Reactions),* 627 (2000).
10. R.L. Augustine, P. Goel, N. Mahata, C. Reyes and S.K. Tanielyan, *J. Mol. Catal., A,* **216**, 189 (2004).
11. C. Reyes, S. Tanielyan and R.L. Augustine, *Chem. Ind.* (Boca Raton, FL USA), **104** (*Catalysis of Organic Reactions*), 59 (2005).
12. R.L. Augustine, S.K. Tanielyan, N. Mahata, Y. Gao, A. Zsigmond and H. Yang, *Appl. Catal. A*, **256**, 69 (2003).
13. R.L. Augustine and S.K. Tanielyan, *Chem. Ind.* (Dekker), **89** *(Catalysis of Organic Reactions)*, 73 (2003).

21. Synthesis of New Ligands for the Preparation of Combined Homogeneous and Heterogeneous Catalysts

Catherine Pinel, Sébastien Noël, Ke Pan, Ciahong Luo and Laurent Djakovitch

IRCELYON, Institut de recherches sur la catalyse et l'environnement de Lyon, UMR 5256 - CNRS - Université de Lyon Bioresources upgrading and green chemistry, 2 avenue Albert Einstein 69626 Villeurbanne Cedex, France

Catherine.pinel@ircelyon.univ-lyon1.fr

Abstract

We studied the asymmetric hydrogenation of β-ketoester in the presence of a chiral homogeneous complex associated with supported metallic catalyst. To induce interactions between both entities, we synthetized chiral ligands bearing a polyaromatic moiety that could adsorb on metallic particles. DIOP-based ligands were elaborated using a Heck coupling reaction between polyaryl or heteroaryl bromides and acrolein derivatives as the key step. The corresponding ruthenium complexes were prepared and used in the hydrogenation of methylacetoacetate. Catalytic activity was not improved upon addition of metallic particles suggesting no synergistic activity between the two entities.

Introduction

The enantioselective catalysis is one of the most elegant methods to prepare optically active products. Asymmetric hydrogenation is largely performed in the presence of homogeneous chiral complexes, mainly ruthenium derivatives. We aimed to develop an alternative approach based on the use of new heterogeneous catalysts involving on the same support grafted/immobilized organometallic complexes and supported metallic particles. This approach was recently reported by several groups for achiral hydrogenation of aromatic moieties (1-5). It seems that the efficiency of such catalysts is linked to the proximity of both entities. In order to maintain a short distance between both moieties, we synthesized chiral diphosphine ligands based on the DIOP [2,3-O-isopropylidene-2,3-dihydroxy-1,4-bis(diphenylphosphino)butane)] structure and bearing a polyaromatic moiety that could adsorb on a metallic surface.

The Heck coupling reaction appeared to be a route of choice to achieve the synthesis of the modified-DIOP ligands. We previously studied the palladium-catalyzed coupling of acrolein and acrolein acetals with several polyaromatic and heteroaromatic bromides either in the presence of homogeneous or heterogeneous catalytic systems (6, 7). After optimization of the reaction conditions, high conversions and selectivities were achieved except with anthracenyl derivatives (8). Based on these results, we developed the synthesis of the desired ligands. The corresponding ruthenium complexes were then synthesized and used as homogeneous catalysts, or associated to supported palladium particles, in the hydrogenation of methyl acetoacetate.

Experimental Section

Typical procedure for one-pot Heck coupling and hydrogenation
In a round bottom flask equipped with a cooling condenser, 4.8 mmol of aryl bromide, 0.048 mmol (2 mol% Pd) palladacycle **1**, 7.2 mmol diethyl 2-vinyl-[1,3]-dioxolane-4,5-diacetate **2**, 7.2 mmol NaOAc were dissolved in 20 mL degassed NMP. The solution was stirred for 2-4h at 140°C. After cooling, 350 mg of activated carbon (CECA, 4S) was added and nitrogen atmosphere was replaced with hydrogen atmosphere. The solution was stirred vigorously until complete hydrogenation of the C=C double bond as analyzed by GC. The mixture was filtered over celite, 200 mL H_2O were added to the filtrate and the aqueous layer was extracted with AcOEt (3 x 100 mL). The organic layers were washed with saturated NaCl (50 mL) and dried over $MgSO_4$. The solvent was evaporated and the product was purified by flash chromatography. Coupling products were characterized by elemental analysis, ^1H and ^{13}C NMR analyses.

Results and Discussion

Heck coupling of diethylacetal derivatives
We have shown that the direct arylation of acrolein toward the synthesis of cinnamaldehyde derivatives was an efficient procedure. Using the palladacycle **1** as catalyst, substituted aldehydes **3** were prepared with up to 87% isolated yield from condensed aryl bromides (Scheme 21.1, Route 1) that was extended successfully to heteroaryl bromides, like bromoquinolines (6). Alternatively, the acrolein diethyl acetal was used as olefin and a selective formation of the saturated ester **4** was attained under the same reaction conditions (Scheme 21.1, Route 2). The expected aldehydes **3** were, however, obtained from most of the aryl halides used under modified conditions. It was shown that the addition of n-Bu$_4$NOAc in the medium

Route 2   Route 3   Route 1

1

Scheme 21.1: Heck arylation of acrolein and acrolein diethylacetal.

was the most important parameter among all those evaluated (i.e.; KCl, solvent...) that affect the selectivity (Scheme 21.1, Route 3). However, moderate activity and selectivity were achieved when using the 9-bromoanthracene whatever the olefin (8). This was attributed to the large steric hindrance of this substrate.

Heck coupling of vinyl dioxolane derivatives

In order to synthesize the DIOP-based ligands, the aldehyde derivatives **3** were treated with commercially available diethyl tartrate in the presence of catalytic amount of PPTS (Scheme 21.2). While good conversions were attained with the naphtyl derivative, they were low with all other aryl compounds and unsuccessful with heteroaromatic moieties (yields < 10%).

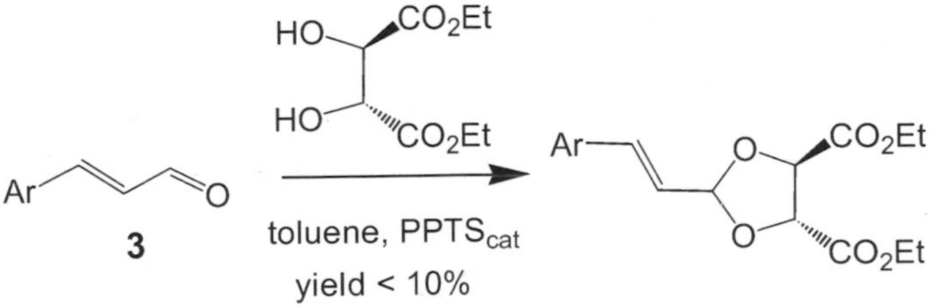

Scheme 21.2: Attempts to acetalize substituted aldehydes.

Table 21.1: Palladium-catalyzed coupling of aryl bromides with dioxolane derivatives. Reaction conditions: **1** (1%), NaOAc, NMP, 140°C.

Entry	Dioxolane	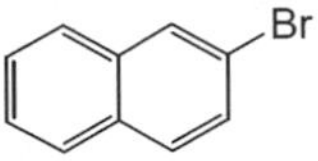 Ar-Br ![naphthyl]	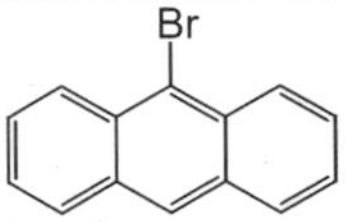 ![anthracenyl]	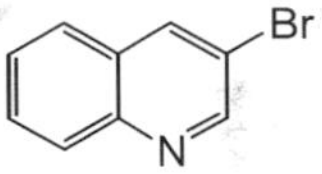 ![quinolinyl]
1	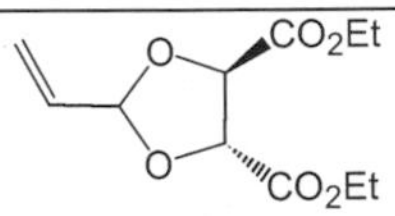 CO₂Et / CO₂Et dioxolane (**2**)	81%[b] (3h)	75%[b] (3h)	84%[b] (3h)
2	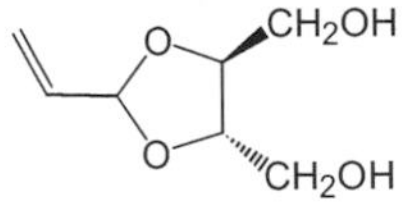 CH₂OH / CH₂OH dioxolane	Mixture	Mixture	
3	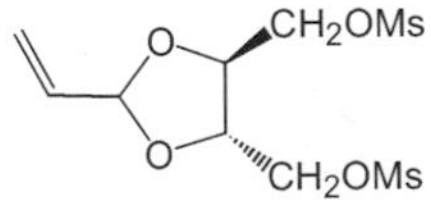 CH₂OMs / CH₂OMs dioxolane	0%	0%	0%
4	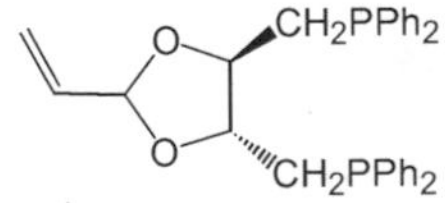 CH₂PPh₂ / CH₂PPh₂ dioxolane	20%[b] (24h)		
5	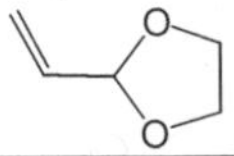 dioxolane [a]	52%[c] (29 h)	73%[c] (29h)	100%[c] (3h)

a: Pd(OAc)₂ (2%), K₂CO₃, DMF, 110°C; b: isolated yield; c: GC conversion

To overcome this drawback, we studied the arylation of diethyl 2-vinyl-[1,3]-dioxolane-4,5-diacetate **2** with several bromo polyaromatic and heteroaromatic substrates (Table 21.1 and Scheme 21.4). In parallel, the Heck coupling of several vinyl dioxolane derivatives with aryl bromides was studied in the presence of homogeneous catalysts (Table 21.1).

The arylation of the unsubstituted acrolein ethylene acetal with the bromopolyaromatics and bromoquinoline was studied in the presence of $Pd(OAc)_2$ as catalyst (reaction conditions: catalyst (2%), K_2CO_3, DMF, 110°C). The reaction was successfully performed with all evaluated poly(hetero)aromatics (Entry 5).

However, the reaction was not selective and a mixture of the acetal **5** and the ester **6** (Scheme 3) was observed, the latter being the major product (56-98% selectivity), especially in the case of the 9-bromoanthracene.

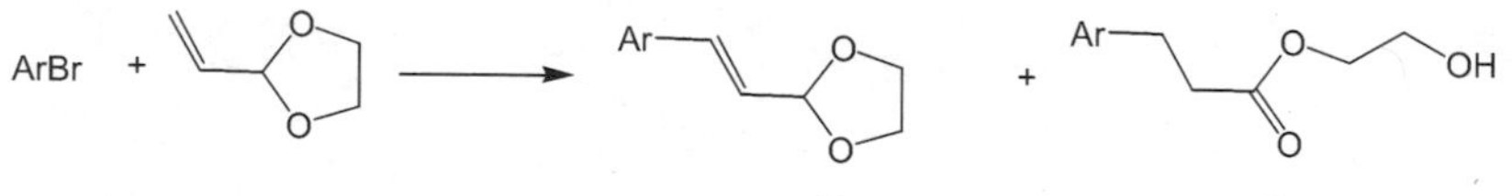

Scheme 21.3: Arylation of acrolein ethylene acetal.

The product **6** results from the syn-β–H-elimination of the palladium with hydrogen borne by the acetal carbon (Scheme 4) as previously reported for corresponding reactions (9, 10). The presence of n-Bu_4NOAc noticeably increases

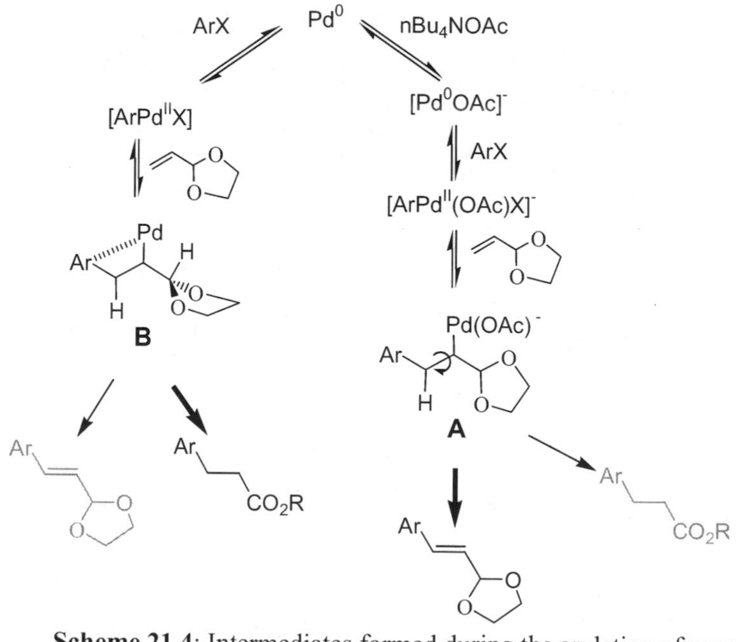

Scheme 21.4: Intermediates formed during the arylation of acrolein ethylene acetal.

the selectivity towards the acetal **5**. As suggested previously with diethyl acetal acrolein we propose that the formation of an anionic palladium intermediate favor the acetal formation.

Surprisingly, diethyl 2-vinyl-[1,3]-dioxolane-4,5-diacetate **2** is very reactive (Entry 1). Whatever the aryl bromide, complete conversions are observed after 3h and high isolated yields (> 73%) toward the expected compound are achieved. No product issued from the alternative syn-β–H-elimination is detected. In that case, we suggest that a specific interaction between the ester group and the palladium center could occur leading to a stabilized 7 membered-ring intermediate **7** avoiding thus the formation of undesired product.

The diol yields a mixture of products, including the expected coupling compound (Table 21.1, Entry 2). The presence of mesityl group inhibits the reaction (Entry 3). In that case, the aryl bromide is recovered but the mesityl derivative decomposes partially under reaction conditions. Interestingly the diphosphine-substituted vinyl dioxolane, a potential substrate for the direct synthesis of the ligands through the Heck arylation, gives encouraging yield, however, lowered due to oxidation of the phosphine moiety during the reaction (Entry 4). Phosphine borane derivatives could be regarded as good candidates to perform this reaction.

In summary, only, the diethyl 2-vinyl-[1,3]-dioxolane-4,5-diacetate **2** is an adequate substrate for the global synthesis of the DIOP-based ligands. Reduction of the coupling product by LiAlH$_4$ to give the expected diol failed while reduction of the corresponding saturated product succeeded. To circumvent this difficulty, we performed directly the one-pot hydrogenation of the coupling product using the palladium catalyst previously involved in the Heck coupling. The coupling of **2** with aryl bromide was monitored by GC analysis. After completion of the reaction (approximately 3h), activated carbon was added to the reaction medium and the nitrogen atmosphere was substituted with hydrogen (1 bar). Former studies showed that the carbon support is crucial to achieve high hydrogenation activity by dispersing finely the palladium as particles onto the support. Complete conversions were achieved within 2-3h, giving the saturated product with high isolated yields after purification (Scheme 21.5).

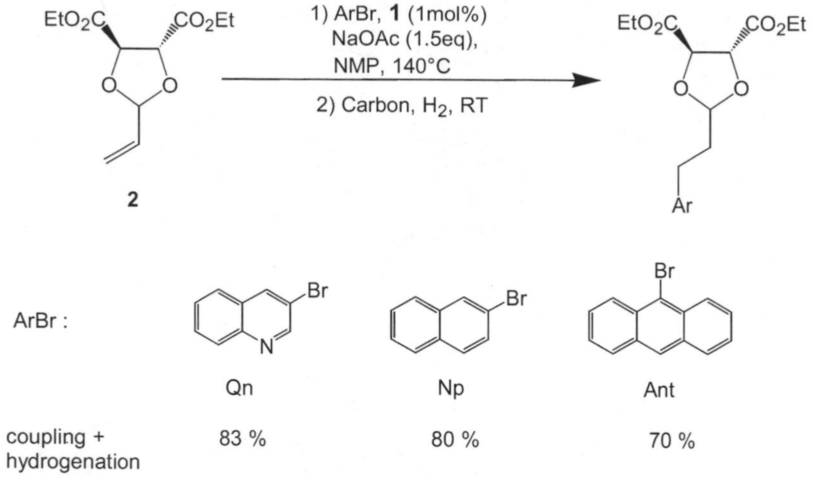

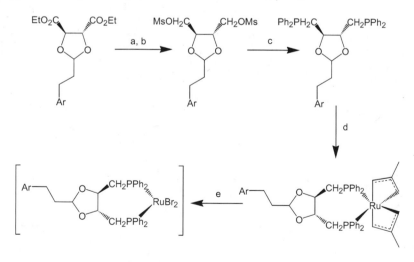

Scheme 21.5: Synthesis of diethyl 2-substituted-[1,3]-dioxolane-4,5-diacetate.

This approach allows the full synthesis of the modified DIOP ligands (Scheme 21.6). Three new ligands were synthesized with 15-20% global yield (6 steps) according to a similar route as described for DIOP synthesis (11). After chemical reduction of the diester to the diol, the mesylated compounds were isolated. Their treatment with diphenylphosphine previously reacted with n-BuLi to yield LiPPh$_2$ gave the expected ligands with 49-55% isolated yield.

a: LiAlH$_4$, THF, reflux, 3-5h, 70-80%; b: MsCl, NEt$_3$, CH$_2$Cl$_2$, reflux, 3h, 73-78%;
c: LiPPh$_2$, THF,0°C, 3h, 49-55%; d: (COD)Ru(Met)$_2$, hexane/toluene (1/1), 50°C, 5h, 54-78%;
e: 2HBr, acetone, RT, 30 mn

Scheme 21.6: Synthesis of ruthenium complexes.

The corresponding ruthenium complexes were prepared from [(COD)Ru(metallyl)$_2$] complex according to the literature with slight modifications (12). In order to solubilize the ligand, the exchange reaction must be performed in hexane/toluene (1/1) mixture at 50-60°C. After cooling, the precipitated complex was washed with degassed hexane. These complexes were fully characterized (^{1}H and ^{13}C NMR, elemental analysis, [α]$_D$). ^{31}P NMR analyses showed that there is no residual free ligand (Table 21.2).

Table 21.2: ^{31}P of free ligands and ruthenium complexes [L$_2$Ru(metallyl)$_2$].

Phosphine	Free ligand δ ^{31}P (ppm)	Yield complex (%)	Complex δ ^{31}P (ppm)
Qn-DIOP	-22.42; -21.56	68	41.77; 42.84
Np-DIOP	-23.67; -22.81	70	41.97; 43.07
Ant-DIOP	-23.17; -22.28	54	42.29; 43.51

These complexes [L$_2$Ru(metallyl)$_2$] are not efficient in C=O hydrogenation and the dibromide complexes associated with the Qn-DIOP and Np-DIOP ligands were synthesized according to the conditions described by Noyori et al. (13). The corresponding bromide complexes were not isolated and used as obtained for the hydrogenation of methyl acetoacetate in MeOH for 6h at 30°C under 20 bar hydrogen pressure (Table 21.3). Under these conditions, 51% conversions were achieved in the presence of [DIOPRuBr$_2$] complex with modest enantioselectivity. The other ruthenium complexes bearing aromatic-substituted DIOP ligands exhibited lower activity (24 and 28% in the presence of [Qn-DIOPRuBR$_2$] and [Np-DIOPRuBr$_2$], respectively).

Whatever the ligand, no improvement was observed by addition of Pd/C, neither on the activity nor on the enantioselectivity indicating that there is no synergetic effect between the two entities.

Table 21.3: Asymmetric hydrogenation of methylacetoacetate.

Catalyst	Conv (%)	ee (%)	+ Pd/C Conv (%)	ee (%)
			4	
DIOPRuBr$_2$	51 (81)^a	10	45	10
Qn-DIOPRuBr$_2$	24	5	17	3
Np-DIOPRuBr$_2$	28	7	23	4

^{a} 24h reaction time

Conclusions

In conclusion, we have shown that the direct arylation of diethyl 2-vinyl-[1,3]-dioxolane-4,5-diacetate is an efficient procedure for the synthesis of cinnamaldehydes dioxolane derivatives. Using the palladacycle as catalyst, the corresponding aryl derivatives are obtained with up to 83% isolated yield. This strategy allows the full synthesis of three new arylated DIOP ligands in 6 steps with 15-20% global yield. The corresponding ruthenium complexes were synthesized and used in asymmetric hydrogenation of methylacetoacetate. Moderate yields and enantioselectivities were achieved that were not improved in the presence of supported palladium suggesting that no cooperation between the two catalytic species to activate the substrate toward hydrogenation exist.

Acknowledgements

S.N. thanks the "Ministère de l'Education Nationale, de l'Enseignement Supérieur et de la Recherche" for a grant.

References

1. H. Gao and R.J. Angelici, *J. Am. Chem. Soc.* **119**, 6937 (1997).
2. H. Yang, H. Gao and R.J. Angelici, *Organometallics* **19**, 622 (2000).
3. R. Abu-Reziq, D. Avnir, I. Milosslavski, H. Schumann and J. Blum, *J. Mol. Catal. A* **185,** 179 (2002).
4. C. Bianchini, V. Dal Santo, A. Meli, S. Monett, M. Moreno, W. Oberhauser, R. Psaro, L. Sordelli and F. Vizza, *Angew. Chem. Int. Ed. Engl.* **42**, 2636 (2003).
5. P. Barbaro, C. Bianchini, V. Dal Santo, A. Meli, S. Moneti, R. Psaro, A. Scaffidi, L. Sordelli and F. Vizza, *J. Am. Chem. Soc.* **128,** 7065 (2006).
6. S. Noël, L. Djakovitch and C. Pinel, *Tetrahedron Letters*, **47,** 3839 (2006).
7. S. Noël, C. Luo, C. Pinel and L. Djakovitch, *Adv. Synth. Catal.* **349**, 1128 (2007).
8. K. Pan, S. Noël, L. Djakovitch and C. Pinel submitted for publication.
9. F. Berthiol, H. Doucet and M. Santelli, *Catal. Lett.* **102**, 281 (2005).
10. M. Lemhadri, H. Doucet and M. Santelli, *Tetrahedron* **60,** 11533 (2004).
11. H.B. Kagan and T.P. Dang, *J. Am. Chem. Soc.* **94,** 6429 (1972).
12. J.P. Genet, S. Mallart, C. Pinel, S. Jugé and J.A. Laffitte, *Tetrahedron: Asymmetry* **2,** 43 (1991).
13. R. Noyori, T. Ohkuma, M. Kitamura, H. Takaya, N. Sayo, H. Kumobayashi and S. Akutagawa, *J. Am. Chem. Soc.* **109**, 5856 (1987).

22. Assessing Catalyst Homogeneity/Heterogeneity via Application of Insoluble Metal Scavengers: Application to Heck and Suzuki Reactions

Christopher W. Jones and John M. Richardson

School of Chemical & Biomolecular Engineering, Georgia Institute of Technology, 311 Ferst Dr., Atlanta, GA 30332

cjones@chbe.gatech.edu

Abstract

Solid palladium scavengers are introduced as tools for the assessment of the heterogeneity/homogeneity of palladium catalyzed Heck and Suzuki coupling reactions. The strengths and limitations of this heterogeneity test are discussed in the context of three examples including coupling reactions of aryl iodides, bromides and chlorides with terminal olefins and aryl boronic acids. In all cases where we have added an excess of solid poison to the reactions using solid precatalysts, we have been able to completely extinguish the catalysis. This has been interpreted as evidence supporting the hypothesis that soluble, leached species are solely responsible for these reactions when using these precatalysts.

Introduction

The use of palladium as a catalyst is common in the development and synthesis of active pharmaceutical ingredients (APIs). Palladium is an expensive metal and has no known biological function. Therefore, there is a need to recover spent palladium, which is driven both by cost and by government regulations requiring residual palladium in APIs to be <5 ppm (1). Thus, much research has been conducted with the aim of heterogenizing active palladium that can then be removed via simple filtration and hopefully reused without significant loss of activity.

A large number of supports and immobilization strategies have been employed. Examples include ionic palladium bound to charged surfaces, palladium entrapped in various matrices (zeolites, polymers, e.g., Pd-EnCat[TM]), metallic palladium particles on oxide or carbon supports and organometallic palladium complexes covalently linked to silica or organic polymeric solids (2). In most cases, these solid precatalysts are assumed to act as active, heterogeneous catalysts. The most common control experiments used to elucidate the nature of the active species are relatively ambiguous methods, such as hot filtration tests and comparisons of reaction yields upon catalyst recycle (2-3). There is a growing consensus that catalysis from immobilized metallic palladium is actually from leached metal in Heck and Suzuki catalysis (2,4-6), perhaps with a common solution phase catalytic species (7-9), although there are still isolated claims of heterogeneous catalysis in these reactions. In no case has direct, experimental proof of Pd(0) surface mediated catalysis been reported. In general, rigorous testing is rarely employed to evaluate catalyst heterogeneity, although new "heterogeneous catalysts" continue to be claimed every year (2,10).

A variety of heterogeneity/homogeneity tests have been developed for liquid phase catalytic reactions. These have been reviewed in a general sense by Finke (11) and specifically for application in palladium catalyzed Heck and Suzuki reactions by Jones (2-3). The most commonly used tests in palladium mediated reactions include the Hg(0) test, the split or filtration test, recycle tests, and 3-phase tests. Each of these control experiments has subtleties in its interpretation and no single experiment can conclusively be used to assign heterogeneity or homogeneity (2-3). Each test is best used in conjunction with others, with the combined data used to shed light on the nature of the truly active species.

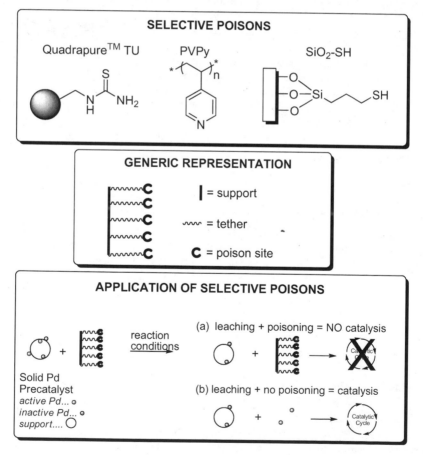

Figure 22.1. Schematic representation of use of insoluble poisons to selectively terminate catalysis from leached palladium species in coupling reactions.

In the course of our work using supported Pd(II) pincer precatalysts in these reactions, we found it was often possible to distinguish whether there were leached, soluble species that were active. However, it was often difficult to exclude the possible parallel participation of catalysis by solid supported Pd(II) pincer complexes. Thus, we sought a reagent that could be used to selectively poison only the soluble catalytic species. To this end, we introduced the use of solid, insoluble poisons that should, in principle, interact only with soluble palladium species in these reactions (Figure 22.1). (12) In this work, we discuss the various solid poisons we have employed in heterogeneity/homogeneity testing [poly(4-vinylpyrdine) (PVPy), (12-15) QuadraPure^{TM} TU (QTU) (3) and mercaptopropyl-modified mesoporous silica (SiO$_2$-SH), (16)] in Heck and Suzuki coupling reactions. In addition, a particular focus is placed on the discussion of subtle details in the application of these and other control experiments to heterogeneity/homogeneity testing.

Experimental Section

Solid palladium scavengers, PVPy, QTU were purchased from commercial sources. The mesoporous silica material, SiO$_2$-SH, was prepared via reaction of SBA-15 (110 Å pore diameter) with 3-mercaptopropyltrimethoxysilane (16). Specifically, a toluene suspension of SBA-15 and 3-mercaptopropyltrimethoxysilane was heated at reflux for two days under Ar. Water was then added to promote the cross-linking and the mixture was heated at reflux for an additional day. The solids were filtered and washed with copious amounts of toluene, hexanes, and methanol to remove un-reacted silanes. The solids were finally Soxhlet extracted with dichloromethane at reflux temperature for 3 days, dried, and stored in a nitrogen dry box. The final solid contained 7.5 wt% sulfur (2.3 mmole S/g solid).

Results and Discussion

First Generation GT Soluble Pd Catalyst Poison - PVPy. In our evaluation of soluble and supported Pd(II) SCS and PCP pincer precatalysts in the Heck coupling of aryl iodides with n-butyl acrylate, we introduced the use of PVPy as a selective poison for leached palladium (12-15). Working in particular with mesoporous silica-supported Pd pincers where the pore size was much smaller than the particle size of the cross-linked PVPy, we hypothesized that this polymer should selectively over-coordinate soluble palladium if the polymer was present in a large excess. Our choice of this polymer stemmed from the low reactivity of the pyridine groups with the reagents used as well as a previous report that suggested that excess PVPy could hamper catalysis in Heck reactions using Pd-loaded PVPy supports (17). In this example and others, the authors intended to prepare a supported precatalyst using supported ligands (e.g., pyrdines (17) or phosphines (18)) but instead found the supported ligand, when used in excess, could poison catalysis. We demonstrated that catalysts covalently tethered within the pores of the mesoporous silica support would not be effected by PVPy using an acid catalysis test reaction. Using mesoporous silica-supported propylsulfonic acid catalyst, the dimerization of α-methylstyrene proceeded smoothly in the presence of the bulky PVPy, whereas the same reaction was poisoned by molecular pyridine, which could enter the catalyst pores (13).

For all the silica catalysts studied, addition of PVPy at a loading of 350 equivalents per Pd effectively extinguished the Heck catalysis (12-14). Thus, we established it as an effective poison for these reactions. Since our early work, several other authors have utilized PVPy in their homogeneity/heterogeneity testing (19-22). However, like all heterogeneity/homogeneity tests, care must be used in the selection and application of a solid poison (2). For instance, the choice of solvent and the degree of solubility of the precatalyst may strongly influence the applicability of a solid poison to various reaction conditions. Furthermore, the solid poisons can degrade in the presence of some reactants or at elevated temperatures and the small molecule degradation products could interact with solid (pre)catalysts in unintended ways.

As an example, consider the use of PVPy as a solid poison in the study of poly(norbornene)-supported Pd-NHC complexes in Suzuki reactions of aryl chlorides and phenylboronic acid in DMF (23). This polymeric precatalyst is soluble under some of the reaction conditions employed and thus it presents a different situation from the work using porous, insoluble oxide catalysts (12-13). Like past studies, addition of PVPy resulted in a reduction in reaction yield. However, the reaction solution was observed to become noticeably more viscous, and the cause of the reduced yield – catalyst poisoning vs. transport limitations on reaction kinetics – was not immediately obvious. The authors thus added a non-functionalized poly(styrene), which should only affect the reaction via non-specific physical means (e.g., increase in solution viscosity, etc.), and also observed a decrease in reaction yield. They thus demonstrated a drawback in the use of the potentially swellable PVPy with soluble (23) or swellable (20) catalysts in certain solvents.

Second Generation GT Soluble Pd Catalyst Poison – Commercial Palladium Scavengers: QTU. In the past years, many new materials have been developed and marketed as heavy metal scavengers for product purification when using transition metal catalysts. These are essentially oxide or organic polymeric solids functionalized with molecular, metal binding groups of a variety of structures. Examples include the QuadraPure[TM] (polymer based) and QuadraSil[TM] (silica based) series of scavengers sold by Reaxa Ltd. and the silica based scavengers sold by Silicycle Inc. Owing to the relatively weak binding affinity of pyridyl groups for palladium, we sought other readily available solids for use as soluble catalyst poisons. QTU, marketed by Reaxa Ltd. and available through Aldrich, is a thiourea-functionalized polymeric material that is well-suited for palladium scavenging.

To assess the utility of this resin, we chose to employ it in the evaluation of the heterogeneity of a commercial polymer-entrapped Pd(OAc)$_2$ precatalyst, Pd-EnCat[TM], also sold by Reaxa. This precatalyst was designed with the goal of providing a heterogeneous catalyst that would allow simple removal of palladium from reactions (24-26). PVPy and QTU were first used as poisons in the Heck reaction of iodobenzene and n-butyl acrylate in DMF using Pd/C as the palladium source. Both poisons were shown to completely extinguish the reaction using this precatalyst, an observation expected since it is known that this precatalyst operates in solely homogeneous mode by leaching of palladium from the solid support in both Heck and Suzuki reactions (2,4,5,27-28).

Application of QTU in Heck reactions of iodobenzene or iodopyridine and n-butyl acrylate using Pd EnCat[TM] in DMF, IPA and toluene resulted in complete cessation of the reaction. Certainly it is unlikely that QTU can poison palladium species encased in a polymeric shell, and thus we interpret these results as being consistent with the hypothesis that all catalysis is associated with leached species using this precatalyst under our conditions. Other authors have also reported convincing evidence for leaching from Pd EnCat (29), but it is the solid phase poisoning results described here that allow one to rule out significant amounts of catalysis possibly occurring within the polymeric shells of Pd EnCat[TM].

Third Generation GT Soluble Pd Catalyst Poison – Mercaptopropyl-modified Mesoporous Silica – SiO₂-SH. QTU, as well as many other effective palladium scavengers, contains sulfur donor ligands. It is thus not surprising that 3-mercaptopropyl-functionalized silica materials can be used to bind Pd(II) species strongly (30,31). Recently, two groups reported the use of mercaptopropyl-functionalized silica materials loaded with Pd(II) salts as catalysts in Heck and Suzuki coupling reactions of a variety of aryl halides (32-34). The solids were used to effectively convert aryl bromides and chlorides in Suzuki reactions, as well as aryl bromides in Heck reactions, all while limiting palladium losses to levels that were low enough for some authors to describe the catalysts as "leach-proof catalysts" (34).

However, given that other, supposedly stable and leach-resistant catalysts that we had previously studied seemed to operate solely by a leaching mechanism, with no supported Pd apparently active for the reaction (vide supra) and that other authors had found precatalysts of this type to operate by solely a leaching mechanism in Heck conversions of aryl iodides (19), we decided to investigate this system further using the solid poison testing method.

Given that mercaptopropyl-modified silica is a good scavenger for palladium (30), we decided to use the unmetallated, bare support as a solid phase poison while studying Heck and Suzuki reactions of aryl iodides, bromides and chlorides under the conditions reported previously (19, 32-34). Whereas we and others have observed problems when using polymeric scavengers such as PVPy that might degrade or swell under reaction conditions, there seemed to be no pathway by which the bare catalyst support, from the same batch as was used to make the catalyst, could effect the reaction by any means other than scavenging leached palladium.

Under all the conditions studied, addition of bare SiO₂-SH to Heck or Suzuki coupling reactions using a variety of bases, aryl halides and solvents resulted in complete cessation of the catalytic activity (35). These results suggest that catalysis with this precatalyst is also associated with labile palladium species that either (i) are already in solution as active species or (ii) can be pulled into solution from a weak interaction on the catalyst surface via a shift in the dynamic surface-solution equilibrium induced by addition of a solution palladium sink (the bare SiO_2-SH) poison. Our interpretation of these results is the former, although the occurrence of the latter cannot be rigorously ruled out with the current data. These new findings have been shared with some of the original authors (33,34) of the previous work on these precatalyst systems and they have subsequently reproduced our results (35).

These observations demonstrate the complexity of the determination of the heterogeneity/homogeneity of palladium catalyzed coupling reactions. Because so little solution phase palladium is necessary to effect the catalysis (even ppb levels of palladium can be active, (2,36-37)) and because of a dynamic equilibrium between solution and supported palladium species that leads to facile palladium redeposition when the system is disturbed (for example by a split test) (2,38), ruling out homogeneous catalysis can be a difficult task. In the case of mercaptopropyl-functionalized silica supported Pd(II), all the early evidence (33), including use of a well-established 3-phase test for assessing catalyst homogeneity (2,39-41), suggested that the catalysis was mostly heterogeneous, although there was some evidence for a homogeneous pathway. This suggests several subtle points about the 3-phase test that have barely been raised in the literature and deserve some discussion here.

Subtleties and Limitations of the Use of Solid Poisons for Soluble, Catalytic Palladium Species. As we have noted previously, every heterogeneity/homogeneity test has subtleties in its interpretation (2). Although researchers often interpret a test as simple as 'result one means "A" and result two means "B"', this simplistic, non-analytical approach can be dangerous and misleading. For example, researchers have for years interpreted an absence of catalysis in the presence of Hg(0) as "proof of heterogeneous catalysis" (2). However, this is overinterpretation, and instead we suggest it is simply evidence for the intermediacy of Pd(0) in the cycle, regardless of whether Pd(0) particles are active for complete catalytic turnovers (2-3). Similarly, the presence of activity in solution after a filtration or split test can be taken as evidence for homogeneous catalysis, but the absence of activity after such a step is not sufficient proof for a lack of catalysis by soluble species (2,3,16), owing to the rapid redeposition that can occur as noted above (2,18). As noted above, the same thing seems to hold true for the 3-phase test.

In the 3-phase test, one reagent is immobilized on a solid support and the catalyst is immobilized on a second, different solid support. The rest of the reagents are then added to the solution. Ideally, if there is no background reaction in the absence of catalyst, the immobilized reagent should only be converted if there is catalyst or substrate leaching. In some instances the lack of conversion or only a very limited conversion of the immobilized reagent is taken as evidence for catalysis by a heterogeneous pathway when there is a concurrent, high activity of a soluble reagent. However, this may not be rigorously correct. Ideally, an additional control experiment should be performed in which a soluble metal precatalyst, known to be active for the specific transformation, is used to catalyze the conversion of the surface bound reagent (e.g., an aryl halide) both in the presence and absence of a soluble reagent (e.g., another aryl halide). Thus, in order to conclude that a lack of activity of surface bound aryl halide is a proof of heterogeneous catalysis, one must also show that the surface bound reagent can be converted at all, under the employed conditions. It is possible and perhaps plausible that the decreased entropy and steric constraints associated with tethering the reagent will affect the rate of the reaction and may even make the tethered reagent unreactive. Assuming the tethered reagent is reactive, the rate of its conversion, especially relative to the rate of conversion of a soluble reagent and to the catalyst lifetime, must be considered. In particular, one must consider whether there is time to diffuse to the bound reagent and react before side product formation, re-deposition of soluble catalyst onto the support, or catalyst deactivation.

With regards to Pd catalyzed reactions, a soluble aryl halide is often added to both demonstrate catalysis and possibly to both promote the leaching pathway and stabilize the soluble Pd via oxidative coupling to the Pd(0) (40). When a soluble reagent is added that competes with the immobilized reagent, then the turnover frequency of the surface-bound reagent could be dramatically retarded until the soluble component is fully converted. Even then the reaction of the surface-bound reagent could be slower than normal due to the reduced concentration of the other coupling reagent (e.g., a boronic acid in Suzuki couplings). When all of these subtleties of the three-phase test are considered, it becomes apparent that it can be difficult to interpret the negative of the test, a lack of conversion of the tethered reagent, as proof of heterogeneous catalysis. We suggest that only the positive of the test can be conclusively used to demonstrate that leaching of active metal does occur, similar to the case of the filtration test (2).

Similarly, there are numerous factors that must be considered when choosing a solid poison and the conditions under which it will be applied. For example:

1. Is the solid poison stable under the proposed conditions?

    Most commercial palladium scavengers are designed for use after a reaction – thus at low temperatures. Is the solid poison being considered stable at the temperatures of the reaction? Many Heck couplings are used at well over 100°C and some polymeric poisons may break down under these conditions. This could lead to ineffective poisoning of solution species as well as unwanted interaction of small molecule degradation products with supported palladium species.

2. What loading of solid poison is needed to bind all the free palladium?

    Of the solid poisons we have used, SiO_2-SH and QTU seem to bind palladium more effectively than PVPy – that is, they have significantly different palladium binding constants. Furthermore, all the solids have different functional group loadings. Thus, determining how much of each poison to add involves some trial and error at the early stages of testing. There is no fixed loading of scavenger to use in all cases.

3.   Does this level of loading drastically change the system?

Given the loading of solid poison that will be used, consider whether this will drastically alter your reaction conditions. Is the system still well-mixed? Are there phase separation problems? Does the viscosity change appreciably when adding the poison, as was observed by one group? (23)

4.   Are all the sites accessible in the media used?

Not all solid poisons will likely be useful in all media. Many polymeric resin-based poisons swell differently in various solvents. In one solvent, all the sites may be accessible, whereas in others, as little as a 1% of the sites, those on the external surface, for example, may be available.

5.   Will the solid poison react with the reagents or solvent under the conditions of use?

Some of the poisons we have used and that others may consider using may react with the reagents in the presence or absence of palladium catalysts. One of the advantages of PVPy, then, is its relative inertness to reactions with typical reagents used in coupling reactions. In contrast, the thiol groups on SiO_2-SH could react with aryl halides in copper or palladium catalyzed C-S bond forming reactions under some conditions. The influence of such side reactions must be ruled out before a solid such as SiO_2-SH can be used as a poison. Similarly, copper and palladium are also useful in promoting C-N and C-O bond forming reactions, so poisons with OH and NH groups need to be considered carefully. Finally, some reagents, such as Grignards (RMgX) in Kumada couplings are highly reactive with such a wide array of functional groups, that use of the solid phase poisoning test for the identification of soluble catalytic species may not be useful under any conditions (43).

Conclusions

Assessing the nature of the true catalyst in palladium catalyzed Heck, Suzuki and related reactions is important for both scientific and practical reasons. Understanding the nature of the truly active species can help in the design of more effective, next generation precatalysts. In addition, knowledge about palladium dynamics and leaching can help the engineer assess what is the best methodology to achieve high product yield in a short time while minimizing cost and metal contamination of the products.

A variety of techniques are available for elucidating the heterogeneity/homogeneity of reactions using supported palladium precatalysts (2). In all cases, it is suggested that a variety of these tests be used together to achieve the clearest possible picture of catalyst heterogeneity/homogeneity. Recently (12), we added the use of solid phase poisons that are ideally selective for only soluble catalytic species to the array of available heterogeneity/homogeneity tests. This approach is unique in its ability, in principle given the caveats noted above, to affect only metal species in solution. We suggest that due to the simplicity of the experiment (e.g., compared to a traditional 3-phase test) use of such solid phase poisons should be adopted as a standard component of heterogeneity/homogeneity testing protocols.

Acknowledgements

We thank the U.S. Department of Energy for funding via Catalysis Science contract # DE-FG02–03ER15459. CWJ also thanks DuPont for a Young Faculty Grant and ChBE at Georgia Tech for support through the J. Carl & Shiela Pirkle Faculty Fellowship.

References

1. C.E. Garrett, K. Prasad, *Adv. Synth. Catal.* **346**, 889 (2004).
2. N.T.S. Phan, M. Van Der Sluys, C.W. Jones, *Adv. Synth. Catal.* **348**, 609 (2006).
3. J.M. Richardson, C.W. Jones, *Adv. Synth. Catal.* **348**, 1207 (2006).
4. K. Kohler, W. Kleist, S. S. Prockl, *Inorg. Chem.* **46**, 1876 (2007).
5. A.F. Shmidt, L.V. Mametova, *Kinet. Catal.* 37, 406 (1996).
6. S. P. Andrews, A. F. Stepan, H. Tanaka, S. V. Ley, M. D. Smith, *Adv. Synth. Catal.* **347**, 647 (2005).
7. J.G. de Vries, *Dalton Trans.* 421 (2006).
8. M. T. Reetz, J. G. de Vries, *Chem. Commun.* 1559 (2004).
9. A.H.M. de Vries, J. Mulders, J.H.M. Mommers, H.J.W. Henderickx, J.G. de Vries, *Org. Lett.* **5**, 3285 (2003).
10. M. Weck, C. W. Jones, *Inorg. Chem.* **46**, 1865 (2007).
11. J.A. Widegren and R.G. Finke, *J. Mol. Catal. A.* **198**, 317 (2003).
12. K.Q. Yu, W. Sommer, M. Weck, C.W. Jones, *J. Catal.* **226**, 101 (2004).
13. K.Q. Yu, W. Sommer, J.M. Richardson, M. Weck, C.W. Jones, *Adv. Synth. Catal.* **347**, 161 (2005).
14. W.J. Sommer, K.Q. Yu, J.S. Sears, Y.Y. Ji, X.L. Zheng, R.J. Davis, C.D. Sherrill, C.W. Jones, M. Weck, *Organometallics* **24**, 4351 (2005).
15. C.W. Jones, M. Holbach, J.M. Richardson, W. Sommer, M. Weck, K.Q. Yu, X. Zheng , in *Catalysis of Organic Reactions* (CRC Press), 3 (2007).
16. J.M. Richardson, C.W. Jones, *J. Catal.* **251**, 80 (2007).
17. S. Klingelhöfer, W. Heitz, A. Greiner, S. Osetreich, S. Förster, M. Antonietti, *J. Am. Chem. Soc.* **119**, 10116 (1997).
18. S. Tasler, B.H. Lipshutz, *J. Org. Chem.* **68**, 1190 (2003).
19. Y.Y. Ji, S. Jain, R.J. Davis, *J. Phys. Chem. B.* **109**, 17232 (2005).

20. A. M. Caporusso, P. Innocenti, L. A. Aronica, G. Vitulli, R. Gallina, A. Biffis, M. Zecca, B. Corain, *J. Catal.* **234**, 1 (2005).
21. O. Aksm, H. Turkmen, L. Artok, B. Cetinkaya, C. Ni, O. Buyukgungor, E. Ozkal, *J. Organomet. Chem.* **291**, 3027 (2006).
22. G. Durgun, O. Aksm, L. Artok, *J. Mol. Catal. A.* **278**, 189 (2007).
23. W. J. Sommer, M. Weck, *Adv. Synth. Catal.* **348**, 2101 (2006).
24. C. K. Y. Lee, A. B. Holmes, S. V. Ley, I. F. McConvey, B. Al-Duri, G. A. Leeke, R. C. D. Santos, J. P. K. Seville, *Chem. Commun.* 2175 (2005).
25. S. V. Ley, C. Ramarao, R. S. Gordon, A. B. Holmes, A. J. Morrison, I. F. McConvey, I. M. Shirley, S. C. Smith, M. D. Smith, *Chem. Commun.* 1134 (2002).
26. *ChemFiles* **4**, (2004).
27. F. Y. Zhao, K. Murakami, M. Shirai, M. Arai, *J. Catal.* **194**, 479 (2000).
28. J. P. Simeone, J. R. Sowa, *Tetrahedron*, **63**, 12646 (2007).
29. S. J. Broadwater, D. T. McQuade, *J. Org. Chem.* **71**, 2131 (2006).
30. V. N. Losev, Y. V. Kudrina, N. V. Maznyak, A. K. Trofimchuk, *J. Anal. Chem.* **58**, 124 (2003).
31. M. Parisien, D. Valette, K. Fagnou, *J. Org. Chem.* **70**, 7578 (2005).
32. K. Shimizu, S. Koizumi, T. Hatamachi, H. Yoshida, S. Komai, T. Kodama, Y. Kitayama, *J. Catal.* **228**, 141 (2004).
33. C.M. Crudden, M. Sateesh, R. Lewis, *J. Am. Chem. Soc.* **127**, 10045 (2005).
34. C.M. Crudden, K. McEleney, S.L. MacQuarrie, B. A., M. Sateesh, J.D. Webb, *Pure Appl. Chem.* **79**, 247 (2007).
35. J. M. Richardson, C. W. Jones, *J. Catal.* **251**, 80 (2007).
36. J. D. Webb, S. MacQuarrie, K. McEleney, C. M. Crudden, *J. Catal.* **252**, 97, (2007).
37. A. S. Gruber, D. Pozebon, A. L. Monteiro, J. Dupont, *Tetrahedron Lett.* **42**, 7345 (2001).
38. R. K. Arvela, N. E. Leadbeater, M. S. Sangi, V. A. Williams, P. Granados, R. D. Singer, *J. Org. Chem.* **70**, 161 (2005).
39. S. Tasler, B.H. Lipshutz, *J. Org. Chem.* **68**, 1190 (2003).
40. I. W. Davies, L. Matty, D. L. Hughes, P. J. Reider, *J. Am. Chem. Soc.* **123**, 10139 (2001).
41. J. Rebek, F. Gavina, *J. Am. Chem. Soc.* **96**, 7112 (1974).
42. J. P. Collman, K. M. Kosydar, M. Bressan, W. Lamanna, T. Garrett, *J. Am. Chem. Soc.* **106**, 2569 (1984).
43. C. Baleizao, A. Corma, H. Garcia, and A. Leyva, *J. Org. Chem.* **69**, 439 (2004).
44. J. M. Richardson, C. W. Jones, manuscript submitted.

23.   Unusual Isolated Pre-Catalyst Systems Using TaniaPhos/MandyPhos Ligands

Oliver Briel[2], Angelino Doppiu[1], Ralf Karch[1], Christophe Le Ret[2], Roland Winde[1] and Andreas Rivas Nass[1]

[1]*Umicore AG & Co. KG, Strategic Research & Development, Rodenbacher Chaussee 4 D-63403 Hanau, Germany*
[2]*Umicore AG & Co. KG, Precious Metals Chemistry, Rodenbacher Chaussee 4 D-63403 Hanau, Germany*

andreas.rivas-nass@eu.umicore.com

Abstract

The catalytic behaviour of MandyPhos (**1**) and TaniaPhos (**2**) ligands (see structures in Scheme 23.1) in chiral hydrogenations was explored in former publications by applying *in situ* catalytic procedures (1).

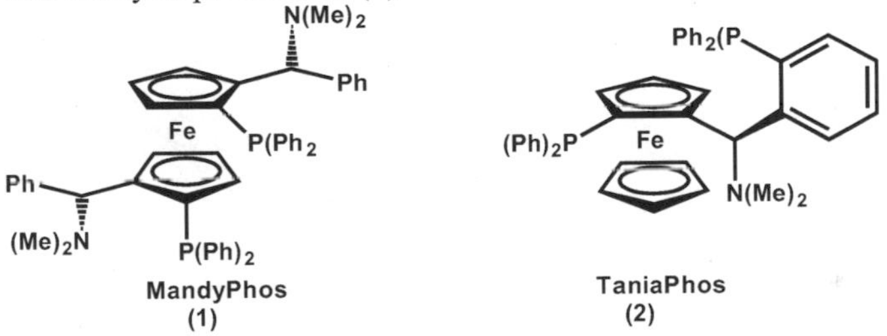

Scheme 23.1: MandyPhos (**1**), Taniaphos (**2**).

The present study investigates the structural properties of the active catalytic species. Emphasis is given to the new TaniaPhos- and MandyPhos-based ruthenium complexes (Scheme 23.2) and their relevance in chiral hydrogenations.

When using a standard [Ru(p-cymene)Cl$_2$]$_2$ precursor both the MandyPhos and TaniaPhos systems generated a variety of complexes. These ligands do not

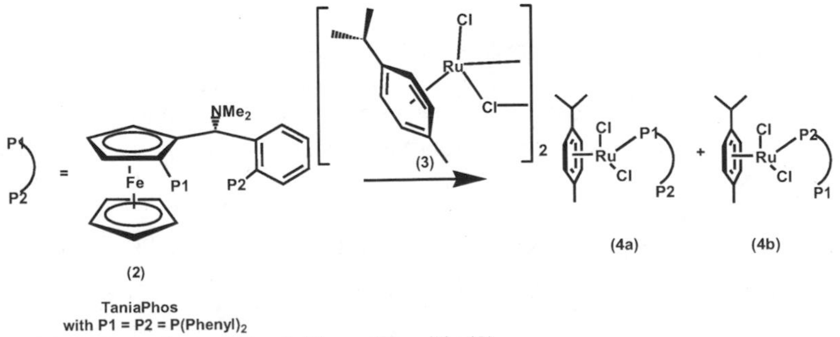

Scheme 23.2: Reaction of TaniaPhos (**2**) with (**3**).

always form phosphine bidentate complexes with ruthenium, even though they are known to react to either bidendate and/or tritendate products, as shown with Pd- and Rh-systems (1, Knochel). Furthermore, unusual transition metal complex structures and chemo- and regio-selective behaviour were observed.

When using commercially available standard precursor catalyst compounds, several complexes were observed. Precise structural identities could not be assigned to these mixtures.

As shown in Scheme 23.3, the use of rather uncommon ruthenium(II) precursors (2) led to formation of defined ruthenium complexes where both phosphorous atoms of (1) and (2) can bind to the metal center:

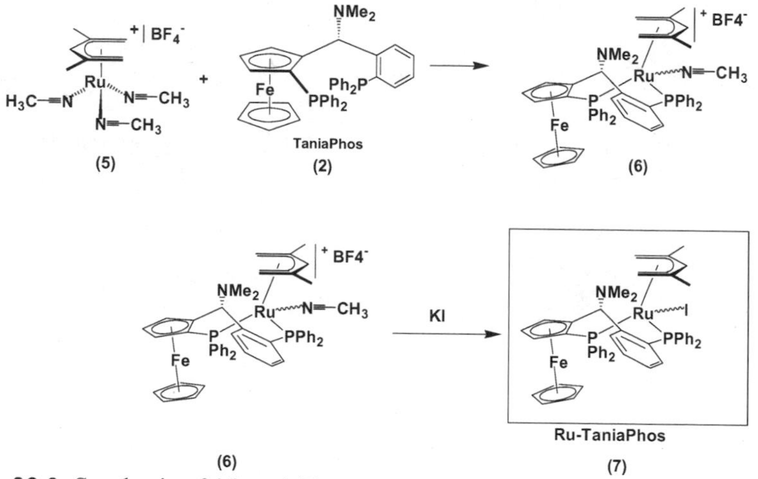

Scheme 23.3: Synthesis of (6) and (7).

Starting with (5) and (2), it was possible to isolate and characterize two stable unusual complexes. This supports the premise of a bidentating structure. In comparison to monodentating complexes, (4a and 4b) those new complexes are expected to behave differently than that of catalysts mentioned in literature (1).

Introduction

Chiral phosphine based transition metal complexes are used as a powerful tool for asymmetric synthesis (3). A fundamental mechanistic understanding is required for rhodium and ruthenium catalyzed reactions. The starting point of those investigations was the clear and detailed structural description of the isolated pre catalyst system.

The coordination chemistry of MandyPhos and TaniaPhos with rhodium and palladium has been explained by Knochel (1, Knochel). The chemistry of ligands such as Duphos or BINAP with all relevant established ruthenium(II) precursors is known and was derived from their bidentating coordination behaviour;

and its implication towards their catalytic behaviour was also investigated (3, 4, 5). In these cases, well-defined structures are known. However, performance data of the chiral ligands MandyPhos and TaniaPhos do not give insights on the active catalytic species in case of ruthenium-based catalytic applications (1). And up to now no defined ruthenium chemistry of the ligands MandyPhos and TaniaPhos was established nor any catalytically active pre-formed species is known, although proposed. This study provides the synthetic routes and characterization of stable ruthenium-based complexes with MandyPhos and TaniaPhos starting from the well-established chemistry of Duphos or BINAP systems and related literature. The synthesis did not yield stable TaniaPhos- and MandyPhos-based complexes. Rather, unusual coordination chemistry. Further investigation provide evidence of stable complexes that have unusual chemistry at industrial scale.

Experimental Section

Reagents and dry solvents were purchased from Aldrich or VWR and used without further treatment. Synthetic preparation of the complexes was done using standard Schlenk techniques (dry argon) unless otherwise stated. Standard ruthenium(II) precursors were prepared as described in the literature or via internal routes.

NMR spectra at r.t. were recorded on a Bruker AV300 spectrometer, with residual protons of deuterated solvents (^{1}H, relative to external $SiMe_4$); ^{31}P data are reported relative to 85% aq. H_3PO_4.

X-ray single crystal analysis: Data collection. Siemens SMART 1K CCD. Diffractometer. Data collection by SMART (Siemens, 1995); cell refinement: SMART; data reduction: SAINT (Siemens, 1995); program(s) used to solve structure: SHELXS97 (Sheldrick, 1997); program(s) used to refine structure: SHELXL97 (Sheldrick, 1997); molecular graphics: XP in SHELXTL (Sheldrick, 1996); software used to prepare material for publication: SHELXL97 (Sheldrick, 1997).

Mass spectra: Finnigan MAT 95, activation energy 70 eV.

Results and Discussion

MandyPhos: Initial attempts to synthesize defined chelating biphosphine complexes by reacting (1) with established Ruthenium(II) precursors such as $[Ru(COD)Cl_2]_x$, $[Ru(benzene)Cl_2]_2$, $[Ru(p\text{-cymene})Cl_2]_2$, $(COD)Ru(Methylallyl)_2$, $(COD)Ru(acetate)_2$ and $(COD)Ru(trifluoroacetate)_2$ were not successful. In all cases, full conversion of the ligand with the precursor was determined via NMR, leading to mixtures of several, partly unidentified species. Furthermore, ^{31}P-NMR experiments proved that in those cases no formation of any PP-coordinated complexes was achieved.

TaniaPhos: In comparison to MandyPhos (1) the structural properties of TaniaPhos (2) gave reason to other synthetic behaviour in those reactions. In spite of that fact, no defined complexes could be observed or isolated when reacting TaniaPhos (**2**) with Ruthenium(II) precursors such as [Ru(COD)Cl$_2$]$_x$, [Ru(benzene)Cl$_2$]$_2$, [Ru(p-cymene)Cl$_2$]$_2$, (COD)Ru(Methylallyl)$_2$, (COD)Ru(acetate)$_2$ and (COD)Ru(trifluoroacetate)$_2$ See Scheme 23.4. In all cases (including when varying solvent, temperature, reaction time) the final mixture consisted of unidentified complex species. Reaction of (**2**) with [Ru(benzene)Cl$_2$]$_2$ or [Ru(p-cymene)Cl$_2$]$_2$, lead to a different picture, as described below for [Ru(p-cymene)Cl$_2$]$_2$: For the first time defined complexes could be isolated with full conversion of the ligand in dichloromethane after several hours at r.t. or under heat (max. 50°C). Full conversion was determined via TLC (ligand as reference). Surprisingly the ^{31}P{^{1}H} NMR spectra showed an unexpected picture: four singlets were observed (^{31}P NMR (CD$_2$Cl$_2$) δ: 35 and –28 ppm for the minor isomer; 12 and –13 ppm for the major isomer) with no coupling of the P atoms via ruthenium giving rise to a monodentating complexation behaviour.

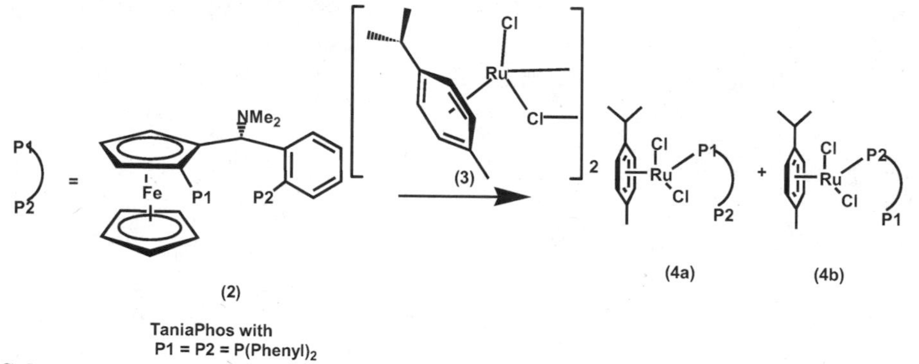

TaniaPhos with
P1 = P2 = P(Phenyl)$_2$

Scheme 23.4: Reaction of (**2**) with (**3**).

It was not possible to correlate specific signals with the proposed structures (**4a**) and (**4b**). The signal in the minus δ$_P$ area could not be dedicated to free unreacted ligand as full conversion via TLC was determined. Therefore, it is reasonable to assign the structure to a mono coordination of the ligand moieties to ruthenium; whereas, the second uncoordinated moiety shows singlets in the minus δ$_P$ area. Interestingly two species could be detected according to the different phosphorous atoms coordinated to the resulting two metal-complexes. Further investigation on a deeper structural description by x-ray single crystal analysis or ^{1}H NMR failed so far.

Precursors synthesis: The observed reaction of TaniaPhos (**2**) with [Ru(benzene)Cl$_2$]$_2$ or [Ru(p-cymene)Cl$_2$]$_2$ confirmed that defined and stable metal-complexes could also be synthesized with this ligand.

Further investigation of complexes using open ruthenocen (2, Salzer) species (Scheme 23.5), which had been noticed as possible precursors for synthesis of BINAP complexes (2) were considered. Several complexes were synthesized through procedures inspired from Salzer's work:

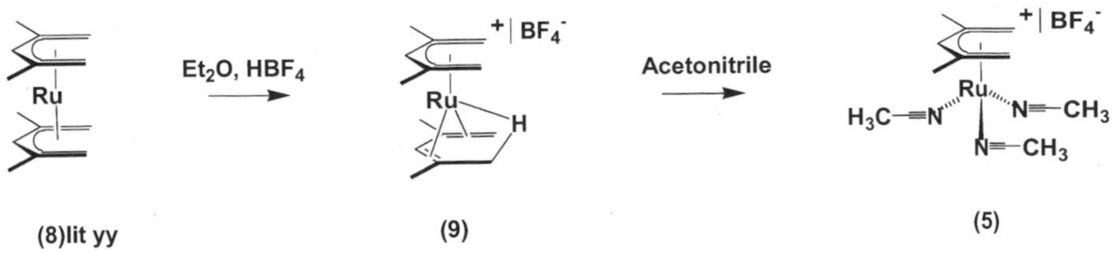

Scheme 23.5: Synthesis of **(9)** and **(5)**.

These precursors were then reacted with MandyPhos and TaniaPhos.

MandyPhos with (9) or (5): Reactions of MandyPhos with **(9)** or **(5)** lead to full conversion of the ligand (determined via TLC), but no defined complex could be observed or isolated. The route for MandyPhos was not investigated further.

TaniaPhos with (9) or (5): Reacting TaniaPhos with **(9)** in THF at r.t. resulted in amine coordination (Scheme 23.6):

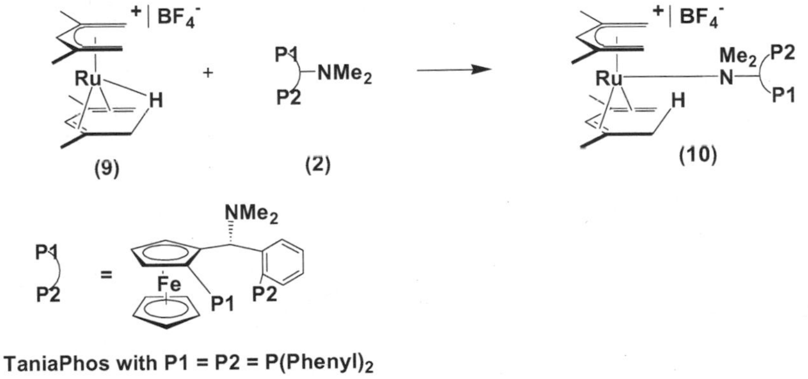

Scheme 23.6: Reaction of **(9)** with **(2)**.

As this result was disappointing TaniaPhos **(2)** was reacted with **(5)**. This reaction, run in THF at r.t., resulted for the first time in the formation with 90% yield of a complex where both phosphorous moieties are coordinated to the metal. Complex was characterized via $^{31}P\{^{1}H\}$ NMR and ^{1}H NMR (2,88 ppm for the first: ^{1}H NMR (CD$_2$Cl$_2$) δ: -1.76 (b, 1H, *pentadienyl*), 0.15 (b, 1H, *pentadienyl*), 0.765 (t, 1H, $J = 6.41$ Hz, *pentadienyl*), 1.24 (b, 3H, *pentadienyl-CH$_3$*), 1.81 (s, 6H, N(CH$_3$)$_2$), 2.34 (m, 3H, *pentadienyl-CH$_3$*), 2.88 (s, 3H, *CH$_3$CN*), 3.63 (b, 1H), 3.66 (m, 1H), 3.76 (s, 5H), 4.39 (t, 1H, $J = 2.4$ Hz), 4.87 (m, 1H), 4.89 (m, 1H), 5.90 (s, 1H), 6.02 (t, 2H), 6.90-8.20 (m, 21H), 8.80 (dd, 1H, $J = 7.6$ Hz, $J = 14.3$ Hz). ^{31}P NMR (CD$_2$Cl$_2$) δ: 36.54 (d, J_{PP}= 34.6 Hz), 31.59 (d, J_{PP}= 34.6 Hz).

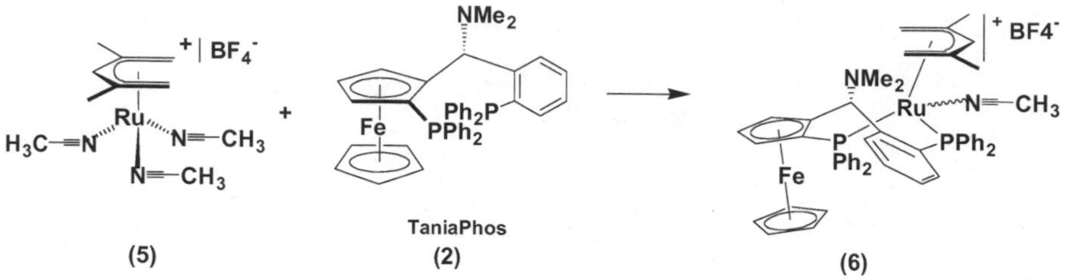

Scheme 23.7: Reaction of (**5**) with (**2**).

As we were interested in halogen containing neutral complexes further treatment of (**6**) with KI in acetone at r.t. during 12 h resulted in another complex where the coordinated acetonitrile was substituted by an iodide as shown in Scheme 23.8.

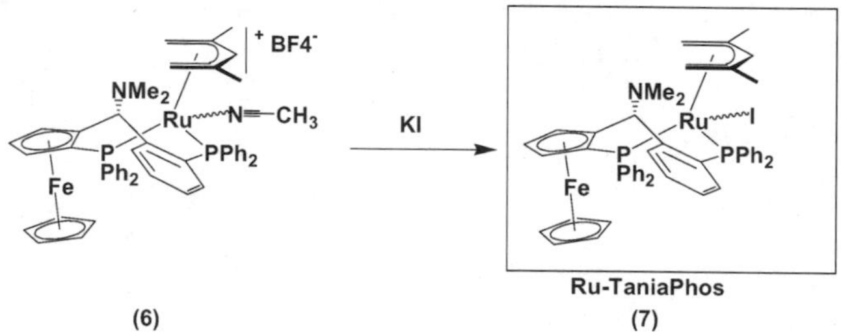

Scheme 23.8: Reaction of (**6**) with KI.

The complex was also analyzed via $^{31}P\{^{1}H\}$ NMR, ^{1}H NMR and mass spectra: ^{1}H NMR (CD$_2$Cl$_2$, major) δ: -2.06 (m, 1H), -0.13 (m, 1H), 0.93 (m, 1H), 1.29 (s, 3H), 1.95 (s, 6H), 2.81 (s, 3H), 3.28 (m 1H), 3.76 (m, 1H), 3.80 (s, 5H), 4.17 (m, 1H), 4.75 (m, 1H), 5.54 (m, 1H), 6.07 (m, 2H), 6.6-7.8 (m, 21H), 9.25 (m, 1H), 9.63 (m, 1H). ^{31}P NMR (CD$_2$Cl$_2$) δ: 39.88 (d, J_{PP}= 40.5 Hz), 29.96 (d, J_{PP}= 40.5 Hz). MS (FAB) *m/z*: 1012 [M$^+$+1], 884 [M$^+$ - I], 788 [M$^+$ - I - pentadienyl].

We could identify that both phosphorous atoms stay coordinated to the metal (PP coupling among metal at 40,5 Hz). Crystallization of the complex in CHCl$_3$/hexane allowed an x-ray single crystal analysis that clearly demonstrates the coordination chemistry as shown above.

TaniaPhos active catalyst discussion: As shown by Salzer (2) such complexes with half sandwich structure result in the catalyst cycle into a hydride species where the pentadienyl moiety can be hydrogenolytically liberated (2, 6). This was verified in the case of BINAP complexes (2, diss. Podewils, Geyser). In accordance to this fact and other mechanistic aspects from Noyori's work (3, 5) it is likely that the pre-catalyst species undergoes the same reaction pathway and that the reactive part of the pre-catalyst, the pentadienyl moiety, will be liberated under hydrogenolytic conditions as shown below in Scheme 23.9:

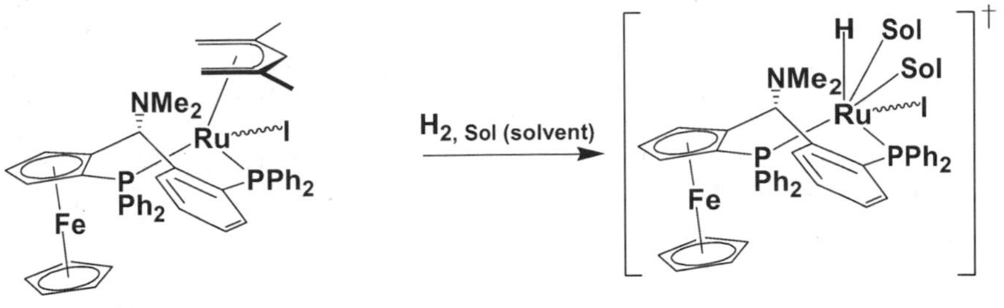

(7)   (11)

Scheme 23.9: Proposal reaction of (**7**) with H_2 according to (2, diss. Podewils, Geyser).

The resulting complex would be a hydride species with two further solvent molecules. Further studies are required to identify the structure of this species (**11**).

Conclusions

It was not possible to isolate defined complexes with clear bidentate coordination for the ligands MandyPhos and TaniaPhos using established ruthenium precursors such as [Ru(COD)Cl_2]$_x$, [Ru(benzene)Cl_2]$_2$, [Ru(p-cymene)Cl_2]$_2$, (COD)Ru(Methylallyl)$_2$, (COD)Ru(acetate)$_2$ and (COD)Ru(trifluoroacetate)$_2$. For the cases where [Ru(benzene)Cl_2]$_2$, [Ru(p-cymene)Cl_2]$_2$ were reacted with TaniaPhos, mixtures of complexes showing mono coordination behaviour for one phosphorous moiety of the ligand were identified.

For ruthenium, special precursors are required to synthesize defined bidentate diphosphine complexes. With Taniaphos for instance, it is possible to synthesize such complexes starting from unusual ruthenium(II) species. The complexes were characterized by NMR and single crystal analysis.

Substantial differences in the catalytic performance of such complexes were expected, when comparing their behaviour with reported catalytic data employing in situ procedures (1). This study focused on the performance of (**6**), (**7**) and the determination the structure of (**11**).

Acknowledgements

The authors would like to thank Professor Dr. A. Salzer (RWTH Aachen, Germany) for cooperating with us in the field of organometallic synthesis of ruthenium complexes.

Additionally, the authors would like to thank Dr. Bats (University of Frankfurt am Main, Germany) for preparing x-ray single crystal analysis.

References

1. Knochel, *Angew. Chem.* 1999, **111**, **Nr. 21** 3397-3400; Schwink, *Dissertation University of Munich*, 1999; Knochel, *Angew. Chem.* 2002, **114, Nr. 24**, 4902-4905; Knochel, *Chem. Eur. J.* 2002, **8, No. 4**, 843-852; Knochel, *Tetrahedron Asymmetry* **15** (2004) 91–102; Spindler *Adv. Synth. Catal.* 2003, **345**, 160-164; Spindler, *Tetrahedron Asymmetry* **15** (2004) 2299–2306.

2. Podewils, F., *Diss. RWTH Aachen*; **2000** Geyser, S., *Dissertation RWTH Aachen*, **2003**; Salzer, A., Geyser, S., Bauer, A., Podewils, F., *Organometallics*, **2000**, *19*, 5471-5476; Cox, D.N., Roulet, R., *Chem.Comm.*, **1989**, 175-176; Ernst, R.D., *Acc. Chem. Res.*, **1985**, *18*, 56-62; Ernst, R.D., Stahl L., Ziegler M.L., *Organometallics*, **1990**, *9*, 2962-2972; Pertici, P., Vitulli, G., Paci, M., Porri, L., *Dalton Trans.*, **1980**, 1961-1964; Cox, D.N., Roulet, R., Lumini, T., Schenk, K., *J.Organometallic Chem.*, **1992**, *434*, 363-385.

3. J. M. Brown, Hydrogenation of functionalized carbon–carbon double bonds. In: E. N. Jacobsen, A. Pfaltz, H. Yamamoto (Eds.), *Comprehensive Asymmetric Catalysis*, Springer, Berlin, 1999, Chapter 5.1, pp. 121–182; U. Nagel, J. Albrecht, *Topics Catal.* **5** (1998); M. J. Burk, F. Bienewald, Unnatural a-amino acids, via asymmetric hydrogenation of enamides. In: M. Beller, C. Bolm (Eds.), *Transition Metals for Organic Synthesis*, **Vol. I**, Wiley-VCH, Weinheim, 1998, Chapter 1.1.2, pp. 13–25; H. Brunner, Hydrogenation. In: B. Cornils, W.A. Herrmann (Eds.), Applied Homogeneous Catalysis with Organometallic Compounds, vols. I and II, VCH, Weinheim, 1996, Chapter 2.2, pp. 201–219; R. Noyori, *Asymmetric Catalysis in Organic Synthesis*, Wiley, New York, 1994, Chapter 2, pp. 16–94; P. A. Chaloner, M. E. Esteruelas, F. Joo, L.A. Oro, *Homogeneous Hydrogenation*, Kluwer Academic Publishers, Dordrecht, 1994; H. Takaya, T. Ohta, R. Noyori, Asymmetric hydrogenation. In: *I. Ojima (Ed.), Catalytic Asymmetric Synthesis*, VCH, New York, 1993, Chapter 1, pp. 1–39.

4. Kagan et al., *JACS*, 1972, **94**, 6429. Knowles, *JACS* 1977, **99**, 5946. Noyori et al., *JACS* 1986, **108**, 7117.

5. Landis, Halpern, *JACS*, 1987, **109**, 1746; Noyori, *Acc. Chem Res.* 1997, **30(2)**, 97 and ref. cited.

6. J. A. Wiles, S. H. Bergens, *Organometallics* **1996**, *15*, 3782–3784; T. Lumin, D. N. Cox, *J. Organomet. Chem.* **1992**, *434*, 363; J. A. Wiles, S. H. Bergens, *Organometallics* **1998**, *17*, 2228–2240; J. A. Wiles, S. H. Bergens, *J. Am. Chem. Soc.* **1997**, *119*, 2940–2041.

24.   New Developments in the Synthesis and Application of Chiral Phospholane Ligands

Renat Kadyrov[1] and John Tarabocchia[2]

[1]*Evonik Degussa GmbH, Rodenbacher Chaussee 4, 63457 Hanau-Wolfgang, Germany*
[2]*Evonik Degussa Corp., 379 Interpace Parkway, Parsippany, NJ 07054*

renat.kadyrov@evonik.com

Abstract

Among electron-rich chiral phosphines, chiral phospholanes have emerged to be one of the most efficient classes of ligands in metal catalyzed enantioselective reactions. We have developed a novel family of bisphospholane ligand namely, catASium M, from laboratory to commercial scale. Trimethylsilylphospholane **1** was employed as a key intermediate to provide access to a large variety of ligands.

The enantiopure 1-chloro-2,5-dimethylphospholane **2** is now available from the corresponding 1-trimethylsilylphospholane **1**. The new phospholane **2** was used as an electrophilic building block in a wide range of coupling reactions giving rise to new phospholanes. These proved to be valuable as chiral ligands in transition metals catalysis with Rh, Ir or Ru complexes.

Introduction

Since the first revolutionary discovery and industrial application of chiral diphosphines in the late 1970s by Knowles, a huge number of chiral phosphine ligands have been synthesized and applied to catalytic asymmetric reactions on industrial scale. Among electron-rich chiral phosphines, chiral phospholanes have emerged to be one of the most efficient classes of ligands in metal catalyzed enantioselective reactions [1]. Development of the bisphospholane DuPHOS by Burk [2] provided a breakthrough and revealed in turn the great potential of these compounds as ligands in asymmetric catalysis. Numerous analogues of related ligands bearing phospholane moiety have been prepared by different academic and industrial groups [3-5]. The synthetic access to most of these ligands is based on the original procedure of Burk, utilizing double nucleophilic substitution of chiral 1,4-sulfates with the corresponding primary diphosphines in the presence of a strong base. This tedious approach is restricted by a low tolerance to functional groups and, therefore, greatly limits the possible variations of the backbone.

EVONIK DEGUSSA GmbH has developed the modular synthesis of diverse arrays of chiral vicinal bisphospholanes in collaboration with the Leibniz-Institut of Organic Catalysis in Rostock [6]. Using trimethylsilylphospholane **1** as a common precursor we generated a large family of catASium® M ligands which are differentiated by varying "bite-angle" and electronic properties of the bridge unit [7].

In this approach 2,5-dimethyl-1-trimethylsilylphospholane **1** acting as nucleophile can be coupled only with activated unsaturated 1,2-dihalogenides. Recently we described the convenient access to the enantiopure 1-chloro-2,5-dimethylphospholane **2** based on the chlorination of the corresponding 1-trimethylsilylphospholane **1** [8]. This new compound **2** available in large amounts can be used as an electrophilic building block for construction of the wide range of new types of phospholanes. We describe herein the synthesis of the new unsymmetrical phospholane ligands and the first application of these ligands in asymmetric catalysis.

Results and Discussion

Lithiation of bromo-substituted aromatic and heteroaromatic derivatives with n-BuLi and subsequent treatment with the chiral chlorophospholane **2** at lower temperature produces the corresponding 2,5-dimethylphospholanes (**3-6**) in good yields (Scheme 24.1). A two-step approach was used to prepare the diphospholane **3** without isolating the intermediate 1-bromo-2-(2,5-(*R*)-dimethylphospholanyl-1)-acenaphthylene.

The efficiency of the new ligands was examined in enantioselective hydrogenation of some prochiral substrates. Itaconic ester hydrogenation using in situ prepared Rh-complexes was the first test reaction chosen. The best results from the screening of various solvents are summarized in Table 24.1. High enantiomeric excesses of 96-99% were reached by employing ligands **3** and **4**. The hydrogenation of this substrate, via catalysts based on ligands **5-6** with thiophene backbone was less selective.

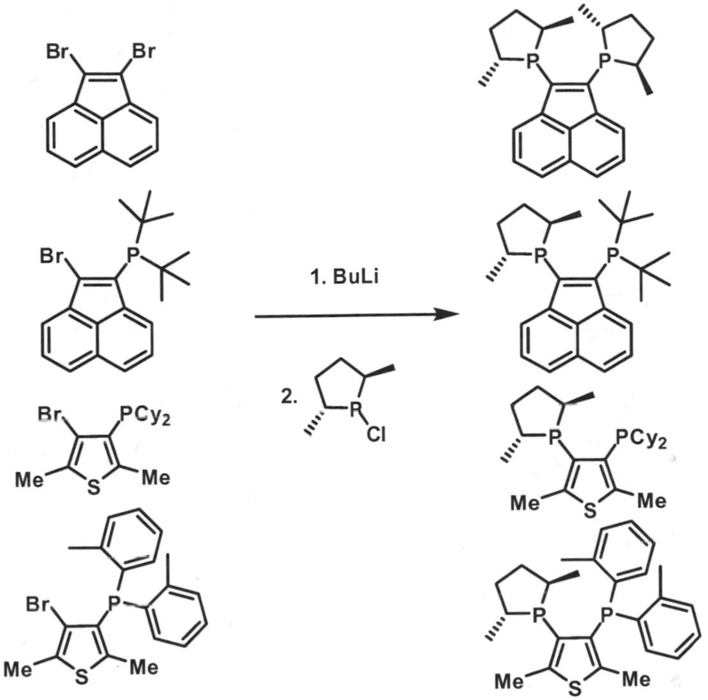

Scheme 24.1

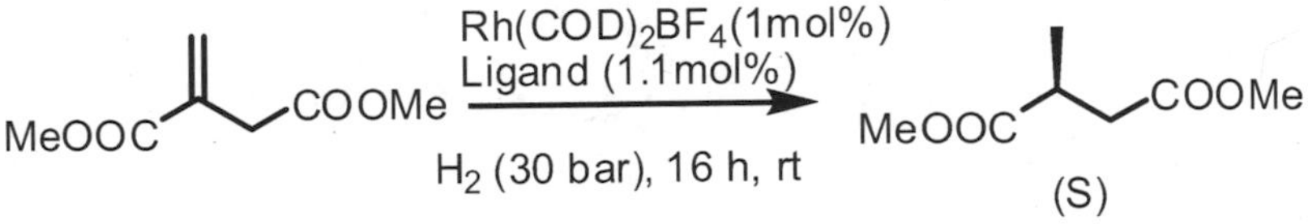

Table 24.1. Enantioselective hydrogenation of dimethyl itaconate using Rh(I)-complexes.

ligand	Solvent	ee (%)[a]
3	MeOH	99
3	THF	98
3	CH$_2$Cl$_2$	98
4	THF	96
5	CH$_2$Cl$_2$	59
6	CH$_2$Cl$_2$	90

[a] Enantiomeric excesses were determined by chiral HPLC on ChiralCel OD (Hexane:2-PrOH 95:5).

The new ligands were also tested in the hydrogenation of methyl (Z)-α-(N-acetamido)-4-chlorocinnamate. The results are summarized in Table 24.2. The symmetrical bisphospholane ligand **3** delivered chiral N-acylaminoacid methyl-ester with excellent enantioselectivities in all solvents studied. The unsymmetrical ligands **4-6** were less selective and sensitive to the choice of the solvent.

Table 24.2. Enantioselective hydrogenation of methyl (Z)-α-(N-acetamido)-4-chlorocinnamate using Rh(I)-complexes.

ligand	Solvent	ee (%)[a]
3	MeOH	98
3	THF	98
3	CH_2Cl_2	98
4	THF	91
5	THF	79
6	THF	74

[a] Enantiomeric excesses were determined by chiral HPLC on ChiralPak AD (Hexane:2-PrOH 75:25).

The most interesting feature of these ligands is their efficiency in the Ru-catalyzed high temperature hydrogenation of β-functionalized ketones. In contrast to the BINAP-type ligands, electron rich P-ligands were rarely used in Ru-catalyzed hydrogenations. There are only a couple of publications that discuss the use of i-Pr-BPE, BisP* and TangPhos in the Ru-catalyzed hydrogenation of substituted β-keto-esters [9].

The selected results of the β-keto ester hydrogenation applying Ru-catalysts with ligands **3-6** are summarized in Table 24.3. In all cases the reaction proceeded smoothly with a substrate to catalyst ratio of S/C = 100 at 80°C to give the corresponding alcohols in almost quantitative yield. High enantiomeric excesses (97-98%) were obtained with ligands **3, 4** and **6**, while the catalyst derived from **5** showed lower asymmetric induction. It is noteworthy that the efficiency of the catalyst is sensitive to the type of Ru-complex preparation. The Ru-complex prepared from metal precursor [Ru(C₄H₉)₂(cod)] generally delivered better enantiomeric excesses in the hydrogenation of methyl acetylacetoate and ethyl benzoylacetoate. No preference was observed in the metal precursor in the hydrogenation of the methyl acetylacetoate using catalyst based on the ligands **6**. On the other hand, the catalyst prepared from ligand **6** and benzene Ru-precursor ([Ru(C₆H₆)Cl₂]₂) leads to better selectivity in the hydrogenation of ethyl benzoylacetoate.

$$R{-}\underset{O}{\overset{\|}{C}}{-}CO_2R' \xrightarrow[50\ bar\ H_2,\ 80\,°C]{\begin{array}{c}[Ru]\ (1mol\%)\\ Ligand\ (1.1mol\%)\end{array}} R{-}\underset{OH}{\overset{|}{C}}H{-}CO_2R'$$

Table 24.3. Enantioselective hydrogenation of β-keto esters using Ru(II)-complexes.

Ligand	[Ru]	R	R'	Solvent	ee (%)[a]
3	[Ru(C$_4$H$_9$)$_2$(cod)]	Me	Me	MeOH	97
3	[Ru(C$_6$H$_6$)Cl$_2$]$_2$	Me	Me	MeOH	88
4	[Ru(C$_4$H$_9$)$_2$(cod)]	Me	Me	MeOH	98
5	[Ru(C$_4$H$_9$)$_2$(cod)]	Me	Me	MeOH	92
6	[Ru(C$_6$H$_6$)Cl$_2$]$_2$	Me	Me	MeOH	97
6	[Ru(C$_4$H$_9$)$_2$(cod)]	Me	Me	MeOH	98
3	[Ru(C$_6$H$_6$)Cl$_2$]$_2$	Ph	Et	EtOH	89
4	[Ru(C$_4$H$_9$)$_2$(cod)]	Ph	Et	EtOH	52
5	[Ru(C$_4$H$_9$)$_2$(cod)]	Ph	Et	EtOH	30
6	[Ru(C$_4$H$_9$)$_2$(cod)]	Ph	Et	EtOH	76
6	[Ru(C$_6$H$_6$)Cl$_2$]$_2$	Ph	Et	EtOH	98

[a] Enantiomeric excesses were determined by chiral GC on Chirsil Dex.

The results clearly show that these novel ligands are able to form a suitable asymmetric environment around the metal resulting in high asymmetric induction. Their catalytic potential has been demonstrated in the highly enantioselective Rh-catalyzed hydrogenation of itaconates and α-enamides and Ru-catalyzed hydrogenation of β-functionalized ketone.

References

1. Recent reviews (a) R. Kadyrov, A. Monsees in *Phosphorus Ligands in Asymmetric Catalysis* , Ed. A. Börner, 2008, Wiley-VCH, Weinheim (b) T. Clark, C. Landis, *Tetrahedron: Asymmetry* **2004**, *15*, 2123-2137; (c) M. Burk, *Acc. Chem. Res.* **2000**, *33*, 363-372; (d) Y. Yamanoi, T. Imamoto, *Rev. Heteroat. Chem.* **1999**, 227-248.
2. (a) M. J. Burk, *J. Am. Chem. Soc.* **1991**, *113*, 8518-8519; (b) M. J. Burk, J. E. Feaster, R. L. Harlow, *Tetrahedron: Asymmetry* **1991**, *2*, 569-592; (c) M. J. Burk, J. E. Feaster, W. A. Nugent, R. L. Harlow, *J. Am. Chem. Soc.* **1993**, *115*, 10125-10138.
3. (a) R. Sablong, C. Newton, P. Dierkes, J. Osborn *Tetrahedron Lett.* **1996**, *37*, 4933-4936; (b) P. Dierkes, S. Ramdeehul, L. Barloy, A. De Cian, J. Fischer, P. C. J. Kamer, P. W. N. M. van Leeuwen, J. A. Osborn, *Angew. Chem.* **1998**, *110*, 3299-3301; (c) J. Holz, M. Quirmbach, U. Schmidt, D. Heller, R. Stürmer, A. Börner, *J. Org. Chem.* **1998**, *63*, 8031-8034; (d) Q. Jiang, Y. Jiang, D. Xiao, P. Cao, X. Zhang, *Angew. Chem.* **1998**, *110*, 1203-1207; *Angew. Chem. Int. Ed. Engl.* **1998**, *37*, 1100-1103; (e) W. Li, Z. Zhang, D. Xiao, X. Zhang,

Tetrahedron Lett. **1999**, *40*, 6701-6704; (f) D. Carmichael, H. Doucet, J. M. Brown, *Chem. Commun.* **1999**, 261-262; (g) W. Li., Z. Zhang, D. Xiao, X. Zhang, *J. Org. Chem.* **2000**, *65*, 3489-3496; (h) E. Fernandez, A. Gillon, K. Heslop, E. Horwood, D. J. Hyett, A. G. Orpen, P. G. Pringel *Chem. Commun.* **2000**, 1663-1664; (i) Y.-Y. Yan, T. V. RajanBabu, *J. Org. Chem.* **2000**, *65*, 900-906; (j) J. Holz, R. Stürmer, U. Schmidt, H.-J. Drexler, D. Heller, H.-P. Krimmer, A. Börner, *Eur. J. Org. Chem.* **2001**, *24*, 4615-4624; (k) C. R. Landis, W. Jin, J. S. Owen, T. P. Clark, *Angew. Chem.* **2001**, *113*, 3540-3542; (l) D. Liu, W. Li, X. Zhang, *Org. Lett.* **2002**, *4*, 4471-4474; (m) G. Hoge, *J. Am. Chem. Soc.* **2003**, *125*, 10219-10227; (n) H. Shimizu, T. Saito, H. Kumobayashi, *Adv. Synth. Catal.* **2003**, *345*, 185-189; (o) G. Hoge, B. Samas, *Tetrahedron: Asymmetry* **2004**, *15*, 2155-2157; (p) A. Zhang, T. V. RajanBabu, *Org. Lett.* **2004**, *6*, 1515-1517; (q) W. Braun, B. Calmuschi, J. Haberland, W. Hummel, A. Liese, T. Nickel, O. Stelzer, A. Salzer, *Eur. J. Inorg. Chem.* **2004**, 2235-2243; (r) T. Benincori, T. Pilati, S. Rizzo, F. Sannicolo, M. J. Burk, L. de Ferra, E. Ulluci, O. Piccolo, *J. Org. Chem.* **2005**, *70*, 5436-5441; (s) D. Qian, C.-J. Wang, X. Zhang, *Tetrahedron* **2006**, *62*, 868-871; (t) U. Berens, U. Englert, S. Geyser, J. Runsink, A. Salzer, *Eur. J. Org. Chem.* **2006**, 2100-2109.

4.   (a) J.-C. Fiaud, J-Y. Legros, *Tetrahedron Lett.* **1991**, *32*, 5089-5092; (b) F. Guillen, M. Rivard, M. Toffano, J.-Y. Legros, J.-C. Daran, J.-C. Fiaud, *Tetrahedron* **2002**, *58*, 5895-5904; (c) C. J. Pilkington, A. Zanotti-Gerosa, *Org. Lett.* **2003**, *5*, 1273-1275; (d) C. Dobrota, M. Toffano, J.-C. Fiaud, *Tetrahedron Lett.* **2004**, *45*, 8153-8156; (e) M. Toffano, C. Dobrota, J.-C. Fiaud *Eur. J. Org. Chem.* **2006**, 650-656.

5.   K. Matsumura, H. Shimizu, T. Saito, H. Kumobayashi, *Adv. Synth. Catal.* **2003**, *345*, 180-184.

6.   J. Holz, A. Monsees, H. Jiao, J. You, I. V. Komarov, C. Fischer, K. Drauz, A. Börner, *J. Org. Chem.* **2003**, *68*, 1701-1707.

7.   (a) J. Holz, O. Zayas, H. Jiao, W. Baumann, A. Spannenberg, A. Monsees, T. H. Riermeier, J. Almena, R. Kadyrov, A. Börner, *Chem. Eur. J.* **2006**, *12*, 5001-5013. (b) T. H. Riermeier, A. Monsees, J. Holz, A. Boerner, J. Tarabocchia, *Chemical Industries* (Boca Raton, FL, United States) **2007**, *115* (Catalysis of Organic Reactions), 455-461.

8.   J. Holz, A. Monsees, R. Kadyrov, A. Boerner, *Synlett* **2007**, 599-602.

9.   (a) M. J. Burk, T.G.P. Harper, C.S. Kalperg, *J. Am. Chem. Soc.* **1995**, *117*, 4423-4424; (b) T. Yamano, N.Taya, M. Kawada, T. Huang, T. Imamoto, *Tetrahedron Lett.* **1999**, *40*, 2577-2580; (c) C.-J. Wang, H. Tao, X. Zhang, *Tetrahedron Lett.* **2006**, *47*, 1901-1903.

25. New Concepts in Designing Ruthenium-Based Second Generation Olefin Metathesis Catalysts and Their Application

Renat Kadyrov[1], Anna Rosiak[1], John Tarabocchia[2], Anna Szadkowska[3], Michał Bieniek[3] and Karol Grela[3]

[1]*Evonik Degussa GmbH, Rodenbacher Chaussee 4, 63457 Hanau-Wolfgang, Germany*
[2]*Evonik Degussa Corp., 379 Interpace Parkway, Parsippany, NJ 07054*
[3]*Institute of Organic Chemistry, Polish Academy of Sciences, Kasprzaka 44/52, 01-224 Warsaw, Poland*

renat.kadyrov@evonik.com

Abstract

Strong enhancement of metathesis reactions using "second-generation" Ru-carbene complexes bearing N-aromatic groups was observed by switching from non-aromatic to aromatic solvents. Fluorinated aromatic solvents as reaction media enabled the completion of very challenging metathesis transformations, such as the formation of tetra-substituted olefins utilizing standard commercially available ruthenium catalysts.

Further improvements in activity of the ruthenium carbene complexes were achieved by incorporation of methyl groups in 3,4-position of imidazol-2-ylidene moiety. Introduction of sulfur in the *trans*-position to the N-heterocyclic carbene leads to increased stability of the resulting ruthenium complexes. The synthesis and the first applications of these new ruthenium complexes are described herein.

Introduction

Olefin metathesis has emerged over the last half century as a powerful tool in chemistry opening unique industrial routes to new petrochemical, polymers and specialty chemicals. Particularly metathesis reactions have found extensive uses in various organic syntheses. In the last decades, thanks to the intensive development of well-defined and stable ruthenium carbenes, there has been an explosion in the number of published syntheses using metathesis as a key step [1]. Following significant breakthroughs in the ruthenium chemistry, a variety of ruthenium carbene catalysts have been prepared and used by many research teams as a convenient strategy in the synthesis of various organic compounds. Among them, "second generation" catalysts bearing N-heterocyclic carbene (NHC) ligands such as 1 exhibit improved activity, stability and an excellent application profile [2]. As an alternative to Ru-benzylidene initiators **1b** and **1c** the 3-phenylindenylidene complex **1a** is commercially available under the trade name catMETium® IMesPCy [3].

1a   **1b**   **1c**

During our efforts in profiling (comparative investigations) of several commercial available metathesis catalysts bearing NHC ligands in different types of metathesis reactions remarkable solvent effects were observed [4]. Interestingly, the efficiency of most transformations studied frequently depended more on solvent and temperature effects rather than on the nature of Ru precursor and NHC ligands.

Results and Discussion

Careful analysis of the screening results for a series of catalysts bearing saturated NHC **1a** and **1b**, and unsaturated NHC **2a-d** applied to a set of metathesis reactions showed remarkably higher activity in aromatic solvents such as benzene and toluene compared to other solvents at the same temperature. It should be noted that such an increased reactivity using aromatic solvents was also observed by Fürstner and Ledoux [5]. Furthermore, as reported by Fürstner, this effect is observable only for complexes bearing N-aryl groups. This was attributed to competing interactions of the N-aryl group and the aromatic solvent reducing the stabilizing effect of the intramolecular π–π interaction with the benzylidene moiety [5a]. An alternative explanation might be a π–π interaction between the N-aryl group that is not coplanar with the benzylidene moiety and the aromatic solvent [5b].

2a   **2b**   **2c**   **2d**

These facts and observations prompted our interest in a systematic investigation of metathesis reactions in fluorinated aromatic solvents as better acceptors in the π–π interaction [6]. Here we will outline some representative examples of this remarkable enhancement of activity in catalytic olefin metathesis applying "second-generation" ruthenium complexes.

The formation of tetra-substituted double bonds is one of the most challenging metathesis transformations with Ru catalysts. This transformation requires the application of "second-generation" catalysts at high loadings typically 5-10 mol%. As can be deduced from the results illustrated in Table 25.1, the replacement of 1,2-dichloroethane with toluene exhibits a strong effect on the turnover numbers (TON). Nevertheless, at the 2 mol% catalyst loading the yields are barely over 50%. By changing to a fluorinated reaction medium, nearly complete conversion can be achieved. Interestingly, for phosphine containing catalysts (**1a-2b**), we observed that TON increases directly with the number of fluorine atoms in the solvent molecule, reaching the maximum with perfluorotoluene. This enhancement is very important for practical applications. For example, in the case of catalyst **2c,** we observed an 18-fold increase in the reaction yield, *by only changing* the solvent from 1,2-dichloroethane to perfluorotoluene.

Table 25.1. GC-yields of **4a** in RCM of **3a** after 3 h at 70°C in presence of 2 mol% Ru

Solvent	1a	1b	2a	2b	2c	2d
ClCH$_2$CH$_2$Cl	38	n.d.	21	28	4	19
toluene	56	49	20	27	33	29
F$_3$-toluene	61	41	36	48	13	40
F$_6$-benzene	74	77	61	91	58	n.d.
F$_8$-toluene	85	n.d.	72	94	72	49

Table 25.2. GC-yields of **4b** in RCM of **3b** after 18 h at 70°C using 5 mol% Ru

Solvent	1a	2b	2c	2d
ClCH$_2$CH$_2$Cl	6	4	13	30
toluene	27	16	35	42
F$_3$-toluene	25	20	52	n.d.
F$_8$-toluene	63	55	58	60

The enyne metathesis transforming an alkene and an alkyne into diene is a valuable new synthetic tool. The enyne cyclo-isomerisation of **3b** has been established in the literature as a challenging transformation in this field. Again we observe a strong increase in catalyst activity by using aromatic solvents instead of chlorinated solvents. The results compiled in the Table 25.2 illustrate that increasing the degree of fluorination of the aromatic solvent generally leads to the higher yields. For example almost a 13-fold enhancement can be observed in the case of **2b** under otherwise identical conditions.

Further improvements in activity of the imidazol-2-ylidene Ru complexes might be attained by the incorporation of a better σ-donor substituents with larger steric requirements. The ligands that most efficiently promote catalytic activity are those that stabilize the high-oxidation state (14 e⁻) of the ruthenium metallacyclobutane intermediate [7]. Both ligand-to-metal σ-donation and bulkiness of the NHC force the active orientation of the carbene moiety and thus contribute to the rapid transformation into metallacyclobutane species [7b]. Both can be realized by incorporation of alkyl groups in 3,4-position of imidazol-2-ylidene moiety.

Starting from diacetyl we prepared and characterized new imidazolium salt **5**. After generation of the N-heterocyclic carbene in situ we formed new complex **1c**. This complex like the parent complexes **1a** and **1b** is stable in the presence of air and moisture.

The efficiency of catalyst **1c** was examined in the ring closing metathesis of challenging substrate **3c**. Full conversion of dimethallylamide (**3c**) was achieved after heating to 80°C for 2.5 h with a catalyst loading of 2.5 mol%. The larger scale run furnished 80% isolated yield of the desired dihydropyrrole **4c**.

The development of perfectly stable and thermally or chemically switchable initiators is of great interest in synthetic and polymer chemistry and related areas. The Ciba-group first developed very slowly initiated 2-pyridylethanyl chelating ruthenium carbene complexes [8]. In an alternative approach originally reported by Hoveyda, a styrenyl ether chelating ligand is only loosely bound resulting in a fast-initiating catalyst [9]. Unexpectedly, we have found that novel complexes formed by replacing the oxygen by sulfur in the Hoveyda-motif exhibit interesting properties.

These new sulfur bearing complexes can potentially be used as a thermo-switchable "latent" catalyst.

The preparation of the new compounds is straightforward, using recently established routes. As shown in the following scheme, the targeted complexes **7a** and **7b** were prepared by a carbene exchange reaction of **1a** or **2a** with a mercaptostyrol derivative **6**.

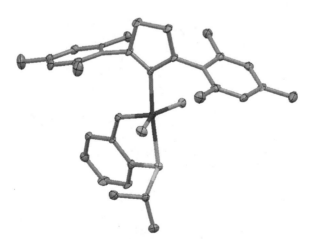

The solid state structure of complex **7b** is shown in Figure 25.1. Similar to parent chelating ether complex [9b], the solid-state structure of **7b** shows a distorted square-pyramidal structure with the benzylidene moiety at the apical position. The N-aryl ring is located above the benzylidene moiety resulting in the relatively close contact of the benzylidene proton with the π-aromatic system of the mesityl group.

Figure 25.1. X-ray crystal structure of **7b**. Hydrogen atoms are omitted for clarity.

The efficiency of the new catalysts **7a** and **7b** was examined in some metathesis transformations. Interestingly, both catalysts can only very slowly initiate typical metathesis reactions when tested at room temperature. At a catalyst loading of 5 mol%, chelating sulfur catalysts **7a** and **7b** achieved 22% and 51% conversion of diallylamide (**3d**), respectively, after heating to 80°C for 24 h. Catalyst **7a** converts

3e only to 13% of trisubstituted olefin **4e** in 24 h at 80°C, whereas catalyst **7b** achieves 28% conversion after 24 h at 80°C. Uninitiated catalyst was found for both **7a** and **7b** even at higher conversions, indicating that only a fraction of added catalyst is engaged in the reaction.

3d, A = NTs, R= H
3e, A = C(CO$_2$Et)$_2$, R = Me

4d, A = NTs, R= H
4e, A = C(CO$_2$Et)$_2$, R = Me

These results clearly show that these novel catalysts exhibit low metathesis activity at the studied temperatures and therefore can operate as "dormant" catalysts. This behavior is very important in the polymer industry.

References

1. Most recent reviews (a) J. C. Conrad, D.E. Fogg, *Current Org. Chem.* **2006**, *10*, 185-202. (b)) K. C. Nicolaou, P. G. Bulger, D. Sarlah, *Angew. Chem. Int. Ed.* **2006**, *45*, 4490–4527; (c) T. J. Donohoe, A. J. Orr, M. Bingham, *Angew. Chem. Int. Ed.* **2006**, *45*, 2664–2670; (d) A. Gradillas, J. Perez-Castells, *Angew. Chem. Int. Ed.* **2006**, *45*, 6086–6101.
2. (a) T. Weskamp, W.C. Schattenmann, M. Spiegler, W.A. Herrmann, *Angew. Chem. Int. Ed.*, **1998**, *37*, 2490-2493; b) T. Weskamp, F. Kohl, W. Hieringer, D. Gleich, W.A. Herrmann, *Angew. Chem. Int. Ed.*, **1999**, *38*, 2416-2419. (c) Fürstner, O. Guth, A. Düffels, G. Seidel, M. Liebl, B. Gabor, R. Mynott, *Chem. Eur. J.* **2001**, *7*, 4811-4820.
3. R. Kadyrov, J. Almena, A. Monsees, T. Riermeier, *Chemistry Today.* **2005**, *23*, 14-15.
4. M. Bieniek, A. Michrowska, D.L. Usanov, K. Grela, *Chem. Eur. J.* **2007** in press.
5. (a) A. Fürstner, L. Ackermann, B. Gabor, R. Goddart, C. W. Lehmann, R. Mynott, F. Stelzer, O.R. Thiel, *Chem. Eur. J.* **2001**, *7*, 3236-3253; (b) N. Ledoux, B. Allaert, S. Pattyn, H. Vander Mierde, C. Vercaemst, F. Verpoort, *Chem. Eur. J.* **2006**, *12*, 4654–4661.
6. B.W. Gung, J.C. Amicangelo, *J. Org. Chem.* **2006**, *71*, 9261-9270.
7. (a) C. Adlhart, P. Chen, *J. Am. Chem. Soc.* **2004**, *126*, 3496–3510; (b) B. F. Straub, *Angew. Chem., Int. Ed.* **2005**, *44*, 5974–5978; (c) G. Occhipinti, H.-R. Bjørsvik, V. R. Jensen, *J. Am. Chem. Soc.* **2006**, *128*, 6952-6964.
8. P. A. van der Schaaf, R. Kolly, H.-J. Kirner, F. Rime, A. Mühlebach, A. Hafner, *J. Organometal. Chem.* **2000**, *606*, 65–74.
9. (a) J.S. Kingsbury, J.P.A. Harrity, P.J. Bonitatebus, A.H. Hoveyda, *J. Am. Chem. Soc.* 1999, 121, 791-799; (b) S.B. Garber, J.S. Kingsbury, B.L. Gray, A.H. Hoveyda, *J. Am. Chem. Soc.* **2000**, *122*, 8168-8179.

26.   Kinetic Implications Derived from Competitive Reactions: Buchwald-Hartwig Amination

Antonio C. Ferretti[1], Jinu S. Mathew[2], Colin Brennan[3] and Donna G. Blackmond[1,2]

[1]*Department of Chemical Engineering and* [2]*Department of Chemistry, Imperial College London SW7 2AZ, United Kingdom,* [3]*Process Studies Group, Syngenta, Huddersfield HD2 1FF, United Kingdom*

d.blackmond@imperial.ac.uk

Abstract

The Buchwald-Hartwig amination reaction has gained great interest in the last decade in both academic and industrial environments. In the work presented herein, we discuss a very interesting effect in the competitive reaction of two amines (benzophenone hydrazone and *n*-hexylamine) with 3-bromobenzotrifluoride.

The reactions involving either benzophenone hydrazone or *n*-hexylamine have been studied by reaction calorimetry. The benzophenone hydrazone reaction presents zero order kinetics, while the hexylamine reaction is first order in the aryl halide and zero order in the amine. Under synthetically relevant conditions, at 90°C, the rate of the hexylamine reaction is about 30-fold higher than the rate of the benzophenone reaction.

Interestingly, when reacting together, the benzophenone reacts first, and the hexylamine does not start to react until the benzophenone is not completely finished. This behavior can be explained by taking into account the mechanism and by observing that the catalyst intermediate which sits as the catalyst resting state in the benzophenone reaction is more stable than the Pd(BINAP) complex, catalyst resting state for the hexylamine reaction, which forms from the dissociation of BINAP from Pd(BINAP)₂ and exists only fleetingly in stoichiometric and catalytic reaction networks. It is the stability of the major intermediate, in this case, which controls selectivity, and not the relative reactivity.

This concept can be used in the study of other parallel reaction networks, and for designing more efficient catalyst systems in kinetic resolutions.

Introduction

The Pd-catalyzed amination of aryl halides had rapidly become one of the most important methods to synthesize substituted arylamines. Since its discovery in 1995, thanks to two independent studies by Buchwald and Hartwig, it has generated a tremendous amount of scientific publications (1-3). The knowledge gained in academia has rapidly been transferred to the industry, where this method is now widely used in the pharmaceutical and agrochemical fields (4). It is a versatile, simple, reliable reaction; several protocols have been developed to be used in mild conditions and with different functional groups.

Several studies were performed in order to establish the mechanism (5-7). The currently accepted mechanism, presented in Scheme 26.1 for the Pd(BINAP) catalyzed amination, involves the formation of a complex, Pd(BINAP)$_2$ from a catalyst precursor (usually Pd(OAc)$_2$ or Pd$_2$(dba)$_3$) and ligand; this complex lies outside the catalytic cycle and undertakes dissociation of one BINAP to form Pd(BINAP); the following steps are the oxidative addition of the aryl halide to the Pd(BINAP), reaction with amine and base, and the reductive elimination of the product to reform Pd(BINAP).

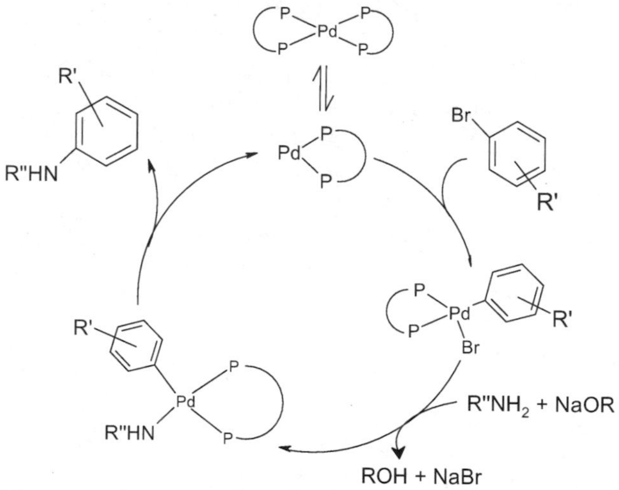

Scheme 26.1: currently accepted mechanism for the Pd/BINAP amination (7).

In our study, we focused our attention on the kinetic behaviour of the two amination reactions represented in Scheme 26.2:

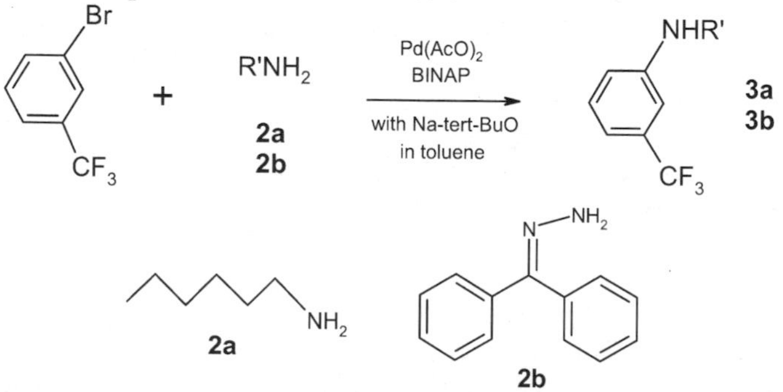

Scheme 26.2: Amination of aryl halides catalyzed by Pd(binap).

The experimental method used for this kinetic study is reaction calorimetry. In the calorimeter, the energy enthalpy balance is continuously monitored; the heat signal can then be easily converted in the reaction rate (in the case of an isothermal batch reactor, the rate is proportional to the heat generated or consumed by the reaction). The reaction orders and catalyst stability were determined with the methodology of reaction progress kinetic analysis (see refs. (8,9) for reviews).

Figure 26.1 represents the heat profile of the benzophenone hydrazone and hexylamine reactions. At the same conditions, at 90°C, the reaction involving hexylamine is considerably faster. The heat profile of the hexylamine reaction at 70°C shows how the reaction has positive order kinetics, while the benzophenone reaction shows overall zero order kinetics.

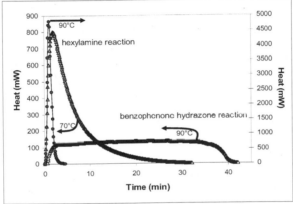

Figure 26.1: Heat profiles for the reactions of 2a and 2b with 1 (Scheme 26.1): 1 (0.3 M); 2a or 2b (0.4 M); base: NaO-*t*-Bu (0.45 M) in toluene using 3 mol% Pd(binap) prepared in situ from 1:1 Pd(OAc)$_2$ and binap. Temperatures: either 70°C or 90°C (see graph).

As shown in previous studies, the catalyst is stable during both reactions (11). The benzophenone hydrazone reaction is zero order in both substrates, while the hexylamine reaction is zero order in the amine and first order in the aryl halide. This change in the kinetics can be explained with a change in the rate limiting step; the oxidative addition is the rate limiting step for the hexylamine reaction, while, during the benzophenone reaction, it is the reductive elimination to be rate limiting. ^{31}P NMR studies where conducted to identify the catalyst resting state. The only species visible during the hexylamine reaction is the Pd(BINAP)$_2$, which lies outside the catalytic cycle. During the benzophenone reaction, a complex showing two doublets was identified (31Pç d 30.1, 19.9; [d, J = 38 Hz]); in light of the kinetics and the accepted mechanism, we assigned these peaks to the BINAP(Pd)Ar(amine) complex which precedes the reductive elimination. Mass measurements carried out with a time of flight electrospray method confirmed our suggestion (the spectrum showed a peak having mass 1069-complex plus H$^+$-).

We then turned our attention on the competitive amination of both these amines and the aryl halide, as in Scheme 26.2.

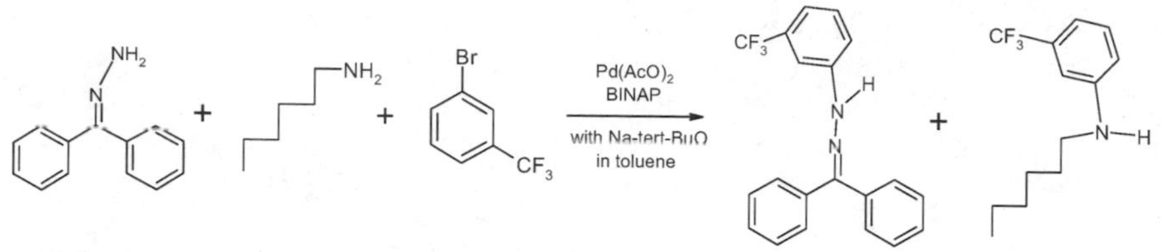

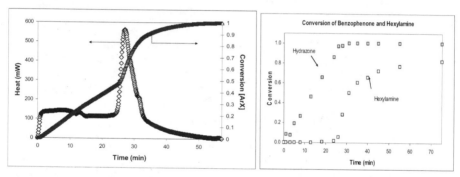

Scheme 26.3: Competitive reaction of benzophenone hydrazone and *n*-hexylamine.

When reacting. alone, the hexylamine is evidently more reactive; on the other hand, the benzophenone hydrazone forms a more stable intermediate (the BINAP(Pd)Ar(amine) complex that we observed by NMR). The interesting question was: which amine will react first?

The answer to this question is presented in Figure 26.2.

Figure 26.2: Heat profile and conversion of the benzophenone hydrazone and of the hexylamine for a competitive amination reaction involving 3-bromobenzotrifluoride (0.5 M), benzophenone hydrazone (0.25 M) and hexylamine (0.25 M). Temperature: 90°C.

The heat profile shows that the reaction has zero order kinetics at first, and then switches to positive order kinetics. The benzophenone hydrazone reacts first; only when it is completely consumed, the reaction involving hexylamine begins. Samples were taken and analyzed by ^{31}P and ^{19}F NMR. One sample was taken when the aryl halide conversion was low, at about 5%, and the profile was overall zero order; the second when the profile had switched to positive order and the conversion of the halide was greater than 50%.

We can see from Figure 26.3 that a complex arises at the beginning, showing peaks in both the phosphorus and fluorine NMR; it is the same previously observed during the benzophenone reaction; when the hexylamine reaction starts, and its product begins to form, this complex disappears and a single peak in the ^{31}P is visible, referring to the Pd(BINAP)$_2$.

These results are very intriguing. It would not be unreasonable to expect the hexylamine to react first, as its reactivity at 90°C is around 30-fold higher. Once again, in order to explain the results, we look at the mechanism.

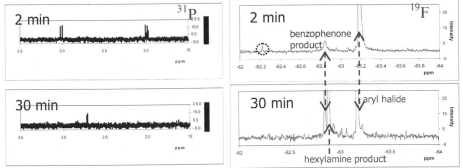

Figure 26.3: ^{31}P and ^{19}F NMR spectra of two samples taken during the competitive experiment after 2 min and 30 min.

The following scheme represents the possible paths which can be followed during a competitive experiment. Path *a* refers to the hexylamine, path *b* to the benzophenone hydrazone.

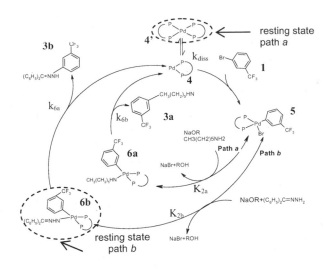

Scheme 26.4: Pathways for the competitive amination reactions of 2a and 2b.

The rates of reaction of path *a* and path *b* are expressed by eqs. 1 and 2. The sum of the rate of these two paths is the total reaction rate, i.e., the total rate at which the aryl halide is consumed (eq. 3):

$$r(pathway\ a) = \frac{k_1 K'_{diss}}{[binap]}[4'][ArX] \qquad \text{eq. 1}$$

$$r\left(pathway\ b\right) = k_{3b}\left[6b\right] \qquad\qquad \text{eq. 2}$$

$$r_{total} = r_{(path\ a)} + r_{(path\ b)} \qquad\qquad \text{eq. 3}$$

In a simple statement of the Curtin-Hammett principle, we know that the rate will depend not only on the separate reactivities of the two pathways, but also on the relative populations of the catalyst species. In the following table we highlight the parameters referring to reactivity and to the population.

	Pathway *a*	Pathway *b*
Reactivity	$k_1\left[ArX\right]$	k_{3b}
Population of catalytic species in the rate-limiting step	$\dfrac{k_1 K_{diss}}{\left[binap\right]}\left[4'\right] = \left[4\right]$	$\left[6b\right]$

Table 26.1: Parameters describing reactivity and population of active catalyst species from eqs. 1 and 2 in the proposed rate limiting step for the mechanism of Scheme 26.2.

The initial dominance of path b in the competitive reaction arises because the concentration ratio $\left[6b\right]/\left[4\right]$ is large enough to overwhelm the reactivity ratio $k_1\left[ArX\right]/k_{3b}$, which would favour the more reactive path *a*. This supports previous suggestions that the Pd(BINAP), which has never been isolated or observed spectroscopically, exists only fleetingly in stoichiometric and catalytic reaction networks.

Interpreting the results in terms of selectivity provides a deeper understanding of these results. We know that we can express selectivity as the ratio of the relative product formation rates, and, in case of pre-equilibrium amine binding and when [2a]=[2b], it takes the form of eq. 4:

$$s = \frac{k_{6b}}{k_{6a}} \cdot \frac{K_{2b}}{K_{2a}} \qquad\qquad \text{eq. 4}$$

Selectivity (*s*) depends on the ratio of the reactivity constant, as well on the ratio of the binding constants, which determines the concentration of the major intermediate.

We can identify three possible scenarios:

1. "Lock and key" kinetics: the more active intermediate is also the more stable one. The major intermediate leads to the major product.

2. "Major-minor": the more active intermediate, even though is less stable, leads to the major product. An example is provided by the asymmetric hydrogenation studied by Landis and Halpern (12).
3. "Monopolizing": the more stable intermediate, even though is less active, leads to the major product, as it "monopolizes" the catalyst (14).

When an intermediate is both less active and less stable, it is always unable to lead to the major product.

Energy diagrams can also be helpful to stress these concepts. Table 26.2 represents the four cases we described. In these diagrams, the intermediates a and b lead to the two products A and B. The difference in the transition state free energies $\Delta\Delta G$ will determine the selectivity. In the first three quadrants, I, II and III, the major product formed is B. In particular, we discussed in this study the "monopolizing" case. We can see that the selectivity, in this case, is dictated by relative intermediate stability.

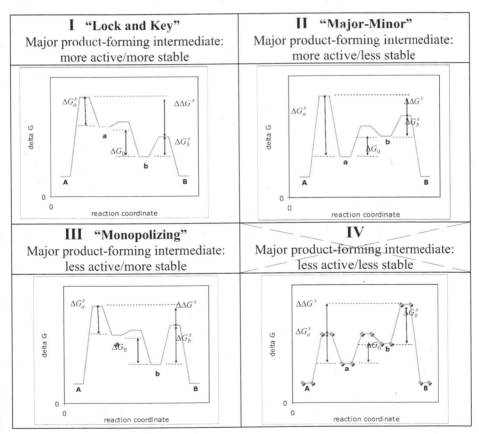

Table 26.2: Classification and free energy diagrams of the four different scenarios in which products *A* and *B* are obtained from the intermediates *a* and *b*.

Buchwald and coworkers also studied how selectivity is influenced by the nature of the different amines during competitive aminations (10). They discussed examples of the "lock and key" kinetics showed in case I and of the "major-minor" kinetics of case II; as they did not present the activities of the single reactions, it is not possible to conclude if any of their systems present the "monopolizing" kinetics shown in this study.

Conclusion

We studied the competitive amination of two amines (benzophenone hydrazone and *n*-hexylamine) and one aryl halide (3-bromobenzotrifluoride), catalyzed by Pd(BINAP). We showed that, when reacting alone at the same conditions, n-hexylamine is considerably more reactive and shows positive order kinetics; benzophenone hydrazone shows zero order kinetics and forms a very stable intermediate, the BINAP(Pd)Ar(amine) we also observed by NMR. During the competitive reaction of the two amines, the benzophenone hydrazone reacts first and only when it is completely consumed, the hexylamine starts to react. In this case it is the stability of the major intermediate, and not the relative reactivity, which dictates the selectivity.

The concept shown in this work, i.e., the fact that strongly differentiated binding can result in high selectivity during competitive reactions, has implications for the study of other parallel reaction networks and for the design of catalysts for kinetic resolutions. It also suggests a cautionary note when investigating single enantiomer reactions to simplify the study of kinetic resolution or during the construction of a Hammett plot: experimental trends of the single reactions studied alone may differ from those obtained during competitive reactions.

Acknowledgements

An EPSRC/CASE Award sponsored by Syngenta is gratefully acknowledged. DGB holds a Royal Society of Chemistry Wolfson Research Merit Award.

References

1. Hartwig, J. F. In *Handbook of Organopalladium Chemistry for Organic Synthesis*; Negishi, E. I., Ed.; Wiley-Interscience: New York, 2002; Vol. 1, p. 1051.
2. Muci, A. R.; Buchwald, S. L., *Top. Curr. Chem.* **2002**, *219*, 131.
3. Hartwig, J. F., *Synlett* **2006**, *9*, 1283.
4. Buchwald, S. L.; Mauger, C.; Mignani, G.; Scholz, U., *Adv. Synth. Catal.* **2006**, *348*, 23.
5. Singh, U. K.; Strieter, E. R.; Blackmond, D. G.; Buchwald, S. L., *J. Am. Chem. Soc.* **2002**, *124*, 14104.
6. Alcazar-Roman, L. M.; Hartwig, J. F.; Rheingold, A. L.; Liable-Sands, L.M.; Guzei, I. A., *J. Am. Chem. Soc.* **2000**, *122*, 4618.

7. Shekhar, S.; Ryberg, P.; Hartwig, J. F.; Mathew, J. S.; Blackmond, D. G.; Strieter, E. R.; Buchwald, S. L., *J. Am. Chem. Soc.* **2006**, *128*, 3584.
8. Blackmond, D. G., *Angew. Chem., Int. Ed.* **2005**, *44*, 4302.
9. Mathew, J. S.; Klussmann, M.; Iwamura, H.; Valera, F.; Futran, A.; Emanuelsson, E. A. C.; Blackmond, D. G., *J. Org. Chem.* **2006**, *71*, 4711.
10. Biscoe, M.R., Barber, T.E., Buchwald, S.L., *Angew. Chemie Int. Ed.,* **2007**, *46*.
11. Ferretti, A.C., Mathew, J.S., Blackmond D.G., *Ind. Eng. Chem. & Res.* **2007,** 46, 8584-8589.
12. Landis C., Halpern J., *J. Am. Chem. Soc.* **1987**, 109, 1746.
13. Ferretti, A.C., Mathew J.S., Ashworth I., Purdy M., Brennan C., Blackmond D.G., *Adv. Synt. Cat.*, **2008**, in press.

27. A Highly Active and Reusable Catalyst for Suzuki Couplings $BaCe_{1-x}Pd_xO_{3-x}$ (0 < x ≤ 0.1)

Xiaoying Ouyang[1], Jun Li[2,3], Ram Seshadri[1,2,3] and Susannah L. Scott[1,4]

[1]Department of Chemistry and Biochemistry, [2]Materials Department, [3]Materials Research Laboratory, and [4]Department of Chemical Engineering University of California, Santa Barbara, CA, 93106

sscott@engineering.ucsb.edu

Abstract

A catalyst consisting of $BaCeO_3$ perovskite with low levels of substitution of Pd(II) on the *B* (i.e., Ce) site and a corresponding number of oxygen vacancies was prepared by a high-temperature synthesis method. The $BaCe_{1-x}Pd_xO_{3-x}$ phase extrudes nanoparticles of *fcc*-Pd when heated in a reducing atmosphere. The oxidized form of the Pd-doped perovskite is highly active for the Suzuki coupling of aryl bromides and iodides with 4-phenylboronic acid (TOF ca. 50,000 at 80°C), provided the solvent system contains 2-propanol. The alcohol is suggested to be a reducing agent for Pd(II), extracting ligand-free Pd(0) from the perovskite into solution. The fully reduced form of the perovskite is a poor catalyst, because of the inaccessibility of the Pd(0). A lower level of cationic Pd substitution in the oxidized phase is associated with a shorter induction period, corresponding to less required reorganization of the remaining perovskite upon extraction of Pd. The catalyst showed little decline in activity after being recycled seven times, suggesting that the perovskite may recapture Pd(II) from the reaction medium. The concentrations of soluble Pd before and after the reaction are also consistent with its dissolution/re-absorption: the Pd level decreased over the course of the reaction, and was undetectable (< 0.1 ppm) at the end.

Introduction

Supported versions of homogeneous catalysts have been studied intensively over the last two decades. In practical terms, solid catalysts facilitate separation from the reaction medium, enabling their recovery and reuse. Consequently, less precious metal is consumed, which is advantageous both economically and environmentally. The search for "green" palladium catalysts for C-C bond-forming processes, such as the Suzuki, Heck and Sonogashira reactions, is motivated by these reasons, as well as by the need to minimize residual Pd in the product. A large number of "heterogeneous" catalysts for Suzuki coupling reactions have been reported, including Pd supported on carbon (1), silica (2), polymers (3) and dendrimers (4). A family of unsupported heterogeneous Pd-containing perovskite catalysts has also been reported (5, 6). Most Pd-containing heterogeneous catalysts show much lower catalytic activities than homogeneous C-C coupling catalysts, and many (perhaps all) involve the formation of soluble Pd during the catalytic cycle (7). Thus removal residual Pd at the end of the reaction remains a problem.

In this contribution, we describe a novel, low surface area (ca. 1 m^2/g) $BaCe_{1-x}Pd_xO_{3-x}$ perovskite catalyst for Suzuki coupling reactions. Recently, we showed that $BaCeO_3$ can incorporate low levels of Pd(II) on the *B* (Ce) sites, and that this Pd moves out of and back into the perovskite lattice in response to heating under reducing (5 % H_2/95 % N_2) or oxidizing (pure O_2) conditions, respectively, based on the reversible change in cell volume observed by XRD during redox cycling, and the appearance and disappearance of reflections due to *fcc*-Pd (8, 9). The XPS spectrum of oxidized $BaCe_{1-x}Pd_xO_{3-x}$ (x = 0.05, 0.10) shows that essentially all of the near-surface Pd is present in cationic form. When reduced, the XPS signals due to cationic Pd are completely suppressed, but the signals due to Pd(0) do not become appreciably more intense. This surprising observation was interpreted as overgrowth of the metal nanoparticles by $BaCeO_3$. Reoxidation of *fcc*-Pd/$BaCeO_3$ resulted in the reappearance of XPS signals corresponding to cationic Pd, confirming that the metal is reabsorbed as Pd(II) into the perovskite lattice (8, 9).

Experimental Section

Synthesis of $BaCe_{1-x}Pd_xO_{3-x}$. $BaCe_{1-x}Pd_xO_{3-x}$ was prepared by grinding together BaO_2 (99 %, Cerac), CeO_2 (99.9 %, Cerac), and PdO (99.95 %, Cerac) in their stoichiometric ratios (1 : 1-x : x), pelletizing the ground powders, and heating them in flowing O_2 at 1000°C for 10 h. BaO_2 was chosen as the Ba source because it is less hygroscopic than BaO, and more easily decomposed than $BaCO_3$. The pellet was then reground, pelletized, and heated for another 10 h at 1000°C in flowing O_2. This fresh sample is called "as-prepared," and has a BET surface area of ca. 1 m^2/g. Reduction of as-prepared $BaCe_{1-x}Pd_xO_{3-x}$ was carried out in flowing 5% H_2/95% N_2 at 1000°C for 1 h, as previously described (8). This treatment causes complete reduction of Pd(II) to Pd(0). Reoxidized $BaCe_{0.95}Pd_{0.05}O_{2.95}$ was prepared by reoxidation of the reduced material in flowing O_2 at 1000°C for 10 h. None of these treatments affect the BET surface area appreciably.

General procedure for Suzuki coupling. 4-Bromoanisole (125 µL, 1 mmol), phenylboronic acid (186 mg, 1.5 mmol), K_2CO_3 (0.55 g, 4 mmol) and the $BaCe_{1-x}Pd_xO_{3-x}$ catalyst were mixed in a 20 mL scintillation vial. A preheated 2-propanol/water solution (IPA/H_2O, 1:1 v/v, 12 mL, 80°C) was added, the vial was immediately placed on a hot plate stirrer and its temperature was maintained at (80 ± 1) °C. The reaction mixture was stirred at 1000 rpm for 3 min, then cooled to room temperature. The 4-methoxybiphenyl product was extracted with diethyl ether (3 × 15 mL). The organic fractions were washed with deionized water and dried with $MgSO_4$. After filtration, volatiles were removed under reduced pressure to yield the isolated product.

To promote phase transfer in the toluene/H_2O solvent system, 1 drop of Aliquat® 336 (tricaprylmethylammonium chloride, Aldrich) was added. For kinetic studies, aliquots (0.1 mL) were withdrawn by syringe at timed intervals (20–360 s) and immediately diluted with 0.2 mL aqueous HCl to quench the reaction. The acid-quenched mixtures were extracted with C_6D_6, and the organic phase was subjected to analysis by ${}^1H$ NMR. For the recycling study, the reaction mixture was prepared according to the method described above. At the end of the reaction, it was cooled to room temperature and diluted with 10 mL acetone to dissolve the solid, organic product. The liquid phase was removed carefully by syringe. Solid $BaCe_{1-x}Pd_xO_{3-x}$ remaining in the scintillation vial was washed with diethyl ether, dried and reused in subsequent reactions.

Analysis of soluble Pd by ICP. A series of reactions was initiated according to the procedure described above. After 20 s, 1 min, 3 min or 1 h, the hot reaction mixture was passed quickly through a 0.45 μm syringe filter to remove solid $BaCe_{1-x}Pd_xO_{3-x}$, then the solvent was removed under reduced pressure. A mixture of $BaCe_{1-x}Pd_xO_{3-x}$ and K_2CO_3 in IPA/H_2O, heated at 80 °C for 1 h, was subjected to the same workup procedure. Each of the solid residues was suspended in 12 mL 10 M aqueous HCl. The mixtures were filtered to produce clear solutions for subsequent analysis on a Thermo Jarrell Ash (TJA) High Resolution ICP spectrometer. A ICP standard solution of Pd (Aldrich) was diluted to 10 ppm for use as the high standard.

Results

Suzuki coupling reactions with aryl halides. Two as-prepared $BaCe_{1-x}Pd_xO_{3-x}$ materials (x = 0.05 and 0.10) were successfully utilized in several Suzuki coupling reactions. Both aryl iodides and aryl bromides react smoothly with 4-phenylboronic acid, eq 1, to yield the corresponding biaryls in high yields ($\geq$ 95%). For both 4-bromoanisole and 4-iodoanisole, the biaryl yields reached nearly 100% in 3 min with $BaCe_{0.95}Pd_{0.05}O_{3-\delta}$ as the catalyst, corresponding to an effective TON of ca. 2,000 and an effective TOF of nearly 50,000 h^{-1}. Results are summarized in Table 27.1.

$$R\text{-}Ar\text{-}X + \text{—B(OH)}_2 \xrightarrow[\text{IPA/H}_2\text{O (1:1), K}_2\text{CO}_3,\ 80\ ^\circ\text{C}]{BaCe_{1-x}Pd_xO_{3-\delta}\,(0.05\text{-}0.1\%\ \text{mol Pd})} R\text{—}$$

$$(1)$$

Kinetic analysis. The kinetics of Suzuki coupling were investigated and compared for $BaCe_{0.95}Pd_{0.05}O_{2.95}$ and $BaCe_{0.90}Pd_{0.10}O_{2.90}$. The results are shown in Figure 27.1a. The reaction between 4-bromoanisole and 4-phenylboronic acid catalyzed by $BaCe_{0.95}Pd_{0.05}O_{2.95}$ shows no induction period. However, a short but measurable induction period (< 1 min) was observed with $BaCe_{0.90}Pd_{0.10}O_{2.90}$. The biaryl yield reached 100% after 160 s for $BaCe_{0.95}Pd_{0.05}O_{2.95}$, and after 240 s for $BaCe_{0.90}Pd_{0.10}O_{2.90}$.

Similar kinetic experiments were carried out to compare the activities of the fully oxidized and fully reduced forms of $BaCe_{0.95}Pd_{0.05}O_{2.95}$, Figure 27.1b. The activity of the reduced catalyst is much lower than that of the as-prepared (i.e., oxidized) catalyst, resulting in only 2% yield after 4 min. However, high activity is completely restored upon reoxidation of the reduced catalyst.

Table 27.1 Suzuki couplings catalyzed by $BaCe_{1-x}Pd_xO_{3-x}$ ($x = 0.05, 0.10$) [a]

entry	substrate	x	reaction time min	isolated yield (%)	TON [c]	TOF [c] h^{-1}
1	Br⟨⟩OMe	0.05	2.7	95 (100) [b]	2,045	45,400
		0.10	4.0	95 (100) [b]	1,015	15,225
2	I⟨⟩OMe	0.05	2.5	96 (100) [b]	2,070	49,680
3	Br⟨⟩Me	0.05	3.0	98	2,110	42,200
4	Br⟨⟩CN	0.05	3.0	96	2,070	41,400
5	Br⟨⟩N	0.05	3.0	95	2,045	40,910

[a] Reagents and conditions: 3.0 mg $BaCe_{1-x}Pd_xO_{3-\delta}$ (0.05 mol % Pd, $x = 0.05$; 0.10 mol % Pd, $x = 0.10$), 1.0 mmol aryl halide, 1.5 mmol $PhB(OH)_2$, 4.0 mmol K_2CO_3, 12 mL IPA-H_2O (1:1, v/v), 80°C. [b] Estimated by ^1H NMR. [c] The TON and TOF effective values, based on the total Pd introduced with the perovskite catalyst.

(a)   (b)

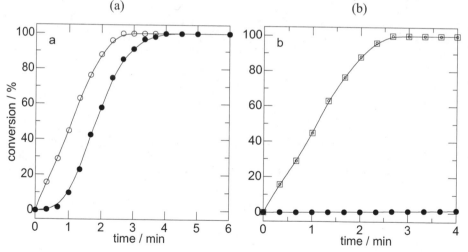

Figure 27.1. Kinetic profiles for the coupling of 4-bromoanisole with phenylboronic acid to yield 4-bromonaphthalene, (a) using as-prepared $BaCe_{0.95}Pd_{0.05}O_{2.95}$ (open circles) or $BaCe_{0.90}Pd_{0.10}O_{2.90}$ (filled circles) as the catalyst; and (b) using $BaCe_{0.95}Pd_{0.05}O_{2.95}$ as-prepared (small open circles); reduced (filled circles); and reoxidized (open squares) as the catalyst. Reagents and conditions: 3.0 mg catalyst (0.05 mol % Pd, $x = 0.05$; 0.10 mol % Pd, $x = 0.10$), 1.0 mmol 4-bromoanisole, 1.5 mmol $PhB(OH)_2$, 4.0 mmol K_2CO_3, 12 mL IPA-H_2O (1:1, v/v), 80°C. Lines are drawn only to guide the eye.

Analysis of soluble Pd. To assess the amount of "ligand-free" soluble Pd present in the reaction mixture, the solution containing 4-bromoanisole/4-

phenylboronic acid reaction and BaCe$_{0.95}$Pd$_{0.05}$O$_{2.95}$ was filtered after 20 s, 1 min, 3 min and 1 h, and the filtrate subjected to ICP analysis. In addition, ICP analysis of Pd was performed on a mixture of BaCe$_{0.95}$Pd$_{0.05}$O$_{2.95}$ and K$_2$CO$_3$ in IPA/H$_2$O, in order to verify whether Pd dissolution occurs only in the presence of the aryl halide. Interestingly, although the Pd levels before, during and after the reaction were all rather low (< 0.2 ppm), the Pd content was highest in the absence of substrate. It decreased progressively during the reaction until it eventually became undetectable (< 0.05 ppm), Figure 27.2.

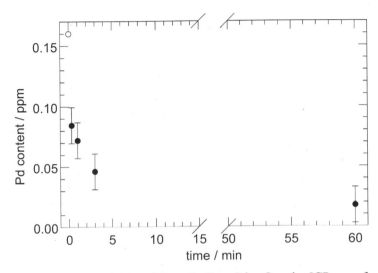

Figure 27.2. Analysis of soluble Pd from BaCe$_{0.95}$Pd$_{0.05}$O$_{2.95}$ by ICP, as a function of reaction time (reagents and conditions: 3.0 mg BaCe$_{0.95}$Pd$_{0.05}$O$_{2.95}$ (0.05 mmol Pd), 4 mmol K$_2$CO$_3$, 1.0 mmol 4-bromoanisole, 1.5 mmol PhB(OH)$_2$, 12 mL IPA-H$_2$O (1:1, v/v), 80°C). The measurement for t = 0 min corresponds to the catalyst stirred with K$_2$CO$_3$ in IPA/H$_2$O (i.e., in the absence of the aryl halide) at 80°C for 1 h.

Maitlis' filtration test. To investigate whether the Pd-doped perovskite is the actual catalyst, or a reservoir of soluble Pd, a Maitlis' filtration test (10) was performed. The reaction of 4-bromoanisole with 4-phenylboronic acid, catalyzed by BaCe$_{0.95}$Pd$_{0.05}$O$_{2.95}$, was interrupted at 20 s and 1 min, corresponding to conversions of 16 and 45%, respectively, by filtering the hot reaction mixture to remove the solid perovskite. The filtrates were allowed to cool to room temperature without stirring. After 3 h, the biaryl yields in both samples were estimated to be 100% by ^{1}H NMR.

Solvent effect. Several different solvent systems were investigated for the coupling of 4-bromoanisole with 4-phenylboronic acid catalyzed by BaCe$_{0.95}$Pd$_{0.05}$O$_{2.95}$. The results are summarized in Table 27.2. A phase-transfer catalyst (Aliquat$^®$ 336) was added in the experiment involving toluene/H$_2$O. However, only the IPA/H$_2$O solvent system was effective, resulting in a biaryl yield of 100%; all other solvents gave rather low yields.

Table 27.2. Effect of solvent system on Suzuki coupling[a] catalyzed by $BaCe_{0.95}Pd_{0.05}O_{3-\delta}$

solvent	additive	time (h)	yield (%)
IPA/H$_2$O (1:1, v/v)	-	30	100 [b]
THF/H$_2$O (1:1, v/v)	-	30	20 [c]
^{t}BuOH/H$_2$O (1:1, v/v)	-	30	5 [b]
toluene/H$_2$O (1:1, v/v)	Aliquat® 336 [d]	30	5 [b]
toluene	-	30	trace [b]

[a] Reagents and conditions: 3.0 mg $BaCe_{0.95}Pd_{0.05}O_{3-\delta}$ (0.05 mol % Pd), 1.0 mmol 4-bromoanisole, 1.5 mmol PhB(OH)$_2$, 4.0 mmol K$_2$CO$_3$, 12 mL solvent, 80°C. [b] Estimated by ^{1}H NMR. [c] Estimated by GC-MS. [d] One drop of the neat liquid.

Catalyst recycling. The solid Pd-doped perovskite catalysts are easily filtered from the reaction mixture for reuse. The activity of the recycled $BaCe_{0.95}Pd_{0.05}O_{2.95}$ catalyst was investigated in the coupling of 4-bromoanisole with 4-phenylboronic acid. The results in Table 27.3 show that high activity was retained even after seven cycles of catalyst use.

Table 27.3. Effect of recycling the $BaCe_{0.95}Pd_{0.05}O_{3-\delta}$ catalyst, on yields in Suzuki coupling [a]

cycle	time (min)	yield (%) [b]
1	5	100
2	5	100
3	5	100
4	5	98
5	5	100
6	5	98
7	5	99

[a] Reagents and conditions: 3.0 mg $BaCe_{0.95}Pd_{0.05}O_{2.95}$ (0.05 mol % Pd), 1.0 mmol 4-bromoanisole, 1.5 mmol PhB(OH)$_2$, 4 mmol K$_2$CO$_3$, 12 mL IPA-H$_2$O (1:1, v/v), 80°C. [b] Estimated by ^{1}H NMR.

Discussion

Formation of the active sites. The results of the Maitlis' filtration test suggest that the active sites for the Suzuki coupling reaction are soluble, "ligand-free" Pd species, rather than Pd present on the solid surface of the perovskite. This is consistent with many previous studies of "heterogeneous" Pd catalysts which act by leaching Pd into solution (7). Furthermore, the requirement for IPA in the solvent system and the observation by ICP of soluble Pd in the absence of the aryl halide substrate suggest that the active site is extracted from the solid reservoir via a reaction with the solvent. According to XPS, only Pd(II) is present on the surface of $BaCe_{0.95}Pd_{0.05}O_{2.95}$ (8, 9). However, Pd(0) is considered essential for entry into the catalytic cycle via its oxidative addition of the aryl halide. Therefore, we propose that reduction of Pd(II) to Pd(0) by basic, aqueous IPA is involved in the extraction of Pd, eq. 2. This is analogous to the extrusion of Pd(0) caused by reduction with H_2 (8), although the reaction conditions are different (80°C for IPA vs. 1000°C for H_2).

$$Pd(II) + \underset{\text{OH}}{\bigwedge\!\!\!\bigwedge} \xrightarrow{\text{base}} Pd(0) + \underset{\text{O}}{\bigwedge\!\!\!\bigwedge} \qquad (2)$$

The level of Pd loading in the perovskite plays a role in the length of the induction period, during which the Pd is extracted from its reservoir. Removal of Pd(II) requires an accompanying rearrangement of the perovskite structure, with formation of a corresponding amount of $Ba(OH)_2$, eq. 3.

$$BaCe_{1-x}Pd_xO_{3-x} + 2x\,H^+ + 2x\,e^- \rightarrow x\,Pd(0) + (1-x)\,BaCeO_3 + x\,Ba(OH)_2 \qquad (3)$$

The greater rearrangement of the perovskite structure in the catalyst associated with the higher level of Pd-doping may be responsible for the longer induction period. After the onset of catalytic activity, the slopes of the two conversion vs. time curves for $x = 0.05$ and $x = 0.10$ in Figure 27.1a are very similar, demonstrating that the two catalysts produce the same soluble active site in similar amounts.

The dramatic suppression of catalytic activity upon H_2 reduction of the catalyst is curious, considering the requirement noted above for formation of Pd(0) in order to enter the catalytic cycle. It is attributed to the unavailability of Pd(0) at the liquid/solid interface. The strong Pd $3d$ signals representing near-surface Pd(II) in the XPS spectra of the as-prepared and reoxidized forms of $BaCe_{1-x}Pd_xO_{3-x}$ disappear completely upon treatment of the catalyst in H_2 at 1000°C (8, 9). Although powder XRD confirmed that *fcc*-Pd(0) is formed (average particle size, 80 nm by line-broadening), it is not present at the perovskite surface. Instead, the metal nanoparticles are overgrown by $BaCeO_3$ and/or by the $Ba(OH)_2$ formed concurrently with Pd(0). This likely prevents the release of soluble Pd(0) into the reaction medium, since the metal is not exposed to the reaction medium. A similar effect of suppressed activity in the reduced form of $LaFe_{0.57}Co_{0.35}Pd_{0.05}O_3$, although no explanation was proposed (6).

Proposed mechanism. Based on the evidence described above, we propose that $BaCe_{1-x}Pd_xO_{3-x}$ acts as a reservoir of "ligand-free" soluble Pd for the Suzuki coupling reaction. Due to the large ionic radii of Ba^{2+} and Ce^{4+}, Pd(II) is very stable at the *B*-site of the host lattice. Reductive extraction under mild conditions (basic, aqueous IPA, 80°C) results in formation of the active, ligand-free Pd(0) species, but at very low levels which do not tend to form inactive Pd colloids by aggregation. Indeed, the effective turnover frequencies reported here are among the highest reported in the literature to date. The soluble, "ligand-free" Pd(0) presumably undergoes oxidative addition of the aryl halide, followed by reductive coupling with the phenylboronic acid.

   Although recapture of soluble Pd by the perovskite has not been demonstrated conclusively, the ability to reuse the catalyst several times without significant loss of activity suggests that reabsorption of Pd may occur. If this were not the case, rapid depletion of near-surface Pd in the low surface area (1 m^2/g) catalyst would be expected to cause a precipitous loss of activity upon catalyst recycling. The dissolution/re-absorption of Pd may also contribute to the stabilization of soluble Pd(0) by preventing the formation of inactive Pd black. The residual Pd in solution at the end of the reaction is significantly lower than that reported for either homogeneous catalyst (e.g., Pd(PPh$_3$)$_4$, residual [Pd] = 40 – 80 ppm (11)) or heterogeneous catalysts (e.g., Pd/C, residual [Pd] < 6 ppm (12); LaFe$_{0.57}$Co$_{0.35}$Pd$_{0.05}$O$_3$, residual [Pd] = 2 ppm (5, 6)). The proposed reaction mechanism is summarized in Scheme 27.1.

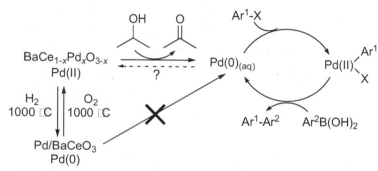

Scheme 27.1. Proposed mechanism for Suzuki coupling by Pd-doped perovskite catalysts.

Conclusions

BaCe$_{1-x}$Pd$_x$O$_{3-x}$ catalysts (x = 0.05 and 0.10), synthesized by a high-temperature solid-state method, are very active in Suzuki coupling (TOF ca. 50,000) and produce very low levels of residual Pd (< 0.05 ppm) in the reaction medium. The as-prepared and reoxidized forms of the catalyst have very similar catalytic activities, while the H$_2$-reduced catalyst in which metallic Pd has been extruded from the perovskite lattice shows much lower activity. Reductive extraction of Pd(II) by basic, aqueous 2-propanol in the reaction medium occurs in the absence of the aryl halide substrate. The low levels of residual Pd at the end of the reaction and the ability to recycle the catalyst without loss of activity suggest that recapture of soluble Pd may also be occurring. The longer induction period exhibited by BaCe$_{0.90}$Pd$_{0.10}$O$_{2.90}$ relative to BaCe$_{0.90}$Pd$_{0.10}$O$_{2.95}$ is attributed to the greater reorganization requirement of the remaining perovskite host.

Acknowledgements

The authors are grateful to Catalytic Solutions, Inc., and the U.S. Department of Energy, Basic Energy Sciences, Grant No. DE-FG02-05ER15725, for funding. This work made use of MRL Central Facilities supported by the MRSEC Program of the National Science Foundation under award DMR05-20020.

References

1. R. G. Heidenreich, K. Kohler, J. G. E. Krauter and J. Pietsch, *Synlett*, 1118-1122 (2002).
2. K. Shimizu, R. Maruyama, S. Komai, T. Kodama and Y. Kitayama, *J. Catal.*, **227**, 202-209 (2004).
3. B. J. Gallon, R. W. Kojima, R. B. Kaner and P. L. Diaconescu, *Angew. Chem. Int. Ed.*, **46**, 7251-7254 (2007).
4. A. K. Diallo, C. Ornelas, L. Salmon, J. R. Aranzaes and D. Astruc, *Angew. Chem. Int. Ed.*, **46**, 8644-8648 (2007).
5. M. D. Smith, A. F. Stepan, C. Ramaro, P. E. Brenan and S. V. Ley, *Chem. Commun.*, 2652-2653 (2003).
6. S. P. Andrews, A. F. Stepan, H. Tanaka, S. V. Ley and M. D. Smith, *Adv. Synth. Catal.*, **347**, 647-654 (2005).
7. N. T. S. Phan, M. Van Der Sluys and C. W. Jones, *Adv. Synth. Catal.*, **348**, 609-679 (2006).
8. J. Li, U. G. Singh, J. W. Bennett, K. Page, J. C. Weaver, J. P. Zhang, T. Proffen, A. M. Rappe, S. L. Scott and R. Seshadri, *Chem. Mater.*, **19**, 1418-1426 (2007).
9. U. G. Singh, J. Li, J. W. Bennett, A. M. Rappe, R. Seshadri and S. L. Scott, *J. Catal.*, **249**, 347-356 (2007).
10. J. E. Hamlin, K. Hirai, A. Millan and P. M. Maitlis, *J. Mol. Catal.* **7**, 543-544 (1980).
11. D. Ennis, J. McManus, W. Wood-Kaczmar, J. Richardson, G. E. Smith and A. Carstairs, *Org. Proc. Res. Dev.* **3**, 248-252 (1999).
12. D. A. Conlon, B. Pipik, S. Ferdinand, C. R. LeBlond, J. R. Sowa, Jr. B. Izzo, P. Collins, G. J. Ho, J. M. Williams, Y. J. Shi and Y. Sun. *Adv. Synth. Catal.* **345**, 931-935 (2003).

28. Biphasic Hydroformylation of Higher Olefins

**Steven D. Dietz, Claire M. Ohman, Trudy A. Scholten, Steven Gebhard and
Girish Srinivas**

TDA Research, Inc., 12345 W. 52nd Ave., Wheat Ridge, CO 80033

sdietz@tda.com

Abstract

A method has been developed for the continuous removal and reuse of a
homogeneous rhodium hydroformylation catalyst. This is done using solvent
mixtures that become miscible at reaction temperature and phase separate at lower
temperatures. Such behavior is referred to as thermomorphic, and it can be used
separate the expensive rhodium catalysts from the aldehydes before they are distilled.
In this process, the reaction mixture phase separates into an organic phase that
contains the aldehyde product and an aqueous phase that contains the rhodium
catalyst. The organic phase is separated and sent to purification, and the aqueous
rhodium catalyst phase is simply recycled.

Introduction

Currently, worldwide production of aldehydes exceeds 7 million tons/year (1). Higher
aldehydes are important intermediates in the synthesis of industrial solvents,
biodegradable detergents, surfactants, lubricants, and other plasticizers. The process,
called hydroformylation or more familiarly, the Oxo process, refers to the addition of
hydrogen and the formyl group, CHO, across a double bond. Two possible isomers
can be formed (linear or branched) and the linear isomer is the desired product for these
applications.

linear   branched

Homogeneous rhodium-phosphine catalysts are used to manufacture lower
aldehydes because they have by far the best activity and selectivity.   In this case the
aldehyde product is distilled out of the reaction solvent and catalyst solvent mixture is
recycled. The process used for conventional hydroformylation of short-chain alkenes
cannot be applied to higher ($>C_5$) olefins because the temperatures need to distill off
the high-boiling aldehyde product are high enough to destroy the catalyst. Therefore
the hydroformylation of higher olefins ($>C_5$) is done with less efficient cobalt catalysts
at high temperatures and pressures.

One approach that has been successfully used to separate the catalyst from the product aldehyde is to use a biphasic system in which the rhodium catalyst is soluble in water and the product is soluble in an organic phase. This approach is used by Hoechst/Rhône-Poulenc to produce more than 600,000 t/year of butyraldehyde (a lower aldehyde) (2). Unfortunately, this process cannot be used to produce higher aldehydes because the water solubility of the higher olefins that are the feedstock is very low, which dramatically reduces the reaction rate.

To eliminate the need to recover the product by distillation, researchers are now looking at thermomorphic solvent mixtures. A thermomorphic system is characterized by solvent pairs that reversibly change from being biphasic to monophasic as a function of temperature. Many solvent pairs exhibit varying miscibility as a function of temperature. For example, methanol/cyclohexane and *n*-butanol/water are immiscible at ambient temperature, but have consolute temperatures (temperatures at which they become miscible) of 125°C and 49°C, respectively (3).

Since these mixtures are immiscible at room temperature but miscible at the higher reaction temperature, there is excellent contact between the rhodium catalyst and the olefin when the reaction is carried out, increasing the reaction rate by orders of magnitude. After the reaction is complete, the reaction mixture is cooled and the phases completely separate. The product can be simply recovered by decantation and the catalyst can be recycled (Figure 28.1).

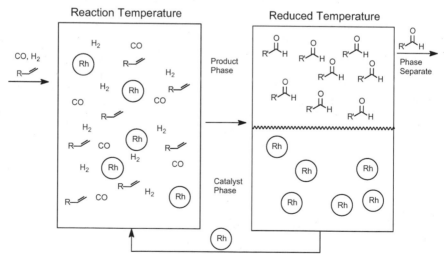

Figure 28.1. Representation of hydroformylation in a thermomorphic system.

The use of thermomorphic systems has recently been studied as a way of achieving catalyst separation in homogeneous catalysis. For example, a biphasic hydroformylation catalyst system was developed to take advantage of the unusual solvent characteristics of perfluorocarbons combined with typical organic solvents (4). Fluorous/organic mixtures such as perfluoromethylcyclohexane

(PFMCH)/toluene) are immiscible at ambient temperature but become a single-phase solution upon heating (5). This characteristic is very useful for homogeneous catalysis because the reaction can be performed at an elevated temperature where the mixture becomes one-phase, eliminating the mass transfer rate limitations between the two phases. Upon reaction completion, the solution is phase separated by cooling and the products are simply decanted. Unfortunately, the expense of the solvents and the ligands make the fluorous biphasic system impractical for industrial application. Other thermomorphic systems studied rely on exotic phosphine ligands for their use. Again, these are not used industrially because of their high cost (6, 7).

Our approach is to use the inexpensive ligands that are already used industrially as well as conventional solvents. The goal of this project is to develop a thermomorphic approach to the rhodium-catalyzed hydroformylation of higher olefins ($>C_6$) that enhances conversion rates and ease of product recovery while minimizing catalyst degradation and loss.

Thermomorphic solvent mixtures have been tested for hydroformylation of 1-octene and 1-dodecene to determine the ease of product recovery and catalyst recycling. Using both batch and continuous reactors, we demonstrated the efficacy of a biphasic, thermomorphic, system that had the following advantages:
  -Suitable for conversion of higher olefins to higher aldehydes.
  -High conversion rates and high selectivity of linear to branch aldehyde.
  -Recovery of the product aldehyde by simple room temperature decantation rather than high temperature distillation.
  -Little to no catalyst leaching and high catalyst recyclability.

Experimental Section

Batch Experimental Apparatus and Methods. The activity of the rhodium catalyst was tested in a 125 mL reactor with a pressure rating of 3000 psi at 350°C and a pressure relief valve that is rated for 1500-2200 psi. If the pressure valve releases, the gaseous contents of the autoclave are safely vented through a 1/4" stainless steel line and the liquid/vapor content in the autoclave is collected in a metal container and the vapor vented out through the hood. The reactor was heated in a silicone oil bath with a digitally controlled heat/stir plate.

In a typical experiment, the catalyst $HRh(CO)(TPPTS)_3$, (TPPTS = tris(3-sulphonatophenyl)phosphine trisodium salt, $Na_3P(C_6H_4SO_3)_3$) was prepared by charging the reactor with $Rh(CO)_2(acac)$ (99%, Strem), $Na_3P(C_6H_4SO_3)_3 \cdot xH_2O$ (10-15% oxide, Strem) and water. CO and H_2 were added in a 1:1 molar ratio and stirred at ambient temperature for 1 h. The reactor was then depressurized and charged with 1-octene (Aldrich) or 1-dodecene (Aldrich) and a nonpolar solvent. Depending on the particular experiment, a surfactant and/or an water miscible cosolvent were also added. The reactor was reassembled, pressurized with 1:1 CO/H_2 and heated in a silicon oil bath to the desired reaction temperature. The mixture was stirred during the entire process with a magnetic stir plate.

After the desired reaction time, the stirring was stopped and the reactor cooled to ambient temperature. The reactor was slowly vented into the hood. The organic phase was typically clear and colorless and the aqueous phase yellow to red. The products were analyzed by gas chromatography using an SRI Instruments gas chromatograph with a flame ionization detector.

To maintain consistency, all the reactions were performed at 100°C using the same amounts of reactants, catalysts and solvents. Under these reaction conditions, only aldehyde products were detected; no alcohol or alkene isomers were formed.

The rhodium loss to the hydrocarbon phase was analyzed by atomic absorption (Hazen Research, Inc., Golden, CO).

Continuous Experimental Apparatus and Methods. Figure 28.2 depicts the P&ID for the apparatus. The apparatus is fully automated using OptoControl software and is designed to operate unattended. It has an Autoclave Engineers reactor with a 300 ml capacity that features a bolted closure with a confined gasket for high temperature sealing. It is designed to withstand a maximum allowable pressure of 3800 psig at 538°C. The contents of the reactor are mixed with an overhead mixer designed specifically for use with the pressure vessel. Two thermocouples measure the internal and external temperature of the reactor wall and a pressure gauge monitors the internal reactor pressure. A rupture disk designed to withstand up to 1000 psig is mounted on the top of the reactor. All process lines to and from the reactor are unheated 1/8" 316 SS tubing with a wall thickness of 0.3".

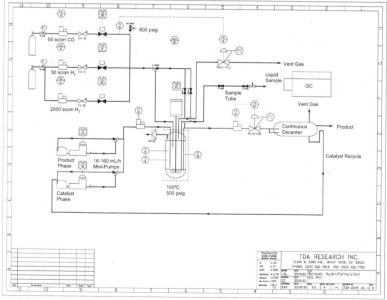

Figure 28.2. P&ID of bench-scale continuous reactor.

Two separate 2.1 L reservoirs contain the catalyst and product phases and the contents are fed into the reactor through a standard liquid mass flow controller. The contents of the reactor can be sampled from a pressure fed sample tube. The pressurized liquid reactor products exit the reactor through a pressure control valve, which reduces the pressure to atmospheric, and the liquid contents are delivered to a continuous decanter where the phases separate. The catalyst phase then settles to the bottom where it is drained for recycle and reuse, while the product phase is collected into a 4.2 L reservoir.

Results and Discussion

Batch Experiments with Thermomorphic Systems. As a reference, we tested the hydroformylation of 1-octene in a completely homogeneous system using the same rhodium triphenylphosphine catalyst that is used for hydroformylation of lower aldehydes. This is sample R39 in Table 28.1, and gives us a baseline to compare the performance of our systems in terms of conversion and selectivity. To maintain consistency, we performed all the reactions at 100°C using the same amounts of reactants, catalysts and solvents. Under these conditions we only detected aldehyde products; no alcohol or alkene isomers were formed.

We based our initial experiments on the work reported by Bergbreiter et al. (6) who showed that 90:10 (v/v) ethanol/water solution is immiscible with heptane at ambient temperature, but becomes a single phase at 70°C. Unfortunately, as shown for sample R13 this combination did not work for us because the catalyst precipitated as a yellow solid from solution. We then increased the water content to 50:50 ethanol/water. As shown for sample R14, this greatly enhances the reaction rate. We added a surfactant to the ethanol/water system, and we found the addition of CTAB to the 90:10 ethanol/water system greatly increased higher olefin conversion indicating that the surfactant help solubilize the catalyst. When we added CTAB to the 50:50 ethanol/water system, we again found that the olefin conversion increased. In addition the linear/branch aldehyde ratio also greatly increased with the addition of CTAB.

Table 28.1. Hydroformylation of 1-octene.

Sample	Surfactant	Aq. Phase	Time (h)	Conversion (%)	L/B[b] ratio
R39[a]	-	-	2.0	20	4.4
R4	CTAB	-	2.5	0.3	-
R13	-	90% EtOH	2.0	0.2	-
R14	-	50% EtOH	2.0	12	1.7
R15A	CTAB	90% EtOH	2.0	19	1.7
R29A	CTAB	50% EtOH	2.0	20	5.8

Reaction conditions: Catalyst H(CO)Rh(TPPTS)$_3$, Rh 0.089 mmol, 1-octene/Rh = 426, CO/H$_2$ (molar ratio) = 1, P(CO/H$_2$) 400 psi, T = 100°C. [a]Single-phase heptane only solvent, catalyst H(CO)Rh(PPh$_3$)$_3$. [b]Linear/branched.

Since it is well known that the higher the olefin, the more difficult it is to hydroformylate, we tried our approach on 1-dodecene. We found that when we used 50:50 ethanol/water (R24) and CTAB only (R33) systems they showed no activity for hydroformylation of 1-dodecene; whereas, the addition of surfactant to the 50:50 ethanol/water mixture (R27) dramatically increased the reaction rate, as well as the selectivity.

Table 28.2. Hydroformylation of 1-dodecene.

Sample	Surfactant	Aq. Phase	Time (h)	Conversion (%)	L/B[b] ratio
R33	CTAB	-	2	0	-
R24	-	50% EtOH	2	0	-
R27	CTAB	50% EtOH	4.6	26	5.3

Reaction conditions: Catalyst H(CO)Rh(TPPTS)$_3$, Rh 0.089 mmol, 1-octene/Rh = 426, CO/H$_2$ (molar ratio) = 1, P(CO/H$_2$) 400 psi, T = 100°C. [b]Linear/branched.

We looked at a number of water soluble cosolvents (**Table 28.3**). In all cases aldehyde products were observed. 1,4-dioxane compares well with ethanol as a co-solvent. The data so far shows that 1,4-dioxane shows slightly lower olefin conversion after two hours than ethanol, but slightly better selectivity.

Table 28.3. Comparison of water-soluble solvents on biphasic hydroformylation of 1-octene.

Aqueous Phase Components	Olefin Conversion (mol. %)	Linear/Branch Aldehyde Ratio
50:50 THF/Water	9.6	5.0
50:50 Acetone/Water	13	3.8
50:50 Acetonitrile/Water	7.3	5.6
50:50 1,4-dioxane/Water	16	6.0
50:50 Ethanol/Water	20	5.8

Reaction conditions: Catalyst = H(CO)Rh(TPPTS)$_3$, CO/H$_2$ (molar ratio) = 1, P(CO/H$_2$) = 500 psi, T = 100°C, Rxn Time = 2h, nonpolar phase = heptane, surfactant = CTAB.

The rhodium loss to the hydrocarbon phase was analyzed by atomic absorption. We found that for a thermomorphic catalyst solution that was cycled three times that the rhodium loss was below the 0.1 ppm detection limit of their instrument.

In summary, what we have found is that the combination of a thermomorphic system and a surfactant is very effective for the hydroformylation of 1-octene and 1-dodecene. We believe that although a 90:10 ethanol/water and heptane system becomes miscible at 70°C, the additional water in a 50:50 ethanol/water and heptane system raises the miscibility temperature to >100°C. When a surfactant is added, the miscibility temperature is lowered and the biphasic solution becomes monophasic below the reaction temperature, resulting in good reaction rates. In addition, the presence of the surfactant also enhances the selectivity compared to the completely homogeneous system from 1.8 to 5.3 L/B ratio in the case of dodecanal. Finally, our rhodium losses into the organic phase are very low, showing that the catalyst can be recycled.

Continuous Experiments with Thermomorphic Systems. For the continuous experiments, we used the best solvent system we identified in the batch reactions, which was 50:50 1,4-dioxane/water. Heptane was the nonpolar solvent and CTAB the surfactant. We chose this system over the 50:50 ethanol/water system because it gave us better selectivity and there is no chance that unwanted acetal side products will be formed by the reaction of ethanol with the aldehyde. We initially used 1-octene as the olefin and after we worked out the process conditions for 1-octene, we tested the higher olefin 1-dodecene.

The first few experiments in the continuous flow reactor yielded inconsistent octene conversions (Figure 28.3). The experiment ran for 218 hours. Initially the conversion was consistent at 3-4% for several hours, then improved significantly to 16% and then rapidly dropped off to less than 2% (Figure 28.3). The selectivity was also very good for this run, with an average normal to branch isomer ratio of 7:1.

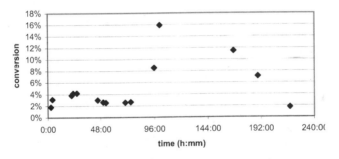

Figure 28.3. Percent octene conversion versus time.

For all of the runs we found that the amount of rhodium catalyst that leached into the product phase is negligible. Initially the reaction rate and selectivity are very good, but after approximately 200 hours the catalyst began to lose activity. In addition, we noticed that the volume of catalyst phase had decreased significantly. It was assumed that this was due to the extraction of dioxane into the organic phase. We speculated that the source of the problem was leaching of the dioxane cosolvent into the organic phase that at first enhances the activity, but then results in the loss of miscibility of the two phases and the conversion rate drops. To address this problem we prepared the catalyst in water only (without a cosolvent) and added the cosolvent to the organic feed so the dioxane is continuously replenished to prevent the loss of catalyst activity. By adding dioxane to the organic phase, we hoped to achieve a steady-state condition in which the catalyst composition would remain constant, rather than a constantly changing composition as the cosolvent is extracted.

In Figure 28.4 is shown an example of the hydroformylation of 1-octene in the continuous bench-scale reaction using biphasic and thermomorphic catalysts. The reaction was carried out with 1:1 molar mixture of H_2/CO with a combined pressure of 500 psi at 100°C for over 720 hours. Initially the polar phase consisted of the water soluble rhodium catalyst $(HRh(CO)(TPPTS)_3)$ was dissolved in water only and the nonpolar phase consisted of 1-octene dissolved in heptane. Under these conditions, the mixture is strictly biphasic. As shown in Figure 28.4, the olefin conversion is very low under biphasic conditions. We then added 10% v/v dioxane to the heptane/1-octene nonpolar feed and the conversion rate remained low. Next the cationic surfactant CTAB was added to the aqueous catalyst phase. This resulted in a dramatic increase in the percent olefin conversion. An observation we made at this point was, to be successful, that the system requires both the use of a water-soluble cosolvent and a surfactant. Adding a surfactant to $HRh(CO)(TPPTS)_3$ dissolved in water (without any cosolvent) causes the catalyst to precipitate from the solution. Precipitation of the catalyst suggests that an exchange reaction occurred between the sodium ions on the phosphines and cationic surfactant reducing the solubility of the catalyst in water. The addition of a water miscible cosolvent such as dioxane or ethanol dissolves the catalyst.

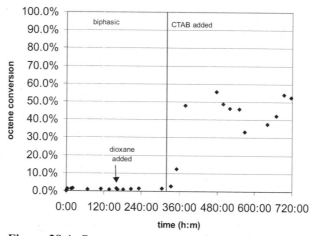

Figure 28.4. Percent octene conversion vs. time.

The 1-octene conversions averaged 50% at the current flow rate (residence time 30 minutes). We believe the scatter in the data is due to the drift in the pump flow rate, which alters the residence time, and not to a change in the catalyst itself. In all cases the linear to branch aldehyde selectivity was very high in the range of 5:1 linear to branch aldehyde. The reaction was run under thermomorphic conditions for over 400 hours and we found that we maintained good conversion and good selectivity.

Figure 28.5 shows another experiment that we did and this case we used 1-dodecene as the olefin. Once again we started under a strictly biphasic condition in which the catalyst was dissolved in water and the olefin dissolved in heptane. Under biphasic conditions the conversion was very low. We then added 10% v/v 1,4-dioxane to the nonpolar olefin-containing phase and the conversion remained low. When the surfactant CTAB was added, the conversion increased to approximately 20% and stayed at that level. The selectivity was greater than 5:1. The same flow rate was used for the octene and dodecene runs and, as expected, the switch to the increasingly higher olefin resulted in a drop in the reaction rate.

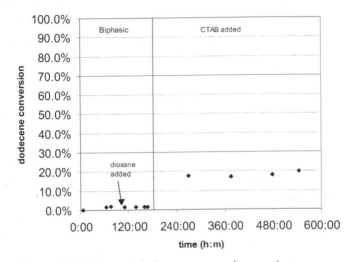

Figure 28.5. Percent dodecene conversion vs. time.

Economics. Comparison of the material and energy balance for our process and the cobalt-based BASF higher olefin process (8), we found that our process reduced the capital investment required by over 50% due to the fact that we require far fewer unit operations, and because the operating pressure is much lower. In summary, the thermomorphic solution developed by TDA allows easy catalyst recycle, which, when coupled with the lower pressure operation possible with Rh catalysts (compared to the cobalt-based process) lowers both capital and operating costs for current oxidation (oxo) plants of similar capacity.

Conclusions

We have developed a thermomorphic catalyst system for the hydroformylation of higher alkenes. We have built a bench-scale continuous reactor and have used it to determine the long-term performance of the thermomorphic catalyst system. Long-term results (>400 h) using 1-octene and 1-dodecene show that the catalyst has high selectivity and no measurable loss in activity.

Acknowledgements

This work was supported by the Department of Energy under SBIR Phase II grant DE-FG03-01ER83307/A002.

References

1. Billig E. and D.R. Bryant in <u>Encyclopedia of Chemical Technology</u> (ed. M. Howe-Grant), Wiley-Interscience, New York, (1995), vol. 17, p. 902.
2. R.T. Baker and W. Tumas, *Science*, **284**, 1477 (1999).
3. A.M. Halpern, and S. Gozashti, *J. Chem. Educ.* **65**, 371 (1988).
4. I.T. Horváth and J. Rábai, *Science* 266, 72-75 (1994); D.J. Adams, D. J. Cole-Hamilton, E.G. Hope, P.J. Pogorzelec and A. M. Stuart, *J. Organomet. Chem.* **689**, 1413 (2004); Y. Huang, E. Perperi, G. Manos and D.J. Cole-Hamilton, *J. Mol. Catal. A: Chem.* **210**, 17 (2004).
5. J.H. Hildebrand and D.R.F. Cochran, *J. Am. Chem. Soc.* **71**, 22 (1949).
6. D.E. Bergbreiter, Y.-S. Liu and P.L., *J. Am. Chem.. Soc.*, **120**, 4250 (1998).
7. C. Feng, Y. Wang, J. Jiang, Y. Yang and Z. Jin, *J. Mol. Catal. A: Chem.* **268**, 201 (2007).
8. C.D. Frohning, C.W. Kohlpaintner and H.-W. Bohnen in <u>Applied Homogeneous Catalysis with Organometallic Compounds</u> (eds. B. Cornils and W.A. Herrmann) Wiley-VCH: Weinheim (2002), vol. 1, p. 31.

29. Hydroformylation of 1-Butene on Rh Catalyst

Andreas Bernas[1], Johan Ahlkvist[1], Johan Wärnå[1], Päivi Mäki-Arvela[1], Juha Lehtonen[2], Tapio Salmi[1] and Dmitry Yu Murzin[1]

[1]Åbo Akademi, Process Chemistry Centre, Laboratory of Industrial Chemistry, FI-20500 Åbo/Turku, Finland
[2]Perstorp Oy, Technology Centre, P.O. Box 350, FI-06101 Borgå/Porvoo, Finland

Tapio.Salmi@abo.fi

Abstract

Kinetics in homogeneously catalyzed hydroformylation of 1-butene was studied in a pressurized semi-batch autoclave. Kinetics was determined for a reaction mixture, which consisted of 1-butene, carbon monoxide, hydrogen, a rhodium-based catalyst and 2,2,4-trimethyl-1,3-pentanediol monoisobutyrate solvent at the temperature 70-100°C, total pressure 1-3-MPa, and catalyst concentration 100-200 ppm, using varied catalyst (Rh)-to-ligand ratio as well as the initial ratio of the synthesis gas (hydrogen and carbon dioxide) components. The solubility of 1-butene, carbon monoxide and hydrogen in the solvent were determined by precise pressure and weight measurements. The solubility was modeled empirically and theoretically. The main reaction products were pentanal (P) and 2-methylbutanal (MB), while trace amounts of *cis*-2- and *trans*-2-butene were detected as reaction intermediates. The kinetic data were combined with solubility models and the parameters of an empirical power-law rate model were determined by non-linear regression analysis.

Introduction

Hydroformylation is the oldest and in production volume the largest homogeneously catalyzed industrial process. The hydroformylation reaction was discovered by Otto Roelen in 1938; the reaction is also called oxosynthesis and Roelen's reaction [1-13].

The catalysts used in hydroformylation are typically organometallic complexes. Cobalt-based catalysts dominated hydroformylation until 1970s; thereafter rhodium-based catalysts were commercialized. Synthesized aldehydes are typical intermediates for chemical industry [5]. A typical hydroformylation catalyst is modified with a ligand, e.g., triphenylphoshine. In recent years, a lot of effort has been put on the ligand chemistry in order to find new ligands for tailored processes [7-9]. In the present study, phosphine-based rhodium catalysts were used for hydroformylation of 1-butene. Despite intensive research on hydroformylation in the last 50 years, both the reaction mechanisms and kinetics are not in the most cases clear. Both associative and dissociative mechanisms have been proposed [5-6]. The discrepancies in mechanistic speculations have also led to a variety of rate equations for hydroformylation processes.

A lot of research has been published on hydroformylation of alkenes, but the vast majority of the effort has been focused on the chemistry of various metal-ligand systems. Quantitative kinetic studies including modeling of rates and selectivities are much more scarce. In this work, we present the approach to modeling of hydroformylation kinetics and gas-solubility. Hydroformylation of 1-butene with a rhodium-based catalyst was selected as a case study.

Experimental Section

Both gas solubility measurements and hydroformylation experiments were carried out in a pressurized autoclave (Parr 4561, 150 ml) at various pressures and temperatures. The flow of the synthesis gas was regulated by a pressure controller (Brooks 5866, Brooks 0154). The temperature was controlled by a heating jacket coupled to a steering unit (Par 4843), which was also used to control the stirring speed. The temperature was measured by a thermocouple and the pressure was detected by a pressure sensor (Keller Type PA21 SR/80520.3-1). In case of solubility measurements, the consumption of 1-butene was recorded by a balance. In the solubility measurements, the liquid-filled reactor was first evacuated by a vacuum pump (Vacuubrand PC 3 Type RZ-2). An analogous procedure was applied on the amount of 1-butene loaded into the reactor in the kinetic measurements. Gas was dispersed with the aid of a sinter (2 μm). The reaction solution was prepared prior to the kinetic experiment. In a tight flask a measured amount of catalyst was dissolved under stirring and moderate heating (50°C). To avoid the oxidation of the ligand and the catalyst the work was performed under nitrogen pressure. The reactor was loaded, rinsed with nitrogen and then exposed to vacuum (10 min). 1-Butene was loaded into the reactor at room temperature under stirring (400 rpm). The amount of 1-butene was measured (about 9.5 g 1-butene) and the pressure was registered. The heating of the reaction mixture to the desired temperature took place under constant agitation (400 rpm). As the reaction temperature was obtained, the stirring speed was adjusted to 1000 rpm. As the system had stabilized itself, the first sample was withdrawn; thereafter synthesis gas (CO/H₂) was introduced into the system. Samples were withdrawn with increasing sampling intervals. The samples were analyzed by gas chromatography. An internal standard method using dodecane was applied. The total amount of butane in the liquid phase was estimated from the total amount of aldehydes.

Results and Discussion

Solubility of 1-butene in 2,2,4-trimethyl-1,3-pentanediol monoisobutyrate. The solubility data were first modeled with the empirical equation [11] giving mole fraction of dissolved gas (x_i)

$$\ln x_i = A' + \frac{B'}{(T/K)} + C \ln(T/K) \tag{1}$$

where i is the component index, T is the absolute temperature (in K) and A', B' and C' are coefficients to be determined from the data by regression analysis. The concentration is given by

$$c_{Li} = p_i x_i c_L \tag{2}$$

where c_L is the total concentration of liquid. The solubility of 1-butene decreases with increasing temperature as expected. As a next step, the empirical equation (1) was fitted to the data by regression giving the following result

$$\ln x_g = -91.417 + \frac{6438,5}{(T/K)} + 12.039 \ln(T/K) \tag{3}$$

The equation is valid at 297 – 373 K. The fit is very good as shown in [12]. The degree of explanation was 99.8%. The solubility of CO and H_2 were reported elsewhere [12].

Kinetic measurements.
All the kinetic experiments were carried out with 1-butene, CO, H_2 and carbonylhydridotris(triphenylphoshine)rhodium(I), $RhH(CO)(TPP)_3$ as the rhodium precursor and triphenylphoshine, TPP as the ligand. H_2 and CO were fed 1:1 mixture in most experiments. Few experiments were carried out with a slight excess of H_2 (H_2 : CO = 55:45). The following process parameters were investigated: temperature (70-100°C), pressure (1-3 MPa), ligand-rhodium ratio (L/Rh=10-100 mol/mol) and rhodium concentration (100-200 ppm).   Based on the experiments, some general statements can be made. Pentanal and 2-methyl butanal were the absolutely dominating products; just trace amounts of plausible reaction intermediates, *cis*-2-butene and *trans*-2-butene were detected. In experiments carried out with a stoichiometric feed of H_2 and CO, no butane was formed, while trace amounts of butane became visible in experiments performed with an excess of H_2. The effect of various process parameters are discussed in detail in the sequel.

Temperature and pressure.
The experiments revealed that the initial rate increases as a function of temperature. The product distribution was not strongly affected by the temperature. The product ratio of pentanal-to-2-methyl butanal is approximately 4. The total pressure did not have any significant effect on the product distribution.

Ligand concentration.
The effect of Rh concentration on the selectivity was negligible. The ligand concentration, on the other hand, is expected to have a considerable effect on the selectivity; thus experiments were carried out with different ligand concentrations. A higher ligand concentration retards the overall rate, but increases the selectivity to pentanal. At the lowest L/Rh ratio (10), the ratio pentanal-to-2-methylbutanal was about 3.4, while it became about 5 for the highest L/Rh ratio (10).

Kinetic modeling principles.
Modeling of the hydroformylation kinetics was started by constructing a stoichiometric scheme for the process. An detailed reaction pattern is shown in Scheme 29.1.

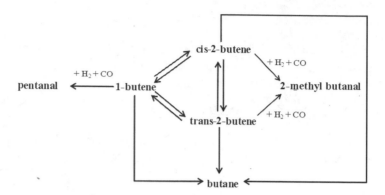

Scheme 29.1. Reaction scheme for hydroformylation of 1-butene.

Experimental data, however, indicated that just trace amounts of *cis*-2-butene and *trans*-2-butene appeared in the experiments; the concentrations of these compounds were typically less than 3 mol%. This implied that their impact on the total balance was very minimal, particularly considering the general accuracy of the data. In the experiments where the stoichiometric ratio $H_2:CO = 1:1$ was used, no butane was detected. Thus in the mathematical modeling of kinetic data, it is possible to use a considerably simplified scheme

$$\text{(1)} \qquad\qquad \text{(2)}$$
$$\text{pentanal} \leftarrow \text{1-butene} \rightarrow \text{2-methyl butanal}$$
$$\text{(P)} \qquad\quad \text{(A)} \qquad\qquad \text{(MB)}$$

The generation rates of the components (A = 1-butene, P = pentanal and MB = 2-methylbutanal) thus, according to the stoichiometry, become

$$r_A = -r_1 - r_2 \tag{4}$$

$$r_P = r_1 \tag{5}$$

$$r_{MB} = r_2 \tag{6}$$

The next step is to find suitable expressions for r_1 and r_2. Considerable simplifications are in any case needed. Because of these reasons, the empirical expressions for r_1 and r_2 listed below were used as a starting approach (L=ligand),

$$r_1 = k_1 c_A^{m_1} c_{cat}^{m_2} c_{CO}^{m_3} c_{H_2}^{m_4} c_L^{m_5} \tag{7}$$

$$r_2 = k_1 c_A^{n_1} c_{cat}^{n_2} c_{CO}^{n_3} c_{H_2}^{n_4} c_L^{n_5} \tag{8}$$

where the constants to be estimated are k_1 and k_2 as well as the exponents (m, n). The ratios of the exponents can be preliminary obtained by checking the product distribution plots. Essentially the product distribution depended on the ligand concentration only, while it was rather independent on the catalyst concentration and total pressure. The change of total pressure (10 MPa, 30 MPa) also implied the change of the partial pressures of CO and H_2 and thus their concentration in the liquid phase. This implies that $m_2 = n_2$, $m_3 = n_3$ and $m_4 = n_4$ in Eqs. (7) and (8).

Consequently, we remain at the first stage with the parameters m_1, m_2, m_3, m_4, m_5 and n_5. To progress with parameter estimation, the mass balances in the reactor will be considered.

Mass balances for the reactor.
The reactor system, where the kinetic experiments were carried out can be described as a semi-batch reactor. Only the synthesis gas (H_2 and CO) was fed into the reactor continuously during the experiments, while 1-butene and the solvent were in the batch mode. All reactions took place in the liquid phase. The mass balance for an arbitrary component in the gas is given by

$$\dot{n}_{0Gi} = N_i A_m + \frac{d n_{Gi}}{dt} \tag{9}$$

where $\dot{n}_{0Gi}$ is the inlet flow and N_i the interfacial flux and A_m is the interfacial area. For the liquid phase, the corresponding balance equation is written as

$$N_i A_m + r_i V_L = \frac{d n_{Li}}{dt} \tag{10}$$

For components which are in batch (1-butene and reaction products) n_{0i} is zero. By introducing the concentrations ($n_{Li} = c_i V_L$ and $n_i = c_{Gi} V_G$) and adding together the balances (for case $n_{oLi} = 0$), we get

$$\frac{V_G}{V_L} \frac{d c_{Gi}}{dt} + \frac{d c_{Li}}{dt} = r_i \tag{11}$$

As the agitation of the reaction mixture was very intensive, interfacial mass transfer resistance is suppressed, and the concentration in gas and liquid are related by the phase equilibrium

$$\frac{c_{Gi}}{c_{Li}} = K_i \quad \Rightarrow \quad c_{Gi} = K_i c_{Li} \tag{12}$$

This is inserted in the balance equation, which gets its final form

$$\frac{d c_{Li}}{dt} = \frac{r_i}{\dfrac{V_G}{V_L} K_i + 1} \tag{13}$$

For non-volatile components $K_i = 0$ the equilibrium parameter (K_i) is expressed by Henry's constant as follows

$$K_i = \frac{H_i}{RT c_L} \tag{14}$$

where c_L is the total concentration of the liquid. Henry's constant was obtained from the solubility model. Since the measured liquid-phase concentration of 1-butene was not completely correct, the concentration of 1-butene in the liquid phase was calculated from the overall balance,

$$c_{AL}(V_L + K_A V_G) = n_{0A} - n_P - n_{MB} \tag{15}$$

i.e., consumed butane is equal to formed products (the concentrations of the products were zero at $t = 0$). The amounts of P and MB are expressed with concentrations, $n_P = c_{PL} V_L$ and $n_{MB} = c_{MBL} V_L$. In case that all 1-butene (A) was reacted, we get the balance

$$\frac{n_{0A}}{V_L} = c_{PL\infty} + c_{MBL\infty} \tag{16}$$

where ∞ denotes the asymptotic concentration of the products. After combining all the information, the concentration of butane in the liquid phase is obtained ($c_{LA} = c_A$),

$$c_A = \frac{c_{P\infty} + c_{MB\infty} - \left(c_P + c_{MB}\right)}{1 + K_A V_G / V_L} \tag{17}$$

This expression was used in the parameter estimation, i.e., just the concentrations c_P and c_{MB} were used in the data fitting, and c_A was calculated from Eq. (17). The rate constants included in the model were described by the modified Arrhenius equation

$$k = k_{ref} e^{\left(-\frac{E_a}{R}\left(\frac{1}{T} \frac{1}{T_{ref}}\right)\right)} \tag{18}$$

where T_{ref} was calculated from

$$k_{ref} = A e^{-\left(\frac{E_a}{RT_{ref}}\right)} \tag{19}$$

T_{ref} was 358.15 K (85°C). The transformation (18) was used to suppress the correlation between A and E_a.

Parameter estimation.
In the parameter estimation, the product components P and MB were included. The objective function was defined as

$$Q = \sum_t \sum_i (c_{it} - c_{it\,exp})^2 \tag{20}$$

where c_{it} denotes the modeled liquid-phase concentration at time t and c_{itexp} is the corresponding experimental concentration. The objective function was minimized with a hybrid Simplex-Levenberg-Marquardt method by using Modest software [13]. The regression analysis revealed that the original rate equations can be considerably simplified. Parameters m_1, n_1, m_2 and n_2 get the value 1, while $m_3 = n_3 = m_4 = n_4 = 0$ in the rate expressions (7–8). This is in accordance with many previous investigations of hydroformylation. Thus the regression procedure ended up with two simple rate equations

$$r_1 = r_P = k_1 \frac{c_A c_{cat}}{c_L^\delta} \tag{21}$$

$$r_2 = r_{MB} = k_2 \frac{c_A c_{cat}}{c_L^\gamma} \tag{22}$$

where $\delta = 0.28$ and $\gamma = 0.44$. The parameters obtained from the parameter estimation are collected in Table 29.1.

Table 29.1. Kinetic parameters for hydroformylation

Parameter	Estimated parameter value	Estimated st. dev.	Estimated rel. st. dev. [%]	St. dev. of the parameter
k_1 $[(dm^3)^{1.72} mol_L^{0.28} g_{Rh}^{-2} min^{-1}]$	8.51E-03	3.93E-04	4.6	21.7
k_2 $[(dm^3)^{1.56} mol_L^{0.44} g_{Rh}^{-2} min^{-1}]$	1.43E-03	1.02E-04	7.1	14.1
E_{a1} [J/mol]	6.49E+04	1.34E+03	2.1	48.3
E_{a2} [J/mol]	7.55E+04	1.88E+03	2.5	40.1
δ	0.278	1.48E-02	5.3	18.8
γ	0.439	2.11E-02	4.8	20.8

As revealed by Table 29.1, the parameters are very well identified and thus statistically relevant. The obtained modeling results can be compared with preliminary results obtained just by checking kinetic data. The product plots c_P vs. c_{MB} should according to the model be linear. The difference between the activation energies E_{a2} - E_{a1} is about 11 kJ/mol, which is a small value. Thus the temperature dependence of the rate constants is very weak, which was confirmed. High temperatures slightly favor the formation of MB. Some selected modeling results are displayed in Figure 29.1. In general the modeling results follow the experimental curves of pentanal (P) and 2-methyl-butanal (MB) rather well except in some few cases. The predicted 1-butene concentration is typically slightly higher than the experimental one, as it should be because of volatilization during the sampling. The degree of explanation became 97-99%, which is very satisfactory. Thus we can conclude that the modeling approach was successful.

Conclusions

Hydroformylation of 1-butene in the presence of the Rh catalyst gave pentanal (P) and 2-methyl butanal as the main products. Just trace amounts of *cis*-and *trans*-2-butene were detected as by-products. No butane was detected in experiments, where a stoichiometric ratio of CO and H_2 were used. Based on preliminary considerations of product distributions, a kinetic model was developed. The kinetic parameters obtained from the model were well identified and physically reasonable. The product concentrations are predicted very well by the kinetic model. The kinetic model can be further refined by considering detailed reaction mechanisms and extending it to the domain of lower partial pressures of CO and H_2.

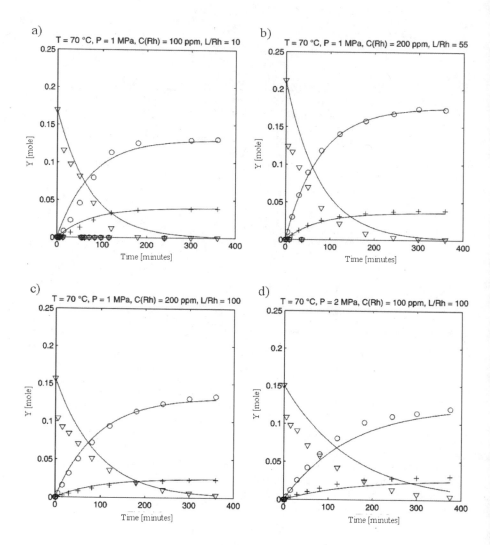

Figure 29.1. Fitting of the model (solid line). (a) T=70°C, p=1 MPa, *c(Rh)*=100 ppm, L/Rh-mole ratio=10, (b) T=70°C, p=1 MPa, *c(Rh)*=200 ppm, L/Rh-mole ratio=55, (c) T=70°C, p=1 MPa, *c(Rh)*=200 ppm, L/Rh-mole ratio=100, (d) T=70°C, p=2 MPa, *c(Rh)*=100 ppm, L/Rh-mole ratio=100. Symbols: (o) pentanal, (+) 2-methyl butanal and (∇) 1-butene.

Acknowledgements

This work is a part of the Finnish Centre of Excellence Programs (2000-2011) financed by the Academy of Finland. Financial support from Perstorp is gratefully acknowledged.

Notation

A	frequency factor
A	gas-liquid interfacial area
A', B', C'	parameters in gas solubility model
c	concentration
E_a	activation energy
H	Henry's constant
K	gas-liquid equilibrium ratio
k	rate constant
k'	transformed rate constant
m, n	empirical exponents in rate expressions
N	interfacial (gas-liquid) flux
n	amount of substance
$\dot{n}$	flow of amount of substance
P	total pressure
p	partial pressure
Q	objective function
R	gas constant (= 8.314 $Jmol^{-1}K^{-1}$)
r	rate
T	absolute temperature (K)
t	time
x	mole fraction
V	volume
$\alpha_1, \alpha_2, \beta_1$	merged parameters
γ, δ	empirical exponents in rate expressions

Subscripts and superscripts

G	gas
L	liquid
i	component index

Abbreviations

A	1-butene
cat	catalyst
L	ligand
MB	2-methylbutanal
P	pentanal
ref	reference
4	asymptotic value

References

1. B. Cornils and W. A. Herrmann, *J. Catal.* **216**, 23 (2003).
2. The new catalyst technical handbook; Johnson Matthey: 2001.

3. B. Cornils and W. A. Herrmann, in *Applied homogeneous catalysis with organometallic compounds: a comprehensive handbook in three volumes*, Wiley-Vch: Weinheim, 2002; volume 1, chapter 1.

4. R. A. Sheldon, Chemicals from Synthesis Gas: Catalytic Reaction of CO and H_2; Reidel: Dordrecht, 1983.

5. C. D. Frohning, C. W. Kohlpaintner, and H. Bohnen, in *Applied homogeneous catalysis with organometallic compounds: a comprehensive handbook in three volumes*, 2nd ed.; Wiley-Vch: Weinheim, 2002; volume 1, chapter 2.

6. H. W. Bohnen, and B. Cornils, *Adv. Catal.* **47**, 1 (2002).

7. G. G. Stanley, *Organometallic Chemistry*, 28.5.2003; http://chemistry.lsu.edu/ stanley/webpub/4571-chap16-hydroformylation.pdf; 4.10.2004.

8. R. L. Pruett and J. A. Smith, *J. Org. Chem.* **34(2)**, 327 (1969).

9. M. Beller, B. Cornils, C. D. Frohning, and C. W. Kohlpaintner, *J. Mol. Catal. A: Chem.* **104**, 17, (1995).

10. R. V. Chaudhari, A. Seayad, and S. Jayasree, *Catalysis Today* **66**, 371 (2001).

11. P. G. T. Fogg and W. Gerrard, Solubility of Gases in Liquids: A Critical Evaluation of Gas/Liquid Systems in Theory and Practice; Wiley: Chichester, 1991.

12. C. Still, T. Salmi, P. Mäki-Arvela, K. Eränen, D. Yu. Murzin, and J. Lehtonen, *Chem. Eng. Sci.* **61**, 3698 (2006).

13. Haario, H. *ModEst*, 6.1; Software for kinetic modeling; ProfMath: Helsinki, 2002.

30. Innovative One-Pot Oxidation Method for the Synthesis of Bioproducts from Renewables

Pierre Gallezot and Alexander B. Sorokin

Institut de recherches sur la catalyse et l'environnement de Lyon 2, avenue Albert Einstein, 69626 Villeurbanne, France

pierre.gallezot@ircelyon.univ-lyon1.fr

Abstract

Fine and specialty chemicals can be obtained from renewable resources via multi-step catalytic conversion from platform molecules obtained by fermentation. An alternative method decreasing the processing cost is to carry out one-pot catalytic conversion to final product without intermediate product recovery. This latter option is illustrated by an innovative oxidation method developed in our laboratory to oxidize native polysaccharides to obtain valuable hydrophilic end-products useful for various technical applications.

Introduction

There are strong incentives for the use of renewables in the production of chemicals. Indeed, bio-based resources are renewable and CO_2 neutral in contrast with fossil fuels. Also, the molecules extracted from bio-based resources are already functionalised so that the synthesis of chemicals may require a lower number of steps than from hydrocarbons whereby decreasing the energy consumption and the overall waste generated. Bio-based products may have unique properties compared to hydrocarbon-derived products, for instance biodegradability and biocompatibility. Products issued from biomass have a potential market differentiation and their marketing is made easier because of their "natural" or "bio" label. However, the processing cost of biomass derivatives to specialties and fine chemicals has to be decreased by designing new value chains and catalytic systems different from those employed for hydrocarbons. Synthesis routes should be adapted to the specific molecular structure of biomolecules.

Two process options to convert biomass derivatives into valuable bioproducts via innovative catalytic routes can be proposed, viz.: (i) Synthesis of chemicals via platform molecules. According to this option biomass is first converted via biotechnological processes to well-identified platform molecules that can be employed as building blocks for chemical synthesis via catalytic routes. (ii) Synthesis of bioproducts via one-pot reaction without recovery of intermediate products. These process options could be integrated in a biorefinery which is a facility to produce various chemicals from renewable resources such as carbohydrates, vegetable oils and terpenes. The underlying idea is to maximize the value derived from biomass via well-integrated processes, valorizing co-products and by-products, optimizing the inputs (feedstock supply, water management) and outputs (energy and product recovery, treatments of waste).

Conversion via platform molecules. As far as carbohydrates are concerned, a number of platform molecules are already well identified and currently employed to synthesize specialties and fine chemicals. Thus, lactic (2-hydroxypropionic) acid obtained by fermentation of glucose and polysaccharides is employed to synthesize polylactide (PLA) a biodegradable polymer industrially produced by NatureWorks®. A number of platform molecules such as ethanol and carboxylic acids are obtained by fermentation of carbohydrates. In the future the challenge will be to produce fermentable sugars from cellulose and hemicelluloses that are available in huge amounts from vegetative biomass. Fatty acid esters, fatty acids and fatty alcohols are all valuable platform molecules derived from vegetable oils. Terpenes such as α– and β–pinenes, which are extracted from turpentine oil, are building blocks for the synthesis of flavor and fragrances. Various examples of catalytic conversion of platform molecules to fine chemicals have been reviewed recently (1).

Conversion via one-pot reaction. The processing cost to convert biomass to valuable products and the amount of wastes can be greatly reduced under the following process conditions: (i) Multistep reactions to targeted molecules carried out by cascade catalysis without intermediate product recovery. Cascade catalysis may involve more than two steps combining enzymatic, homogeneous, and heterogeneous catalysis. Several examples of this approach were given by Schoevaart and Kieboom (2). (ii) One-step conversion to a mixture of products that can be used as such for the further synthesis of end-products. In the food industry there is usually no requirement to prepare specific molecules to be used as nutrients, but rather a mixture of triglycerides and/or carbohydrates. This could well be extended to prepare commodities such as paints, paper, construction materials, cosmetics, etc. This innovative approach will be illustrated in the present paper by the chemical modification of polysaccharides into valuable products. Figure 30.1 gives a simplified scheme summarizing these process options.

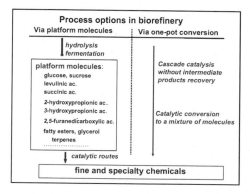

Figure 30.1 Process options for the catalytic conversion of biomass.

Oxidized polysaccharides are already used in different technological applications such as paper, textile, and detergent industries as well as in pharmaceutical and cosmetic applications (3-5). However, the potential of starch and other polysaccharides can still be improved to meet the specifications such as adhesive power, viscosity, water binding capacity and sequestering ability needed for various applications. The oxidation of native biopolymers needs to be tailored to specifications by developing clean, general and practical oxidation methods. Current methods of preparation of oxidized polysaccharides, mainly starch, are often based on stoichiometric oxidation with NaOCl (6-9) or N_2O_4 (10) to introduce carboxyl groups or with $NaIO_4$ (6, 11) to obtain aldehyde functions. These oxidations are efficient but lead to large amounts of inorganic wastes and/or toxic chlorinated by-products. Catalytic approaches have been proposed to improve starch oxidation using 2,2,6,6-tetramethyl-1-piperidinyloxy (TEMPO) in combination with NaOCl/NaBr (12-16) or Fe, Cu and W salts in combination with H_2O_2 (17-20). However, high amounts of metal (typically 0.01–0.1%) are necessary to achieve an efficient oxidation. Since oxidized starch has good chelating properties, heavy metals can be retained by carboxyl functions in the modified starch. The high metal content in modified starch causes undesirable coloration thus preventing its use in applications where brightness and/or low metal content are required. Recently, we have developed an efficient and practical method for a waste-free oxidation of starch (21-23). In this chapter, we describe the application of this method for the oxidation of other polysaccharides.

Experimental Section

The oxidation of insoluble polysaccharides was performed by H_2O_2 in the presence of small amounts of iron tetrasulfophthalocyanine, a cheap blue dye used as catalyst, to introduce aldehyde and carboxyl functions in polymer chains (Figure 30.2).

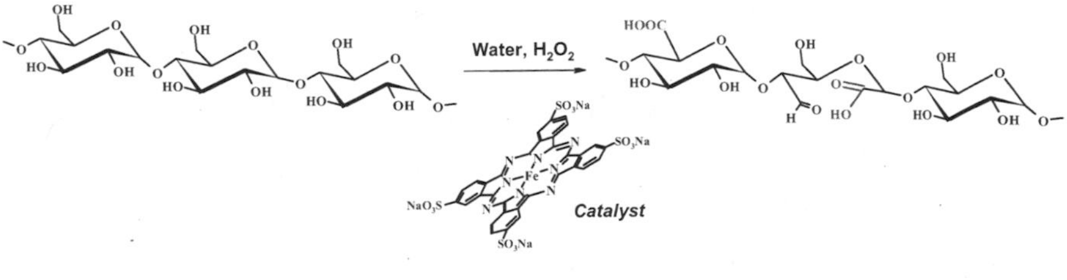

Figure 30.2 Schematic representation of starch oxidation.

The oxidation can be performed heterogeneously with the insoluble polysaccharide suspended in an aqueous solution of catalyst. In this case the reaction conditions should be carefully controlled because of possible gelatinization. The isolation of product can be performed either by water elimination, which is a time and energy consuming procedure, or by addition of organic solvents such as EtOH to precipitate the product. Another more practical approach was to perform the oxidation in semi-dry conditions because the pores in starch granules are able to absorb up to 0.5 g of water per gram of starch. Accordingly, small amounts of aqueous solution containing 0.00004–0.0002 mole per anhydroglucose unit (AGU) of catalyst were added to the mechanically mixed polysaccharide powder. The mixing resulted in an even distribution of the catalyst in the powder as indicated by the homogeneous blue color of the mixture. Then, required amounts of H_2O_2 to obtain the desired level of oxidation was added drop wise, usually 0.1–0.4 molar equiv. to polysaccharide, while the reaction mixture was mixed and heated at 60°C. After complete H_2O_2 consumption determined by the iodometric method, a colorless solid mixture was obtained because the catalyst was oxidized. Trace amounts of bleached catalyst can be easily removed by washing with cold water. After drying, the oxidized polysaccharides exhibited low residual iron content, typically 6-40 ppm, i.e., only slightly more than the iron content of native potato starch. Due to the very small amounts of catalyst, the modified polysaccharides prepared by this method are pure enough to be used directly in most applications without further water washing.

The degree of starch oxidation was evaluated by measuring the degree of substitution by carboxyl (DS_{COOH}) and carbonyl (DS_{CO}) groups. The degree of substitution DS_{COOH} and DS_{CO} expressed as the number of carboxyl and carbonyl groups per 100 AGU were determined by Smith's method (17). Modified polysaccharides were also characterized by scanning electron microscopy, thermogravimetric analysis and ^{13}C NMR (Figure 30.3).

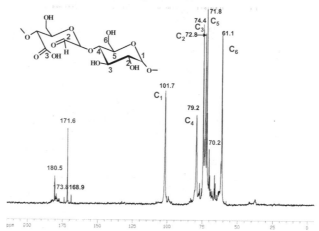

Figure 30.3 ^{13}C NMR spectrum of oxidized starch. (Signals at 171.6 and 180.5 ppm are attributed to carboxyl and aldehyde groups, respectively).

Results and Discussion

We have developed an efficient and practical method for clean oxidation of starch (21-23) resulting in the oxidation of primary alcohol function in C_6 position and the cleavage of vicinal diols in C_2 and C_3 position (Figure 30.2). We used small amounts of cheap iron tetrasulfophthalocyanine catalyst, pure water as reaction medium and H_2O_2 as clean oxidant to achieve a one-pot conversion of starch resulting in the introduction of aldehyde and carboxyl functions in polymer chains. The iron content in the final product was only slightly higher than that of native starch because of the small amounts of catalysts used and because part of the iron (0.004–0.008 mol %) was not retained in starch after washing with cold water.

The scope of the method was extended to the oxidation of other polysaccharides with very different structures and properties (Figure 30.4). Cellulose is the most abundant polysaccharide, but the most difficult to modify because of β-*1,4*-glycosidic linkages. Hydroxyethyl cellulose, a soluble, non-ionic polymer containing 0.8–3 ethyleneglycol residues per anhydroglucose unit (AGU), is mainly used in pharmaceutical and cosmetic industries (annual production: 42.5 million pounds, 1979). Carboxymethyl cellulose (annual production: 700 million pounds, 2001) is used in food, cosmetic, personal care products and drug applications. Guar gum possesses a linear backbone composed of β-*1,4*-linked d-mannopyranosyl units with irregularly distributed α-*1,6*-linked galactopyranosyl residues (mannose/galactose ratio ~ 2:1). Inulin from plants (low molecular weight, 480–10,000) or from fungal and bacterial origin (high molecular weight, $2·10^7$) is composed of linear *1,2*-polyfructosides with a terminal glucose unit.

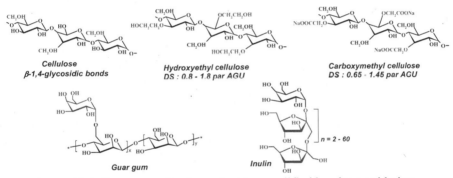

Figure 30.4. Structures of polysaccharides modified by clean oxidation.

The introduction of aldehyde and carboxyl groups via oxidation of these polysaccharides can give access to new materials with increased hydrophilic character. All these polysaccharides have been successfully modified. Substitution degrees and ratio COOH/CHO are listed in Table 30.1. Depending on the structure of polysaccharide different COOH/CHO ratios were obtained. Thus, depending on the choice of starting polysaccharides and the level of oxidation, modified starch materials with very different properties were obtained. The presence of carboxyl functions improves hydrophilic properties that are important for numerous technological applications, while the presence of carbonyl groups allows further chemical modifications to obtain materials with specific properties. We believe that this catalytic method, which is easy to handle and fulfill all the principles of green chemistry, might have a real future in using the renewable materials substituting feedstock derived from fossil fuels. The natural origin of these materials brings a high added value to the final products and allows an easier marketing.

Table 30.1. Oxidation of polysaccharides with H_2O_2 catalyzed by iron tetrasulfophthalocyanine.

Polysaccharides	DS_{COOH}*	DS_{CHO}*	COOH/CHO
Inulin	14.8	17.9	0.82
α-cellulose	5.2	4.4	1.18
Potato starch	19.4	15.6	1.24
Guar gum	13.7	9.2	1.49
Hydroxyethyl cellulose	11.4	7.2	1.58
Carboxymethyl cellulose	15.5	4.4	3.52

*Degree of substitution in carboxyl (DS_{COOH}) and carbonyl (DS_{CHO}) expressed as the number of carboxyl and carbonyl groups per 100 anhydroglucose units (AGU).

A tentative oxidation mechanism was proposed in Figure 30.5. In a first step the oxidation could occur at C_2/C_3 position resulting in a ketone group adjacent to the hydroxyl group in C_2 position. Then, this intermediate can undergo a nucleophilic attack at keto-position by iron (III) peroxo complex $PcFe^{III}\text{-}O\text{-}O^-$ to form a six-membered transition state. The C_2–C_3 bond is then cleaved as indicated in Figure 30.5 via Grob type fragmentation to form aldehyde and carboxyl functions. Aldehyde groups can be further oxidized to carboxyl groups. This mechanism is akin to that previously proposed for the oxidation of chlorinated phenols by iron phthalocyanine–H_2O_2 system (24-27). The oxidation of the primary alcohol function at C_6 position could also occur, but ^{13}C NMR data are in agreement with the proposal that the principal pathway is the cleavage of C_2–C_3 bond.

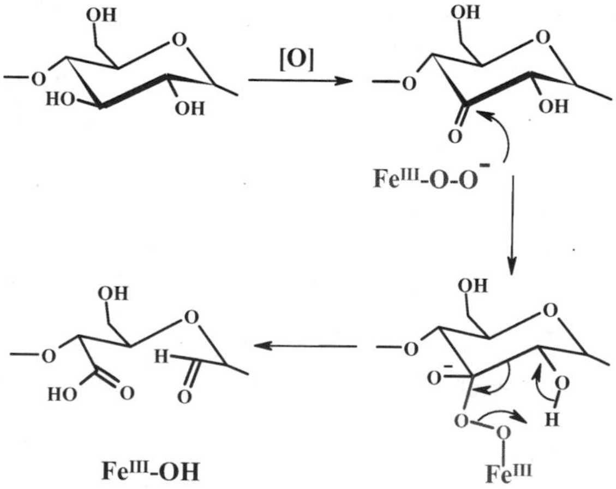

Figure 30.5. Proposed oxidation mechanism of glucoside unit by iron phthalocyanine-H_2O_2 system.

Conclusion

A practical method of modification of polysaccharides by clean oxidation using H_2O_2 as oxidant and cheap iron phthalocyanine as catalyst has been developed. Since no acids, bases or buffers and no chlorinated compounds were used, a pure product can be recovered without additional treatment. Importantly, this flexible method provides materials with a wide range of DS_{CHO} and DS_{COOH} just by an appropriate choice of the reaction conditions. Oxidized polysaccharides thus obtained possess various, tailormade hydrophilic/hydrophobic properties which have been tested successfully in cosmetic and other applications.

References

1. P. Gallezot, *Catal. Today*, **121**, 76 (2007).
2. R. Schoevaart, T. Kieboom, *Topics Catal.*, **27**, 3 (2004).
3. H.W. Maurer, R.L. Kearney, *Starch/Stärke*, **50**, 396 (1998).
4. D.J. Thomas, W.A. Atwell, *Starches* (Eagan Press, St. Paul, MN, 1999).
5. H. Röper, *Starch/Starke*, **54**, 89 (2002).
6. M.S. Nieuwenhuizen, A.P.G. Kieboom, H. van Bekkum, *Starch/Stärke*, **37**, 192 (1985).
7. M. Floor, A.P.G. Kieboom, H. van Bekkum, *Starch/Stärke*, **41**, 348 (1985).
8. A. Teleman, K. Kruus, E. Ämmälahti, J. Buchert, K. Nurmi, *Carbohydr. Res.*, **315**, 286 (1999).
9. D. Kuakpetoon, Y.-J. Wang, Starch/Stärke, **53**, 211 (2001).
10. H. Kochkar, M. Morawietz , W.F. Hölderich, *Appl. Catal. A*, **210**, 325 (2001).
11. S. Veelaert, D. de Wit, K.F. Gotlieb, R. Verhé, *Carbohydr. Polym.*, **33**, 153 (1997).
12. A.E.J. de Nooy, A.C. Besemer, H. van Bekkum, *Carbohydr. Res.*, **269**, 89 (1995).
13. P.S. Chang, J.F. Robyt, *Carbohydr. Lett.*, **3**, 31 (1998).
14. P.L. Bragd, A.C. Besemer, H. van Bekkum, *J. Mol. Catal. A: Chem.*, **170**, 35 (2001).
15. Y. Kato, R. Matsuo, A. Isogai, *Carbohydr. Polym.*, **51**, 69 (2003).
16. P. Bragd, A.C. Besemer, H. van Bekkum, *Carbohydr. Polym.*, **49**, 397 (2002) and references therein.
17. P. Parovuori, A. Hamunen, P. Forssell, K. Autio, K. Poutanen, *Starch/Stärke*, **47**, 19 (1995).
18. A.M. Sakharov, N.T. Silakhtarnyan, I.P. Skibida, *Kinet. Catal.*, **37**, 368 (1996).
19. R. Manelius, A. Buléon, K. Nurmi, E. Bertoft, *Carbohydr. Res.*, **329**, 621 (2000).
20. M. Floor, K.M. Schenk, A.P.G. Kieboom, H. van Bekkum, *Starch/Stärke*, **41**, 303 (1989).
21. A.B. Sorokin, S.L. Kachkarova-Sorokina, C. Donzé, C. Pinel, P. Gallezot, *Top. Catal.*, **27**, 67 (2004).
22. S.L. Kachkarova-Sorokina, P. Gallezot, A.B. Sorokin, *Chem. Commun.*, 2844 (2004).

23. S.L. Kachkarova-Sorokina, P. Gallezot, A.B. Sorokin, WO Patent 2004/007560 A1 (2004).
24. A.B. Sorokin, J.-L. Séris, B. Meunier, *Science*, **268,** 1163 (1995).
25. A.B. Sorokin, B. Meunier, *Chem.Eur. J.*, **2**, 1308 (1996).
26. B. Meunier, A.B. Sorokin, *Acc. Chem. Res.*, **30,** 470 (1997).
27. A.B. Sorokin, S. De Suzzoni-Dezard, D. Poullain, J.-P. Noël, B. Meunier, *J. Am. Chem. Soc.*, **118**, 7410 (1996).

31.   A Process for the Production of Biodiesel from Non-Conventional Sources

Paolo Bondioli[1], Laura Della Bella[1],
Nicoletta Ravasio[2] and Federica Zaccheria[2]

[1]*Stazione Sperimentale Oli e Grassi, via G. Colombo79,*
20133 Milano, Italy
[2]*CNR –ISTM and Dip. CIMA, University of Milano, via G.Venezian 21, 20133*
Milano, Italy

federica.zaccheria@istm.cnr.it

Abstract

Good quality biodiesel can be prepared starting from highly unsaturated starting materials, that means high iodine value oils, through a selective hydrogenation process promoted by heterogeneous copper catalysts. Thus, the process described allows improving the oxidative stability of the starting material maintaining good cold properties.

Introduction

Biodiesel may be represented chemically as a mixture of fatty acid methyl esters. It is a naturally derived liquid fuel, produced from renewable sources which, in compliance with appropriate prescriptions, may be used in place of diesel fuel for both internal combustion engines and for producing heat in boilers. The advantages, especially environmental, which can potentially result from the widespread use of biodiesel, are manifold:

- Its combustion does not contribute towards increasing the net atmospheric carbon dioxide concentration, one of the major factors responsible for the greenhouse effect;
- Considering the fact that the triglyceride oils used for the production of biodiesel are sulphur free, the use of biodiesel does not contribute towards the phenomenon of acid rain;
- Due to its particular composition, biodiesel is biodegradable and allows reduced emissions, in terms of particulates and polycyclic aromatic hydrocarbons. Instead, the results of the combustion of biodiesel are contentious in relation to so-called NO_x emissions, where it has been observed that such emissions are more or less increased, with respect to conventional diesel, depending on the characteristics of the engine in which it is used;
- For economies like the European economy, this is an excellent opportunity to have available and autonomous, as well as renewable energy source. The European Union strongly encourages this initiative, which allows the use of

marginal land, not dedicated to food production, with undoubted advantages for safeguarding and increasing the workforce and protecting the environment.

The production of this alternative fuel has seen enormous developments over the past fifteen years, during which production has progressed from the trial stage to annual global production of 5,800,000 tons (2007 data), the majority of which of European origin.

One of the most important barriers to the utilization of biodiesel is the price and availability of starting material. Vegetable oils that commonly are used as a feedstock for biodiesel production are responsible for more than 80% of the total biodiesel production costs. During this time used frying oils, as well as animal fats or other non-conventional oils were studied to assess their aptitude for biodiesel preparation. However, the use of alternative feedstocks can influence very much the quality of biodiesel obtained, this often obliges the manufacturer to prepare biodiesel blends in order to fulfill international standards such as EN 14214 or ASTM D 6751. This chapter is a description of a process that, starting from several high iodine values (IV) vegetable or animal oils, allows the preparation of a good quality biodiesel.

Experimental Section

Esterification reactions were carried out using 100 parts by weight of tall oil fatty acids and 100 parts by weight of methyl alcohol, in the presence of 0.5 g (approx. 0.3 ml) of concentrated sulphuric acid or 1 g of para-toluenesulphonic acid monohydrate. At the end of reaction, after lower phase separation containing the excess methanol, the water formed during the esterification reaction and the main part of acidic catalyst, the esterification mixture was washed several times with distilled water and finally dried under vacuum. After drying the ester fraction was distilled under reduced pressure (1 mbar at 180°C) in order to obtain a purified distilled fraction to be used for hydrogenation experiments. This unit operation allows to remove the main part of residual rosin acids that are not esterified in the above mentioned conditions and to remove the trace of eventually present sulphonated products. According to our experience the sulphonation of double bond of fatty acids does not take place under the described experimental conditions. Nevertheless the distillation step represents a safety guard for these unwanted side-products.

Hydrogenation reactions were carried out in a stainless steel autoclave at 180°C, under 2-8 atm H_2, in the presence of powdered supported Cu catalysts (10% wt) with a 3-15% copper loading. SiO_2, Al_2O_3 and TiO_2 were used as the catalyst support and the catalysts prepared as already reported[1]. These reactions were carried out on the esterified and distilled fraction of tall oil fatty acids.

Reaction mixtures, separated by simple filtration, were analyzed by GC (HP-6890) using a non-bonded, bis-cyanopropylpolysiloxane (100 m) capillary column. Control tests done by ICP-AES did not show any catalyst contamination on the final product.

The Nickel catalyst used for comparison was Calsicat E472D from Mallinckrodt Specialty Chemicals containing 22% Ni.

Results and Discussion

The degree of unsaturation of the fatty acids is normally expressed as the iodine number (IV), i.e., the number of grams of iodine that have reacted with 100 g of product analysed. The higher the index (number), the greater the degree of unsaturation. For example, for biodiesel intended for haulage use, the most remunerative use, a maximum iodine number limit is envisaged of 120 gI₂/100g. Biodiesel is normally produced by starting from vegetable oils having an iodine number of less than or equal to 130, i.e., having a low unsaturation index, such as rape seed oil (IV=112), sunflower oil (IV=123), soybean oil (IV=130).

The major fatty acids present in plant-derived fatty substances are oleic acid (9-octadecenoic, C18.1), linoleic acid (9,12-octadecadienoic, C18:2) and the conjugated isomers thereof and linolenic acid (9,12,15-octadecatrienoic, C18:3) (Scheme 31.1). Their rates of oxygen absorption are 100:40:1, respectively[2], hence partial hydrogenation with consequent lowering of the iodine number would lead to a significant increase in oxidative stability, particularly when C18:3 is reduced.

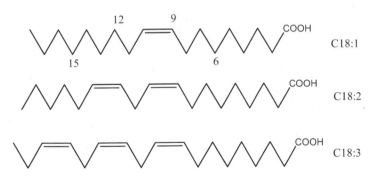

Scheme 31.1

On the other hand, in order to preserve the cold properties of the fuel (Cloud Point, Pour Point and low-temperature filterability), it is mandatory not to increase the melting point, that in turn depends on both the saturated compound (stearic acid, C18:0) content and the extent of *cis/trans* and positional isomerization as the difference in melting point between the *cis* and *trans* isomer is at least 15°C according to double bond position as shown in Table 1.

The use of copper hydrogenation catalysts in the fat and oils industry mainly relies on copper chromites for the production of fatty acid alcohols from esters[3]. However, copper-based systems have long been known in edible oils hydrogenation as the most selective ones for the reduction of linolenate C18:3 to oleate C18:1 without affecting linoleate, the valuable component from the nutritional point of view. Monoenes are not reduced; therefore, the percentage of saturated is scarcely changed during the hydrogenation process[4,5]. Moreover, some of us already showed that Cu/SiO_2 catalysts prepared by chemisorption hydrolysis and pre-reduced ex situ are effective and very selective in the hydrogenation of rapeseed oil methylesters allowing to obtain up to 88% of C18:1 derivative without modifying the amount of C18:0 and with a *trans*-content of about 20%[6].

Table 31.1 Influence of double bonds number, position and geometry on the melting points of the 18 carbon atom fatty acids

Fatty Acid	N° of double bonds	MP (°C)
9,12,15 C18:3	3	-13
9,12 C18:2	2	-7
9 C18:1	1	+16
C18:0	0	+70
Monounsaturated fatty acid:influence of double bond position		
	MP*cis* (°C)	MP*trans* (°C)
Δ^6	28	53
Δ^9	16	45
Δ^{12}	27	52
Δ^{15}	40	58

This prompted us to investigate the possibility of selectively hydrogenate highly unsaturated oils, unsuitable for the production of Biodiesel, in order to improve their oxidative stability while keeping the cold properties.

We focused our attention on Tall oil, a by-product of the paper industry, whenever this is prepared according to the KRAFT process. Said material consists of a mixture of highly unsaturated fatty acids (many of which with conjugated diene systems) and terpene derived rosin acids. The rosin acids have the molecular formula $C_{20}H_{30}O_2$ and thus belong to the diterpenes (pimaric and abietic acids). Tall Oil has an iodine number equal to approximately 170 $gI_2/100$ g.

The peculiarity of this material is that it consists of a mixture of acids and not triglycerides; therefore, its transformation in biodiesel requires only an esterification reaction instead of a *trans*-esterification one and therefore does not produce glycerol, making the total economy lighter and independent of the critical marketing of this polyalcohol. However, reports on the use of Tall Oil to produce biodiesel are very rare. Liu et al.[7] developed a process for producing a diesel oil additive, not biodiesel, from pine oil. That process produced a diesel oil cut which was blended with a base diesel fuel.

Hydrotreating has been proposed by Arbokem Inc. in Canada[8] as a means of converting Crude Tall Oil into biofuels and fuel additives. However, this process is a hydrogenation process which produces hydrocarbons rather than biodiesel. Recently a process for making biodiesel from crude tall oil has been proposed.[9] It relies on the use of an acid catalysts or of an acyl halide for the esterification reaction, but no information is given on the properties of this fuel, particularly concerning the oxidative stability.

In the process we are describing here[10], the esterification reaction was carried out in the homogeneous phase and followed by distillation under vacuum of the resinic acids. Hydrogenation of Tall Oil methylesters obtained in this way gave the results reported in Table 31.2.

Table 31.2: Selective hydrogenation of Tall Oil methylesters

catalyst	P atm	C18:2	C18:2$_{conj}$	C18:1	C18:0	IV	PourP (°C)
Initial Composition		44	11	37	0.4	145	
Cu/SiO$_2$	6	22	2	68	0.5	117	-18
Cu/Al$_2$O$_3$	6	24	4	63	1	112	-12
Cu/Al$_2$O$_3$	3	19	4	68	1	-	-
Ni/SiO$_2$	4	25	0.5	56	11	104	+3

The reaction was very fast and led to nearly complete suppression of conjugated dienes and significant reduction of linoleic acid derivative leaving unaffected the stearic acid content. Owing to this very high selectivity, not only the iodine value (IV) can be reduced to meet the European standard (EN 14214 requires IV$_{max}$=120), but also excellent cold properties can be obtained (Figure 31.1). The hydrogenated oil is completely colorless. Moreover the treatment improves very much the Conradson carbon residue (CCR) (Figure 31.2), that is to say the tendency for the fuel to form carbon deposits when used with stoichiometric quantities of comburent, such as for example in diesel cycle engines.

IV, CCR and oxidation stability are three strictly co-related parameters. As a general rule, the reduction of IV (on the same feedstock) dramatically improves the oxidation stability. On the contrary the distillation step removes the main part of naturally occurring antioxidants. For this reason, even after hydrogenation the Rancimat induction time (as measured according to the EN 14112 standard) of the hydrogenated sample does not fulfill the EN 14214 requirement for oxidation stability (6 hours at 110°C), 4 hours being the measured induction period.

Nevertheless the non-hydrogenated, distilled Tall Oil methylester has a Rancimat induction time lower than 1 hour, demonstrating

Figure 31.1 – Reaction profile and product distribution obtained in the hydrogenation reaction carried out by using Cu/SiO_2

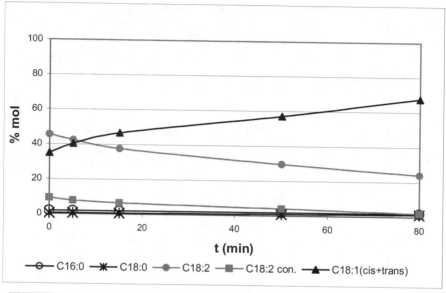

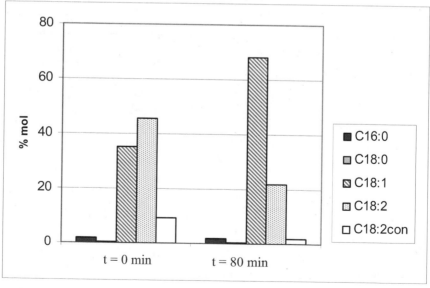

an evident stabilization effect provided by partial selective hydrogenation. To fulfill the EN requirement, it is common practice the use of synthetic antioxidants[11]. A further demonstration of the quality improvement can be obtained from the visual evaluation of tubes used for CCR evaluation. It is well known that the CCR represents the tendency of a fuel to produce coke residues when burnt using a stoichiometric amount of oxygen. For FAME fuels, CCR correlates with the respective residual amount of glycerides, free fatty acids, soaps and catalyst polyunsaturated residues[12]. Moreover the parameter is also influenced by high concentration of fatty acid methylesters and related polymers[13]. In our case the first mentioned factor has no impact, as the feedstock origin is the same in terms of major and minor constituents. On the contrary the important influence of selective hydrogenation reaction on CCR is clearly demonstrated in Figure 31.2, where CCR obtained from Tall Oil FAMEs non-hydrogenated and selectively hydrogenated are shown.

No significant differences were observed among the different catalysts used. On the contrary a catalyst with the same Cu loading but prepared by conventional incipient wetness technique did not show any activity even after long reaction time. The sharp difference in activity between catalysts prepared by our technique and the IW one has been deeply investigated in the case of Cu/TiO_2 systems and attributed to both very high dispersion of the metallic phase and morphology of Cu crystallites[14]. A comparison test carried out over a commercial Ni catalyst shows that early formation of stearic acid (C18:0) results in a very high pour point.

Figure 31.2 – Conradson carbon residue of Tall Oil (A), hydrogenated Tall Oil (B) compared with a standard oil (C)

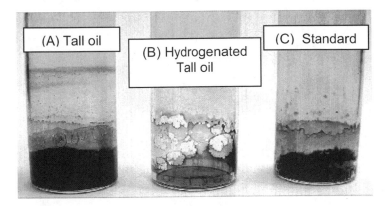

Conclusions

Selective hydrogenation over low-loading, supported Cu catalysts has shown to be a valuable tool for the production of high quality biodiesel from Tall Oil, a byproduct of the Pulp & Paper industry. These results allow planning the use of a great variety of non-conventional oils with high iodine value for the production of biodiesel.

References

1. T. F. Boccuzzi, A. Chiorino, G. Martra, M. Gargano, N. Ravasio and B. Carrozzini, *J. Catal.*, **165**, 129 (1997).
2. Frankel E.N. Ed., Lipid Oxidation, 2nd Ed., The Oily Press, Bridgewater (UK), (2005).
3. K. Noweck, H. Ridder, in: W. Gerhartz (Ed.), Ullmann's Encyclopaedia of Industrial Chemistry, 5th Edition,Vol. A10, VCH, Weinheim, 277 (1987).
4. L.E. Johansson, *J. Am. Oil. Chem. Soc.*, **57,** 16 (1980).
5. S. Koritala and H.J. Dutton, *J. Am. Oil. Chem. Soc.* **46**, 265 (1969).
6. N. Ravasio, F. Zaccheria, M. Gargano, S. Recchia, A. Fusi, N. Poli and R. Psaro, *Applied Catal. A: General* **233**, 1 (2002).
7. D.D. S.Liu, J. Monnier, G.Tourigny, *Perol Sci. & Tech.*, **16**, 597 (1998).
8. http://www.arbokem.com/nat_r/cetane_2.html.
9. S.G. Chatterjee, S. Omori, S.Marda and S.Shastri, US Pat. Appl. US2007130820 A1 (2007).
10. P.F. Bondioli, M.N. Ravasio and F. Zaccheria, PCT/IT2006/000258 (2006) WO2006/111997 A1; EP 06745284.7 (2007).
11. S. Schober and M. Mittelbach, *European Journal of Lipid Science and Technology* **106**, 382-389, (2004).
12. M. Mittelbach, *Bioresource Technology* **56**, 7-11, (1996).
13. M. Mittelbach and H. Enzelsberger, *J. Am. Oil Chem. Soc.* **78** (5), 545-550, (1999).
14. F. Boccuzzi, A. Chiorino, M. Gargano and N. Ravasio, *J. Catal.*, **165**, 140 (1997).

32. Heterogeneous Catalyst and Process for the Production of Biodiesel from High Free-Fatty Acid-Containing Feedstocks

Rajiv Banavali[1], Robert T. Hanlon[1], Karel Jerabek[2] and Alfred K. Schultz[1]

[1]*Rohm and Haas Company, LLC, Spring House, PA 19477*
[2]*Institute of Chemical Process Fundamentals, Academy of Sciences of the Czech Republic, Rozvojova 135, 165 02, Prague 6, Czech Republic*

rbanavali@rohmhaas.com

Abstract

We will present research describing our newly developed polymeric catalyst technology which enables the production of biodiesel from feedstocks containing high levels (> 1 wt %) of free fatty acids (FFAs). Current biodiesel manufacture via alkali-catalyzed transesterification of an oil or fat is now limited by both cost and availability of refined feedstocks. While non-refined feed stocks, such as crude palm oil, rendered animal fat, and yellow and brown greases, are inexpensive and readily available, the high FFA contents of these feedstocks limits their use since the acids unfavorably react with the base catalysts employed in transesterification. Because of this, conventional biodiesel technology typically limits feedstock FFA to < 0. 1 wt%. We will present our work using a novel polymer catalyst for esterifying the FFA present in greases (1-100 wt %) to their corresponding methyl esters in quantitative yields. The resulting ester-oil stream can then be readily converted to biodiesel by base-catalyzed transesterification. The novel catalyst overcomes the drawbacks of traditional catalysts such as limited catalyst life time, slow reaction rates, and low conversions. We will also discuss the chemistry of catalyst functionality, and how these properties relate to improved catalytic activity, reaction rates, kinetics, and mechanisms. We will also describe a continuous process with reactor design optimized for conversion, longevity, ease of use, and economic impact.

Introduction

Biodiesel is a fuel derived from renewable natural resources such as soybean and rapeseed and consists of alkyl esters derived from transesterification of triglycerides with methanol. In spite of all the advantages of biodiesel, such as low emissions, biodegradability, non-toxicity, and lubricity, the major hurdle in penetration of biodiesel is its high cost because of the expensive food grade refined vegetable oil feedstock.

To produce biodiesel, refined vegetable oils are reacted with methanol in the presence of alkali catalysts such as sodium hydroxide, potassium hydroxide, and sodium methylate. The overall base-catalyzed process has several problems that also translate into high production costs. Strict feedstock specifications are a main issue with this process. In particular, the total free fatty acid (FFA) content associated with the lipid feedstock should not exceed 0.5 wt %. Removing the FFA from the fat is important; otherwise soap formation seriously hinders the production of fuel quality biodiesel. Soap forms when the metal hydroxide catalyst reacts with FFA in the feedstock. Soap production gives rise to the formation of gels, increases viscosity, and greatly increases product separation cost.[1]

Use of low quality feedstocks such as waste cooking oils, crude oils, and animal fats which are available cheaply, instead of refined vegetable oil will help in improving the economical feasibility of biodiesel. The amount of such feedstocks generated in each country varies, depending on the use of oils and fats. In the United States alone, there are approximately 3 billion pounds of waste recyclable restaurant grease, 12 billion pounds of animal derived fat, and 24 billion pounds of vegetable oil generated annually.[2] Yellow grease, which is obtained from rendered animal fats and waste restaurant cooking oil, has FFA levels of 6–15 wt % and sells for $0.09–0.20/lb. Brown grease, also known as trap grease, refers to grease collected from traps installed in commercial, municipal or industrial sewage facilities which separate water from oil and grease in wastewater. This waste grease has FFA levels greater than 15 wt %.

The production of biodiesel from low quality oils such as animal fats, greases, and tropical oils is challenging due to the presence of undesirable components especially FFA and water. A pre-treatment step is required when using such high fatty-acid feedstock. Generally, this "esterification" pre-treatment employs liquid sulfuric acid catalyst which must subsequently be neutralized and either disposed of or recycled. However, requirement of high temperature, high molar ratio of alcohol to FFA, separation of the catalyst, environmental and corrosion related problems make its use costly for biodiesel production.

There is a real opportunity to reduce biodiesel production costs and environmental impact by applying modern catalyst technology, which will allow increased process flexibility to incorporate the use of low-cost high-FFA feedstock, and reduce water and energy requirement. Solid catalysts such as synthetic polymeric catalysts, zeolites and superacids like sulfated zirconia and niobic acid have the strong potential to replace liquid acids, eliminating separation, corrosion and environmental problems. Lotero et al. recently published a review that elaborates the importance of solid acids for biodiesel production.[3]

Apart from a few reports[4,5] on solid acid catalyzed esterification of model compounds, to our knowledge utilization of solid catalysts for biodiesel production from low quality real feedstocks have been explored only recently.[6] 12-Tungstophosphoric acid (TPA) impregnated on hydrous zirconia was evaluated as a solid acid catalyst for biodiesel production from canola oil containing up to 20 wt % free fatty acids and was found to give ester yield of 90% at 200°C.[7] Propylsulfonic acid-functionalized mesoporous silica catalyst for esterification of FFA in flotation beef tallow showed a superior initial catalytic activity (90% yield) relative to a

Nafion NR50 catalyst at 120°C. However, the performance of the recycled acidic mesoporous catalyst was negatively affected by the presence of polar impurities in the beef tallow.

Very recently, Nolan et al.[8] studied several acidic polymeric catalysts for the esterification of fatty acids. The highest FFA conversion (45.7%) was obtained over strong acidic macroreticular polymer catalysts; Amberlyst™ 15 at 60°C compared with Amberlyst 35, Amberlyst 16, and Dowex™ HCR-W2.

Reports have shown solid catalysts for esterification of FFA have one or more problems such as high cost, severe reaction conditions, slow kinetics, low or incomplete conversions, and limited lifetime. We will present research describing our newly developed polymeric catalyst technology which enables the production of biodiesel from feedstock containing high levels (> 1 wt %) of FFAs. The novel catalyst, named Amberlyst™ BD20, overcomes the traditional drawbacks such as limited catalyst life time, slow reaction rates, and low conversions.

Catalyst Description

Amberlyst BD20 is a patent-pending catalyst manufactured by Rohm and Haas Company. Its key properties are listed in Table 32.1. Figure 32.1 below shows a micrograph of the actual catalyst.

Table 32.1. Properties of Amberlyst™ BD20.

Property	Value
Average Particle Diameter	750 um
Particle Size Uniformity Coefficient (UC = d_{60} / d_{10})	1.2
Modulus	3-5 GPa
Color	Amber
Crystalinity	0%
Solubility	Insoluble
Density	0.73 g/mL
Functionality	Dual

Figure 32.1. Amberlyst™ BD20; Amber, spherical polymeric beads of ~750 μm diameter.

Experimental

Esterification over Amberlyst BD20 was evaluated by processing a model mixture in a fixed-bed reactor. The model reaction mixture was prepared by dissolving 10 wt.% of pure stearic acid (> 97%, Fluka, Germany) in a low-acid vegetable oil (0.04 %) bought in the supermarket. Methanol (> 99.5%) was used without any preliminary treatment.

A fixed-bed reactor system was employed (Figure 32.2). Each of the two reactors was charged with 38 cc of Amberlyst BD20 catalyst. Sample ports located at the exit of each reactor enabled increased acquisition of residence time data. Pressure was maintained by a back pressure control valve to maintain methanol in the liquid phase. After charging, the 1st and then 2nd reactors were connected to the pumps and filled with the reaction mixture while vapor was released from each through the top vent valve. Once each reactor was filled with liquid and emptied of vapor, the pressure regulator was connected to the output and both reactors were immersed into the water bath.

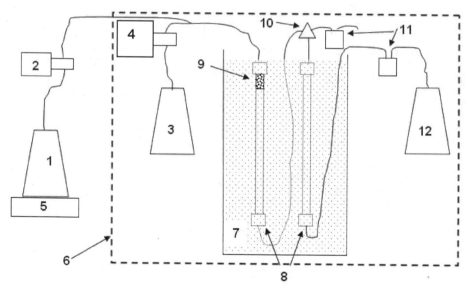

1-MeOH container, 2-MeOH pump (computer controlled flow range 0.1–2.5 ml/min), 3-Oil container, 4-Oil pump (computer controlled flow range 0.1–2.5 ml/min), 5-Digital balances, 6-Heated enclosure maintained at 60°C, 7-Thermostatic water bath, 8-Reactors with inner diameter 10 mm and length 50 cm (volume ml = 0.785 ml x bed length) connected in series, 9- 3 cm long mixing zone filled with glass beads 0.2 mm diam., 10- 3-way valve to collect samples after the 1st reactor, 11-Zero dead volume SSI Flow-Through Back-Pressure Regulator maintaining in the reactor pressure about 4 bars, 12- container for products.

Figure 32.2 - Experimental setup for kinetic experiments using fixed-bed reactors.

Oil and methanol streams were pumped by separate pumps to the 1st reactor. Both flows were merged in a 3-way T-connector and then homogenized by passing through a 3-cm long mixing zone filled with 0.2 mm glass beads. The head of the oil pump was inserted into the heated compartment to ensure good flow while the MeOH pump was located outside. The MeOH bottle stood on an electronic scale and the weight difference recorded in regular intervals. Total flow rate was determined by weighing samples collected at the output from the pressure regulator. The Liquid Hourly Space Velocity (LHSV) was then calculated as:

LHSV = Total Liquid Flow Rate (cc/hr) / Total Amberlyst BD20 Catalyst Volume (cc)

For each condition change, product samples were collected in regular (usually 1 hour) intervals at the output of the pressure regulator. Sampling time was adjusted to obtain about 10 g of the reaction mixture. After weighing, the sample was diluted with 50 ml iPrOH to achieve a homogeneous clear liquid. Acid content was determined by titration with NaOH solution in iPrOH using phenolphthalein as indicator.

Stearic acid conversion was calculated in two steps. First, the result of the titration of the whole sample was converted into a value related to the weight of the oil part of the reaction mixture only:

$$B = A * (F_R - F_{MeOH})/F_R$$

Where,
A = alkali solution consumption per gram of the whole sample,
F_R = flow rate of the reaction mixture (determined by weighing of samples collected at the reactor output),
F_{MeOH} = methanol flow rate (determined from the decrease of the MeOH container weight),
B = alkali solution consumption per gram of the oil part of the reaction mixture.

Stearic acid conversion was then computed according to the following formula: SA conversion (%) = 100* B/C, in which C is the consumption of the alkali solution during the titration of the starting oil + SA mixture determined the same day.

Difference in stearic acid conversion between two consecutive samples of lower than 0.1% was considered as evidence of reaching the steady state operation. This typically required about 5 hours time as reflected in Table 32.2 below which shows change in SA conversion over time after a condition change (T = 63°C, MeOH flow rate = 15 g/h, oil flow rate = 67 g/h).

Table 32.2. Conversion vs. Time after Condition Change.

Duration of the experiment (hours)	1	2	3	4	5	6	7	8
SA conversion (%)	97.83	97.53	96.59	95.71	94.81	94.42	94.37	94.46
* Reaction run in continuous fixed-bed configuration 63°C, 2.8 LHSV, 80:20 Oil:MeOH								

Results and Discussion

The traditional catalyst used for esterification of acids to methyl esters is sulfuric acid. Homogeneous sulfuric acid catalysis has many downsides.[9] When using sulfuric acid, much capital expense is required for Hastalloy and/or other specialty metals of construction. Homogeneous catalysis results in the contamination of the product by sulfur containing species. Therefore, neutralization and removal of acid is required to meet biodiesel specifications and to protect the downstream transesterification reactor. Inevitably, when using sulfuric acid, organic sulfur compounds will be produced. These products will cause the resultant biodiesel to fail specification tests.

Comparison of Amberlyst BD20 with sulfuric acid shows virtually identical behavior when employing low FFA feedstocks, but when employing oils with higher FFA content, sulfuric acid catalysis becomes sluggish, and lower overall yields are achieved. This is due to the hydration of the sulfuric acid by water produced in the esterification reaction. The graph below shows reaction profiles of sulfuric acid vs. Amberlyst BD20 catalyzed esterification using two different FFA content oils.

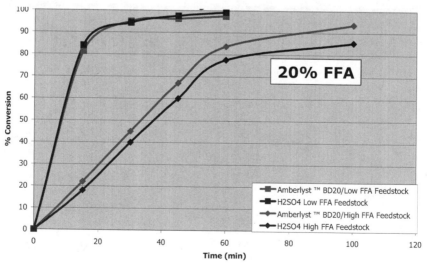

Figure 32.3. Reaction Profile Comparing Heterogeneous Catalysis to Homogeneous Catalysis; Differing FFA Content.

There are numerous potential feedstocks for use in biodiesel production, including crude vegetable oils, animal fats, and recycled oils. These materials differ in quantity and composition of fatty acids along with differing levels of impurities. We have found that fats and oils containing shorter chain fatty acids (e.g., palmitic) tend to react faster than oils containing longer chain fatty acids (e.g., stearic). More than 20 oils were tested, and in each case, the catalyst was effective at converting the fatty acids to the corresponding methyl esters. Table 32.2 summarizes the overall conversion and relative rate constants of some representative oil samples.

Table 32.3. Comparison of Various Oils Tested.

"Commercial" Oil	% Free Fatty Acid (in Feedstock)	* Conversion (%)	Rate Constant (K_{obs}) (min^{-1})
Tallow from Canada	0.9	96	8.5×10^{-2}
Used Cooking oil from PA	1.6	96	4.7×10^{-2}
Animal Fat from USA	2.2	98	10.0×10^{-2}
Coconut Oil from Phillipines	3.1	96	11.0×10^{-2}
Palm Oil from Thailand	5.3	98	11.0×10^{-2}
Tallow from Holland	20.1	96	4.4×10^{-2}
Palm Oil Fatty Acid Distillate	79.5	96	2.0×10^{-2}

* Reaction run in batch configuration (in closed reactor) without water removal 85°C, 50:50 oil:MeOH; 5% catalyst, 60 min.

The first series of experiments in the fixed-bed continuous reactor examined the influence of reaction temperature (63–93°C) at a constant MeOH:SA molar ratio of 20.

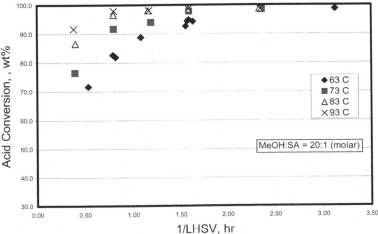

Figure 32.4. Conversion vs. Residence Time at Differing Temperatures.

As anticipated, SA conversion increases with increasing residence time (1/LHSV) and with increasing temperature to a maximum of about 98%. This limit is most likely caused by equilibrium. This limit and thus the equilibrium constant were not affected by the temperature range studied, consistent with a low heat of reaction. The sum of the molar heats of combustion of stearic acid (11320 kJ/mol) and methanol (720 kJ/mol) is almost the same as the heat of combustion of methyl stearate (12010 kJ/mol), meaning that the change in enthalpy of this reaction is nearly zero and that the equilibrium constant is essentially temperature independent.

The next series of experiments were run with the same feed and catalyst charge to assess the influence of the molar ratio MeOH/SA (10–30) at constant temperature of 83°C (Figure 32.5).

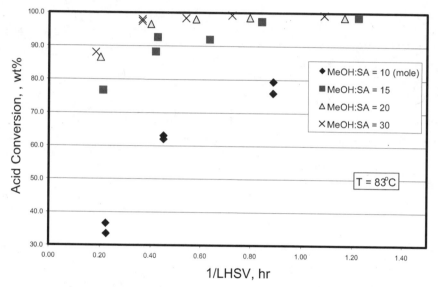

Figure 32.5. Conversion vs. Residence Time at Differing MeOH/Oil Concentrations.

SA conversion increases with increasing residence time, and with increasing MeOH:SA to a maximum of about 98%. It appears that the maximum conversion increases to 99% for the highest MeOH:SA studied (30), consistent with an equilibrium limited reaction. The esterification reaction rate was strongly suppressed by lowering MeOH/SA ratio below 10%.

Evidently, the increase of methanol concentration in the reaction mixture above the molar ratio MeOH:SA = 20 has no effect on the attainable conversions. On the other hand, decrease of the methanol concentration below this level has a very detrimental effect. It is probably connected with the influence of the reaction mixture composition on the polymer catalyst.

In this study of the fatty acid esterification reaction mixture, only methanol swells the catalyst. And the catalyst is maximally swollen when the methanol concentration is above the saturation level in the oily part of the mixture, which we visually observed to occur between 10 and 20 methanol wt% on total feed. As suggested in Figure 32.5, once methanol falls below this saturation level, the catalyst shrinks, decreasing accessibility to its acid sites, thereby decreasing catalytic activity. In addition to swelling the catalyst and increasing activity, the presence of separate methanolic phase at higher methanol levels in the feed probably also greatly assists in removal of water from the catalyst, which helps shift the reaction equilibrium towards the product's side.

The presence of two phases in the reaction mixture may seem to be a mass-transfer engineering problem, but even moderate stirring of the mixture produces an emulsion, which greatly facilitates the phase transfer steps of the reaction mechanism. In our fixed-bed reactor, the turbulence resulting from the flow rates used seemed to suffice to eliminate external mass transfer limitations. At MeOH:SA of 20 and identical LHSV values, similar acid conversions were observed for two linear flow velocities differing by a factor of two.

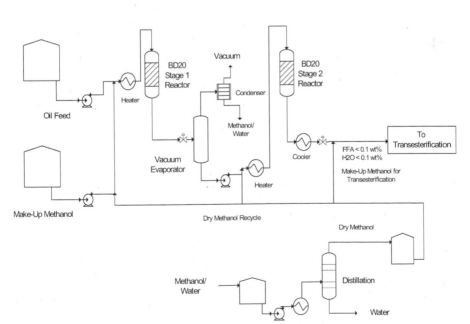

Figure 32.6. Continuous Reactor Design.

The esterification process can be carried out in either batch or continuous mode, the final decision depending most likely on the size of the flow rates involved. For most commercial sizes of 15 MM gal/yr or higher, the continuous process is probably more cost effective and for this option, two additional options are available: continuous stirred tank reactor (CSTR) or a fixed-bed reactor (FBR).

Regardless of the reactor design mode, equilibrium limitations will have to be addressed should high overall FFA conversions (>99%) be desired. For example, in this study, we have observed maximum conversions of about 98% for fixed-bed reactors employing co-current flow of oil and methanol. To effect even higher conversions would require the use of additional reaction stages together with the inter-stage removal of water as shown in an example process design below. The equilibrium limitation can also be addressed by designing for high methanol flows relative to oil and, more specifically, relative to the FFA content in the oil. This methanol stream would then require distillation prior to recycle in order to remove product water. Both design features, i.e., high methanol and low water, aid in achieving overall FFA conversions greater than 99%.

Conclusions

The new polymeric catalyst, Amberlyst BD20 offers highly specialized morphology providing excellent accessibility of the supported catalytic sites even for sterically demanding molecules such as fatty acids. Its working-state porosity is optimized for achieving high production per unit of reactor volume and its mechanical properties allow for its use in fixed-bed reactors as well as stirred batch reactors without danger of the catalyst bed deformation and clogging. Using Amberlyst BD20, it is possible to design continuous process able to reduce free fatty acid concentration in low-cost feed stocks to levels acceptable for their further processing in biodiesel production.

Acknowledgements

We thank the Rohm and Haas Company for financial support.

References

1. Ma, F. R. and Hanna, M. A., *Bioresour. Technol.* **1999**, *70*, 1-15.
2. Talley, P., *Render Magazine*, **2004**, 16.
3. Lotero, E; Liu, Y.; Lopez, D. E.; Suwannakarn, K.; Bruce D. A. and Goodwin, J. G., *Ind. Eng. Chem. Res.,* **2005**, *44*, 5353.
4. Kiss, A. A.; Dimian, A. C. and Rothenberg, G., *Adv. Synth. Catal.*, **2006**, *348*, 75.
5. Lopez, D. E.; Goodwin Jr., J. G.; Bruce, D. A. and Lotero, E., *Appl. Catal., A,* **2005**, *295*, 97.
6. Kulkarni, M. G.; Gopinath, R.; Meher, L. C. and Dalai, A. K., *Green Chem.*, **2006**, *8*, 1056.
7. Mbaraka, I. K.; McGuire, K. J. and Shanks, B. H., *Ind. Eng. Chem. Res.* **2006**, *45*, 3022.

8. Nolan, O'zbay N.; Oktar, N. and Tapan, N., *Fuel* **2008**, Internet Preprint.
9. Canakci, M. and Van Gerpen, J., *Trans. ASAE*, **1999**, *42*, 1203.

33.   Sustainable Biodiesel Production by Catalytic Reactive Distillation

Anton A. Kiss and Gadi Rothenberg

van 't Hoff Institute for Molecular Sciences, University of Amsterdam, Nieuwe Achtergracht 166, 1018 WV Amsterdam, the Netherlands

gadi@science.uva.nl

Abstract

This chapter outlines the properties of biodiesel as renewable fuel, as well as the problems associated with its conventional production processes. The synthesis via fatty acid esterification using solid acid catalysts is investigated. The major challenge is finding a suitable catalyst that is active, selective, water-tolerant and stable under the process conditions. The most promising candidates are metal oxides. A novel sustainable process based on catalytic reactive distillation is proposed as base case and a heat-integrated design as alternative. Significant energy savings of ~45% are possible compared to conventional reactive distillation designs.

Introduction

The increased energy demand, depletion of petroleum reserves, as well as major concerns of rising greenhouse gas emissions make the implementation of alternative and renewable sources of energy a crucial issue worldwide.

Biodiesel is a very attractive alternative fuel because it is sustainable and combines high performance with environmental benefits. Unlike petroleum diesel, biodiesel consists of a mixture of fatty acids alkyl esters. It can be produced from vegetable oils, animal fat or even recycled grease from the food industry (1). Biodiesel has several advantages over petroleum diesel (1-4): it is safe, renewable, non-toxic and biodegradable; it contains no sulfur and is a better lubricant. Moreover, it has a higher cetane number and an eco-friendly life cycle. Biodiesel is the only fuel with an overall positive life cycle energy balance. Despite the chemical differences, these two fuels have similar properties. The presence of oxygen in biodiesel improves combustion and reduces CO, soot and hydrocarbon emissions (~70% less gas pollutants and ~50% less soot particles), while slightly increasing the NOx emissions. Biodiesel brings also additional benefits to the society: rural revitalization, less global warming, energy supply security. Its production is increasing rapidly as biodiesel can be distributed using the current infrastructure.

Fatty acid methyl esters (FAME) are currently manufactured mainly by trans-esterification with an alcohol, using a homogeneous base catalyst (NaOH/KOH). Methanol is more suitable for biodiesel manufacturing, but other alcohols can in principle also be used, depending on the feedstock available. The base catalyst must be neutralized afterwards; hence, it forms salt waste streams. Also due to the presence of free fatty acids (FFA) it reacts to form soap as an unwanted by-product, requiring expensive separation. Another method to produce fatty esters is by esterification of fatty acids, catalysed by H_2SO_4. The problem is the batch operation mode that again involves costly neutralization and separation of the homogeneous catalyst (5).

The current biodiesel manufacturing processes have several disadvantages: shifting the equilibrium to fatty esters by using an excess of alcohol that must be separated and recycled, making use of homogeneous catalysts that require neutralization hence causing salt waste streams, expensive separation of fatty esters products from the reaction mixture, high production costs due to relatively complex processes involving one or two reactors and several separation units. Hence, the current biodiesel manufacturing is an energy intensive process that consumes large amounts of energy, primarily from fossil sources. Production costs and pollution produced by the present process may outweigh advantages of using biodiesel. Thus, biodiesel remains an attractive but still costly alternative fuel.

During the last decade many industrial processes shifted towards using solid acid catalysts (6). In contrast to liquid acids that possess well-defined acid properties, solid acids contain a variety of acid sites (7). Solid acids are easily separated from the biodiesel product; they need less equipment maintenance and form no polluting by-products. Therefore, to solve the problems associated with liquid catalysts, we propose their replacement with solid acids and develop a sustainable esterification process based on catalytic reactive distillation (8). The alternative of using solid acid catalysts in a reactive distillation process reduces the energy consumption and manufacturing pollution (i.e., less separation steps, no waste/salt streams).

This chapter presents the findings of the experimental work, highlights the pros and cons of several solid catalyst types, and the results of the rigorous simulations of the reactive distillation process using AspenTech AspenPlus™. The heat integrated design proposed in this work overcomes the shortcomings of conventional processes, by combining reaction and separation into one unit. Compared to classic reactive distillation, the energy requirements in this heat-integrated design are further decreased with −43% and −47% for heating and cooling, respectively.

Results and Discussion

At an industrial scale, the esterification catalyst must fulfill several conditions that may not seem so important at lab-scale. This must be very active and selective as by-products are likely to render the process uneconomical, water-tolerant and stable at relatively high temperatures. In addition, it should be an inexpensive material that is readily available on an industrial scale. In a previous study we investigated metal oxides with strong Brønsted acid sites and high thermal stability. Based on the literature reviews and our previous experimental screening, we focus here on application of metal oxide catalysts based on Zr, Ti, and Sn.

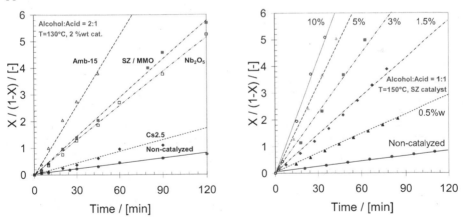

$$C_{10}H_{23}-\overset{O}{\overset{\|}{C}}-OH \; + \; HO-\overset{}{\underset{}{C}}-\textit{n-Bu} \quad \underset{120 \text{ min}}{\overset{160 \text{ °C}}{\rightleftharpoons}} \quad C_{10}H_{23}-\overset{O}{\overset{\|}{C}}-O-\overset{}{\underset{}{C}}-\textit{n-Bu} \; + \; H_2O \quad (1)$$

The following experimental results are presented on the use of solid acid catalysts in esterification of dodecanoic acid with 2-ethylhexanol and methanol. In the next figures, conversion is defined as: $X \, [\%] = 100\cdot(1 - [\text{Acid}]_{\text{final}} / [\text{Acid}]_{\text{initial}})$, and the amount of catalyst used is normalized: $W_{\text{cat}} \, [\%] = 100\cdot M_{\text{cat}} / (M_{\text{acid}} + M_{\text{alcohol}})$. Several alcohols were used to show the range of applicability. The selectivity was assessed by testing the formation of side products in a suspension of catalyst in alcohol. Under the reaction conditions, no products were detected by GC analysis.

Zeolites gave an insignificant increase in acid conversion compared to the non-catalysed reaction. This agrees with previous findings suggesting that the reaction is limited by the diffusion of the bulky reactant into the zeolite pores (9). Amberlyst-15 exhibits high activity (Figure 33.1, left), but regrettably it not stable at temperatures higher than 150°C, making it unsuitable for industrial RD applications (10). The cesium salt of tungstophosphoric acid (Cs2.5) is super acidic and its mesoporous structure has no limitations on the diffusion of the reactants. However, Cs2.5 exhibits only low activity per weight; hence, it is not suitable for industrial applications.

Figure 33.1. Esterification of dodecanoic acid with 2-ethylhexanol (left); non-catalysed and catalysed (0.5-5 wt% SZ catalyst) reaction profiles (right).

Out of the metal oxides, sulfated titania and tin oxide performed slightly better than the sulfated zirconia (SZ) catalyst and niobic acid (Nb_2O_5). However, SZ is cheaper and readily available on an industrial scale. Moreover, it is already applied in several industrial processes (7,8). Zirconia can be modified with sulfate ions to form a superacidic catalyst, depending on the treatment conditions (11-16). In our experiments, SZ showed high activity and selectivity for the esterification of fatty acids with a variety of alcohols, from 2-ethylhexanol to methanol. Increasing the amount of catalyst leads to higher conversions (Figure 33.1, right). This makes SZ suitable for applications where high activity is required over short time spans. SZ is also easily regenerable and thermally stable. The reusability and robustness of the SZ catalyst was also tested and in five consecutive runs, the activity dropped to about 90% of the original value, then it remained constant thereafter. Note that the re-calcination of the used catalyst restored the original activity.

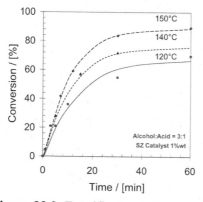 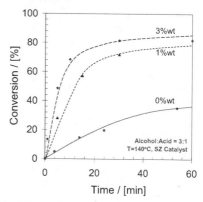

Figure 33.2. Esterification of dodecanoic acid with methanol, using an alcohol:acid ratio of 3:1 and sulfated zirconia (SZ) as catalyst.

Considering the promising results with 2-ethylhexanol, we tested the applicability of SZ also for esterification with methanol. High conversions can be reached even at 140°C providing that an increased amount of catalyst is used (Figure 33.2). It is worthwhile noting that the esterification with methanol takes place at the highest rates. This can be explained by the alcohols' relative sizes.

The pore size of the catalyst plays an important role as the reactants and the products must be able to fit inside the catalyst to take full advantage of the total surface area available. The pore size of metal oxides are sufficiently large (>2 nm) to facilitate the mass transfer into and from the catalyst pores. This compensates for their lower acidity compared to other solid acids. Table 33.1 gives an overview of the tested catalysts, showing their pros/cons with respect to the fatty acid esterification reaction.

Table 33.1. Advantages and disadvantages of the tested catalysts.

Catalyst	Advantages	Disadvantages
H_2SO_4	Highest activity	Liquid catalyst
Ion-exchange resins	Very high activity	Low thermal stability
$H_3PW_{12}O_{40}$	Very high activity	Soluble in water
$Cs_{2.5}H_{0.5}PW_{12}O_{40}$	Super acid	Low activity per weight
Zeolites (H-ZSM-5, Y and Beta)	Controlable acidity and hydrophobicity	Small pore size Low activity
Sulfated metal oxides (zirconia, titania, tin oxide)	High activity Thermally stable	Deactivates in water, but not in organic phase
Niobic acid (Nb_2O_5)	Water resistant Thermally stable	Medium activity

The reaction mechanism for the heterogeneous and homogeneous acid-catalysed esterification were reported to be similar (17). However, there is a major difference concerning the surface hydrophobicity. *Reaction pockets* are created inside a hydrophobic environment, where the fatty acid molecules can be absorbed and react further. Water molecules are unlikely to be absorbed on sites enclosed in hydrophobic areas.

Reactive distillation design - In this part we present the design for fatty acids esterification using metal oxides such as SZ as 'green' solid acid catalyst. The problem is highly complex, as it involves chemical and phase equilibria, VL and VLL equilibria, catalyst activity and kinetics, mass transfer in gas-liquid and liquid-solid, adsorption on the catalyst and desorption of products (8,18-20). The design of the process is based on a reactive distillation column (RDC) that integrates reaction and separation into a single operating unit. An additional flash and a decanter or a distillation column, are used to guarantee the high purity of the products (21-23). RDC consists of a core reactive zone completed by rectifying and stripping sections, whose extent depends on the separation behavior of the reaction mixture. Since methanol and water are much more volatile than the fatty ester and acid, these will separate easily in the top. The flowsheet of the process is shown in Figure 33.3.

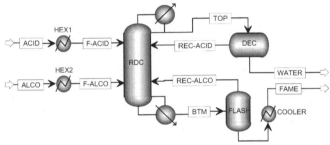

Figure 33.3. AspenPlus flowsheet of the catalytic reactive distillation process.

RDC is operated in the temperature range of 70–210°C, at ambient pressure. Out of the 15 stages of the reactive distillation column, the reactive zone is located in the middle of the column (stages 3-10). The fatty acid stream is fed on top of the reactive zone while the alcohol is fed as saturated liquid, below the reactive zone. The reflux ratio is relatively low (0.1 kg/kg) since a higher reflux ratio is detrimental as it brings back water by-product into the column, thus decreasing the fatty acids conversion by shifting the equilibrium back to reactants. High purity products are possible, but due to the thermo-stability and high boiling points of the fatty esters (i.e., high temperature in the reboiler) this should be avoided. By allowing ~0.2% of alcohol in the bottom stream, the reboiler temperature can be limited to ~200°C.

The base-case design (Figure 33.3) is amenable to heat integration, as the feed stream could be pre-heated using the fatty ester product stream. Obviously, a feed-effluent heat exchanger (FEHE) should replace each of the two heat exchangers HEX1 and HEX2 (Figure 33.4). The hot bottom product of the column, a mixture of fatty esters, is used to pre-heat both reactants: the fatty acid and alcohol feed streams.

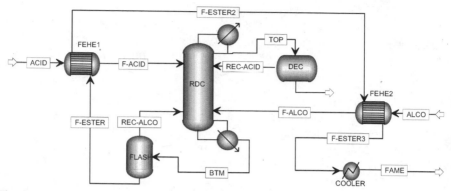

Figure 33.4. Flowsheet of biodiesel production by heat-integrated reactive distillation.

Figure 33.5 shows the composition, temperature and reaction rate profiles in the reactive distillation column. The ester product with traces of methanol is the bottom product, whereas a mixture of water and fatty acid is the top product. This mixture is then separated in the additional distillation column and the acid is refluxed back to the RDC. The fatty ester is further purified in a small evaporator and methanol is recycled back to the RDC (Figures 33.3 and 33.4).

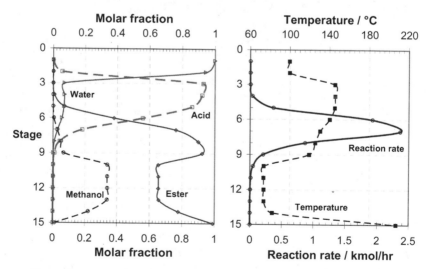

Figure 33.5. Liquid composition profiles (left) and temperature profile (right) in RDC.

The mass balance of these designs is given in Table 33.2, while Table 33.3 shows a comparison between the base case and the heat-integrated alternative, in terms of energy requirements. Compared to the conventional reactive distillation design, the energy requirements in the heat-integrated case are further decreased with

−43% and −47% for heating and cooling, respectively. Note that both design alternatives are suitable for a large range of fatty acids and alcohol feedstocks.

Table 33.2. Mass balance of the biodiesel production based on reactive-distillation.

	F-ACID	F-ALCO	BTM	REC-ALCO	FAME	TOP	WATER
Temperature K	418.1	338.6	480.4	480.4	303.1	372.8	323.1
Mass Flow kg/hr							
METHANOL	0	188.631	1.883	0.391	1.492	0.011	0.011
ACID	1167.607	0	0.144	0	0.144	0.11	0.015
WATER	0	0	0.005	0.001	0.003	104.988	104.986
ESTER-M	0	0	1249.195	0.834	1248.361	0.01	0
Mass Frac							
METHANOL	0	1	0.002	0.319	0.001	0	0
ACID	1	0	0	0	0	0.001	0
WATER	0	0	0	0.001	0	0.999	1
ESTER-M	0	0	0.998	0.68	0.999	0	0

Table 33.3. Comparison of energy consumption: base case vs. heat-integrated design.

Operating unit	Base case	Heat integration	Difference
Heating requirements	KW	KW	
1. RDC reboiler	136	136	
2. HEX1 / FEHE1	95	0	
3. HEX2 / FEHE2	8	0	−43 %
Cooling requirements	KW	KW	
1. RDC condenser	−72	−72	
2. Decanter	−6	−6	
3. Cooler	−141	−38	−47 %

Conclusions

The systematic study of reaction rates under controlled process conditions in terms of temperature, pressure, and reactants ratio, is indeed a suitable method for screening catalyst candidates for fatty acids esterification. The hydrophobicity of the catalyst surface and the density of the acid sites are of paramount importance in determining the activity and selectivity. Catalysts with small pores, such as zeolites, are not suitable for making biodiesel because of diffusion limitations of the large fatty acid molecules. Ion-exchange resins are very active strong acids, but have low thermal stability, which is problematic as relatively high temperatures are required for increased reaction rates. Heteropolyacids are super-acidic compounds but due to the high molecular weight, their activity per weight of catalyst is not sufficient for industrial applications. However, of the mixed metal oxides family, sulfated zirconia and niobic acid were found as good candidates. These are active, selective, and stable under the process conditions hence suitable for industrial reactive distillation applications.

Biodiesel can be produced by a sustainable continuous process based on catalytic reactive distillation, powered by metal oxides as solid acid catalysts for fatty acids esterification. The integrated design ensures the removal of water by-product that shifts the chemical equilibrium to completion and preserves the catalyst activity. By adding heat integration to the conventional reactive distillation setup, significant energy savings of ~45% are possible. This heat integrated alternative improves the HSE benefits and economics of conventional biodiesel processes, and reduces dramatically the number of downstream processing steps by replacing the liquid catalysts with solid acids. The major benefits of this integrated approach are:

1. No excess of alcohol is required, as both reactants are fed in stoichiometric ratio.
2. No catalyst neutralization step; hence, no salt waste streams are being produced.
3. Environmental friendly catalyst functional over a large range of temperatures.
4. Sulfur-free fuel, since solid acid catalysts do not leach into the biodiesel product.
5. Very high conversions, as chemical equilibrium is shifted towards completion.
6. Increased unit productivity, up to 5-10 times higher than conventional processes.
7. Reduced investment costs, due to very few operating units required.
8. Smaller plant footprint compared to conventional biodiesel processes.
9. Minimum energy consumption, due to the heat integrated RD design.
10. Multifunctional plant suitable for a large range of alcohols and fatty acids.

Experimental Section

The experimental results are presented for the esterification of dodecanoic acid ($C_{12}H_{24}O_2$) with 2-ethylhexanol ($C_8H_{18}O$) and methanol (CH_4O), in presence of solid acid catalysts (SAC). Reactions were performed using a system of six parallel reactors (Omni-Reacto Station 6100). In a typical reaction 1 eq of dodecanoic acid and 1 eq of 2-ethylhexanol were reacted at 160°C in the presence of 1 wt% SAC. Reaction progress was monitored by gas chromatography (GC). GC analysis was performed using an InterScience GC-8000 with a DB-1 capillary column (30 m × 0.21 mm). GC conditions: isotherm at 40°C (2 min), ramp at 20°C min^{-1} to 200°C, isotherm at 200°C (4 min). Injector and detector temperatures were set at 240°C.

Chemicals and catalysts - Double distilled water was used in all experiments. Unless otherwise noted, chemicals were purchased from commercial companies and were used as received. Dodecanoic acid 98 wt% (GC), methanol, propanol and 2-ethylhexanol 99+ wt% were supplied by Aldrich, niobic acid by Companhia Brasileira de Metalurgia e Mineração (CBMM), zirconil chloride octahydrate 98+ wt% by Acros Organics, 25 wt% NH_3 solution and H_2SO_4 97% from Merck. Zeolites beta, Y and H-ZSM-5 were provided by Zeolyst, and ion-exchange resins by Alfa.

Preparation of mixed metal oxides - The sulfated metal oxides (zirconia, titania and tin oxide) were synthesized using a two-step method. The first step is the hydroxylation of metal complexes. The second step is the sulfonation with H_2SO_4 followed by calcination in air at various temperatures, for 4 h, in a West 2050 oven, at the temperature rate of 240°C h^{-1}. **Sulfated zirconia:** $ZrOCl_2.8H_2O$ (50 g) was dissolved in water (500 ml), followed by precipitation of $Zr(OH)_4$ at pH = 9 using a 25 wt% NH_3 soln. The precipitate was washed with water (3×500 ml) to remove the chloride salts (Cl^- ions were determined with 0.5 N $AgNO_3$). $Zr(OH)_4$ was dried (16 h at 140°C, impregnated with 1N H_2SO_4 (15 ml H_2SO_4 per 1 g $Zr(OH)_4$), calcined at 650°C. **Sulfated titania:** HNO_3 (35 ml) was added to an aqueous solution of $Ti[OCH(CH_3)_2]_4$ (Acros, >98%, 42 ml in 500 ml H_2O). Then 25% aqueous ammonia was added until the pH was raised to 8. The precipitate was filtered, washed and dried (16 h at 140°C). The product was impregnated with 1N H_2SO_4 (15 ml H_2SO_4 per 1 g $Ti(OH)_4$). The precipitate was filtered, washed, dried and then calcined. **Sulfated tin oxide:** $Sn(OH)_4$ was prepared by adding a 25% aqueous NH_3 solution to an aq. sol. of $SnCl_4$ (Aldrich, >99%, 50 g in 500 ml) until pH 9-10. The precipitate was filtered, washed, suspended in a 100 ml aq. sol. of 4% CH_3COONH_4, filtered and washed again, then dried for 16 h at 140°C. Next, 1N H_2SO_4 (15 ml H_2SO_4 per 1 g $Sn(OH)_4$) was added and the precipitate was filtered, washed, dried and calcined.

Preparation of Cs2.5 catalyst [$Cs_{2.5}H_{0.5}PW_{12}O_{40}$] - Cs_2CO_3 (1.54 g, 10 ml, 0.47 M) aqueous solution were added dropwise to $H_3PW_{12}O_{40}$ (5 ml, 10.8 g, 0.75 M aq. sol.). Reaction was performed at room temperature and normal pressure while stirring. The white precipitate was filtered and aged in water for 60 hours. After aging, the water was evaporated in an oven at 120°C. White solid glass-like particles of $Cs_{2.5}H_{0.5}PW_{12}O_{40}$ (9.0375 g, 2.82 mmol) were obtained.

Table 33.4. Catalyst characterization.

Catalyst sample	Surface area	Pore volume	Pore diameter max./mean/calc.	Sulfur content
$Cs_{2.5}H_{0.5}PW_{12}O_{40}$	163 m^2/g	0.135 cm^3/g	2 / 5.5 / 3 nm	N/A
ZrO_2/SO_4^{2-} / 650°C	118 m^2/g	0.098 cm^3/g	4.8 / 7.8 / 7.5 nm	2.3%
TiO_2/SO_4^{2-} / 550°C	129 m^2/g	0.134 cm^3/g	4.1 / 4.3 / 4.2 nm	2.1%
SnO_2/SO_4^{2-} / 650°C	100 m^2/g	0.102 cm^3/g	3.8 / 4.1 / 4.1 nm	2.6%
$Nb_2O_5 \cdot nH_2O$	110 m^2/g	0.120 cm^3/g	– / – / – nm	N/A

Catalyst characterization - Characterization of mixed metal oxides was performed by atomic emission spectroscopy with inductively coupled plasma atomisation (ICP-AES) on a CE Instruments Sorptomatic 1990. NH_3-TPD was used for the characterization of acid site distribution. SZ (0.3 g) was heated up to 600°C using He (30 ml min^{-1}) to remove adsorbed components. Then, the sample was cooled at room temperature and saturated for 2 h with 100 ml min^{-1} of 8200 ppm NH_3 in He as carrier gas. Subsequently, the system was flushed with He at a flowrate of 30 ml min^{-1} for 2 h. The temperature was ramped up to 600°C at a rate of 10°C min^{-1}. A TCD was used to measure the NH_3 desorption profile. Textural properties were established from the N_2 adsorption isotherm. Surface area was calculated using the BET equation and the pore size was calculated using the BJH method. The results given in Table 33.4 are in good agreement with various literature data.

Catalyst leaching - The mixture may segregate leading to possible leaching of sulfate groups. The leaching of catalyst was studied in organic and in aqueous phases.

First, a sample of fresh SZ catalyst (0.33 g) was stirred with water (50 ml) while measuring the pH development in time. After 24 h, the acidity was measured by titration with KOH. The suspension was then filtered and treated with a $BaCl_2$ solution to test for SO_4^{2-} ions. In a second experiment, the catalyst was added to an equimolar mixture of reactants. After 3 h at 140°C, the catalyst was recovered from the mixture, dried at 120°C and finally stirred in 50 ml water. The pH was measured and the suspension titrated with a solution of KOH. SO_4^{2-} ions in the suspension were determined qualitatively with $BaCl_2$. In a third experiment, the procedure was repeated at 100°C when the mixture segregates and a separate aqueous phase is formed. From the leaching tests it can be concluded that SZ is not deactivated by leaching of sulfate groups when little water is present in the organic phase but it is easily deactivated in water or aqueous phase. There are several methods to prevent aqueous phase formation and leaching of acid sites: 1) use an excess of one reactant, 2) work at low conversions, and 3) increase the temperature exceeding the boiling point of water to preserve the catalyst activity and drive reaction to completion.

Selectivity and side reactions - Typically, the alcohol-to-acid ratio inside an RD unit may vary over several orders of magnitude. Especially for stages with an excess of alcohol, the use of a SAC may lead to side reactions. Selectivity was assessed by testing the formation of side products in a suspension of SZ in pure alcohol under reflux for 24 h. No ethers or dehydration products were detected by GC analysis.

Acknowledgements

We thank M.C. Mittelmejer-Hazeleger and J. Beckers for the technical support.

References

1. F. Maa and M. A. Hanna, *Bioresource Technol.*, **70**, 1 (1999).
2. B. Buczek and L. Czepirski, *Inform*, **15**, 186 (2004).
3. A. Demirbas, *Energy Exploration & Exploitation*, **21**, 475 (2003).
4. W. Körbitz, *Renewable Energy*, **16**, 1078 (1999).
5. M. A. Harmer, W. E. Farneth and Q. Sun, *Adv. Mater.*, **10**, 1255 (1998).
6. J. H. Clark, *Acc. Chem. Res.*, **35**, 791 (2002).
7. K. Wilson, D. J. Adams, G. Rothenberg and J. H. Clark, *J. Mol. Catal. A:Chem.*, **159**, 309 (2000).
8. F. Omota, A. C. Dimian and A. Bliek, *Chem. Eng. Sci.*, **58**, 3159 & 3175 (2003).
9. T. Okuhara, *Chem. Rev.*, **102**, 3641 (2002).
10. M. A. Harmer and V. Sun, *Appl. Catal. A: Gen.*, **221**, 45 (2001).
11. S. Ardizzone, C. L. Bianchi, V. Ragaini and B. Vercelli, *Catal. Lett.*, **62**, 59 (1999).
12. H. Matsuda and T. Okuhara, *Catal. Lett.*, **56**, 241 (1998).
13. M. A. Harmer, Q. Sun, A. J. Vega, W. E. Farneth, A. Heidekum and W. F. Hoelderich, *Green Chem.*, **2**, 7 (2000).
14. G. D. Yadav and J. J. Nair, *Micropor. Mesopor. Mater.*, **33**, 1 (1999).
15. M. A. Ecormier, K. Wilson and A. F. Lee, *J. Catal.*, **215**, 57 (2003).

16. Y. Kamiya, S. Sakata, Y. Yoshinaga, R. Ohnishi and T. Okuhara, *Catal. Lett.*, **94**, 45 (2004).
17. R. Koster, B. van der Linden, E. Poels and A. Bliek, *J. Catal.*, **204**, 333 (2001).
18. H. G. Schoenmakers and B. Bessling, *Chem. Eng. Prog.*, **42**, 145 (2003).
19. R. Taylor and R. Krishna, *Chem. Eng. Sci.*, **55**, 5183 (2000).
20. H. Subawalla and J. R. Fair, *Ind. Eng. Chem. Res.*, **38**, 3696 (1999).
21. A. A. Kiss, A. C. Dimian, G. Rothenberg, *Adv. Synth. Catal.,* **348**, 75 (2006).
22. A. A. Kiss, G. Rothenberg, A. C. Dimian, *Top. Catal.*, **40**, 141 (2006).
23. A. A. Kiss, A. C. Dimian, G. Rothenberg, *Eng & Fuels*, **22**, 598 (2008).

34.　　New Catalysts for the Hydrogenolysis of Glycerol and Sugar Alcohols

Johnathan E. Holladay[1], James F. White[1], Thomas H. Peterson[1,2], John G. Frye[1], Aaron A. Oberg[1], Lars Peerboom[3], Dennis J. Miller[3] and Alan H. Zacher[1]

[1]*Pacific Northwest National Laboratory P.O. Box 999, Richland, WA 99352*
[2]*The Dow Chemical CompanyMidland, MI 48674*
[3]*Department of Chemical Engineering and Materials Science Michigan State University, East Lansing, MI 48824*

Alan.Zacher@pnl.gov

Abstract

Development of value-added products from glycerol can help the total economics of an oilseed biorefinery. Propylene glycol is one such product. This chapter will present the development of catalysts that can convert glycerol to propylene glycol in high yields. Our work has focused on a class of catalysts based on Re, which as a co-metal imparts important character to the catalysts.

Introduction

Recent interest in production of biodiesel provides an opportunity for developing new products from glycerol. Domestic biodiesel production is expected to grow to one to two billion gallons by 2010. One billion gallons would result in the production of 770 million pounds of glycerol (Figure 34.1). The U.S. glycerol market is only 320 million pounds. Developing value-added products from glycerol will not only help the total economics of an oilseed biorefinery but also further the environmental stewardship goal of replacing products that have historically been derived from petroleum sources.

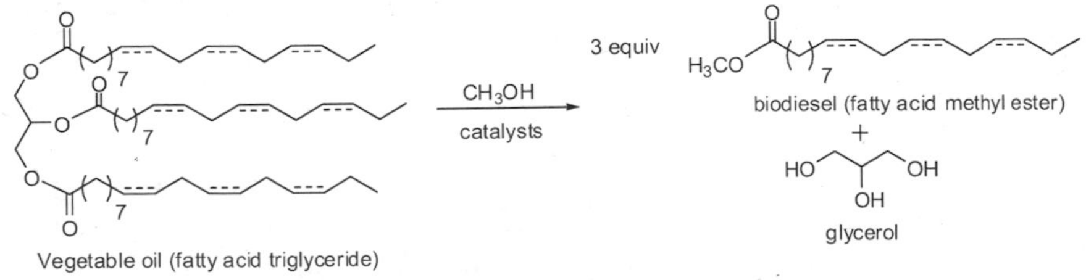

Figure 34.1. Biodiesel production

Numerous companies have announced plans for converting glycerol to propylene glycol (PG), including UOP, Dow, Huntsman, Ashland, Cargill, Archer Daniels Midland (ADM), Senergy Chemical, and Virent.[1] A process that has received significant attention in the press is from Galen Suppes.[2,3] His technology employs copper-chromite catalyst in a two-step process in which glycerol undergoes dehydration to form acetol which subsequently is hydrogenated to PG. Another industrially interesting process comes from Davy Process Technology, LTD. This process also employs a copper-based catalyst system operating in the gas phase at 205–220°C at very high H_2 to glycerol ratio of 500:1 and a pressure of 290 psig.[4]

PNNL has a long history studying hydrogenolysis as a means to form value-added products from sugar alcohols including glycerol.[5-7] In this paper we will report on a subset of this work, focused on rhenium-based multi-metallic catalysts supported on carbon.

A significant volume of literature relates to our work. Concerning choice of support, Montassier et al. have examined silica-supported catalysts with Pt, Co, Rh Ru and Ir catalysts.[8] However, these systems are not stable to hydrothermal conditions.[9] Carbon offers a stable support option. However, the prior art with respect to carbon-supported catalysts has generally focused on Ru and Pt as metals.[2,10-17] Additionally, unsupported catalysts have also been reported effective including Raney metals (metal sponges).[8,18] Although the bulk of the literature is based on mono-metallic systems, Maris et al. recently reported on bimetallic carbon-supported catalysts with Pt/Ru and Au/Ru.[11] In contrast, our work focuses primarily on the development of a class of rhenium-based carbon supported catalysts that have demonstrated performance equal to or better than much of the prior art. A proposed reaction mechanism is shown in Figure 34.2[10].

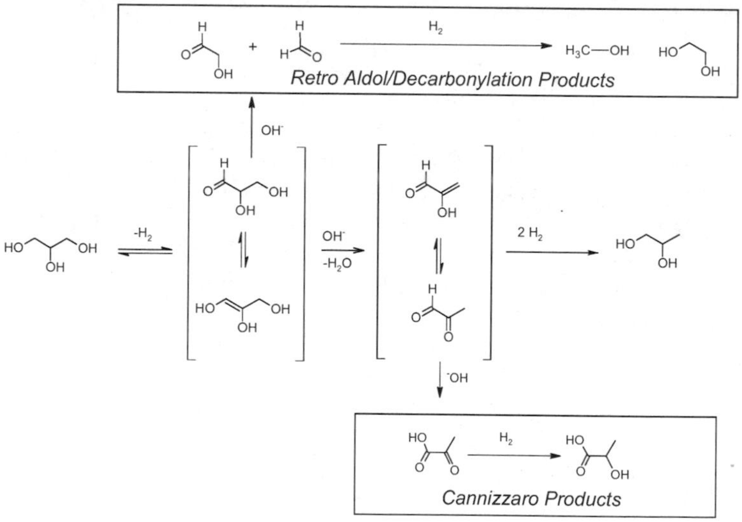

Figure 34.2. Hydrogenolysis of glycerol

Experimental Section

Catalysts were prepared on various supports. One example was a Calgon 120% CTC coconut carbon. The impregnation volume of metal solution was calculated using the measured incipient wetness of the support; 0.85 cc liquid per gram of carbon for the as received Calgon support. Nickel was added as a nickel nitrate solution, rhenium was added as perrhenic acid, ruthenium was added as ruthenium chloride or nitrosyl nitrate and other metals were added as various salts. Impregnation solutions were added dropwise and the carbon shaken thoroughly at about each 10% addition interval. After the metal addition, the vessel was capped for 15 minutes after which the catalyst was dried at 80°C under 510 mmHg vacuum.

In batch experiments catalyst, normally 1.25–2.5 g, was added into a 300 ml Parr autoclave. The reactor was purged with nitrogen and the catalyst reduced under hydrogen atmosphere at elevated temperature. After cooling to room temperature and purging the headspace of products from the catalyst reduction, an aqueous solution of glycerol and sodium hydroxide was added. The reactor was pressurized to the desired pressure with hydrogen and heated to the desired temperature. Hydrogen was continuously added to the reactor as it was consumed to maintain the desired pressure. Liquid samples were pushed out intermittently through a valved dip tube.

In continuous flow experiments, catalyst was packed into a downflow trickle-bed reactor of 30 cc bed volume. Hydrogen was passed slowly over the catalyst at atmospheric pressure and the temperature was slowly raised to the desired reduction/activation temperature and held for at least four hours. After activation, the reactor was cooled to the desired reaction temperature, the pressure was raised, and flow of an aqueous feed of glycerol and sodium hydroxide initiated along with a corresponding amount of hydrogen. A large set of reaction conditions was tested.

Results and Discussion

In this paper we will briefly discuss the role of both support and metals as both impact catalysis performance and stability. In our early work we discovered that Re imparts important character to Ni in producing an effective catalyst. A large study was completed using a design of experiments. A full paper describing that design is planned. In this paper we provide a summary of the results. Figure 34.3 shows

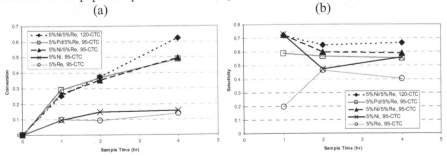

Figure 34.3a and b. Ni, Pd, and Re in catalyst compositions on carbon.

comparative data for Ni, Re monometallic catalysts and Ni/Re, Pd/Re bimetallic catalysts. In the reactions 2.5 g of catalyst charge was used with 105 g of feed containing 25 wt% glycerol and 0.5 wt% NaOH in H_2O. The reaction temperature was 230°C, and pressure was 1250 psi H_2.

The conversion was typically quite low when using the monometallic catalysts, but significantly improved for Ni/Re bimetallic catalysts. We do note that the total metal loading on the bimetallic catalyst was double (on a wt%) that of the monometallic catalysts; however, this data still demonstrates that the combined activity of the individual metal catalysts does not approach the activity of the bimetallic catalyst. Of interest is that Ni/Re resulted in equivalent conversion as was observed with the more expensive Pd/Re catalyst. One other point of importance can be noted in this work: Ni/Re was prepared on two carbon supports in these tests. The initial carbon support was a granular activated carbon having a CTC number of 95. CTC number is a measurement of porosity of the activated carbon as a measure of adsorption of saturated vapor of carbon tetrachloride and is a commonly used indirect indicator of surface area. A second activated granular carbon was also tested having a CTC number of 120. The catalyst on this support led to a higher conversion over the 4 h run time. The role of the support will be discussed later.

Selectivity data is also shown for the experiments described in Figure 34.3. Although selectivity information is given for each catalyst, we will only compare the data for the bimetallic catalysts due to the need to compare selectivities only at similar conversions to make accurate performance comparisons. There was little distinction between the selectivity for the Ni/Re and Pd/Re catalyst on the CTC-95 granular carbon. However, there was a significant improvement for the Ni/Re catalyst on the CTC-120 support, which again highlights the importance of selecting an appropriate support, as there are significant performance differences even among similar carbon supports used in catalysts for this reaction.

A second set of data is shown in Figure 34.4 comparing Ni/Re with Cu/Re and Ni/Cu each on the same granular carbon support (CTC-120) and also a comparison to Ni monometallic catalyst, however, the monometallic catalyst being on a different support. The reaction conditions for this set were 200°C, 1200 psi, 2.5 g of catalyst, and 105 g of feed containing 25 wt% glycerol and 0.25 wt% NaOH in H_2O. Although the conversion level was low (12-14%), the conversion with the Re-based bimetallic catalysts were significantly better than conversion of a Ni/Cu catalyst on

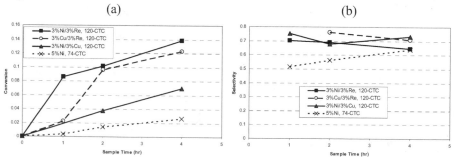

Figure 34.4a and b. Ni, Cu and Re catalyst compositions on carbon.

the same support or the Ni monometallic catalyst (albeit different support) but run as part of the same series in the design. Selectivity data is also shown. The differences in selectivity are minor, and due to the low conversion level the inherit error in the numbers prevent us from confidently ranking the catalysts as to selectivity performance. Based on the catalyst metal work we settled on a Ni/Re metal composition. Additional efforts have focused on optimizing the support.

Under hydrothermal conditions, alumina suffers from increased rates of hydration and hydrolysis of the oxides, resulting in the weakening of conventional supports according to Equation 1. Silica can suffer a similar hydrothermal fate. Process pH can exacerbate the problem (Fig. 34.5). Hence hydrothermal stability is a paramount criterion.

$$Al_2O_3 + H_2O \longrightarrow 2AlO(OH) \qquad (Eq.\ 1)$$

Granular and extruded carbon supports are stable to hydrothermal conditions as measured by crush strength unlike silica or alumina. There are certain metal oxides, such as rutile titania that are also stable. The data shown in Figure 34.5 shows the retention of original crush strength following 24 h soak in pH adjusted water at 200°C.

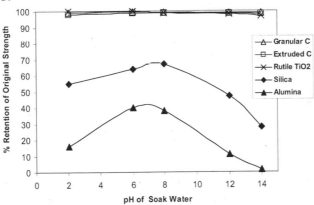

Figure 34.5. Crush strength of various supports after soaking for 24 h at 200°C in pH adjusted water.

Based on stability we have limited our study of supports to the following (surface areas shown in parentheses): carbon (<10 to < 2,000 m^2/g), rutile TiO$_2$ (1 – 40 m^2/g), monoclinic ZrO$_2$ (15-60 m^2/g), Nb$_2$O$_5$, and more esoteric supports such as SnO$_2$ and BaSO$_4$. Carbon offers good chemical stability, mechanical robustness and ease of catalyst metal recovery. Additionally, many carbon supports have higher porosity and surface area than oxides which can lead to high catalyst activity.

Of particular note we found improved selectivity with certain extruded carbon supports versus granular carbons, graphitic carbons, metal oxides (titania and zirconia), and carbon treated with zirconia.[19] Data shown in Figures 34.6 and 34.7 were generated in batch experiments using Ni/Re as catalytic metals. In Figure 34.6a, glycerol conversion and selectivity is graphed versus support for a series of batch experiments run under identical conditions (1.25 g catalyst, 100 g of 30% glycerol/2.1% NaOH aqueous solution, T = 190°C, t = 1.5 h, pressure = 1200 psig). In these experiments the same Ni/Re metal composition was used for each catalyst. Note in Figure 34.6a the striking difference in conversion for the various supports. The differences in selectivity were less striking. Because the selectivity data are at different levels of conversion, comparison can be difficult. In Figure 34.6b, we show selectivity versus conversion (omitting data for low conversion, which are less reliable under the batch conditions). Of interest is that all of the lowest selectivity data points were obtained using the catalysts on the "alternative" (i.e., metal oxide) supports.

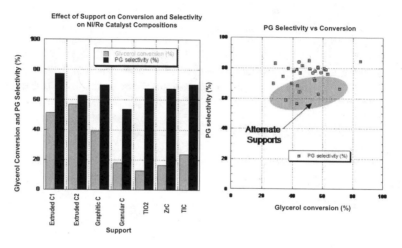

Figure 34.6a and b. Effect of support on conversion and selectivity for Ni/Re catalyst composition.

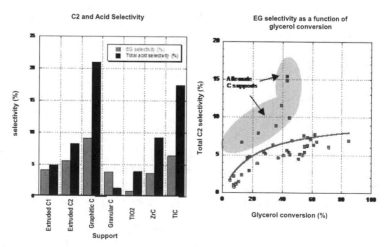

Figure 34.7a and b. EG and organic acid selectivity as a function of conversion with different supports.

In Figure 34.7a ethylene glycol (EG) selectivity is shown as a function of conversion for the same reactions discussed previously. Note that all of the highest selectivities to the two-carbon byproduct occurred on "alternative" supports.
In Figure 34.7b, the relative selectivity to byproducts such as EG and organic acids is shown (primarily acetic, lactic and glyceric acids). Not all carbon supports are equivalent, as there are a wide variety of source materials that are used in their production. Note that the highest acid selectivity is shown with the catalyst based on a graphitic carbon and on a carbon support first treated with titania.

At this point we are not able to explain the results on a mechanistic basis. More fundamental work is needed to correlate the characteristics of various carbons (surface area, pore distribution, point of zero charge, and surface activation) with observed activity and selectivity. The first step in the mechanism is likely dehydrogenation of glycerol to form glyceraldehyde, presumably bound to the metal of the catalyst. In the desired reaction, glyceraldehyde-metal adduct undergoes dehydration followed by hydrogenation to give PG (see Figure 34.2).[10] However, the intermediate can also undergo decarbonylation or retro-Diels-Alder to form two-carbon adducts that are hydrogenated to ethylene glycol. The adducts can alternately undergo base-catalyzed chemistry, resulting in organic acid salts perhaps under Cannizzaro-type chemistry. Organic salts are not reduced under the reaction conditions.[20]

Although carbon has many important qualities for a support material, it also appears to play a role in the access of the substrate to the active sites. Activated carbon preferentially adsorbs organic material from aqueous solutions. Thus the local concentration of reactants and products can be quite different at the catalyst surface than in the bulk solution.

(a)   (b)

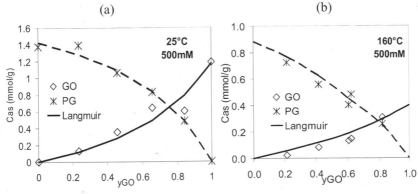

Figure 34.8. Competitive adsorption studies of glycerol (GO) and PG on carbon.

Initially we examined the adsorption of the primary individual species, glycerol and PG. The amount of material adsorbed was determined by differences in solution concentration over extended exposure of solutions to activated carbon support. The results indicate that the PG pore concentration can be three- to five-fold higher than the concentration in the bulk solution. Comparatively, glycerol pore concentration can be only 25% higher than the bulk solution.

Competitive adsorption on carbon was also studied. The results are shown in Figure 34.8. The product PG competitively adsorbs to carbon more readily than the starting material. This can have implications in reaching full conversion and on product stability. The impact of the relative adsorption is alleviated under continuous flow reactor conditions where we are able to achieve high conversion and high yield. A full accounting of the adsorption work will be the subject of a later publication.

From the catalyst metal and catalyst support studies PNNL settled on a preferred catalyst composition. We joined UOP in a DOE funded program to commercialize the process. A simplified process flow diagram is shown in Figure 34.9.

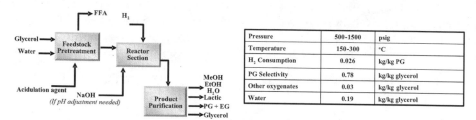

Pressure	500-1500	psig
Temperature	150-300	°C
H₂ Consumption	0.026	kg/kg PG
PG Selectivity	0.78	kg/kg glycerol
Other oxygenates	0.03	kg/kg glycerol
Water	0.19	kg/kg glycerol

Figure 34.9. Simplified process flow diagram (courtesy of UOP).

In our process development efforts PNNL demonstrated the life of the Re-containing catalyst using both pristine feed and feed from biodiesel desalted glycerin (see Figure 34.10).

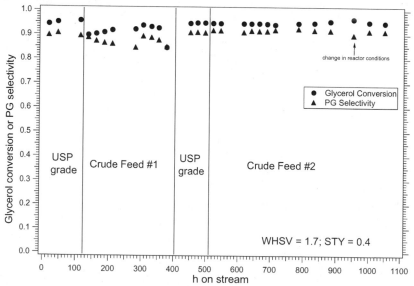

Figure 34.10. Continuous flow reactor.

Conclusions

Advances in catalysis are crucial to improving the economics and the environmental benefits of biorefineries, as well as encouraging the use of renewable feeds. An effective catalyst has been demonstrated for aqueous phase processing of sugar alcohols, like glycerol, to propylene glycol and other glycols. Rhenium-containing multimetallic catalysts result in superior conversion and selectivity and will represent an important part in the commercialization of technologies designed to turn byproduct glycerol derived from commercial scale bio-diesel production into value added products.

Acknowledgements

This work was supported by the United States Department of Energy at the Pacific Northwest National Laboratory (PNNL), a multiprogram national laboratory operated by Battelle for the U.S. Department of Energy under Contract DE-AC06-76RL01830. We acknowledge Tom Binder and Paul Bloom at ADM and Maureen Bricker, James Vassilakis, Simon Bare, Sharry Lynch, Laura Leonard, Todd Kruse, and Jennifer Holmgren at UOP with special thanks.

References

1. S. Shelly, http://www.aiche.org/uploadedFiles/SBE/Publications/Renewable%20PG.pdf, accessed 3/12/2008.
2. M. A. Dasaria, P.-P. Kiatsimkula, W. R. Sutterlin and Galen J. Suppes, *Appl. Catal. C.* **281**, 225 (2005).
3. C. W. Chin, M. A. Dasari, G. J. Suppes and W. R. Sutterlin, *AIChE J.* **52**, 3543 (2006).
4. M. W. M. Tuck and S. N. Tilley, *US Patent application* 20070149830, (2007).
5. T. A. Werpy, J. G. Frye Jr., A. H. Zacher and D. J. Miller, *US Patent* 7,038,094 (2006).
6. T. A. Werpy, J. G. Frye Jr., A. H. Zacher and D. J. Miller, *US Patent* 6,479,731 (2006).
7. T. A. Werpy, J. G. Frye, Jr., A. H. Zacher and D. J. Miller *US Patent* 6,841,085 (2005).
8. C. Montassier, D. Giraud and J. Barbier, Heterogeneous Catalysis over Metals, Elsevier Science Publishers, Amsterdam, 1988, p. 165.
9. E. P. Maris, W. C. Ketchie, V. Oleshko and R. J. Davis, *J. Phys. Chem. B* **110**, 7869 (2006).
10. E. P. Maris and R. J. Davis, *J. Catal.* **249**, 328 (2007).
11. E. P. Maris, W. C. Ketchie, M. Murayama and R. J. Davis, *J. Catal.* **251**, 281 (2007).
12. D. G. Lahr and B. H. Shanks, *Ind. Eng. Chem. Res.* **42**, 5467 (2003).
13. D. G. Lahr and B. H. Shanks, *J. Catal.* **232,** 386 (2005).
14. C. Montassier, J. C. Menezo, L. C. Hoang, C. Renaud and J. Barbier, *J. Mol.Catal.* **70**, 99 (1991).

15. J. Chaminand, L. Djakovitch, P. Gallezot, P. Marion, C. Pinel and C. Rosier, *Green Chem.* **6**, 359 (2004) 359.
16. Y. Kusunoki, T. Miyazawa, K. Kunimori and K. Tomishige, *Catal. Commun.* **6**, 645 (2005).
17. T. Miyazawa, Y. Kusunoki, K. Kunimori and K. Tomishige, *J. Catal.* **240**, 213 (2006).
18. A. Perosa and P. Tundo, *Ind. Eng. Chem. Res.* **44**, 8535 (2005) 8535.
19. T. A. Werpy, J. G. Frye Jr., Y. Wang and A. H. Zacher, *US Patent* 6,670,300 (2003).
20. F. T. Jere, D. J. Miller and J. E. Jackson, *Organic Lett.* **5**, 527 (2003).

35. Catalytic Hydrogenolysis of Glycerol

Michèle Besson, Laurent Djakovitch, Pierre Gallezot, Catherine Pinel, Alain Salameh and Matevz Vospernik

Institut de recherches sur la catalyse et l'environnement de Lyon, UMR 5256 – CNRS - Université de Lyon, 2 avenue Albert Einstein, 69626 Villeurbanne Cedex, France

Catherine.pinel@ircelyon.univ-lyon1.fr

Abstract

Rhodium-based catalysts were evaluated for the hydrogenolysis of glycerol. The study was focused on the influence of pH and metal additives on catalyst activity and selectivity. Under basic conditions, 1,2-PDO was formed with high selectivity (> 90%). Traces of acetol and ethylene glycol were detected and lactic acid was observed at high pH. Under acidic conditions (H_2WO_4), or in the presence of metal salts (i.e., $FeCl_2$, $ErCl_3$...), 1,2-PDO; 1,3-PDO and 1-propanol were formed. The highest 1,3-PDO/1,2-PDO ratio was 0.9.

Introduction

Biomass is a renewable resource from which various useful chemicals and fuels can be produced. Glycerol, obtained as a co-product of the transesterification of vegetable oils to produce biodiesel, is a potential building block to be processed in biorefineries (1,2). Attention has been recently paid to the conversion of glycerol to chemicals, such as propanediols (3, 4), acrolein (5, 6), or glyceric acid (7, 8).

Propylene glycol, i.e., 1,2-propanediol (1,2-PDO), is an important commodity chemical. It is used as biodegradable functional fluids and as precursors for the syntheses of unsaturated polyester resins and pharmaceuticals (9-10). Propylene glycol is currently produced from petroleum-derived propylene via oxidation to propylene oxide and subsequent hydrolysis (9, 11). However, the rising cost of propylene provides an incentive to find a substitute to propylene for this

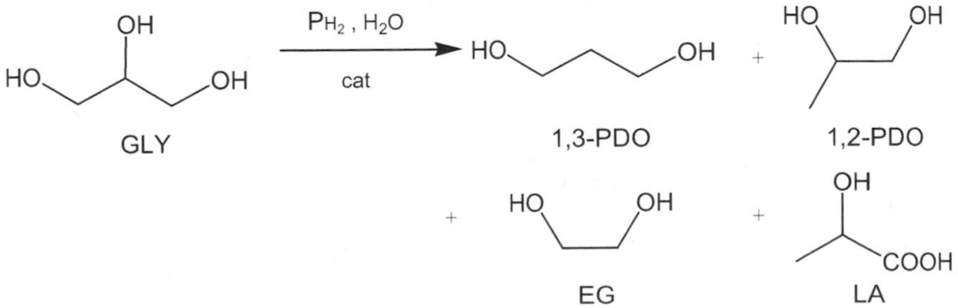

Scheme 35.1: Main products obtained from the conversion of glycerol.

application. In this respect, bio-renewable glycerol has emerged as a promising candidate. 1,3-propanediol (1,3-PDO) found applications as monomer in the preparation of polyesters SORONA® from Dupont or CORTERRA® from Shell, which are used in the manufacture of carpet and textile fibers exhibiting unique properties (11). 1,3-PDO is currently produced from petroleum derivatives (12, 13) or via enzymatic transformation (14). Alternatively, 1,2- and 1,3-propanediol can be obtained by hydrogenolysis of glycerol in the presence of heterogeneous catalysts (10, 11, 15-20). In this paper, the effects of pH and metallic salts additives on the selectivity of glycerol hydrogenolysis have been investigated.

Experimental Section

The reactions were performed either in an automated slurry reactor system SPR-16 (Amtec) consisting of 16 Hastelloy C22 15 mL reactors, or in a 150 mL stirred autoclave constructed in stainless steel and lined with graphite-stabilized Teflon. Temperature and hydrogen pressure were controlled all over the reaction. In a typical experiment, glycerol dissolved in water (20 wt%), supported metal catalyst, and optional additives, were introduced in the autoclave which was purged with nitrogen and pressurized with hydrogen. The reaction mixture was then heated and pressurized under stirring.

Samples were analyzed by GC or HPLC. For GC analysis (17), samples were prepared by mixing 200 µL of the reaction medium with 800 µL of a solution of the internal standard (xylene) in MeOH, and then analyzed (1 µL) over a GC type Agilent Technologies 6890N equipped with FID detector and a capillary column MTX-WAX (30 m × 280 µm × 0.25 µm) working at a maximum temperature of 250°C. For HPLC analysis (18), samples were diluted to 1/40 and then analyzed (20 µl) over an HPLC type Shimadzu CTO-6A equipped with a CARBOSep COREGEL-87H3 column (eluent H_2SO_4 0.005M, 70°C, 0.8 mL/mn) and refractive index detector (Shimadzu RID-A10).

Results and Discussion

Several supported metallic catalysts were evaluated for the selective hydrogenolysis of glycerol. Initially, the reactions were performed under acidic conditions in order to promote the formation of 1,3-PDO. Ruthenium-based catalysts were found to be the most active catalysts but significant amount of undesired products resulted from C-C cleavages were detected. On the contrary, Rh/C catalysts were found selective to C-O cleavages. As far as the selectivity to 1,3-PDO was concerned, we previously reported that the addition of iron salts in the medium improved the 1,3-PDO/1,2-PDO selectivity (11). A systematic study on the influence of additives was therefore carried out in the present investigation. Mineral and organic acids were evaluated for this purpose (Table 35.1).

Acetic or trifluoroacetic acids gave a very low glycerol conversion and no traces of 1,3-PDO could be detected. When the reaction was performed in the presence of tungstic or phosphomolybdic acids, higher conversions were achieved and some 1,3-PDO was formed.

Table 35.1: Influence of acids on the conversion of glycerol and 1,3-PDO/1,2-PDO selectivity.

Acid	%mol (acid/pure metal)	Conv (%)	1,3-PDO (%)	1,3-PDO / 1,2-PDO
H_2WO_4	0.1	36	4	0.30
$[H_3(Mo_{12}PO_{40})]$		36	2.6	0.23
CF_3COOH	1	6	0	
H_3BO_4	0.1	12	1.0	0.2

Reaction conditions: catalyst 5%Rh/C, 180°C, 80bar H_2, 24h

In order to improve the selectivity toward the formation of 1,3-PDO, we studied the influence of metal salt additives. While the addition of calcium or copper salts exhibited a moderate influence, the presence of iron salts played a significant role on the rate and selectivity of the reaction (Figure 35.1). The metal additives reduced noticeably the activity of the rhodium catalysts suggesting that they acted as a surface poison, but they modified the selectivity of the glycerol hydrogenolysis, probably through selective diol chelation.

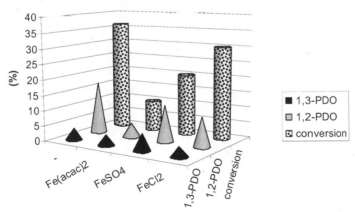

Figure 35.1: Influence of Fe-salts on the hydrogenolysis of glycerol.

At this stage, the influence of several lanthanide salts was evaluated (Figure 35.2) (19). In the presence of lanthanide nitrates, the conversion and the selectivity towards 1,3-PDO decreased compared to the reference, i.e., the reaction run without any additives, except H_2WO_4. In contrast, the addition of lanthanide chlorides had a positive effect on the selectivity to 1,3-PDO. However, low conversions were observed in all cases. Under these conditions, the highest ratio 1,3-PDO/1,2-PDO reached 0.9.

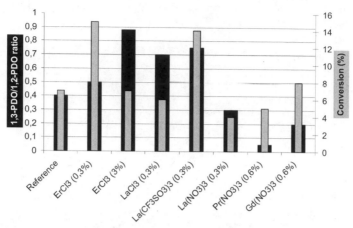

Figure 35.2: Influence of lanthanide salts addition on the hydrogenolysis of glycerol.

To the best of our knowledge, the highest selectivity towards 1,3-PDO (1,3-PDO/1,2-PDO = 15) was achieved by Bullock et al. (20) who performed the reaction in the presence of homogeneous $\{[Cp*Ru(CO_2)_2(\mu H)]^+OTf^-\}$ complex in sulfolan; however, the catalytic activity reported by these authors was very low (TON = 7).

The hydrogenolysis of glycerol was studied over a large range of pH in order to determine the optimal pH value at which the higher reaction rate and selectivity to 1,2-PDO are obtained. Selected results are summarized in Table 35.2. In the presence of Rh/C and under neutral conditions (pH_i = 5.5), a low conversion of glycerol was achieved (< 4%) after 48 h. The main product was 1,2-PDO, but EG, 1-propanol, ethanol and 1,3-PDO were also detected. Increasing the initial pH from 5.5 to 12.0 had beneficial effects both on the conversion and selectivity. The selectivity to EG decreased from 13% under neutral condition to less than 1% while the selectivity toward the desired 1,2-PDO increased from 52% to 96%.

Table 35.2. Influence of the initial pH value on the conversion of glycerol and selectivity to 1,2-PDO.

Solvent (pH_i)	Conv. %	Selectivity, %			
		1,2-PDO	EG	LA	other
H_2O (5.5)	3.6	52	13	0	35
Phosphate buffer (8.0)	<2	~ 100	0	0	/
0.4M Na_2CO_3 (11.0)	13	95	0.9	0.6	/
0.2 M NaOH (12.0)	11	96	0	0.4	/
1M NaOH (14.0)	53	38	0.4	0.8	/

Reaction conditions: catalyst 5%Rh/C, 180°C, 50bar H_2, 48h

At higher pH, lactic acid can be formed via internal Cannizzaro rearrangement of the intermediate ketoaldehyde (Scheme 35.2). Similar observations were reported by Davis in the presence of ruthenium-based catalysts, lactic acid becoming the predominant product under alkaline conditions (18). When reactions were performed at higher pH values (> 13.0) some compounds formed during the reaction were neither analyzed by GC nor by HPLC. This mass balance defect could be attributed to the formation of polyethers under the basic conditions.

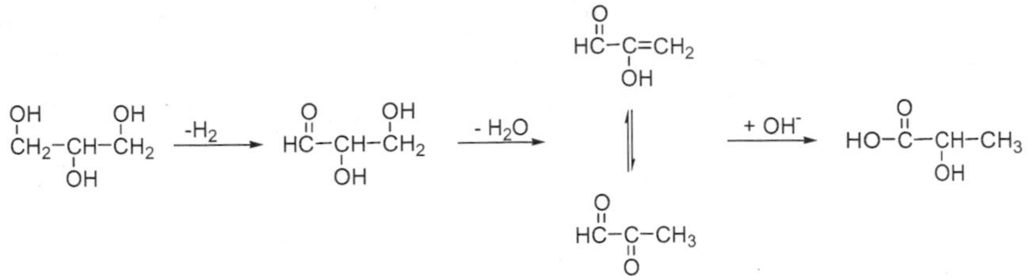

Scheme 35.2: Formation of lactic acid via Cannizaro rearrangement.

Based on literature data and on our own work, a general scheme for the hydrogenolysis of glycerol can be proposed (Scheme 35.3). Both dehydration and hydrogenation steps occur during the reaction. Under acidic conditions, the protonation of hydroxyl groups and dehydration yield acetol that can be easily hydrogenated to 1,2-PDO. Under basic conditions, pyruvaldehyde is formed transiently before being hydrogenated to 1,2-PDO (21). In the presence of metal additives further chelation modify the selectivity towards 1,2-PDO or 1,3-PDO.

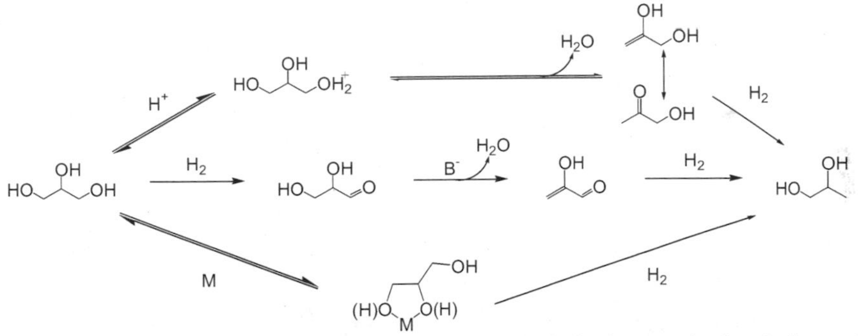

Scheme 35.3: Proposed routes for the catalytic hydrogenolysis of glycerol.

In conclusion, it was shown that the hydrogenolysis of glycerol in the presence of heterogeneous rhodium-based catalysts yielded mainly either 1,2-, or 1,3-propane diol. Many parameters influenced the activity and the selectivity of the catalysts, particularly the presence of metal additives and the initial pH value. 1,2-propanediol can be obtained nearly quantitatively at high pH. Further work is currently under progress in order to optimize this reaction.

Acknowledgements

We are indebted to Agrice (program 02.01.006) and to EU (Topcombi Contract No: NMP2-CT2005-515792) for financial support. Johnson Matthey is gratefully acknowledged for the gift of catalysts.

References

1. For a recent review A. Behr, J. Eitling, K. Irawadi, J. Leschinski, F. Lindler, *Green Chem.* **10**, 13 (2008).
2. G. W. Huber, S. Iborra, A.Corma, *Chem. Reviews* **106**, 4044 (2006).
3. T. Haas, B. Jaeger, R. Weber, S.F. Mitchell, C.F. King, *Appl. Catal. B* **280**, 83 (2005).
4. I. Furikado, T. Miyazawa, S. Koso, A. Shimao, K. Kunimori, K. Tomishige, *Green Chem.* **9**, 582 (2007).
5. M. Watanabe, T. Iida, Y. Aizawa, T.M. Aida, H. Inomata, *Bioresour. Technol.* **98**, 1285 (2006).
6. E. Tsukuda, S. Sato, R. Takahashi, T. Sodesawa, *Catal. Commun.* **8**, 1349 (2007).
7. S. Demirel, K. Lehnert, M. Lucas, P. Claus, *Appl. Catal. B* **70**, 637 (2007).
8. N. Dimitratos, C. Messi, F. Porta, L. Prati, A. Villa, *J. Mol. Catal. A*, **256**, 21 (2006).
9. R.D. Cortright, M. Sanchez-Castillo, J.A. Dumesic, *Appl. Catal. B* **39**, 353 (2002).
10. J. Chaminand, L. Djakovitch, P. Gallezot, P. Marion, C. Pinel, C. Rosier, *Green Chem.* **6**, 359 (2004).
11. M.A. Dasari, P.P. Kiatsimkul, W.R. Sutterlin, G.J. Suppes, *Appl. Catal. A* **281**, 225 (2005).
12. D. Arntz, T. Haas, A. Müller, N. Wiegand, *Chem. Ing. Tech.* **63**, 733 (1991).
13. *WO Patent 97 16250* (1997).
14. *Int. Patent 01/11 070 A2* (2001).
15. *US Patent 5,214,219* (1993).
16. C. Montassier, J.M. Dumas, P. Granger, J. Barbier, *Appl. Catal. A* **121**, 231 (1995).
17. A. Perosa, P. Tundo, *Industrial & Engineering Chemistry Research* **44**, 8535 (2005).
18. Y. Kusunoki, T. Miyazawa, K. Kunimori, K. Tomishige, *Catal. Commun.* **6**, 645 (2005).
19. C.W. Chiu, M.A. Dasari, W.R. Sutterlin, G.J. Suppes, *Industrial & Engineering Chemistry Research* **45**, 791 (2006).
20. T. Miyazawa, Y. Kusunoki, K. Kunimori, K. Tomishige, *J. Catal.* **240**, 213 (2006).
21. E.P. Maris, R.J. Davis, *J. Catal.* **249**, 328 (2007).
22. A.A. Neverov, T. McDonald, G. Gibson, R.S. Brown, *Can. J. Chem.* **79**, 1704 (2001).
23. M. Schlaf, P. Ghosh, P.J. Fagan, E. Hauptman, R.M. Bullock, *Angew. Chem. Int. Ed.*, **40**, 3887 (2001).
24. C. Montassier, J.C. Menezo, L.C. Hoang, C. Renaud, J. Barbier, *J. Mol. Catal.* **70**, 99 (1991).

36.   Development of Commercially Viable Thermomorphic Catalysts for Controlled Free Radical Polymerization

Christina M. Older[1], Soley Kristjansdottir[1], Joachim C. Ritter[1], Wilson Tam[1] and Michael C. Grady[2]

[1]DuPont Central Research & Development, Experimental Station, Wilmington, DE 19880
[2]DuPont Performance Coatings, Marshall Laboratories, Philadelphia, PA 19146

christina.m.older@usa.dupont.com

Abstract

Controlled radical polymerization to build complex polymer architectures is of great commercial interest to DuPont. Using cobalt-mediated catalytic chain transfer technology (CCT) allows for precise control of molecular weight in the free-radical polymerization of methacrylate esters. The polymers produced via CCT are highly functionalized with vinyl end groups, making them ideal macromonomer precursors for a variety of specialty polymer architectures. However, the homogeneous cobalt catalysts currently used for CCT are intensely colored and difficult to remove from the product resin, thus limiting potential applications. Eliminating the metal and color contamination would significantly enhance the commercial feasibility of CCT technology. Heterogeneous catalyst supports for CCT have proved untenable due to the increased polydispersity of the product resins so homogenous polymer-tethered catalysts were therefore evaluated as an alternative to traditional solid supports. Several classes of thermomorphic cobalt catalysts were demonstrated to be efficient recoverable catalysts. Polyethylene-tethered cobalt phthalocyanine and porphyrin catalysts were chosen as leading candidates for commercial implementation due to their inherent stability and facile separation from the product resin.

Introduction

Catalytic chain transfer (CCT) polymerization utilizing cobalt(II) macrocycles is currently the most effective means of controlling molecular weight in free-radical polymerizations of methacrylates.[1] This technology allows for the synthesis of low molecular weight polymers with terminal alkene groups (Scheme 36.1). These *macromonomers* are of particular interest as precursors to specialty polymer architectures and are of increasing commercial interest. Block, graft, telechelic and dendrimeric structures used in a wide variety of specialty applications are all available through these precursors. A challenge for CCT polymerization has been minimizing the amount of residual catalyst in the final polymer, since CCT catalysts are intensely colored and difficult to remove from the product resin. The color issue was previously mitigated by developing ever more active catalysts, which could be used in smaller quantities.

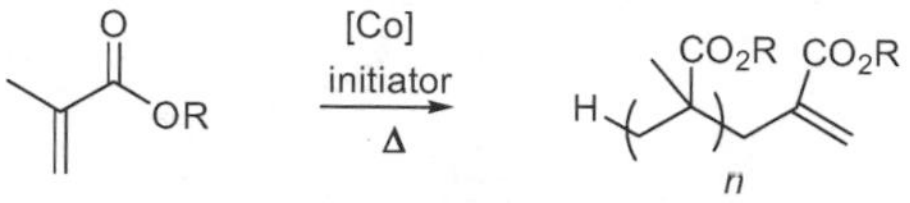

Scheme 36.1. Co^{II}-catalyzed chain transfer of methacrylate free radical polymerization.

Controlled synthesis of low molecular weight polymers is assuming greater importance as the drive to lower volatile organic content (VOC) necessitates replacing volatile solvents with oligomeric resins that become incorporated in the final application. High catalyst loadings are required to achieve these low molecular weights, and the resin color consequently becomes more intense. Previous efforts have been made to develop solid-supported CCT catalysts in an attempt to reduce color and to recycle the catalysts, but results have generally been disappointing.[2]

An alternative approach to heterogeneous catalysts of much recent interest is the use of soluble polymer supports. To minimize solvent handling and purification processes, we decided to focus our attention on thermoresponsive polymer supports. These catalysts rely on temperature-dependent solubilities and are designed to be homogeneous under reaction conditions, but to phase-separate from the product during work-up.[3] While extensively developed for small molecule synthesis, the use of polymer-supported catalysts in polymerization reactions poses special challenges.[4,5] Of the thermoresponsive polymer-supports currently in use, polyethylene offers practical advantages since it is chemically inert, inexpensive and, although soluble in a broad range of organic solvents at higher temperatures (>80°C), the linear ethylene oligomers are insoluble in all solvents at room temperature.[5]

While cobaloximes are the most active CCT catalysts known, they are sensitive to air and moisture, so the more robust porphyrins were considered a better choice for a recoverable catalyst. To better understand how to design an optimal thermomorphic catalyst we initiated an investigation to learn how polyethylene length, number, and the covalent linker influence the catalyst activity and the final color of the methacrylate resin. A series of polyethylene-supported CCT catalysts were thus prepared for study (Scheme 36.2).

To prepare the thermomorphic CCT catalysts, we originally sought to graft commercially available functionalized ethylene oligomers to the porphyrin ligand core. A range of polyethylene monoalcohols from Baker Petrolite (Unilin® Alcohols 460, 700 and 2000 Da) were used in the preliminary screen. Placing a single polyethylene tail on the porphyrin was accomplished by starting with pure mono(4-hydroxyphenyl)-tritolyl porphyrin. However, most of the polymer-tethered ligands were prepared from tetra(4-hydroxyphenyl) porphyrin, via alkylation by the ethylene oligomer mesylate derivative. By carefully controlling reaction conditions, it was possible to alkylate an average of 2 or 3 tails, or to alkylate all 4 phenols of the porphyrin. The polyether-linked catalyst was formed similarly using commercially available functionalized polyethylene-poly(ethylene glycol) diblock. The ester-

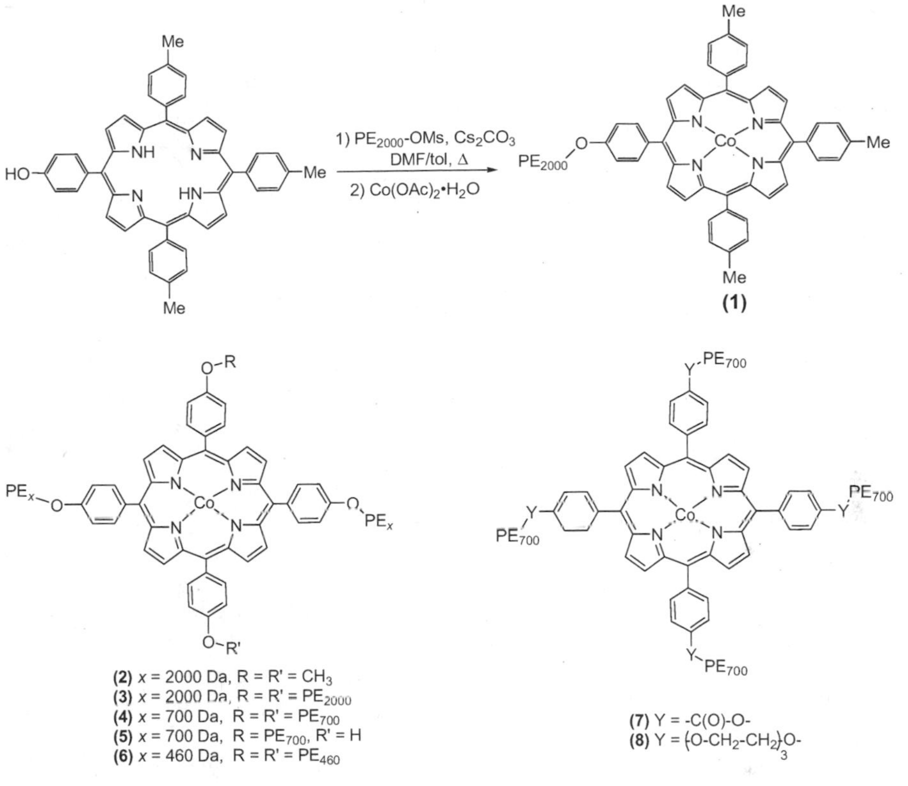

Scheme 36.2. Polyethylene-supported cobalt porphyrin catalysts.

linked polyethylene porphyrins were readily prepared via treatment of the carboxylic acid chloride porphyrin with polyethylene monoalcohols. In each case, the polymer-supported porphyrin ligand was metallated with cobalt acetate in a subsequent step. Each polymer-supported porphyrin complex was purified extensively through multiple Soxhlet extractions since it is crucial to remove any colored impurities, particularly any soluble porphyrins, before evaluating the catalyst. The rigorous washing procedures also removed most polyethylene-derived byproducts such as polyethylene monoalcohol, vinyl-terminated polyethylene, as well as some low-temperature soluble catalyst. All of the isolated polymer-bound porphyrin catalysts were highly pure materials and could be characterized using standard spectroscopic methods in solution (at >80°C), a distinct advantage over heterogeneous catalysts.

Experimental Section

Below are selected examples of catalyst synthesis and a description of the polymerization screening process (patent pending).[6] Polyethylene monoalcohols (Unilin® brand with an average M_n of 460, 700 and 2000 Daltons, respectively) were provided by Baker Petrolite. These materials consisted of the primary alcohol as a statistical distribution of molecular lengths and unfunctionalized hydrocarbon (~15-25% by weight). 5,10,15,20-Tetrakis(4-octadecyloxyphenyl)porphyrin[7] and mono(4-hydroxyphenyl)-tritolyl porphyrin[8] were prepared following literature procedures. Polyethylene methylsulfonic esters (PE-OMs) were prepared as described by Bergbreiter.[9] All other materials were used as received from Aldrich without further purification.

Synthesis of 5,10,15,20-Tetrakis(4-(polyethyleneoxy)phenyl)) porphyrin. A slurry of polyethylene methylsulfonic ester (PE$_{700}$-OMs) (20.0 g, 69% functionalized, M$_n$ ~780 Daltons) and anhydrous cesium carbonate (Cs$_2$CO$_3$) (9.05 g, 27 mmol) in dry toluene (75 mL) was prepared and to this mixture a purple solution of 5,10,15,20-tetrakis(4-hydroxyphenyl)-21H,23H-porphine (2.9 g, 4.27 mmol) in 75 mL *N,N*-DMF was added. The reaction mixture was warmed to 95°C with stirring for 18 hours, then the temperature increased to 130°C for a further 5 hours before cooling. The purple-brown solid was collected by filtration, washed thoroughly with methanol and dried under vacuum to yield 21.0 g of the crude ligand. ^{1}H NMR (toluene-d$_8$, 80°C): δ 8.96 (s), 8.10 (d), 7.22 (d), 4.05 (t), 1.88 (quin.), 1.58 (quin.), 1.31 (br. s), 0.88 (t).

Synthesis of 5,10,15,20-Tetrakis(4-(polyethyleneoxy)phenyl) porphyrin Cobalt (II) (4). A slurry of 21.0 g tetrakis(4-(polyethyleneoxy)phenyl)) porphyrin in 100 mL of a 1:1 mixture of DMF and toluene was prepared and Co(OAc)$_2$•4H$_2$O (1.72 g, 6.9 mmol) was added and the reaction warmed to 120°C with stirring. After 5 hours the reaction mixture was cooled to room temperature and the red-brown precipitate collected by filtration, then thoroughly washed with methanol, water, and finally acetone. The crude product was dried under vacuum to yield 21.43 g of a red powder. Crude tetrakis(4-(polyethyleneoxy)-phenyl)porphyrin cobalt complex (21.43 g) was transferred to a jacketed Soxhlet apparatus and washed with MEK until the washings were colorless. The Soxhlet thimble containing the washed cobalt complex was then dried overnight and the product extracted using hot xylenes in a jacketed Soxhlet apparatus. The precipitate was collected by filtration and washed a second time with MEK in a Soxhlet apparatus until the washings were again colorless. The final weight of dried product was 13.4 g of an orange-red powder. ^{1}H NMR (toluene-d$_8$, 80°C): δ 14.93 (v. br. s), 12.51 (br. s), 9.16 (s), 5.22 (s), 2.66 (s), 2.27 (s), 1.97 (s), 1.84 (s), 1.32 (br. s), 0.86 (t). UV-vis (toluene, 80°C) λ$_{max}$ 415, 531 nm. Elemental analysis: 1.24% cobalt.

Polymerization Screening Protocol: A 100 mL stock solution of degassed methyl methacrylate (MMA)/n-butyl methacrylate (n-BMA)/methyl ethyl ketone (MEK) (33 g: 33 g: 35g) and 1.5 g butyl acetate (internal GC standard) was prepared prior to the run and stored in the freezer until required. In a glovebox, each 20 mL vial was charged with catalyst, a magnetic stir-bar and 5.0 g of the stock solution. The vials were sealed and preheated at 100°C for at least 5 minutes to dissolve the catalyst before adding 0.5 mL of a stock solution of 2,2'azobis(2-methyl) butanenitrile (VAZO-67®) (30 mg / mL of MEK) via syringe. The polymerization reactions were stirred at this temperature for 1 hour. Upon cooling to room temperature, an aliquot was removed from each vial via pipette for SEC and GC analysis. The remaining sample was then filtered using syringe filters (25 mm Acrodisc® with PFTE membrane, 0.2 μL pore size) to remove any precipitated material. The sample was diluted to 7.0 g using MEK and the filtrate analyzed for color. Measurement of the yellow resin colors was initially determined via direct comparison by eye of the product samples to Platinum-Cobalt Color standards, prepared according to ASTM D5386-93B. For more detailed analytical measurements of low color resin samples a spectrophotometer (Varian Cary50 UV/Visible Spectrophotometer) was used to determine APHA color values. *Molecular weight determination:* An integrated multidetector SEC system, Waters 150CV™ model with two online detectors, differential refractometer (DR) and capillary viscometer (CV) was used to determine MN, MW and polydispersity (PDI). Two SEC columns, PL Gel Mixed C and PL Gel 500A columns from Polymer Laboratories were used for separation in THF stabilized with BHT.

Table 36.1. Representative screening of MMA/n-BMA using $(PE_{700})_4Co^{II}$ porphyrin (**4**).

Catalyst (ppm)	$M_N^{\pm}$	$PDI^{\pm}$	Color	$APHA^{\pm}$
6.0×10^2	2861	2.13	colorless	22
20×10^2	1383	1.74	colorless	27
50×10^2	842	1.48	colorless	26
100×10^2	601	1.30	colorless	23
200×10^2	484	1.21	colorless	25

$^{\pm}$ Data represents an average of two polymerization runs.

Synthesis of Polyethylene-tethered Phthalonitrile. Polyethylene monoalcohol (~700 Daltons, 80% functionalized) in pellet form (6.0 g, 8.0 mmol) and anhydrous THF (62 mL) were added to a flask equipped with a magnetic stirring bar. This was allowed to stir for 2 hours before 4-nitrophthalonitrile (2.25 g, 13.0 mmol) was added to the mixture and the suspension heated to reflux. After the polymer had dissolved the Cs_2CO_3 (4.22 g, 13.0 mmol) was added in three equal portions over 3 hours. After 24 hours the reaction was cooled to 60°C then poured into 500 mL of water with stirring. The precipitate was filtered and washed thoroughly with water until pH neutral. The product was then rinsed with *N,N*-DMF, THF and then dried under vacuum to yield 5.03 g of material. ^{1}H NMR (toluene-d$_8$, 100°C, 71% conversion): δ 6.86 (d), 6.59 (s), 6.41 (m), 3.41 (t), 3.37 (t), 1.31 (s), 0.88 (t).

Synthesis of Tetrakis(polyethyleneoxy)phthalocyanine cobalt (II) (**9**). The polyethylene-tethered phthalonitrile (0.827 g, 1.0 mmol), 1,5-diazabicyclo [4.3.0] non-5-ene (0.062 g, 0.50 mmol) and cobalt acetate (0.044 g, 0.25 mmol) were added to a reaction vial equipped with a magnetic stir bar. The reaction was then heated to 175°C for 2 hours before reducing the temperature to 95°C and adding toluene. The reaction mixture was then poured into methanol, filtered, and the solid washed further with methanol. The collected product was then dried under vacuum to yield

0.62 g of a dark green solid. ^{1}H NMR (toluene-d$_8$, 100°C): δ 7.60-6.5 (br. m), 3.38 (t), 1.31 (v. br. s), 0.89 (t). UV-vis (toluene, 80°C) λ$_{max}$ 672, 704 nm.

Results and Discussion

Most examples of thermomorphic polyethylene-supported catalysts have had a single-appended chain of ~1200-2000 Da length.[3,10] The modular design of the porphyrin catalysts gave us an opportunity to study the effect of varying length and number of polyethylene "tails" on catalyst activity and separation. For the purposes of screening CCT catalysts, a representative polymerization reaction for a typical methacrylate coatings application was developed: a monomer solution consisting of a mixture of methyl methacrylate and *n*-butyl methylacrylate esters (1:1 by weight) at high solids loading (only 33% MEK solvent) was warmed to temperature in the presence of dissolved catalyst before adding initiator. Each catalyst was subjected to two polymerization runs at 5 different catalyst loadings in order to generate catalyst activity curves (Figure 36.1). The number average molecular weight (M$_N$), weight average molecular weight (M$_w$), polydispersity (PDI) and percent monomer conversion was determined for each polymer sample. The color of the resin solution after filtration was then determined in APHA units; a value of >100 units was considered unacceptably high for commercial application (Figure 36.2). Tetrakis(4-octadecyloxy-phenyl)porphyrin was used as a soluble control catalyst of known activity. All porphyrin examples in the polymer screen resulted in >85% monomer conversion (as measured by SEC and GC analysis).

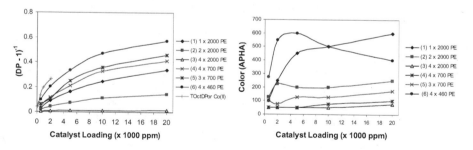

Figure 36.1. Catalytic activity of PE-supported cobalt porphyrins.

Figure 36.2. Measured color of MMA/*n*BMA polymer solution (after catalyst removal).

While the control resins were deep red in color due to the presence of soluble porphyrin complexes, the methacrylate resins obtained after removal of the polyethylene-supported catalysts varied from light yellow to nearly water-white (APHA < 25). UV-Vis spectrophotometric analysis of the yellow resins indicated an absorption signal for the cobalt porphyrin complex Soret band (wavelength of cobalt(II) porphyrin species appears at ~415 nm; free porphyrin ligand is found at ~423 nm). Resin samples that visually appear as water-white show little or no porphyrin species present in the spectrum. Measured catalyst activity and PDI of the polyethylene-supported porphyrin complexes are in the expected range for soluble porphyrin CCT catalysts (PDI = M$_w$/M$_N$ ~ 1.2- 2.0).[1b] The screening results clearly demonstrate a relationship between the polyethylene polymer supports and resin color: (i) color control improves as the polyethylene support increases in length and (ii) color control improves dramatically with multiple tails. For this mixed methyl methacrylate/n-butyl methacrylate CCT polymerization the optimal configuration for color control was four polyethylene tails with a minimum tether length of ~700 Da. Shorter tails resulted in more intense resin color due to more residual catalyst in the filtrate. A single 2000 Da polyethylene tail, catalyst (1), gave surprisingly poor color control despite repeated attempts to remove trace soluble fractions. Multiple long tails yielded very good resin color control but the catalyst activity was lower, even after adjusting for the difference in molecular weight.

Polymerization screening runs using porphyrin catalysts with different linker groups between the 700 Da polyethylene tail and the porphyrin core showed little variability. The catalytic activity for the 4 x 700 PE porphyrin catalysts with ether (5), ester (7) or polyether (8) linkers was nearly identical, with only small differences in the resin colors being noted (Figures 36.3 and 36.4). These results are consistent with literature findings that porphyrin CCT catalysts are relatively insensitive to functional groups on the equatorial plane.[1b]

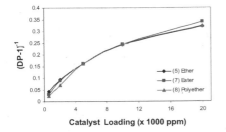

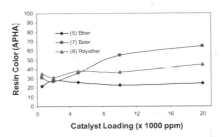

Figure 36.3. Catalyst activity of $(PE_{700})_4 -$ Co porphyrins as function of linker group.

Figure 36.4. MMA/BMA polymer solution color as function of linker group.

For a recoverable CCT catalyst to be commercially viable, it must be both robust and efficiently recycled. Little loss of catalyst activity per cycle can be tolerated, whether through physical losses during processing or through chemical deactivation. Although porphyrin ligands are generally thermally stable, the cobalt catalysts can be deactivated through a variety of pathways.[1b] To determine the stability of our optimal thermomorphic porphyrin catalyst (5), a small scale (10 g) recycle study was carried out using a combination of centrifugation and syringe filtration to minimize physical losses. After nine recycles of the thermomorphic cobalt porphyrin catalyst, no decrease in catalyst activity was observed within experimental error (Figure 36.5). Low color and narrow polydispersity were also maintained during the recycle study.

Although porphyrin-derived thermomorphic complexes were demonstrated to be robust and recyclable CCT catalysts, barriers to implementation still existed. One obstacle to scaling the process was the cost of the tetrakis(4-hydroxyphenyl)porphyrin starting material; porphyrins are expensive due to their notoriously low synthetic yields. Another concern was the rigorous purification that was required to remove all traces of soluble porphyrin impurities from the polyethylene-tethered materials. These two factors added significantly to the cost of catalyst manufacture.

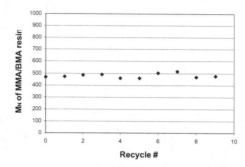

Figure 36.5. Recycle study of $(PE_{700})_4$ cobalt porphyrin (**4**) (10 g scale).

While not as well known in catalysis as porphyrin systems, the closely related cobalt phthalocyanines have also been reported as CCT catalysts.[1b] These compounds are extremely robust, being insensitive to both air and moisture and easier to prepare at commercial scale than porphyrins. A simple two-step convergent synthetic route to the polyethylene-tethered cobalt phthalocyanine was developed from known procedures,[11] although extensive optimization was necessary. An unexpected advantage of the convergent method was soon discovered; unlike the grafting approach used with the porphyrins, there is little in the way of soluble colored impurities formed during the convergent pathway. Simple washing of the crude material is all that is necessary to obtain a catalyst capable of producing colorless methacrylate resins. While analysis of the isolated material indicates that a significant amount of polyethyelene-tethered oligomeric impurities are present (~40% by weight), these impurities do not chelate cobalt and do not interfere with the subsequent polymerization reactions. A recycle study similar to the one conducted on the polyethylene-tethered porphyrin was carried out to confirm that catalyst activity and color control were stable over multiple cycles.

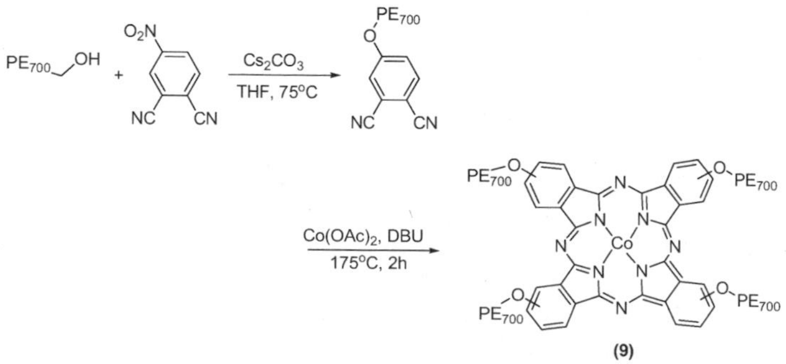

Scheme 36.3. Preparation of polyethylene-supported cobalt phthalocyanine catalyst.

Conclusions

We have demonstrated a new class of effective, recoverable thermormorphic CCT catalysts capable of producing colorless methacrylate oligomers with narrow polydispersity and low molecular weight. For controlled radical polymerization of simple alkyl methacrylates, the use of multiple polyethylene tails of moderate molecular weight (700 Da) gave the best balance of color control and catalyst activity. Porphyrin-derived thermomorphic catalysts met the criteria of easy separation from product resin and low catalyst loss per batch, but were too expensive for commercial implementation. However, the polyethylene-supported cobalt phthalocyanine complex is more economically viable due to its greater ease of synthesis.

Acknowledgements

We gratefully acknowledge the use of research samples of Unilin® monoalcohols provided by Baker Petrolite. We also thank Prof. Bergbreiter of Texas A&M University for helpful discussions.

References

1. For general reviews on catalytic chain transfer catalysis, see: (a) J. P.A. Heuts, G. E. Roberts, J. D. Biasutti *Aust. J. Chem* **55**, 381 (2002). (b) A. A. Gridnev and S. D. Ittel , *Chem. Rev.* **101(12)**, 3611 (2001), and references therein. (c) A. A. Gridnev *J. Polym. Sci., Polym. Chem.* **38**, 1753 (2000).
2. W. Tam and C. M. Older, unpublished results.
3. For recent reviews on soluble polymer supports, see: (a) D. E. Bergbreiter *Chem. Rev.*, **102**, 3345 (2002). (b) T. J. Dickerson, N. N. Reed, K. D. Janda, *Chem. Rev.* **102**, 3325 (2002). (c) D. E. Bergbreiter *Journal of Polymer Science: Part A*, **39**, 2351 (2001). (d) D.E. Bergbreiter, P.L. Osburn, and J. D. Frels, *J. Am. Chem. Soc.*, **123**, 11105 (2001).
4. D. E. Bergbreiter, L. B. Chen, and R. Chandran, *Macromolecules,* **18**, 1055 (1985).
5. (a) Y. Shen and S. Zhu, *Macromolecules*, **34**, 8603 (2001). (b) Y. Shen, S. Zhu, and R. Pelton, *Macromolecules* **34**, 3182 (2001). (c) S. Liou, J. T. Rademacher, D. Malaba, M. E. Pallack and W. J. Brittain, *Macromolecules,* **33**, 4295 (2000).
6. H. E. Bryndza, M. C. Grady, S. S. Kristjansdottir, C. M. Older, M. A. Page, J. Ritter, W. Tam WO 2007/024634 (E.I. DuPont de Nemours and Co.) "Recoverable polymer-bound homogeneous catalysts for catalytic chain transfer processes."
7. D. A. Lerner, M. B. Quintal, P. Maillard, C. Giannotti, *J. Chem. Soc., Perkin Trans. 2,* **7**, 1105 (1990).
8. R. G. Little, J. A. Anto, P. A. Loach, J. A. Ibers *J. Heterocyclic Chem.* **12**, 343 (1975).

9. D. E. Bergbreiter, J. R. Blanton, R. Chandran, M. D. Hein, K.-J. Huang, D. R. Treadwell, and S. A. Walker, *Journal of Polymer Science: Part A*, **27**, 4205 (1989).
10. M. Karabörk and S. Serin, in *Synthesis and Reactivity in Inorganic and Metal-Organic Chemistry* (Dekker), **32 (9),** 1635 (2002).
11. D. Wöhrle, G. Schnurpfeil and G. Knothe, *Dyes and Pigments,* **18**, 91 (1992).

37.   Rhodium Catalyzed Carbonylation of Ethylene and Methanol in the Absence of Alkyl Halides Using Ionic Liquids

Joseph R. Zoeller, Mary K. Moore, Andrew J. Vetter, Ashley Quillen and
Theresa Barnette

*Eastman Chemical Company, Research Laboratories, P.O. Box 1972,
Kingsport, TN 37660*

jzoeller@eastman.com

Abstract

Rhodium catalyzed carbonylations of olefins and methanol can be operated in the absence of an alkyl iodide or hydrogen iodide if the carbonylation is operated in the presence of iodide-based ionic liquids. In this chapter, we will describe the historical development of these non-alkyl halide containing processes beginning with the carbonylation of ethylene to propionic acid in which the omission of alkyl halide led to an improvement in the selectivity. We will further describe extension of the non-alkyl halide based carbonylation to the carbonylation of MeOH (producing acetic acid) in both a batch and continuous mode of operation. In the continuous mode, the best ionic liquids for carbonylation of MeOH were based on pyridinium and polyalkylated pyridinium iodide derivatives. Removing the highly toxic alkyl halide represents safer, potentially lower cost, process with less complex product purification.

Introduction

Historically, the rhodium catalyzed carbonylation of methanol to acetic acid required large quantities of methyl iodide co-catalyst (1) and the related hydrocarboxylation of olefins required the presence of an alkyl iodide or hydrogen iodide (2). Unfortunately, the alkyl halides pose several significant difficulties since they are highly toxic, lead to iodine contamination of the final product, are highly corrosive, and are expensive to purchase and handle. Attempts to eliminate alkyl halides or their precursors have proven futile to date (1).

Recently, Eastman Chemical Company reported that ionic liquids can be successfully employed in a vapor take-off process for the carbonylation of methanol to acetic acid in the presence of rhodium and methyl iodide (3). While attempting to extend this earlier work to the carbonylation of ethylene to propionic acid, we discovered that, when using ionic liquids as a solvent, acceptable carbonylation rates could be attained in the absence of any added alkyl iodide or hydrogen iodide (4). We subsequently demonstrated that the carbonylation of methanol to acetic acid could also be operated in the absence of methyl iodide when using ionic liquids (5).

In this manuscript, we will chronicle the discovery and development of these non-alkyl halide containing processes for the rhodium catalyzed carbonylation of ethylene to propionic acid and methanol to acetic acid when using ionic liquids as solvent.

Experimental Section

All ionic liquids were made by alkylation of their parent amine or phosphine with methyl iodide or ethyl iodide. The following abbreviations for the various ionic liquids will be used throughout the manuscript:
  DMII = N,N-dimethyl imidazolium iodide
  DEII = N,N-diethyl imidazolium iodide
  BMII = N-butyl-N-methyl imidazolium iodide
  [Etpy]I = N-ethyl pyridinium iodide
  [pyMe]I = N-methyl pyridinium iodide
  MTOPI = methyl trioctyl phosphonium iodide
  [Bu₃PMe]I = methyl tributyl phosphonium iodide.

General Procedure for Batch Carbonylation of Ethylene in the Absence of Ethyl Iodide. A complete set of procedures appears in ref. 4 but the following procedure may be regarded as representative. To a 300 mL Hastelloy® C-276 autoclave equipped with a condenser to return liquid to the autoclave and a gas purge at the top of the condenser, was added 0.789 g (3.0 mmol) of $RhCl_3 \cdot 3H_2O$, 122 g (0.545 mol) of N,N'-dimethyl imidazolium iodide, 27.0 g (1.5 mol) of water. The autoclave was sealed, flushed with nitrogen, and then pressurized to 250 psi (1.72 MPa) with 5% hydrogen in CO. A purge of 1.0 mol/hr was established through the condenser and gas purge and the condenser cooled to 10°C mixture. The mixture was heated to 190°C maintaining the gas purge pressure at 250 psi (1.72 MPa) of 5% hydrogen in carbon monoxide. Upon reaching temperature the gas feed was switched to 100% CO and the pressure adjusted to 450 psi (3.10 MPa) using a mixture of 50 mol% CO: 45 mol% ethylene; 5 mol% hydrogen. The temperature and pressure were maintained for 8 hours using the 50 mol% CO: 45 mol% ethylene; 5 mol% hydrogen gas mixture as needed. After 8 hours, the reaction was cooled, vented, and the product transferred to a sample bottle. GC analysis of the product indicated that the mixture contained 1.04 wt.% ethyl propionate, and 39.53 wt. % propionic acid. This represents 1.153 moles of propionic acid and 0.018 moles of ethyl propionate. No ethyl iodide was detected in the product by GC analysis. Reaction rates could be obtained by further equipping the reactor with a liquid sampler and removing samples for GC analysis every 30 minutes.

General Procedure for Batch Carbonylation of Methanol in the Absence of Methyl Iodide. A complete set of procedures appears in ref. 5 but the following procedure is representative of a methanol carbonylation. To a 300 mL Hastelloy® C-276 autoclave was added 0.396 g (1.5 mmol) of $RhCl_3 \cdot 3H_2O$, 112.0 g (0.507 mol) of N-methyl pyridinium iodide, 30.0 g (0.5 mol) of acetic acid, and 64.0 g (2.0 mol) of methanol. The mixture was heated to 190°C under 250 psi (1.72 MPa) of 5% hydrogen in carbon monoxide. Upon reaching temperature the gas feed was switched to 100% CO and the pressure adjusted to 750 psi (5.17 MPa) using 100% CO. The temperature and pressure were maintained for 5 hours using 100% CO as needed. After 5 hours, the reaction was cooled, vented, and the product transferred to a sample bottle. GC analysis of the product indicated that the mixture contained 0.25 wt. % methyl acetate, 0.04 wt. % methanol, and 55.84 wt. % acetic acid. This represents 2.33 moles of acetic acid (Net production of acetic acid = 1.83 moles after accounting for acetic acid in the original solution) and 0.008 mol of methyl acetate along with 0.035 moles of unreacted methanol. No methyl iodide was detected in the product by GC analysis.

General Procedure for Continuous Vapor Take-Off Reaction in the Absence of Methyl Iodide. (Condensed from reference 5.) A baffled 300 mL Hastelloy® C-276 autoclave equipped with (1) a dip tube for gas and methanol introduction and (2) a mechanical stirrer was fitted with a band heater on the top of the autoclave modified to maintain a temperature control over the entire reactor, and then connected via a U tube to a high pressure Hastelloy® C-276 condenser such that the vapors from the autoclave fed to the top of the chilled (20°C) condenser. To collect the condensate, the end of the condenser was connected to a Hastelloy® C-276 high pressure receiver which was equipped with a backpressure regulator at the top of the receiver to allow pressure control in the system and a valve at the bottom to allow the receiver to be drained. The reactor was filled with 2.0 g of solid $RhCl_3 \cdot 3H_2O$ (7.60 mmol) followed by a solution of 125 g of quaternary ammonium or phosphonium salt in 60 g of acetic acid. The autoclave was sealed and flushed first with nitrogen and then with 5% hydrogen in carbon monoxide. After flushing with 5% hydrogen in carbon monoxide, the feed gas was switched to pure carbon monoxide and fed at a rate of 0.90 mol/hr (336 sccm) with the backpressure set to maintain a pressure of 250 psi (1.72 MPa) in the reactor. The reactor was heated to 190°C and upon reaching temperature, methanol was fed at a rate of 0.40 mL/min. (24 ml/hr, 0.59 mol/hr, CO/MeOH ratio = 1.5/1.) The condensate was collected in the high pressure receiver and the receiver was drained every 6 hours. The condensate was weighed and analyzed for methyl iodide, methyl acetate, methanol, water, and acetic acid using gas chromatography. The reaction was normally continued until it was evident that there was no longer any significant weight gain in the product. Results appear in Figures 37.1 and 37.2.

Results and Discussion

Ethylene Carbonylation. The classical rhodium catalyzed carbonylation of ethylene to propionic acid (Eqn. 1) used ethyl iodide or HI as a co-catalyst (6). In the presence of excess ethylene and CO the process could proceed further to propionic anhydride (Eqn. 2). While additional products, such as ethyl propionate (EtCO₂Et), diethyl ketone (DEK), and ethanol were possible (See Eqns. 3-5), the only by-product obtained when using a rhodium–alkyl iodide catalyst was small amounts (ca. 1-1.5%) of ethyl propionate. (See Eqns. 3-5.)

$$H_2C=CH_2 + CO + H_2O \rightarrow EtCO_2H \qquad (1)$$
$$H_2C=CH_2 + CO + EtCO_2H \longrightarrow (EtCO)_2O \qquad (2)$$

$$H_2C=CH_2 + EtCO_2H \longrightarrow EtCO_2Et \tag{3}$$
$$2\ EtCO_2H \longrightarrow DEK + CO_2 + H_2O \tag{4}$$
$$H_2C=CH_2 + H_2O \longrightarrow EtOH \tag{5}$$

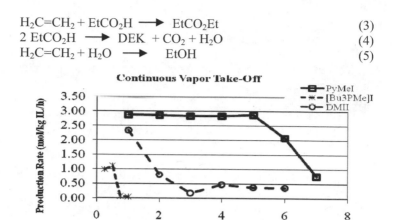

Figure 37.1. Continuous Vapor Stripped Process for the Rh Catalyzed Carbonylation of MeOH in the absence of MeI.

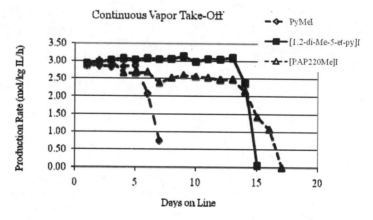

Figure 37.2. Continuous Vapor Stripped Process for the Rh Catalyzed Carbonylation of MeOH in the absence of MeI using Higher Boiling Pyridine Derivatives.

Unfortunately, when the carbonylation of ethylene with a rhodium-ethyl iodide catalyst was operated in ionic liquid media generated the product mixture now contained a significant amount of $EtCO_2Et$ (15-35%). (See Table 37.1.) Unless this selectivity issue was resolved, the carbonylation of ethylene in ionic liquids would have been untenable.

Closer examination of the mechanism for the Rh catalyzed carbonylation of ethylene provides a rationale for the poor selectivity. The mechanism for the carbonylation of ethylene (Scheme 37.1) is well known (6) and proceeds via two simultaneously operating mechanisms which generate a common $EtRh(CO)_2I_2$ intermediate which rapidly reacts with iodide (Eqn. 10) to generate $EtRh(CO)_2I_3^-$. The first, and predominant, mechanism is a hydride mechanism (Eqns. 6-8 below) in which the proton required for the formation of $HRh(CO)_2I_2$ and initiation of the

Table 37.1. Carbonylation of Ethylene in Ionic Liquids in the Presence of EtI.[a]

Ionic Liquid	Temperature (°C)	Total Propionyl Generation Rate[a] $(mol/kg_{ionic\ liquid}/h)$	% Ester (ester/total propionate)
DMII	190	2.13^b	*31%*
DEII	175	2.60	*36%*
DEII	190	2.35	*19%*
MTOPI	175	3.16	*17%*

[a] Conditions: $RhCl_3 \cdot H_2O$: 3.0 mmol; H_2O: 2.0 mol; gas feed: 50% CO, 45% ethylene, 5% H_2 (purge rate 1 mol/h); 190°C; 450 psi (3.10 Mpa), 5 hrs. Rates obtained by liquid sampling. [b] Rate data for DMII were estimated from a batch reactor.

hydride mechanism is provided by the reversible elimination of HI from ethyl iodide. The second mechanism proceeds via a nucleophilic (S_N2) displacement of iodide from ethyl iodide by $Rh(CO)_2I_2^-$ (Eqn. 7). The hydride mechanism, which normally predominates, is inhibited by iodide salts since the iodide competes with ethylene for vacant coordination site via the reversible formation of $HRh(CO)_2I_3^-$. Therefore, in the presence of iodide containing ionic liquids, the predominant hydride mechanism is suppressed by the high iodide levels.

However, if we consider the alternative nucleophilic displacement, it is known that nucleophilic processes are accelerated by ionic liquids, but more pertinent is the fact that the S_N2 displacement of iodide from alkyl iodide (MeI) by $Rh(CO)_2I_2^-$ is slightly accelerated by ionic liquids (7). Unfortunately, ionic liquids would also be expected to accelerate the nucleophilic displacement of iodide from ethyl iodide by propionic acid to form ethyl propionate (Reaction 8). In fact, as an S_N2 Type II displacement (the interaction of two neutral species), the ester formation from propionic acid and ethyl iodide would be expected to be significantly increased compared to the reaction of $Rh(CO)_2I_2^-$ with EtI. Therefore, by operating in iodide containing ionic liquids, we had set up a situation in which we suppressed the normally predominant hydride mechanism, slightly accelerated the alternative nucleophilic mechanism, but dramatically increased the ethyl propionate by-product forming pathway.

$$EtI \rightleftarrows H_2C=CH_2 + HI \qquad (6)$$

Hydride mechanism:
$$Rh(CO)_2I_2^- + H^+ \longrightarrow HRh(CO)_2I_2 \qquad (7)$$
$$HRh(CO)_2I_2 + H_2C=CH_2 \longrightarrow EtRh(CO)_2I_2 \qquad (8)$$

Nucleophilic mechanism:
$$Rh(CO)_2I_2^- + EtI \longrightarrow EtRh(CO)_2I_2 + I^- \qquad (9)$$

Steps common to both mechanisms:
$$EtRh(CO)_2I_2 + I^- \longrightarrow EtRh(CO)_2I_3^- \qquad (10)$$
$$EtRh(CO)_2I_3^- \longrightarrow EtC(=O)Rh(CO)I_3^- \qquad (11)$$
$$EtC(=O)Rh(CO)I_3^- + CO \longrightarrow EtC(=O)I + Rh(CO)_2I_2^- \qquad (12)$$
$$EtC(=O)I + H_2O \longrightarrow EtCO_2H + HI \qquad (13)$$

Scheme 37.1. Mechanism for the carbonylation of ethylene with EtI or HI.

It appeared that, we needed to limit or omit the ethyl iodide if we were going to operate the ethylene carbonylation in ionic liquids. Unfortunately, the previous literature indicated that EtI or HI (which are interconvertible) represented a critical catalyst component. Therefore, it was surprising when we found that, in iodide based ionic liquids, the Rh catalyzed carbonylation of ethylene to propionic acid was still operable at acceptable rates in the absence of ethyl iodide, as shown in Table 37.2. Further, we not only achieved acceptable rates when omitting the ethyl iodide, we also achieved the desired reduction in the levels of ethyl propionate. More importantly, when the reaction products were analyzed, there was no detectable ethyl iodide formed in situ. However, we should note that we now observed traces of ethanol which were normally undetectable in the earlier EtI containing experiments.

Table 37.2. Carbonylation of Ethylene in the Absence of Ethyl Iodide.[a]

Ionic Liquid	Time (h)	$EtCO_2H$ Produced (mol)	$EtCO_2Et$ Produced (mol)	% $EtCO_2Et$ (of total Propionyl)	EtOH Produced (mol)
DEII	6	0.994	0.039	3.7%	0.014
DMII	8	1.153	0.018	1.5%	n.d.
[Etpy]I[b]	5	0.555	0.020	3.5%	0.002
MTOPI	5	0.253	n.d.	0%	n.d.

[a] Condensed from ref. 4. Conditions: $RhCl_3 \cdot H_2O$: 1.5 mmol; H_2O: 2.0 mol; gas feed: 50% CO, 45% ethylene, 5% H_2 (purge rate 1 mol/h); 190°C; 450 psi (3.10 Mpa), 5 hrs. [b] Operated at 750 psi (5.17 Mpa).

While the mechanism in the absence of EtI or HI is still a matter of conjecture, it is unlikely that a hydride mechanism was operable since, whereas we could possibly envision an imidazolium salt donating a hydrogen via carbene formation, there is no corresponding viable source of hydride when using pyridinium and phosphonium salts which are also effective solvents for the process. Therefore, by process of elimination, it was more likely that the process was operating via a nucleophilic process.

This likelihood that a nucleophilic process was more likely was reinforced by comparing initial rate data for dimethyl imidazolium iodide (DMII) and diethyl imidazolium iodide (DEII) based processes. Before discussing the rate data further, we should point out that, unlike the EtI containing processes, the non-EtI containing processes slowed as we consumed water (i.e., with increased conversion) so batch data is difficult to use in determining rates. Removing liquid samples every 30 minutes allowed us to deduce the initial rate. When measured using the liquid sampling procedure, the initial carbonylation rate using DMII as solvent was 1.86 mol propionyl/kg of DMII/h vs 0.42 mol propionyl/kg of DEII/h when using DEII as solvent. This is consistent with the nucleophilic displacement of one of the alkyl groups on the imidazolium salt providing the initiating event in catalysis since we would expect the displacement of a methyl group via an S_N2 process to be faster than the displacement of an ethyl group. Had the hydride mechanism been operable, it is unlikely that there would be a significant difference between the two imidazolium salts. Further strengthening the case for the nucleophilic process was the observation (by NMR) that the methyl groups on DMII were being replaced by ethyl groups.

Methanol Carbonylation without MeI. While resolving the selectivity issue in ethylene carbonylation was exciting, the observations indicating that the reaction was likely proceeding via a nucleophilic reaction between Rh and the ionic liquid and did not require EtI provided an even more exciting opportunity. If a nucleophilic mechanism is operative, it is likely that we could extend the technology to the much more commercially important carbonylation of methanol.

Traditional rhodium catalyzed methanol carbonylation to acetic acid (Eqn. 14) requires 15-20 wt.% MeI as co-catalyst (MeOH:MeI normally 10:1) and proceeds via an S_N2 reaction between MeI and $Rh(CO)_2I_2^-$ (8). Consistent with the S_N2 behavior, the reaction is 1st order in both MeI and Rh. (See Scheme 37.2 for the mechanism.)

Overall Reaction:
$$MeOH + CO \rightarrow AcOH \tag{14}$$
Mechanism for the Carbonylation of Methanol:

$$Rh(CO)_2I_2^- + MeI \rightarrow MeRh(CO)_2I_2 \quad \text{(slow, rate determining)} \tag{15}$$
$$MeRh(CO)_2I_2 + I^- \rightarrow MeRh(CO)_2I_3^- \tag{16}$$
$$MeRh(CO)_2I_3^- \rightarrow AcRh(CO)I_3^- \tag{17}$$
$$AcRh(CO)I_3^- + CO \rightarrow AcI + Rh(CO)_2I_2^- \tag{18}$$
$$AcI + H_2O \rightarrow AcOH + HI \tag{19}$$
$$HI + MeOH \rightarrow MeI + H_2O \tag{20}$$

Side reactions (DME and MeOAc are in equilibrium and ultimately consumed.)

$$MeOH + AcOH \rightleftarrows MeOAc + H_2O \tag{21}$$
$$2\ MeOH \rightleftarrows DME + H_2O \tag{22}$$

Scheme 37.2. Mechanism for the Rh-MeI Catalyzed Carbonylation of Methanol.

Unfortunately, MeI is toxic, corrosive, leads to product contamination, and is expensive to purchase and handle. While a substitute for MeI has been sought since the discovery of the Rh catalyzed carbonylation nearly 40 years ago, to date, no adequate replacement has been identified due to the integral role it played in the catalysis.

Based on our success with the carbonylation of ethylene and the evidence supporting the nucleophilic process, there was reason to believe we might also be able to conduct Rh catalyzed carbonylation of methanol in the absence of MeI. As displayed in Table 37.3, the Rh catalyzed carbonylation of methanol was achievable in the absence of MeI with a variety of ionic liquids. More importantly, at most, only traces of MeI could be detected in the medium. The lack of any MeI was significant since we were concerned that we could generate significant levels of MeI *in situ* via the nucleophilic displacement of the methyl group by iodide (Eqn. 23.)

$$MeOAc + I^- \rightarrow MeI + AcO^- \tag{23}$$

Table 37.3. Rh Catalyzed Carbonylation of MeOH in the Absence of MeI.[a]

Ionic Liquid	Gas Feed	MeOH Conv. (%)	Net Acetyl Formed in 5 h (mol)	% MeI in Product (mmol)
DMII*	100% CO	97%	1.82	0.05% (0.83 mmol)
DMII	95:5 CO:H$_2$	96%	1.44	n.d.
[pyMe]I	95:5 CO:H$_2$	100%	2.00	n.d.
[pyMe]I (run 1)	100% CO	98%	1.82	n.d.
[pyMe]I (run 2)	100% CO	100%	2.00	0.15% (2.8 mmol)
[Etpy]I	100% CO	92%	1.64	n.d.
BMII	95:5 CO:H$_2$	100%	1.94	n.d.
[Bu3NMe]I	100% CO	98%	1.38	n.d.
MTOPI	100%	66%	0.61	0.19% (4 mmol)

[a] Condensed from ref. 5. Conditions: 1.5 mmol RhCl$_3$·H$_2$O: MeOH: 2.0 mol; gas feed: 50% CO, 45% ethylene, 5% H$_2$ (purge rate 1 mol/h); 190°C; 750 psi (51.7 Mpa), 5 hrs.

Given the absence (or near absence) of MeI in these reactions, we sought to demonstrate the process on a continuous process. We purposely chose to employ the same vapor take-off (vapor stripped) reactor we used in our earlier studies of the Rh/MeI co-catalyzed carbonylation of methanol (3) since the vapor stripping procedure used in the process would force *any* MeI formed in situ overhead where we could detect it in the effluent. This would represent the most rigorous test of the new non-MeI process we could contrive.

Figure 37.1 compares results for a continuous vapor stripped (vapor take-off) process using ionic liquids selected from three different classes of salts: phosphonium ([Bu$_3$PMe]I), imidazolium (DMII), and pyridinium ([MePy]I). With [Bu$_3$PMe]I, the reaction was both slow and became inactive in less than a day and DMII deactivated in less than 3 days. The fastest, and longest lasting, reaction utilized N-methyl pyridinium iodide ([MePy]I), although the reaction rapidly slowed after about 5 days. The processes generated a mixture of acetic acid and methyl acetate as expected (MeOAc is generated by in situ esterification.) and analysis of the effluent indicated that we were generating only a very low level of MeI. (There was < 0.3 wt.% MeI in the effluent despite the use of a vapor stripped reactor which intentionally biased the process toward producing MeI.)

Upon opening the autoclave the cause of the deactivation was clear. The volume in the reactor had decreased until it now occupied too little volume to reach the stirrer in the reactor. After evaporating a portion of the effluent and examining the residue by NMR, we found that the residue contained a ca. 3:2 mixture of MePy$^+$:pyH$^+$ indicating that we had lost a portion of ionic liquid to dealkylation. No further effort to analyze this effluent was attempted and at this stage it was unclear whether the volume loss was to aspiration or distillation.

To suppress the loss of ionic liquid we examined two higher boiling methylpyridinium iodide salts, 1,2-dimethyl-5-ethyl pyridinium iodide and N-methylated PAP-220 iodide. (The parent pyridines 2-methyl-5-ethyl pyridine and PAP-220 are available from Aldrich Chemical. The PAP-220 is a mixture of polyalkylated pyridines (PAP) isomers boiling above 220°C. The presence of a substituent at the 2 position may also inhibit dealkylation of the pyridinium salt.) In Figure 37.2, it is clear that this strategy was successful at extending the lifetime of reaction, allowing us to triple the time before the ionic liquid was depleted. (Again, opening the autoclave revealed that we had lost sufficient volume to fall below the level of the stirrer, so the reaction ceased.)

As in the earlier examinations, the amount of methyl iodide detected in the purged product was still averaged ca. 0.3 wt.%. However, unlike the earlier run with [MePy]I, we took a close look at the effluent from the operation with 1,2-dimethyl-5-ethyl-pyridinium iodide ([DMEpy]I). All the product was distilled overhead leaving a residue that upon examination by NMR contained a ca. 3:1 acetate:[DMEpy]$^+$ ratio. Closer examination by NMR revealed that only about 3% of [DMEpy]I in the overhead distillate had been dealkylated to 2-methyl-5-ethyl pyridium hydroiodide.

More interesting was the elemental analysis of the residue. Whereas a 2:1 AcOH:[DMEpy]I should have contained 33% iodine, the elemental analysis indicated the residue contained only 0.7% iodine. This clearly indicated that we no longer had an iodide salt, but more likely had an acetate salt, most likely a 2:1 mixture of AcOH: :[DMEpy][OAc]. (The formation of a 2:1 salt would be typical of our experience with ionic liquids. In practice they normally tenaciously retain ca. 2 mol AcOH/mol of ionic liquid, a phenomena we noted in our earlier reports. (3) Closer comparison of the salt obtained and low levels of MeI detected in the effluent indicated that the amount of [DMEpy][OAc] generated closely matched the total MeI (ca. 90-95% yield of MeI based on [DMEpy][OAc].) Further, the elemental analysis was unable to detect any Rh in the effluent, so we could conclude that there was no aspiration occurring. This clearly indicated that our ionic liquid loss was due to metathesis of the ionic liquid from the iodide to the acetate salt, likely due to reaction (23) which likely sublimed overhead. In principle, the miniscule amount of MeI and ionic liquid could be returned to the reactor to maintain the process.

This successfully demonstrated the Rh catalyzed carbonylation of methanol in the absence of MeI. Even under the most forcing of conditions experienced in the vapor take-off reactor, the effluent contained, at most, ca. 0.3 wt% MeI which represents a >50X reduction from current commercial practice.

Conclusions

The rhodium catalyzed carbonylation of ethylene and methanol can be conducted in the absence of added alkyl halide if the reactions are conducted in iodide based ionic liquids or molten salts. In the case of ethylene carbonylation, the imidazolium iodides appeared to perform best and operating in the absence of ethyl iodide gave improved selectivities relative to processes using ethyl iodide and ionic liquids. In the case of methanol carbonylation, the optimal performance was obtained using pyridinium iodide based salts. The methanol carbonylation can be conducted on a continuous basis in a vapor take-off reactor.

This process provides an option to avoid the expense and handling of toxic alkyl halides in the course of conducting Rh based carbonylations to carboxylic acids. The mechanism is still not completely clear and future work will be dedicated to clarifying the key chemical pathways that permitted us to omit the alkyl halides which were previously regarded as indispensable to the reaction.

Acknowledgements

We thank Eastman Chemical Company for permission to publish this work, Theda Grimm for conducting GC analyses, and Chris Meade for experimental assistance.

References

1. For reviews see: M. J. Howard, M. D. Jones, M. S. Roberts and S. A. Taylor, *Catalysis Today*, **18**, 325 (1993); J. R. Zoeller, *Acetic Acid and its Derivatives*, Marcel-Dekker, NY, NY 1993, p. 35; A. Haynes, *Top. Organomet. Chem,* **18**, 179 (2006).
2. W. Bertloff, *Ullmann's Encyclopedia of Industrial Chemistry,* 6th edition, Vol. 6, Wiley-VCH Verlag GmbH & Co., KGaA, Weinheim, Germany, p. 473 (2003).
3. G.C. Tustin, R. M. Moncier and J. R. Zoeller, *ACS Symp. Ser.*, **975**, 128 (2007).
4. J. R. Zoeller and M. K. Moore, *US Pat. Pub.,* US 2007/0299280 A1 (2007).
5. J. R. Zoeller, M. K. Moore and A. J. Vetter, *US Pat. Pub.,* US 2007/0293695 A1 (2007).
6. D. Forster, A. Hershman, and D. E. Morris, Donald E., *Catalysis Reviews - Science and Engineering,* **23(1)**, 89 (1981).
7. S.J. Pool and K.H. Shaughnessy. "Effects of ionic liquids on oxidative addition to square planar iridium and rhodium complexes." Abstracts of Papers, 231st ACS National Meeting, Atlanta, GA, USA, March 26-30, 2006 (2006).
8. A. Haynes, B. E. Mann, D. Gulliver, G. E. Morris, and P. M. Maitlis, *J. Amer. Chem. Soc.* , **113**, 8567 (1991); A. Haynes, B. E. Mann, G. E. Morris, and P. M. Maitlis, *J. Amer. Chem. Soc.*, **115**, 4093 (1993); T. Griffin, D. B. Cook, A. Haynes, J. M. Pearson, D. Monti, and G. E. Morris, *J. Amer. Chem. Soc.* , **118**, 3029 (1996).

38. Preparation and Characterization of Supported Amine Catalysts

Brittni A. Scruggs[1], Suzanna L. Kilgore[1], Sarah L. Hruby[2], Brent H. Shanks[2] and Bert D. Chandler[1]

[1]Department of Chemistry, Trinity University, San Antonio, TX 78212
[2]Department of Chemical Engineering, Iowa State University, Ames, IA 50011

*bert.chandler@trinity.edu*T

Abstract

Anchored amine materials can be prepared through a number of synthetic methodologies. Because of the potential importance of these materials to organic synthesis, a ninhydrin assay was developed as a rapid laboratory determination of available surface amines. The assay agreed well with expected values for aminopropyltriethoxysilane grafted onto commercial silica. The assay also distinguished between reactive amines and protonated or poisoned surface amines on co-condensed SBA-15 materials.

Introduction

Supported amine materials are being investigated as catalysts for a number of important organic reactions including Michael additions,[1,2] Knoevenagel condensation,[3] aldol condensations,[4,5] and cyano-O-ethoxycarbonylation.[4] For many reactions, controlling the spacing between surface amine groups, or between amine groups and complementary functional groups, is important. A variety of synthetic methodologies are now emerging to exact this control; in many cases, substantial reactivity enhancements are observed when amine spacing can be carefully controlled.[5-9]

We have been developing methods to prepare and characterize supported amine catalysts using readily available commercial supports. One potential means of depositing amines on oxide surfaces is shown in Scheme 38.1, in which the micelle's role is to space the amines on the surface. Current work is directed towards characterizing these samples, particularly applying fluorescence resonance energy transfer (FRET) techniques.

Determining the number of reactive surface amines is important for FRET measurements and for evaluating catalytic activities of supported amine catalysts. Elemental analyses give total nitrogen content, but an alternative that would only measure available amine groups would be more relevant for characterization catalytic materials. Traditional aqueous[10,11] and non-aqueous[6] acid-base titrations are complicated by the buildup of surface charges, which shift pKa values and require extremely long (24+ hours) equilibration times.[10,11] For organic reactions that deal with potentially large substrates or reaction products (e.g., from coupling reactions), it is also important to distinguish amines that can readily react with larger substrates from those that may be occluded in micropores or are otherwise inaccessible. As a first step in evaluating new supported base materials, we set out to develop a simple, fast assay for surface amines that could be used to quickly quantify the number of accessible amines in a typical research laboratory.

Experimental Section

Materials. 1,1,3,3-tetramethyldisilazane (Gelest),

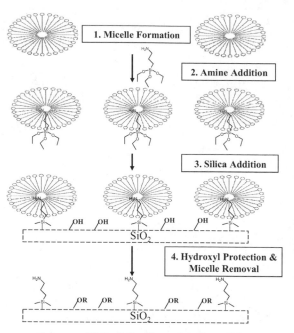

Scheme 38.1. Micelle amine deposition

3-indolepropionic acid (Aldrich Chemical Co.), aminopropyltriethoxysilane (Gelest), cetyltrimethylammonium bromide (Aldrich Chemical Co.), and ninhydrin (Aldrich Chemical Co.) were all used without further purification. Davicat SI-1403 silica powder (245 m^2/g) was supplied by Grace-Davison. Water was purified to a resistivity of 17-18 MΩ-cm with a Barnstead Nanopure system. TRIS buffer (5 mM) was prepared using Tris-HCl and nanopure water, adjusting the buffer to pH 8.2 using dilute sodium hydroxide solution. Phosphate buffer (100 mM) was prepared using NaH$_2$PO$_4$ and Na$_2$HPO$_4\cdot$7H$_2$O, adjusting the pH to 6.5. Solution UV-visible absorption spectra were collected using a Jasco V-530 spectrophotometer.

Preparation of Grafted Amine-Functionalized Silica. Silica was pressed, crushed, and sieved to 40-60 mesh particles and calcined at 550°C overnight. To anhydrous toluene (40 mL), silica (1.0 g) was added under N$_2$ and stirred for an hour. Aminopropyltriethoxysilane (APS; 1.0 g, 5.58 mmol) was syringed into mixture and stirred for 24 hours at room temperature under nitrogen. The functionalized silica was then filtered, washed with toluene three times, and dried under vacuum at 50°C.

Silica Alkylation. To tetramethyldisilizane (TMDS, 1.0 g) in anhydrous toluene (30 mL), amine-functionalized silica was added and stirred overnight at room temperature under N$_2$. Then it was then filtered, washed with toluene three times, and dried under vacuum at 50°C. Ethanol (190 mL) and nanopure water (10 mL) were mixed, and the pH was adjusted to 4.7 with acetic acid. Isobutyltrimethoxysilane (4.0 mL) was added to the solution, stirred for five minutes, and the modified silica was added. After 10 minutes, the solution was decanted, the solid was washed with ethanol, cured at 110°C for ten minutes, and dried under vacuum.

APS-SBA-15 synthesis. SBA-15 materials were prepared as described in the literature.[12,13] The structure-directing agent, Pluronic P123 (BASF Co.), was dissolved in 125 ml deionized water and 25 ml hydrochloric acid (12.1 N) with stirring. Tetraethyl orthosilicate (TEOS) was added as the silica precursor (98%, Acros Organics) at 40°C. 3-aminopropyltriethoxysilane (APS) (99%, Aldrich) was added after a TEOS prehydrolysis period of one hour. The resulting mixture (1 TEOS: 0.1 APS/IPTES: 7.76 HCl: 171 H_2O molar ratio) was stirred at 40°C for 20 h and aged at 90°C for 24 h before being filtered. The surfactant template was removed by refluxing in ethanol with 10% hydrochloric acid for 24 h. The catalyst was then filtered and washed with ethanol. Excess protons from the acidic synthesis conditions were removed with 5 ml tetramethylammonium hydroxide (TMAH) solution (25 wt. % in methanol, Acros Organics) in 45 ml methanol with stirring for 30 min. The solid was filtered, washed 3x with methanol, and dried under vacuum.

Ninhydrin Assays. Ninhydrin tests were performed using a modified procedure of Taylor et al.[14] APS Silica (10-75 mg) of various loadings (0.857, 0.571, and 0.343 mmol NH_2/g Silica) was added to phosphate buffer (5 mL, 100 mM, pH 6.5), and 1 mL of a 5% w/v solution of ninhydrin in ethanol was added to the slurry. After stirring for an hour in a boiling water bath, the mixture was allowed to cool slowly to room temperature. The silica was then filtered and washed three times with 70°C distilled water. The filtrate was collected, added to a volumetric flask, diluted to 100 mL, and the absorbance of this solution at 565 nm was measured using a UV-visible spectrophotometer. The reference solution was prepared as above with unmodified amine-free silica. Calibration standards were prepared with aliquots of a 1 mg/mL solution of APS in ethanol.

Results and Discussion

Ninhydrin Assays of Grafted Silicas. The reaction between ninhydrin and a primary amine, based on the work by Taylor and Howard, is shown in Scheme 38.2 (assay details can be found in the experimental section). For a surface titration, this reaction is advantageous because it cleaves the C-N bond of primary amines, resulting in a soluble highly colored analyte. The production of an extended aromatic product, along with three equivalents of water provides the strong driving force necessary to cleave the C-N bond. Calibration curves prepared with APS were linear and could be prepared daily for direct comparison with anchored amine materials. Figure 38.1 shows the results of ninhydrin assays of grafted, alkylated silica samples with various amine loadings. The first sample was repeated on different days (4 times each day) with freshly prepared solutions to show day-to-day reproducibility.

Ninhydrin   APS Silica   Colored product

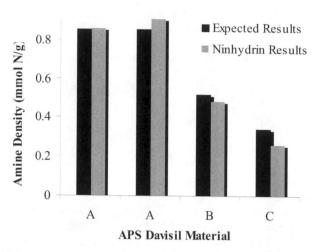

Figure 38.1. Ninhydrin tests on grafted APS silica materials. The two samples with 0.857 mmol/g loadings were tested using different solutions on different days to test reproducibility from day-to-day.

Table 38.1. Ninhydrin assays for surface amines using grafted silicas. [a]Determined from grafting synthesis and elemental analysis. [b]Typical standard deviations were 0.08 mmol N/g.

Grafted Davisil	Expected Loading[a] (mmol N/g)	Experimental Results[b] (mmol N/g)	Error (mmol N/g)	Percent Error
Davisil A	0.86	0.86	0	0.4%
Davisil A	0.86	0.90	0.04	5%
Davisil B	0.52	0.48	0.04	7%
Davisil C	0.34	0.26	0.08	23%

Ninhydrin assay data for the grafted silicas is compiled in Table 38.1. The typical errors in the ninhydrin assays were within one standard deviation of the actual value, and the % errors were generally less than 10%, indicating that the assay gave reasonable results. The lowest amine loading had slightly larger errors, which may be due to the lower total nitrogen content, or possibly to partial protonation (*vida infra*).

Ninhydrin Assays of Amine Functionalized SBA-15 Materials. The ninhydrin assay can also be used to enhance traditional elemental analysis, as it can distinguish between total N content and available/free amines. The latter are, of course, of primary concern for catalysis, and the assay offers an opportunity to distinguish between amines that readily react with organic molecules and those that may be protonated, occluded in the material, trapped in micropores, or poisoned by remnants of synthetic templates. SBA-15 materials, which are prepared using an oxide polymer template, make for a good test of the ninhydrin assay because they require polymer removal. For APS-SBA-15 materials, where the amine is incorporated during the SBA-15 synthesis, the polymer cannot be calcined without destroying the amine functionality. Less forcing conditions are therefore necessary, and it is important to confirm the availability of the remaining surface amines.

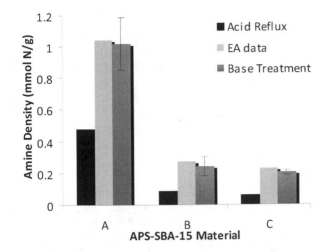

Figure 38.2. Ninhydrin tests on APS-SBA-15 performed after treatment in refluxing ethanol and after an additional treatment with $N(CH_3)_4OH$ in refluxing methanol. Ninhydrin tests are compared to the expected results determined from elemental analysis.

Table 38.2. Ninhydrin assays for surface amines on APS-SBA-15 materials. [a]all materials were alkylated after polymer removal; [b]determined from elemental analysis; [c]refluxed in 10% HCl in ethanol for 24 hours; [d]stirred with 25 wt% $N(CH_3)_4OH$ in methanol for 30 minutes.

SBA Material[a]	N Loading[b] (mmol N/g)	Acid Reflux[c] (mmol N/g)	Base Treatment[d] (mmol N/g)
APS-SBA-15 A	1.0	0.48	$1.0 \pm .17$
APS-SBA-15 B	0.27	0.09	$0.24 \pm .06$
APS-SBA-15 C	0.23	0.06	$0.21 \pm .02$

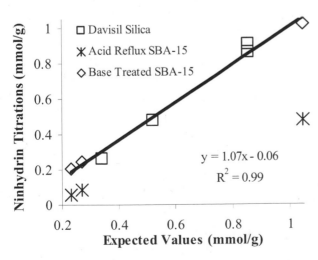

Figure 38.3. Parity plot for ninhydrin titrations showing measured amine densities plotted against the expected values.

The ninhydrin assay clearly shows that only a fraction of the total amines react with ninhydrin, suggesting that most of the amines are either protonated (and therefore unreactive) or poisoned by polymer remnants. Treating the materials with $N(CH_3)_4OH$ removes any excess protons and may help to dislodge the remaining polymer from the solid. After this treatment, the SBA-15 materials behave similarly to the amorphous silicas, as shown in Figure 38.3. Additionally, this parity plot shows the consistency of the titration over these materials.

Conclusions

A relatively simple, fast assay for accessible surface amines which can be readily applied without scrupulously excluding water (e.g., with glove box or schlenk techniques) was developed using ninhydrin to oxidize an anchored primary amine and cleave the C-N bond. The ninhydrin assay was generally in good agreement with known values for amines grafted onto commercial silicas and mesoporous SBA-15 materials that had amines incorporated into the mesopores during synthesis. Ninhydrin assays of the SBA-15 materials also distinguished between functional and unreactive amines, and highlighted the importance of a base treatment after polymer removal if active amines are desired. This assay will aid in future characterization schemes and provides a rapid means of evaluating catalytic activity on a per amine basis.

Acknowledgements

The authors gratefully acknowledge the National Science Foundation (CTS-0455965) for financial support of this work. BAS also thanks Trinity University and the Mach Research Scholarship Program for a Mach Research Scholarship.

References

1. Fuerte, A.; Corma, A.; Sanchez, F. *Catal. Today.* **107-108**, 404, (2005).
2. Gruttadauria, M.; Riela, S.; Lo Meo, P.; D'Anna, F.; Noto, R. *Tet. Lett.* **45**, 6113, (2004).
3. Bass, J. D.; Anderson, S. L.; Katz, A. *Angew. Chem. Intl. Ed.* **42**, 5219, (2003).
4. Motokura, K.; Tada, M.; Iwasawa, Y. *J. Am. Chem. Soc.* **129**, 9540 (2007).
5. Zeidan, R. K.; Hwang, S.-J.; Davis, M. E. *Angew. Chem. Intl. Ed.* **45**, 6332, (2006).
6. Hicks, J. C.; Jones, C. W. *Langmuir* **22**, 2676, (2006).
7. Bass, J. D.; Solovyov, A.; Pascall, A. J.; Katz, A. Smart, J. L.; McCammon, J. A. *J. Am. Chem. Soc.* **128**, 3737 (2006).
8. McKittrick, M. W.; McClendon, S. D.; Jones, C. W. Chemical Industries, **104**, (*Catal. Org. React.*), 267 (2005).
9. McKittrick, M. W.; Jones, C. W. *Chem. Mater.* **17**, 4758 (2005).
10. Walcarius, A.; Delacote, C. *Chem. Mater.* **15**, 4181 (2003).
11. Smart, J. L.; McCammon, J. A. *J. Am. Chem. Soc.* **118**, 2283 (1996).
12. Wang, X.; Lin, K. S. K.; Chan, J. C. C.; Cheng, S. *J. Phys. Chem. B* **109**, 1763 (2005).
13. Chong, A. S. M.; Zhao, X. S. *J. Phys. Chem. B* **107**, 12650 (2003).
14. Taylor, I.; Howard, A. G. *Anal. Chimica Acta* **271**, 77 (1993).

39.   One-Pot Gas Phase Synthesis of 1,2-Methylenedioxybenzene

Dario F. Impalà[1], Oreste Piccolo[2] and Angelo Vaccari[1]

[1] *Dip. di Chimica Industriale e dei Materiali, ALMA MATER STUDIORUM –
Università di Bologna, Viale Risorgimento 4, 40136 BOLOGNA BO, Italy*
[2] *Studio di consulenza scientifica (www.scsop.it)
Via Bornò 5, 23896 SIRTORI LC, Italy*

angelo.vaccari@unibo.it

Abstract

A careful investigation of the one-pot gas phase synthesis of 1,2-methylenedioxybenzene (MDB), without using halogenated reagents, was carried out by reacting pyrocatechol (PYC) with formaldehyde acetals on Ti-silicalite (TS-1), TiO_2/SiO_2 or ZrO_2/SiO_2. The catalysts were previously characterized by using different techniques including BET surface area, XRD, FT-IR and TGA. Promising results were achieved with the ZrO_2/SiO_2 catalyst, although activity and selectivity proved lower than those obtained with TS-1. The latter showed results significantly better than those of TiO_2/SiO_2 highlighting the role of Ti coordination. Among the formaldehyde acetals used (dimethoxymethane, diethoxymethane or 1,3-dioxolane) diethoxymethane was the most active and selective reagent, giving rise to selectivity values higher than 80%. Lastly, a possible reaction pathway was proposed indicating the role of 2-(ethoxymethoxy)phenol as intermediate and explaining the origin of by-products.

Introduction

1,2-Methylenedioxybenzene (MDB) is a useful small-volume commodity, used in the synthesis of agrochemicals (for example Piperonyl Butoxide, an insecticide synergist), drugs (Tadafil[TM], Anolignan A[TM], etc.) and fragrances (Piperonal, Helional[TM]) [1-9]. Many of these products can also be obtained starting from natural sources, although the latter are becoming increasingly scarce and consequently more and more expensive, causing interest to shift toward alternative synthetic methods. The synthesis of MDB is usually carried out in the liquid phase by a reaction of pyrocatechol (PYC) with methylene dihalide (mainly chloride) under alkaline conditions [10], operating in a batch reactor and using aprotic solvents such as dimethylsulfoxide (DMSO) [11,12], N-methylpyrrolidone (NMP) [13] or dimethylformamide (DMF) [14]. A process using a phase transfer catalyst was also reported [15]. Lastly the KF-catalyzed methylenation of PYC was also studied, although only at the research level [16]. All of these synthetic routes present some significant drawbacks, such as the use of toxic solvents, the production of equimolar amounts of halogenated wastes, or an excess of inorganic salts, which may be key factors as far as environmental restrictions and the high cost of waste water treatment are concerned.

Nowadays, few data have been reported in the literature [17] on the one-pot gas phase synthesis of MDB, using Ti-silicalite (TS-1) as a catalyst and

1,3,5 trioxane as the source of formaldehyde. Low-yield values were obtained by feeding a large molar excess of PYC compared to the formylating agent (3:1), resulting in high recycling costs. The aim of this study was to shed light on the key factors in the gas phase synthesis of MDB from PYC and a few formaldehyde acetals (Fig. 39.1) and to try to identify a possible reaction mechanism in order to develop, in the near future, custom-made catalysts capable of improving yield and selectivity values.

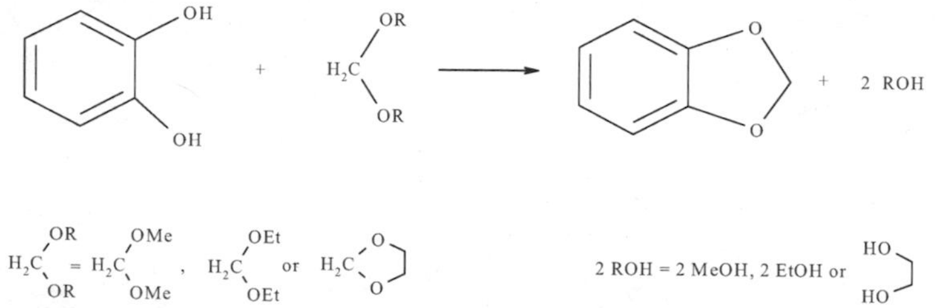

Figure 39.1. Reaction between PYC and formaldehyde acetals to form MDB.

Experimental

PYC, MDB, diethoxymethane (DEM), dimethoxymethane (DMM), 1,3-dioxolane (DOX), 1,3,5-trioxane (TOX), and Zr- and Ti-acetylacetonate (Zr-Ac and Ti-Ac) were purchased from Aldrich Chemicals ($\geq$ 99,0 wt.%) and used without any further purification. The synthesis of 2-(ethoxymethoxy)phenol (2-EMP) was carried out in the liquid phase under reflux at 353 K for 2 h, using 4.0 g of PYC (0.036 mol), 11.0 g of DEM (0.141 mol), and 0.5 g of SiO_2 powder (Si-1803 by Engelhard, surface area = 300 m^2/g, porosity = 0.5368 cm^3/g). The reaction mixture was filtered and the ethanol and residual DEM were removed under reduced pressure (< 100 Pa). The TS-1 (3.45 wt. % of Ti) was prepared according to literature [18-22]; TiO_2/SiO_2 (3.45 wt % of Ti) or ZrO_2/SiO_2 (6.56 wt % of Zr) were prepared by incipient wetness impregnation of the SiO_2 (Si-1803), using solutions of Zr-Ac or Ti-Ac in acetic acid, followed by drying at 373 K for 8 h and calcination at 773 K for 5 h.

X-ray diffraction (XRD) analyses were carried out with a Philips PW 1050/81 diffractometer (40 kV, 25 mA) equipped with a PW 1710 Unit, and Cu-K$_\alpha$ radiation (λ = 0.15118 nm). A 2 theta (2ϑ) range from 10° to 80° was studied at a scanning rate of 0.10°/s. BET surface area values were determined by physical adsorption of N_2 at 77 K using a Micromeritics AUTOCHEM 2910, pre-treating the samples under a vacuum at 373 K for 10 min to eliminate the adsorbed water. The thermogravimetric analyses were carried out in a TGA 2050 by TA Instruments, under an air flow and with a heating rate of 10 K/min. The surface acidity was determined by FT-IR spectroscopy using a Perkin-Elmer 1750 spectrometer. First, the self-supporting wafer of pure catalyst was evacuated at 500 K and 10^{-4} Pa, then adsorption of pyridine was carried out at 303 K, recording the FT-IR spectra after outgassing steps carried out at 323, 373 and 423 K, respectively.

The catalytic tests [temperature = 543-703 K; liquid hourly space velocity (LHSV) = 2,0 h^{-1}; gas hourly space velocity (GHSV) (if not differently reported) = 510 h^{-1}; PYC/formaldehyde acetal = 1:1 mol/mol were carried out in a fixed-bed glass micro-reactor (i.d. 7 mm, length 400 mm), placed in an electronically controlled oven, operating at atmospheric pressure and using 2 cm³ of pellettized catalyst (40-60 mesh). The isothermal axial temperature profile of the catalytic bed during the tests was determined with a 0.5-mm J-type thermocouple, sliding in a glass capillary tube. Before the tests, the catalyst was maintained at the desired temperature under N_2 flow for 1 h. The mixture of PYC and alkylating agent was fed into the top of the reactor by means of a Precidor model 5003 infusion pump. The products were analyzed using a Carlo Erba 4300 gas chromatograph, equipped with FID and a wide-bore OV1 column (length 30 m, i.d. 0.53 mm, film width 0.5 μm) and tentatively identified by GC-MS with a Hewlett-Packard GCD 1800 system equipped with a HP5 column (length 25 m, i.d. 0.25 mm, film width 1.5 μm). The identifications were subsequently confirmed by comparison of the experimental GC-MS patterns with those obtained for pure reference compounds.

Results and discussion

Role of the organic feed XRD powder patterns and FT-IR spectra confirmed that pure MFI-type Ti-silicalite (TS-1) was obtained [18-23], with a surface area of 530 m² g⁻¹. FT-IR spectra of adsorbed pyridine (Fig. 39.2) showed the presence only of weak Brönsted and Lewis sites [24,25], as confirmed by the complete evacuation from the surface at 373 K.

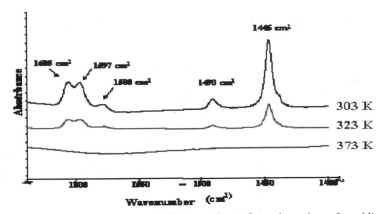

Figure 39.2. FT-IR spectra of TS-1 catalyst after adsorption of pyridine and following evacuation at increasing temperature.

In a preliminary catalytic test, performed at 603 K and feeding an equimolar mixture of PYC/TOX, a yield value was obtained higher than that previously reported [17] (34.5% versus 28.0%, referring to TOX), notwithstanding the lower amount of alkylating agent present, showing the good activity of the TS-1 catalyst prepared. Thus, equimolar mixtures of PYC/alkylating agent were fed in the following tests, to avoid the recycling costs. Table 39.1 shows the results obtained by feeding DMM, including the formation of 2-(methoxymethoxy)phenol (2-MMP) and 3-methylcatechol (3-MC), especially up to 623 K. The selectivity in MDB was relatively low at all the temperatures studied, while the C-balance was usually higher than 90 %.

Table 39.1. Activity as a function of temperature for TS-1 in the synthesis of MDB by reaction of pyrocatechol (PYC) and dimethoxymethane (DMM). (* Yield and selectivity referring to PYC).

Reaction temperature (K)	Convers. of PYC (%)	Yield * (%)			Selectivity in MDB * (%)
		MDB	2-(Methoxy methoxy) phenol	3-Methyl catechol	
573	6.5	3.8	1.0	0.8	57.1
603	17.5	5.6	6.9	3.0	36.5
623	13.3	1.3	8.1	2.9	9.8
643	8.2	1.4	3.0	3.2	18.5

Significantly higher yield and selectivity values were obtained by feeding DEM (Table 39.2), showing an optimum temperature of about 623 K and C-balance values higher than 90%. However, at temperatures higher than 663 K, TS-1 showed significant loss of activity due mainly to the deposition of heavy by-products (tar) as evidenced by the dramatic decrease in the C-balance. Also, the amount of the by-product of 3-MC formed was much lower than that observed with DMM, whereas no appreciable amount of ethylcatechol was detected. Finally, by feeding DOX (i.e., a cyclic acetal of formaldehyde) surprisingly low-yield values were obtained in MDB, regardless of the temperature studied, with significantly higher amounts of by-products formed (Table 39.3).

Table 39.2. Activity as a function of temperature for TS-1 in the synthesis of MDB by reaction of pyrocatechol (PYC) and diethoxymethane (DEM). (* Yield and selectivity referring to PYC).

Reaction temperature (K)	Convers. of PYC (%)	Yield * (%)			Selectivity in MDB * (%)
		MDB	2-(Ethoxy methoxy) phenol	3-Methyl catechol	
543	3.9	0.0	3.5	0.0	0.0
573	8.4	1.1	6.9	0.0	13.1
603	10.3	7.7	2.1	0.0	74.7
623	21.2	18.1	1.2	0.0	85.4
643	25.0	17.5	2.1	0.0	70.0
663	11.4	6.3	3.4	< 0.1	55.3
683	4.1	2.1	0.7	0.2	51.2
703	5.7	1.4	0.9	0.8	24.5

Table 39.3. Activity as a function of temperature for TS-1 in the synthesis of MDB by reaction of pyrocatechol (PYC) and 1,3 dioxolane (DOX) (* Yield and selectivity referring to PYC).

Reaction temperature (K)	Convers. of PYC (%)	Yield * (%)			Selectivity in MDB * (%)
		MDB	Non-identified compound	3-Methyl catechol	
573	8.0	4.1	1.6	1.0	51.3
603	17.1	4.1	6.1	5.1	23.9
623	13.7	3.3	3.0	4.9	24.0
643	16.7	3.0	3.3	7.1	17.9

To shed light on these significantly different reactivities, the thermal stability of each formaldehyde acetal in the absence of PYC was investigated under the same reaction conditions (Fig. 39.3). Worthy of note is the very low thermal stability of the DMM (Fig. 39.3a), which decomposed significantly to formaldehyde and methanol at 523 K in the feed line (0-60 min), while at 603 K an almost quantitative decomposition of DMM occurred in the reactor, at first to CH_3OH and HCHO and afterwards to water and dimethylether. Thus, the high thermal decomposition of DMM suggests a low availability to react with the PYC during the catalytic tests, justifying the low yield values in MDB previously reported.

On the contrary, DEM (Fig. 39.3b) showed a good thermal stability: a partial dehydration of ethanol occurred and about 20% of the starting reagent remained undecomposed. The higher thermal stability of DEM and, consequently, its higher availability during the catalytic tests are in keeping with the higher yield values obtained in MDB (Table 39.2), while the absence between the by-products of those due to the ethylation reaction on the aromatic ring, despite the presence of a significant amount of ethanol in the reactor, is noteworthy.

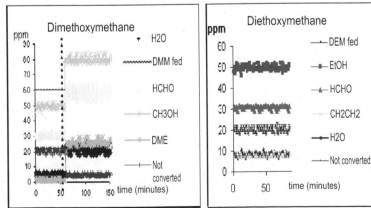

Figure 39.3. Thermal stability at 603 K of (a) dimethoxymethane (DMM) and (b) diethoxymethane (DEM).

Lastly, the very poor catalytic results observed when feeding DOX (Table 39.3) may be attributed to its very high stability in the reaction conditions, in which only about 10% is decomposed to formaldehyde and glycol, the latter being partially dehydrated to acetaldehyde and water. However, in addition to its high stability, it cannot be ruled out that DOX might be not able to interact properly with PYC due to its steric hindrance.

Role of the reaction parameters

In the following tests, devoted to optimize the reaction parameters, an equimolar PYC/DEM mixture was fed, since in previous experiments it showed the best values of activity and selectivity in MDB. The maximum yield and selectivity in MDB was found for a GHSV value of 510 h^{-1} (Table 39.4), while the increase of the GHSV value gave rise to a progressive decrease in the conversion, although the selectivity remained relatively high (always with C-balance values higher than 90%).

By decreasing the GHSV values, the selectivity dramatically decreased due to the presence of the side-reaction of C-alkylation on the aromatic ring, giving rise to relevant amounts of 3-MC and a not-fully-identified methyl-MDB derivative (however, the 3-methyl isomer is the most probable candidate). Lastly, the lowest GHSV value was conducive to the condensation of PYC, with a formation of heavy by-products, a dramatic decrease of C-balance, and resulting catalyst deactivation.

Table 39.4. Activity as a function of GHSV value at 623 K for TS-1 in the synthesis of MDB by reaction of pyrocatechol (PYC) and diethoxymethane (DEM) (* Yield and selectivity referring to PYC).

GHSV (h^{-1})	Convers. of PYC (%)	Yield * (%)				Selectivity in MDB * (%)
		MDB	2-(Ethoxy methoxy) phenol	3-Methyl catechol	Methyl -MDB	
3600	8.0	4.0	3.9	0.0	0.0	50.0
1200	11.0	8.0	2.2	0.0	0.0	72.7
720	19.0	16.6	2.1	0.0	0.0	87.4
510	21.0	18.7	1.2	0.1	0.0	89.0
390	27.0	15.9	1.1	5.9	2.5	58.9
270	26.0	8.0	1.8	6.5	4.9	30.8

Since it is reported in the literature that small amounts of water may improve the catalyst life with time-on-stream, reducing the formation of tar [26], a small percentage of water was added in the feed (Table 39.5), producing a very negative effect on the selectivity in MDB. Furthermore the formation of a new by-product, 2-methoxyphenol (2-MP), was observed. On the contrary, no increase was found in the C-balance values.

Table 39.5. Activity at 623 K for TS-1 in the synthesis of MDB by reaction of pyrocatechol (PYC) and diethoxymethane (DEM) in presence or absence of water (* Yield and selectivity referring to PYC).

Water amount (%)	Convers. of PYC (%)	Yield (%) *				Selectivity in MDB * (%)
		MDB	2-(Ethoxy methoxy) phenol	3-Methyl catechol	2-Methoxy phenol	
0.0	20.9	18.6	1.4	0.2	0.0	89.0
10.0	20.4	14.1	0.4	1.1	4.1	69.8

After 5 h of time-on-stream the TS-1 catalyst showed a significant decrease in activity (Fig. 39.4), mainly related to the formation of heavy by-products by PYC condensation on the surface [27,28]. The TG analysis carried out on the catalyst after the tests, showed the presence of two signals corresponding to a weight loss: the first one, around 523 K, may refer to residual reagents and/or products adsorbed on the catalyst, while the second one at 743 K may be attributed to the combustion of heavy by-products deriving from PYC condensation, which is the main cause of deactivation.

Therefore, we attempted to regenerate the catalyst by calcination at 773 K for 6 h; the calcined catalyst showed a slight decrease in the surface area (450 m^2 g^{-1} versus 530 m^2 g^{-1} of the fresh catalyst) and recovered its activity significantly, although proving less stable with time-on-stream than the fresh one.

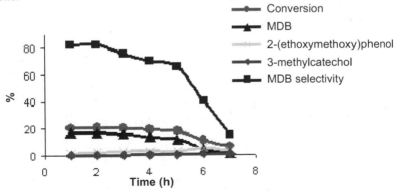

Figure 39.4. Activity with time-on-stream at 623 K for TS-1 catalyst in the synthesis of MDB by reaction of pyrocatechol (PYC) and diethoxymethane (DEM) (* Yield and selectivity referring to PYC).

More information on the nature of active sites was obtained using some model catalysts obtained by incipient wetness impregnation of a commercial silica (Si-1803 with surface area = 300 m^2 g^{-1}). A preliminary test performed using the support (Table 39.6) showed a very low selectivity to MDB, with the preferential formation 2-EMP, indicating that acid sites alone are not able to promote the cyclization of the intermediate.

The TiO_2/SiO_2 catalyst (surface area = 290 m^2g^{-1}), for which XRD patterns indicated the presence of crystalline TiO_2, was conducive mainly to the formation of by-products, such as 3-MC and phenol (deriving from the dehydroxylation of PYC), indicating that octahedrally coordinated Ti^{4+} ions are not useful, giving rise to low selectivity in MDB together with a rapid catalyst deactivation.

In keeping with that already reported for the gas phase synthesis of alkylindoles [29,30], ZrO_2/SiO_2 catalyst (surface area = 283 m^2g^{-1}) was also investigated. The XRD pattern did not show any difference from that of the support, indicating the good dispersion of ZrO_2 on the surface. This catalyst showed an improvement of the catalytic performances in comparison to the support, although they remained lower than those obtained with TS-1 (Table 39.2), suggesting that reaction conditions for Zr-containing catalysts need to be custom-made later on.

Table 39.6. Activity at 623 K of some model catalysts in the synthesis of MDB by reaction of pyrocatechol (PYC) and diethoxymethane (DEM) (* Yield and selectivity referring to PYC).

CAT	Convers. of PYC (%)	Yield (%) *				Selectivity in MDB * (%)
		MDB	2-(Methoxy methoxy) phenol	3-Methyl-catechol	Phenol	
SiO_2	22.7	2.6	13.1	2.1	0.0	11.5
TiO_2/SiO_2	14. 5	2.3	4.9	4.2	1.7	15.8
ZrO_2/SiO_2	19.3	6.1	6.1	2.9	0.0	31.6

Proposed reaction pathways

On the basis of the previous data, possible reaction pathways may be proposed (Fig. 39.6). To confirm this hypothesis, the key intermediate 2-(ethoxymethoxy)phenol (2-EMP) was synthesized and fed, alone or together with water (Table 39.7). As expected, MDB was formed up to 47.0%, together with a smaller amount of 3-MC, confirming that 2-EMP is intermediate for both the above products. Worthy of note is the formation of significant amounts of PYC, showing that the formation of 2-EMP is at equilibrium with starting reagents. Lastly, the presence of water not only increased the amount of PYC, but also aided the formation of 2-MP.

Table 39.7. Activity at 603 K of TS-1 in the synthesis of MDB from 2-(ethoxymethoxy)phenol (2-EMP) (* Yield and selectivity referring to 2-EMP).

Water amount (%)	Convers. of 2-(Ethoxymethoxy) phenol (%)	Yield (%)*				Selectivity in MDB * (%)
		MDB	PYC	3-Methyl catechol	2-Methoxy phenol	
0.0	80.4	47.0	22.0	10.0	0.0	58.5
10.0	87.0	40.1	25.9	8.9	8.0	46.1

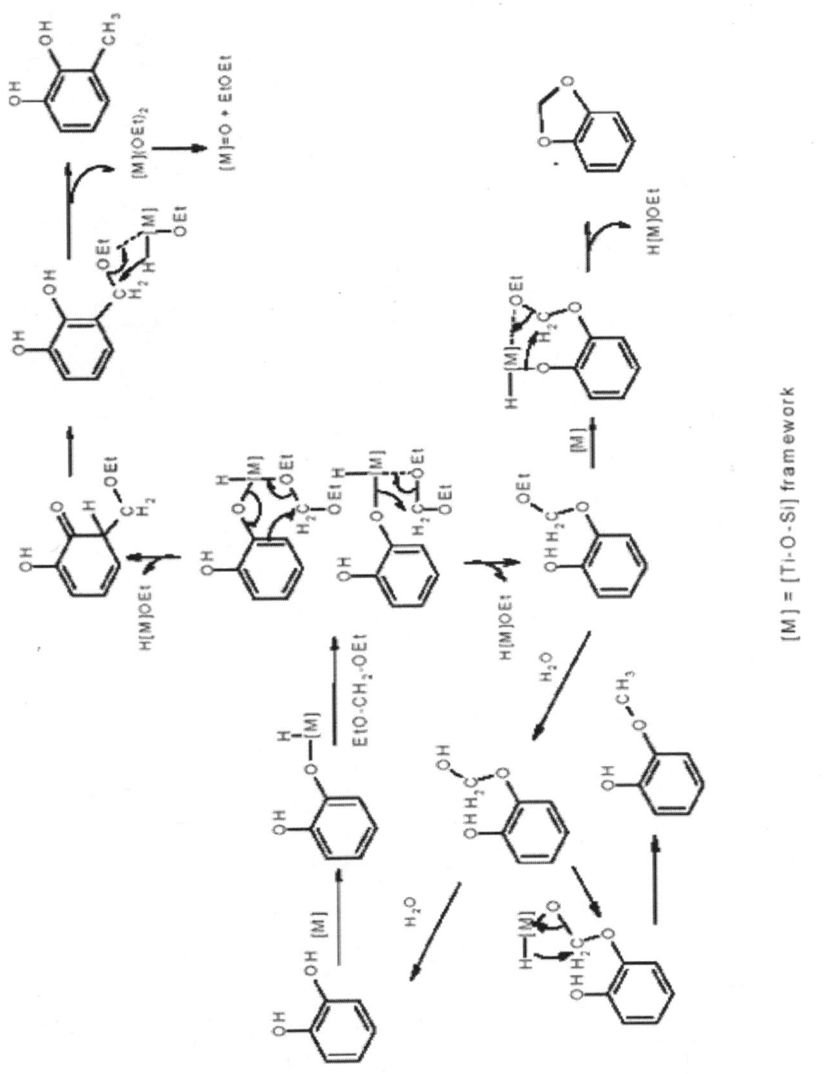

Figure 39.5. Possible reaction pathway in the synthesis of MDB by reaction of pyrocathechol (PYC) and diethoxymethane (DEM).

Conclusions

An environmentally friendly synthesis of 1,2-methylenedioxybenzene (MDB) can be efficiently carried out in the gas phase, by feeding pyrocatechol (PYC) and formaldehyde acetals and using a catalyst containing weak acid sites and redox sites. The Ti-silicalite (TS-1) was identified as the most active and selective catalyst, indicating the role of well-dispersed octahedrally-coordinated Ti^{4+} ions in comparison with some model catalysts.

Catalyst deactivation was mainly due to the deposition of heavy by-products from PYC condensation. However, the catalytic activity may be partially recovered by calcination at 773 K for 4 h. Among the different formaldehyde acetals tested, the best values were obtained by feeding diethoxymethane (in the temperature range 603-643 K and for a GHSV value of 510 h^{-1}), the different reactivity being correlated to the reagent thermal stability. 2-(Ethoxymethoxy)phenol (2-EMP) is the reaction intermediate, as confirmed when feeding pure 2-EMP, with two side reactions forming 3-methylcatechol or 2-methoxyphenol, the latter being aided by the presence of water.

Acknowledgements

The financial support from the Ministero per l'Università e la Ricerca (MUR, Roma) is gratefully acknowledged.

References

1. V. Borzatta, C. Gobbi, E. Capparella and E. Poluzzi, World Patent 042,512 (2005) to Endura.
2. K.G. Fahlbusch, F.J. Hammerschmidt, J. Panten, W. Pickenhagen, Di. Schatkowski, K. Bauer, D. Garbe and H. Surburg, in *"Ullmann's Encyclopedia of Industrial Chemistry"* (B. Elvers, M. Bohmet, S. Hawking, W.E. Russey Ed.s), 6th ed., Vol. 14 , Wiley-VCH, Wenheim (D) , 2003, p. 73.
3. Michael Zviely in *"Kirk-Othmer Encyclopedia of Chemical Technology"* (J. I. Kroschwitz Ed.) 5th ed., Vol. 3, Wiley Interscience, New York (USA), 2007, p. 226.
4. M. Kushiro, T. Masaoka, S. Hageshita, Y. Takahashi, T. Ide and M. Sugano, *J. Nutr. Biochem.* **13**, 289 (2002).
5. http://www.erowid.org
6. G. Salmoria, E. Dall'Oglio and C. Zucco, *Synth. Commun.* **27**, 4335 (1997).
7. P. Vanelle, J. Meuche, J. Maldonado, M. P. Crozet, F. Delmas and P. Timon-David, *Eur. J. Med. Chem.* **35**, 157 (2000).
8. Flavours and Fragrances of Plant Origin, FAO Report M37 (1995).
9. M. Shirai, Y. Yoshida and S. Sadaike, World Patent 054,997 (2004) to Ube Industries.
10. P. Maggioni, U.S. Patent 4,183,861 (1980) to Brichima.
11. J. W. Cornforth, U.S. Patent 3,436,403 (1969) to Shell Oil.
12. J. A. Kirby, U.S. Patent 4,082,774 (1978) to Eli Lilly.
13. P. Panzeri, G. Castelli and V. Messori, European Patent 0,877,023 (1998) to Borregaard Italia.
14. V. Borzatta and D. Brancaleoni, U.S. Patent 6,342,613 (2002) to Endura.

15. B. Jursic, *Tetrahedr.* **44**, 6677 (1988).
16. J.H. Clark, H. L. Holland and J. M. Miller, *Tetrahedr. Letters* **17**, 3361 (1976).
17. J. Roland and J. Desmurs, World Patent 108,385 (2005) to Rhodia Chimie
18. M. Taramasso, G. Perego and B. Notari, U.S. Patent 4,410,501 (1983) to Enichem.
19. U. Bellussi, A. Esposito, M. Clerici and U. Romano, US 4,701,428 (1987) to Enichem Sintesi.
20. X. Wang and X. Guo, *Catal. Today*, **51**, 177 (1999).
21. J. P. Catinat and M. Stebelle, U.S. Patent 6,169,050 (2001) to Solvay.
22. H. Liu, G. Lu, Y. Guo and Y. Guo, *Appl. Catal.* **A293**, 153 (2005).
23. http://www.iza-structure.org.
24. J. Dakta, A. M. Turek, J. M. Jehng and I. E. Wachs, *J. Catal.* **135**, 186 (1992).
25. C. A. Emeis, *J. Catal.* **141**, 347 (1993).
26. E.M. Miller, U.S. Patent 4,001,282 (1977) to General Electric.
27. S. Sanchez-Cortes, O. Francioso, J.V. García-Ramos, C. Ciavatta and C. Gessa, *Colloids and Surfaces A: Physicochemical and Engineering Aspects* **176**, 177 (2001).
28. D. F. McMillen, R. Malhotra, S.Chang and S. E. Nigenda, *Fuel* **83**, 1455 (2004).
29. M. Campanati, F. Donati, A. Vaccari, A. Valentini, O. Piccolo, in *Catalysis of Organic Reaction* (M.E. Ford Ed.), Dekker, New York, 2001, p. 157.
30. M. Campanati, S. Franceschini, O. Piccolo, A. Vaccari, *J. Catal.* **232**, 1 (2005).

40.   The Gas-Phase Ammoxidation of *n*-Hexane to Unsaturated C_6 Dinitriles, Intermediates for Hexamethylenediamine Synthesis

Nicola Ballarini[1], Andrea Battisti[1], Alessandro Castelli[1], Fabrizio Cavani[1], Carlo Lucarelli[1], Philippe Marion[2], Paolo Righi[3] and Cristian Spadoni[1]

[1]*Dipartimento di Chimica Industriale e dei Materiali, Università di Bologna, Viale Risorgimento 4, 40136 Bologna, Italy. INSTM, Research Unit of Bologna: a Partner of NoE Idecat, FP6 of the EU*
[2]*Rhodia Operations, Centre de Recherches et Technologies, 85, Rue des Frères Perret 69190 Saint Fons, France: a Partner of NoE Idecat, FP6 of the EU*
[3]*Dipartimento di Chimica Organica A. Mangini, Università di Bologna, Viale Risorgimento 4, 40136 Bologna, Italy*

fabrizio.cavani@unibo.it

Abstract

This chapter reports about an investigation on the catalytic gas-phase ammoxidation of *n*-hexane aimed at the production of 1,6-C_6 dinitriles, precursors for the synthesis of hexamethylenediamine. Catalysts tested were those also active and selective in the ammoxidation of propane to acrylonitrile: rutile-type V/Sb and Sn/V/Nb/Sb mixed oxides. Several *N*-containing compounds formed; however, the selectivity to cyano-containing aliphatic linear C_6 compounds was low, due to the relevant contribution of side reactions such as combustion, cracking and formation of heavy compounds.

Introduction

Hexamethylenediamine (HMDA), a monomer for the synthesis of polyamide-6,6, is produced by catalytic hydrogenation of adiponitrile. Three processes, each based on a different reactant, produce the latter commercially. The original Du Pont process, still used in a few plants, starts with adipic acid made from cyclohexane; adipic acid then reacts with ammonia to yield the dinitrile. This process has been replaced in many plants by the catalytic hydrocyanation of butadiene. A third route to adiponitrile is the electrolytic dimerization of acrylonitrile, the latter produced by the ammoxidation of propene.

These processes are used commercially for many years. However, an interest exists for the development of alternative technologies for HMDA production, that may offer advantages with respect to conventional ones, such as: (i) the use of less dangerous reactants, and (ii) a better overall economics, also achieved by the use of highly selective catalytic systems. One possible synthetic pathway is the gas-phase ammoxidation of *n*-hexane to 1,6-C_6 dinitriles, the latter containing either a saturated or an unsaturated aliphatic chain; these dinitriles can be hydrogenated to obtain the diamine.

$$C_6H_{14} + 2\ NH_3 + (6+x/2)/2\ O_2 \rightarrow NC\text{-}(C_4H_{8-x})\text{-}CN + (6+x/2)\ H_2O \qquad (x = 0\ to\ 4)$$
$$NC\text{-}(C_4H_{8-x})\text{-}CN + (4+x/2)\ H_2 \rightarrow NH_2\text{-}(CH_2)_6\text{-}NH_2 \quad (HMDA)$$

The catalytic gas-phase ammoxidation of alkanes to aliphatic nitriles has been the object of several investigations in recent years, but the great majority of papers deal with the direct transformation of propane to acrylonitrile, as an alternative to the commercial process of propylene ammoxidation. Two catalytic systems have been proposed in literature for propane ammoxidation: (i) rutile-type metal antimonates, based on V/(Al,W)/Sb/O (1,2) and (ii) multicomponent molybdates (Mo/V/Nb/Te/O) (3). A few papers and patents report about the ammoxidation of C_4 hydrocarbons (butenes, *n*-butane, butadiene) to maleonitrile and fumaronitrile (4,5). The best results were obtained with a catalyst made of TiO_2-supported oxidic active phase containing V, W, Cr and P, and with butadiene as the reactant. With *n*-butane, the highest yield to nitriles was 26%. Some papers deal with the ammoxidation of C_6 hydrocarbons (6,7). In ref. 6, the ammoxidation of *n*-hexane is carried out with a catalyst based on V/P/Sb/O. At 425°C, 40% adiponitrile selectivity and 30% hexanenitrile selectivity were obtained at 12% conversion. In ref. 7, the catalyst described is a Ti/Sb mixed oxide; from cyclohexane, the main products obtained in a pulse reactor were adiponitrile and benzene, with an overall selectivity of more than 90%. An important aspect pointed out by the authors was the strong interaction that develops between the catalyst surface and the nitriles and that favours the occurrence of consecutive reactions and hence the formation of heavy by-products.

In the present chapter, we report about an investigation of the catalytic performance of rutile-type V/Sb and Sn/V/Sb/Nb mixed oxides in the gas-phase ammoxidation of *n*-hexane. These catalysts were chosen because they exhibit intrinsic multifunctional properties; in fact, they possess sites able to perform both the oxidative dehydrogenation of the alkane to yield unsaturated hydrocarbons, and the allylic ammoxidation of the intermediate olefins to the unsaturated nitriles. These steps are those leading to the formation of acrylonitrile in propane ammoxidation. The Sn/V/Sb/(Nb)/O system is one of those giving the best performance in propane ammoxidation under hydrocarbon-rich conditions (8,9).

Experimental

Sn/V/Nb/Sb/O catalysts were prepared with the co-precipitation technique, developed for the synthesis of rutile SnO_2-based systems claimed by Rhodia (8). The preparation involved the dissolution of $SnCl_4 5H_2O$, $VO(acac)_2$, $SbCl_5$ and $NbCl_5$ in absolute ethanol, and by dropping the solution into a buffered aqueous solution maintained at pH 7. The precipitate obtained was separated from the liquid by filtration. The solid was then dried at 120°C and calcined in air at 700°C for 3 hours. The V/Sb/O catalyst was prepared by means of the "slurry" method that consists in a redox reaction between Sb_2O_3 and NH_4VO_3 in water medium, for 18h at 95°C. The solvent is then evaporated and the solid obtained is dried and then calcined in air at 650°C.

Catalytic tests were performed in a gas-phase continuous-flow reactor. The outlet flow of the reactor was either sampled for the analysis of the gaseous components, or condensed in a dry frozen trap, for the analysis of the solid and liquid products. Two liquid layers formed: an organic layer containing the unconverted *n*-hexane, and an aqueous layer; some products dissolved preferentially in the organic layer, others in the aqueous one. Both layers were analyzed by gas chromatography.

Results and Discussion

Figure 40.1 reports the flammability diagrams for the ternary mixtures: ammonia/oxygen/inert and *n*-hexane/oxygen/inert. The boundary conditions for the catalytic tests were the following:
1. Due to the narrow safe area in the hydrocarbon-lean zone, operation in the *n*-hexane-rich zone was preferred. In the latter case, however, feed compositions having high hydrocarbon concentration had to be avoided, in order to limit the contribution of radical-chain reactions, favoured at high temperature under aerobic conditions. Therefore, operation with a diluted feed was preferred; the ballast used for the reaction was helium, in order to allow evaluation of the amount of N_2 produced by ammonia combustion.
2. Hydrocarbon-rich conditions imply that oxygen is the limiting reactant, due to the high oxygen-to-hydrocarbon stoichiometric ratio in *n*-hexane ammoxidation. Therefore, the conversion of the hydrocarbon is low; this should favour, in principle, the selectivity to products of partial (amm)oxidation instead of that to combustion products.

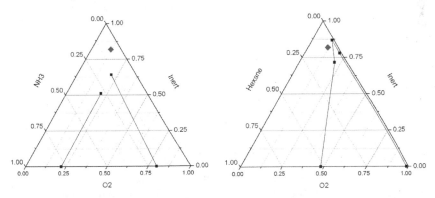

Figure 40.1. Flammability areas for NH_3/O_2/inert and *n*-hexane/O_2/inert mixtures. ◆: Experimental feed composition used for catalytic tests.

The feed composition chosen was: 6 mol% *n*-hexane, 6 mol% ammonia, 12 mol% oxygen and remainder helium, with an overall gas residence time of 2.5 s. Due to the low temperature of *n*-hexane self-ignition (T 234°C), a relevant contribution of homogeneous, radical reactions was expected. Tests made in the absence of catalyst and with the empty reactor, evidenced that at 310°C the conversion of *n*-hexane was 4%, and that of oxygen was 13%. However, already at 330°C the conversion of *n*-hexane and of oxygen were higher than 20 and 30%, respectively. Under these conditions, the main products of the reaction were CO, CO_2 but also a relevant amount of heavy compounds formed downstream the catalytic bed; these compounds led to the blockage of the reactor exit. These problems were in part avoided by minimizing the post-catalyst void fraction in the reactor; the latter was filled with packed quartz-wool, while the void space up-stream the catalytic bed was filled with corundum beads. Under these conditions, the conversion of *n*-hexane in the absence of catalyst was lower than 10% up to 450°C; therefore, this made possible enlarge the temperature zone for catalytic tests, while minimizing the undesired contribution of homogeneous, radical reactions.

Catalysts tested for the reaction of *n*-hexane ammoxidation are reported in Table 40.1. Samples with composition Sn/V/Nb/Sb (atomic ratios between components) equal to x/0.2/1/3 were prepared and characterized. The atomic ratio between V, Nb and Sb was fixed because it corresponds to the optimal one for the active components when these catalysts are used for propane ammoxidation (10).

Table 40.1. Catalysts prepared, and value of specific surface area.

Composition, atomic ratios	Surface area, m^2/g
Sn/V/Nb/Sb 1/0.2/1/3	74
Sn/V/Nb/Sb 3/0.2/1/3	76
Sn/V/Nb/Sb 5/0.2/1/3	82
SnO_2	92
V/Sb 1/1	10

Figure 40.2 shows the X-ray diffraction patterns of Sn/V/Nb/Sb/O catalysts, after calcination at 700°C. All samples showed the reflections typical of the rutile-type compounds, with crystallite size ranging from 7 to 10 nm. There was no relevant effect of the composition on the catalyst surface area. The low crystallinity degree of samples was the reason for the high value of the surface area (reported in Table 40.1), remarkably higher than that typically reported for rutile-type mixed oxides prepared with conventional methodologies like the slurry-redox method. With the latter method, surface area is usually of 10 m^2/g or less. This difference was due to the preparation procedure of our samples, the co-precipitation from the alcoholic medium. This method avoided the segregation of the single metal oxides, and allowed obtaining the rutile mixed oxide by a thermal treatment to be carried out even at temperatures as low as 500°C. Higher catalyst surface areas should allow carrying out the ammoxidation reaction at milder reaction temperatures, and limit the combustion reactions.

In Sn/V/Nb/Sb/O catalysts, different compounds form (10): rutile SnO_2 (also incorporating Sb^{5+}), Sb/Nb mixed oxide and non-stoichiometric rutile-type V/Nb/Sb/O; the latter segregates preferentially at the surface of the catalyst. Tin oxide (cassiterite) provides the matrix for the dispersion of the active components; therefore, a variation of the value of x in Sn/V/Nb/Sb x/0.2/1/3 catalysts implies a change of the ratio between the catalytically inert tin oxide and the components of the active phase. A higher dispersion of the active components should allow minimizing consecutive reactions occurring on both reactants and products, which may be favoured at high concentration of the oxidizing sites.

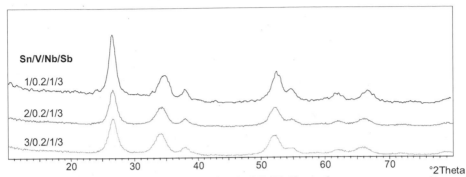

Figure 40.2. X-ray diffraction patterns of Sn/V/Nb/Sb/O catalysts.

The results of reactivity tests performed with catalyst Sn/V/Nb/Sb 1/0.2/1/3 are summarized in Figures 40.3 and 40.4, which report the conversion of reactants and the selectivity to the main products as a function of temperature, respectively. The conversion of *n*-hexane was 20% at 475°C, with a 60% oxygen conversion and 45% ammonia conversion. The reaction yielded many products; besides CO and CO_2, several *N*-containing compounds with 1 to 6 C atoms formed (the overall selectivity to these products is given in Figure 40.4): cyanhydric acid, acetonitrile, acrylonitrile, methacrylonitrile, 2-butenenitrile, isobutyronitrile, butyronitrile, fumaronitrile, maleonitrile, cyclopropanecarbonitrile, 2,4-pentadienenitrile, 2-furancarbonitrile, furandicarbonitrile, pyridine, pyrole, 2-methylpyridine, 3-methylpyridine, benzonitrile, 2-pyridinecarbonitrile, 3-pyridinecarbonitrile and 4-pyridinecarbonitrile. Small amounts of linear aliphatic C_6 nitriles formed: muconodinitrile, hexanenitrile, 2,4-hexadienenitrile and adiponitrile. However, the overall selectivity to these compounds was less than 2%.

Undesired side reactions, i.e., combustion, cracking to light *N*-containing compounds and cyclization of C_6 unsaturated dinitriles, were the reason for the low selectivity to the desired products. The selectivity to C_1-C_5 nitriles increased when the reaction temperature was increased, and that to C_6 nitriles (mainly cyclic compounds) correspondingly decreased. The highest selectivity, amongst nitriles, was to fumaronitrile (17% at 475°C). Another side reaction was the combustion of ammonia; however, the yield to N_2 (also reported in Figure 40.4) decreased when the reaction temperature was increased; at the same time, in fact, the selectivity to *N*-containing compounds became higher. Figure 40.4 also reports the carbon balance, as calculated from the ratio between the overall C atoms found in the products identified, compared to C atoms transformed due to converted *n*-hexane. It is shown that the balance was lower than 50% for reaction temperatures below 400°C, and it increased when the reaction temperature was increased. A carbon balance much lower than 100% indicates that a considerable amount of the formed products has not been analyzed. This was due to the formation of heavy by-products that accumulated in the condenser after the reactor and were not eluted in the GC column.

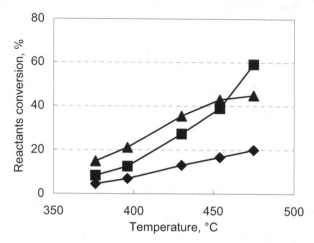

Figure 40.3. Conversion of reactants in *n*-hexane ammoxidation as a function of the reaction temperature. Symbols: conversion of *n*-hexane (◆), ammonia (▲) and oxygen (■). Catalyst Sn/V/Nb/Sb 1/0.2/1/3.

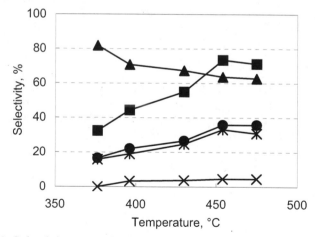

Figure 40.4. Selectivity to products in *n*-hexane ammoxidation as a function of the reaction temperature. Symbols: selectivity to CO (✕), CO_2 (✱), *N*-containing compounds (●, calculated with respect to *n*-hexane converted) and N_2 (▲, calculated with respect to ammonia converted). Carbon balance (■). Catalyst Sn/V/Nb/Sb 1/0.2/1/3.

Figure 40.5 compares the conversion of ammonia and *n*-hexane for catalysts with composition Sn/V/Nb/Sb x/0.2/1/3, while Figure 40.6 reports the overall selectivity to *N*-containing compounds. Surprisingly, the samples containing the greater amount of Sn were the most active. This result was quite unexpected, because tin oxide was catalytically inert when used as one component in catalysts for propane ammoxidation.

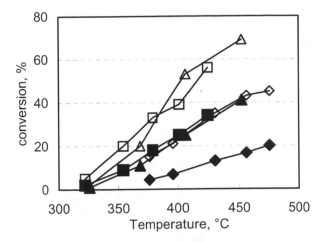

Figure 40.5. Conversion of *n*-hexane (full symbols) and of ammonia (open symbols) as a function of temperature for catalysts with composition Sn/V/Nb/Sb x/0.2/1/3. Symbols: x = 1 (◆◇), 3 (■□) and 5 (▲△).

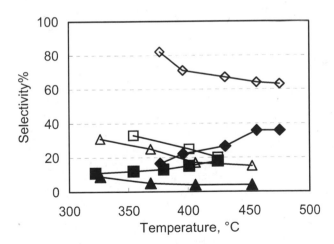

Figure 40.6. Selectivity to *N*-containing products (full symbols, calculated with respect to *n*-hexane converted) and to N_2 (open symbols, calculated with respect to ammonia converted) as a function of temperature for catalysts with composition Sn/V/Nb/Sb x/0.2/1/3. Symbols as in Figure 40.5.

The selectivity to *N*-containing compounds was the lower with catalysts having the greater amount of Sn (x=3 and 5). The yield of products having more than 2 C atoms was very low; in fact the prevailing products were cyanhydric acid and acetonitrile. The main product of the reaction was CO_2, with negligible formation of

CO, while the carbon balance was between 40 and 60% for the catalyst with x=3, and lower than 40% for the catalyst with x=5. On the other hand, the selectivity to N_2, formed by ammonia combustion, was much lower than that obtained with Sn/V/Nb/Sb 1/0.2/1/3. Therefore, in catalysts with the greater amount of Sn, most of N was contained in products that were not detected and that contributed to the relevant loss in carbon balance.

The results indicate that tin oxide plays a direct role in the reaction, addressing the reaction towards the formation of heavy compounds, carbon oxides and cyanhydric acid. This was confirmed by carrying out reactivity tests with the sample made of SnO_2 only; its activity was just the same as that one of samples Sn/V/Nb/Sb x/0.2/1/3 having x=3 and 5. The conversion of ammonia and the selectivity to N_2 were low, and the prevailing products were carbon oxides and heavy compounds. Furthermore, a non-negligible amount of cyanamide formed, that in part oligomerized to form melamine; these compounds, however, were not quantified. Therefore, in the gas-phase ammoxidation of *n*-hexane to linear C_6 dinitriles the main limitation to the selectivity is the concomitant occurrence of several undesired side reactions: (a) The cracking of the reactant and of the unsaturated aliphatics intermediately formed to lighter hydrocarbons. The latter undergo ammoxidation to yield saturated and unsaturated C_1-C_5 nitriles. (b) The combustion of *n*-hexane and of intermediates to CO_2. (c) The combustion of ammonia to N_2, a reaction in competition with the insertion of N to yield *N*-containing compounds. (d) The formation of cyclic C_6 compounds (pyridinecarbonitriles); the latter may occur either by cyclization of unsaturated linear nitriles, or by condensation of lighter nitriles. (e) The formation of heavy compounds and tars.

The reaction mechanism is likely similar to that one commonly accepted for the ammoxidation of propane to acrylonitrile (11). The activation of *n*-hexane may occur via formation of radical species, that either oxidehydrogenate to yield unsaturated C_6 olefins and diolefins, or undergo fragmentation to yield lighter aliphatics. Therefore, the high selectivity to the several C_1-C_5 *N*-containing products obtained is due to the higher reactivity of *n*-hexane as compared to propane, and to the several possible fragmentations of the molecule. In fact, in propane ammoxidation the Sn/V/Nb/Sb/O catalyst yields cyanhydric acid, acetonitrile and acrylonitrile, but the selectivity to acrylonitrile is greater than that to the C_1 and C_2 nitriles. In *n*-hexane ammoxidation, cracking of the C_6 intermediates is quicker than N insertion, and finally the overall selectivity to *N*-containing linear C_6 compounds is lower than the selectivity to lighter compounds.

The catalyst composition has a role in the control of selectivity. The rutile-type V/Sb/(Nb) mixed oxide activates the hydrocarbon and ammonia. However, most of the ammonia is burnt to N_2, rather than being inserted on the hydrocarbons; this likely occurs because the catalyst is not very efficient in the generation of the selective Me=NH species when reaction temperatures lower than 400°C are used (11). In fact, with all catalysts the selectivity to *N*-containing compounds increased when the reaction temperature was increased, and the selectivity to N_2 correspondingly decreased (Figure 40.6). The dilution of the active phase with tin oxide, a component that we supposed to be catalytically inert, decreased the extent of ammonia combustion to nitrogen, but on the other hand it unexpectedly contributed to the conversion of *n*-hexane and to the formation of by-products, such as heavy compounds, cyanamide, cyanhydric acid and CO_2. Therefore, tin oxide is able to activate *n*-hexane, generate radical species and finally favour the fragmentation into lighter compounds. In samples with the greater amount of Sn (x=3 and 5), the low concentration of the active sites able to perform the selective insertion of N led to an enhancement of side reactions occurring at the surface of tin oxide, such as combustion and cracking of the radical intermediates to CO_2 and to light *N*-containing compounds, but also coupling or condensation reactions to yield tars.

In order to confirm the hypothesis made on the role of catalyst components, we carried out the reaction with a rutile-type V/Sb/O catalyst, having V/Sb atomic ratio equal to 1/1 (Table 40.1). This catalyst was prepared with the conventional slurry method, and therefore had a surface area of 10 m^2/g, lower than that obtained with the Sn/V/Nb/Sb/O catalysts prepared with the co-precipitation method. However, despite this difference, with V/Sb/O the conversion of *n*-hexane was similar to that one obtained with Sn/V/Nb/Sb/O. This is shown in Figure 40.7, which reports the conversion of *n*-hexane, the selectivity to CO_2, to *N*-containing compounds and the carbon balance as a function of the reaction temperature.

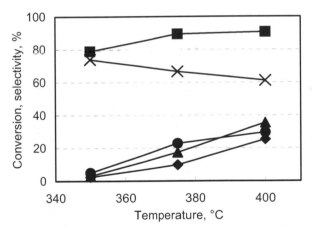

Figure 40.7. Conversion of *n*-hexane (♦) and of ammonia (▲), selectivity to CO + CO_2 (✗), to *N*-containing compounds (●) and carbon balance (■) as a function of the reaction temperature. Catalyst V/Sb 1/1.

The activation energy for *n*-hexane conversion was 37±5 kcal/mole, while it was 17±2 kcal/mole for the Sn/V/Nb/Sb 1/0.2/1/3 catalyst (Figure 40.3). Therefore, with the high-surface-area Sn/V/Nb/Sb/O catalyst, the particle efficiency was largely lower than 1, while this was not the case for the non-porous low-surface-area V/Sb/O. This is likely due to the high intrinsic reactivity of *n*-hexane. For what concerns the selectivity, the main characteristics of the V/Sb/O catalyst were the following (Figure 40.7): (a) a carbon balance close to 80-90%, that indicates the low formation of heavy compounds; (b) a selectivity to CO_2 + CO higher than 70%; and

(c) a relevant degree of ammonia combustion to N_2. The overall selectivity to N-containing compounds increased when the reaction temperature was increased. At 350°C, the prevailing N-containing compounds were C_6 nitriles and dinitriles. The selectivity to N-containing C_4-C_6 products became nil when the reaction temperature was increased, whereas the selectivity to C_1-C_3 nitriles increased. This confirms the hypothesis of an enhanced contribution of cracking reactions and of a higher efficiency in N-insertion at temperatures higher than 350°C.

Conclusions

The gas-phase ammoxidation of n-hexane to nitriles was investigated using rutile-type Sn/V/Nb/Sb and V/Sb mixed oxide catalysts. Several N-containing C_1-C_6 compounds formed; however, the selectivity to cyano-containing aliphatic linear C_6 compounds was low, due to the relevant contribution of side reactions such as combustion, cracking, combustion of ammonia and formation of tars. The ratio between catalyst components influenced the catalytic performance; catalysts having the greater amount of Sn were the least selective, because SnO_2 promoted the formation of CO_2, of light N-containing compounds and of heavy compounds.

References

1. F. Cavani and F. Trifirò, in *Basic Principles in Applied Catalysis* (M. Baerns Ed.), Springer, Berlin, Series in Chemical Physics 75, 2003, p. 21.
2. R.K. Grasselli, *Topics Catal.* **21**, 79 (2002).
3. T. Ushikubo, K. Oshima, A. Kayou, M. Vaarkamp and M. Hatano, *J. Catal.* **169**, 394 (1997).
4. I. Furuoya, *Stud. Surf. Sci. Catal.* **121**, 343 (1999).
5. A. Peters and P.A. Schevelier, WO 2006/053786 A1 (2006), assigned to DSM.
6. B.M. Reddy and B. Manohar, *J. Chem. Soc. Chem. Comm.* 330 (1993).
7. O.Yu. Ovsitser, Z.G. Osipova and V.D. Sokolovskii, *React. Kinet. Catal. Lett.* **38**, 91 (1989).
8. G. Blanchard, P. Burattin, F. Cavani, S. Masetti and F. Trifirò, WO Patent 97/23,287 A1 (1997), assigned to Rhodia.
9. S. Albonetti, G. Blanchard, P. Burattin, F. Cavani, S. Masetti and F. Trifirò, *Catal. Today* **42**, 283 (1998).
10. N. Ballarini, F. Cavani, M. Cimini, F. Trifirò, J.M.M. Millet, U. Corsaro and R. Catani, *J. Catal.* **241**, 255 (2006).
11. G. Centi, S. Perathoner and F. Trifiro, *Applied Catal. A* **157**, 143 (1997).

41.   TUD-1: A Generalized Mesoporous Catalyst Family for Industrial Applications

Philip J. Angevine[1], Anne M. Gaffney[1], Zhiping Shan[2], Jan H. Koegler[1] and Chuen Y. Yeh[1]

[1]Lummus Technology, 1515 Broad Street, Bloomfield, NJ 07003
[2]Huntsman Chemical, 8600 Gosling Road, The Woodlands, Houston, TX 77381

cyeh@cbi.com

Abstract

This chapter discusses the synthesis, characterization and applications of a very unique mesoporous material, TUD-1. This amorphous material possesses three-dimensional interconnecting pores with narrow pore size distribution and excellent thermal and hydrothermal stabilities. The basic material is Si-TUD-1; however, many versions of TUD-1 using different metal variants have been prepared, characterized, and evaluated for a wide variety of hydrocarbon processing applications. Also, zeolitic material can be incorporated into the mesoporous TUD-1 to take the advantage of its mesopores to facilitate the reaction of large molecules, and enhance the mass transfer of reactants, intermediates and products. Examples of preparation and application of many different TUD-1 are described in this chapter.

Introduction

Porous materials have a successful history in heterogeneous catalysis. While both microporous and macroporous materials have been used in industry for many decades, academic and industrial scientists have been searching for materials with intermediate pore sizes between the microporous and the macroporous range. These "mesoporous" materials are defined as having pore diameters between 2 and 50 nm. In addition, materials with hierarchical pore structures, i.e., going from larger mesopores to smaller micropores, are also of practical interest. Both types of materials are anticipated to be useful for conversions of higher molecular weight materials. Catalysts based on these materials should have significant benefits for reactions where mass transfer plays a role.

Since the discovery by researchers at Mobil of a new family of crystalline mesoporous materials (1), a large effort has been expended on synthesis, characterization, and catalytic evaluation (2). MCM-41 is a one-dimensional, hexagonal structure. MCM-48 is a cubic structure with two, nonintersecting pore systems (3). MCM-50 is a layered structure with silica sheets between the layers (4). Many scientists also looked into other mesoporous materials, of note the HMS (Hexagonal Molecular Sieve) family (5) and SBA-15 (acronym derived from Santa Barbara University) (6), but to date few materials have been both catalytically significant and inexpensive to synthesize.

This paper describes the unique properties of TUD-1, a new, amorphous mesoporous material. We will highlight several organic synthesis applications. (We discuss petrochemical and refining applications of TUD-1 in more detail elsewhere (7).)

TUD-1

A joint research project between Lummus Technology and the Delft University of Technology led to the discovery of a new mesoporous material, named TUD-1 (8). TUD-1 is a three-dimensional amorphous structure of random, interconnecting pores. The original emphasis was on the silica version, which has since been extended to about 20 chemical variants (e.g., Al, Al-Si, Ti-Si, etc.).

Key common properties of TUD-1 are:
- Random, three-dimensional interconnecting pores
- Tunable porosity (pore volumes of 0.3-2.5 cc/g and diameters of 4-25 nm)
- High surface area: 400-1000 m^2/g
- Excellent thermal, hydrothermal, and mechanical stability

TUD-1 is an amorphous material. Unlike crystalline structures, it has no characteristic x-ray diffraction pattern. Figure 41.1 illustrates the pore diameter of TUD-1 in comparison to some major molecular sieves - ZSM-5, Zeolite Y, and MCM-41. It is important to note that the pore diameter of TUD-1 can be varied from about 40Å to 250 Å.

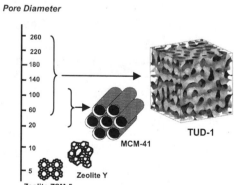

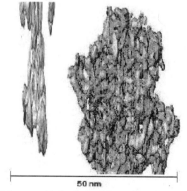

Figure 41.1: Pore diameters of TUD-1, MCM-41 and zeolites.

Figure 41.2: 3-D TEM showing the irregular pore structure of TUD-1.

One of the early questions raised on TUD-1 dealt with its pore structure: did it have intersecting or nonintersecting pores? At the University of Utrecht, one conclusive characterization was carried out with a silica TUD-1 with Pt inserted, which was analyzed by 3-D TEM (transmission electron microscopy) (9). The Pt anchors (not shown) were used as a focal point for maintaining the xyz orientation. As shown in Figure 41.2, the TUD-1 is clearly amorphous. While not quantitatively measured for this sample, the pores appear rather uniform, consistent with all porosimetry measurements on TUD-1 showing narrow pore size distributions.

General Method of Synthesis

The original synthesis route (i.e., Si-TUD-1) involves a monomeric silica source TEOS (tetraethylorthosilicate), mixing with TEA (triethanolamine), and optionally TEAOH (tetraethylammonium hydroxide). The TEA serves as a template for the mesopore formation. Desirable properties of the TUD-1 template are: physically stable at elevated temperatures (200-250°C), chemically interactive with the inorganic phase, and inexpensive. The TEAOH serves as both a source of quaternary cation (to generate some micropores, if required) and a basic environment to accelerate TEOS hydrolysis. The reaction rate increases with pH; to a large extent, this acceleration can also be achieved by increasing the temperature. The second step involves an aging/drying phase to establish the primary pore structure. The last step – calcination – is required to remove the large quantities of organics. An optional step, between drying and calcination, is a pore modification step employing elevated temperature.

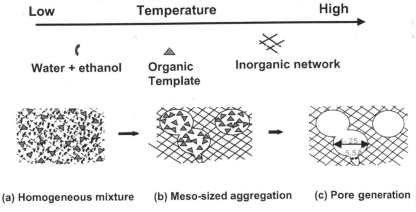

Figure 41.3: Schematic representation of the TUD-1 formation.

Figure 41.3 highlights the major structural transformations that take place in TUD-1 synthesis (10). The three major steps are: (a) formation of a homogeneous mixture, (b) migration of the template to achieve meso-sized aggregation, and (c) pore generation. Additionally, if inorganic bases such as TEAOH are used, some micropores are formed in addition to the mesopores. This is another key differentiator from many other crystalline mesoporous materials. In terms of unit operations, the major steps can be divided into six steps: (a1) mixing, (a2) hydrolysis, (b1) aging, (b2) drying, (c1) heat treatment [optional], and (c2) calcination.

The flexibility of the synthesis method provides for opportunities to incorporate other elements besides silicon. One of the first elements added was Al (11). Soon after the discovery of the silicon based TUD-1, it was found that adding suitable aluminum sources to the above procedure yielded very similar Al-Si-TUD-1 structures. Since then, many other TUD-1 variants have been prepared. Most TUD-1 variants are either Si-TUD-1 or an M-Si version, where M is another element (e.g.,

Ti, Al, Cr, Fe, Zr, Ga, Sn, Co, Mo, V, etc.). Non-siliceous versions of Al- and Ti-TUD-1 have also been made. The typical SiO_2/M_xO_y molar ratio is 20-∞.

Figure 41.4 shows a typical XRD (X-Ray Diffraction) pattern of TUD-1, along with a TEM image (12). Similar to other mesoporous materials, TUD-1 has a broad peak at low 2θ. However, it has a broad background peak, commonly called an "amorphous halo," and lacks any secondary peaks that are evident for example in the hexagonal MCM-41 and cubic MCM-48 structures. The TEM shows that the pores have no apparent periodicity. In this example the pore diameter is about 5 nm.

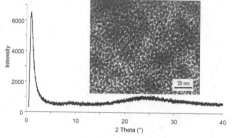
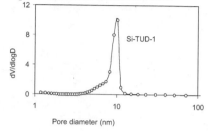

Figure 41.4: XRD pattern and TEM image of TUD-1.

Figure 41.5: Typical pore size distribution of TUD-1.

The narrow pore size distribution of TUD-1 is illustrated in Figure 41.5 by the single peak derived from the nitrogen desorption isotherm. Moreover, an important feature of the material is the easy tunability of the pore sizes over a wide range while maintaining a narrow pore size distribution.

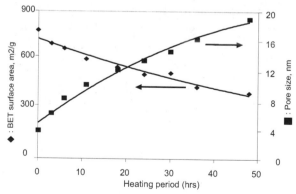

Figure 41.6: Tuning porosity of TUD-1 by varying heat treatment times.

Figure 41.6 illustrates the typical capabilities in tailoring Si-TUD-1 pore diameter using different hydrothermal treatment times (10). The pore diameter can be varied from 4 to 19 nm by increasing the hydrothermal heating time from 0 to 48 hours. The sample with 48 hours of hydrothermal treatment still had an appreciable surface area of 400 m^2/g.

Synthesis of TUD-1 with Embedded Zeolite

The combination of mesoporous materials with embedded zeolites is one of the most promising concepts for catalysts with enhanced activity. When the zeolite is distributed evenly throughout TUD-1, the synergistic benefits are (a) high accessibility to the internal zeolite crystal (achieving higher effective activity) and (b) easy product removal from the zeolite surface (resulting in fewer secondary reactions to form unwanted by-products, and reduced coking, pore-mouth plugging, and associated aging). The zeolite embedding method is described in detail elsewhere (7,13). A high resolution TEM study of TUD-1 with embedded zeolite Beta (14), showed the electron diffraction pattern of zeolite Beta domains within the amorphous mass of the TUD-1, indicating retention of zeolite crystallinity within the composite material.

Applications of TUD-1

Ti-Si-TUD-1 for Epoxidation of Cyclohexene

The synthesis of Ti-Si-TUD-1 is analogous to the silica version; a portion of the reactant is a titanium alkoxide, such as titanium (IV) n-butoxide. One of the early comparative catalytic tests of TUD-1 versus MCM-41 was for cyclohexene epoxidation.

The Ti was loaded using two methods: direct incorporation into the synthesis mixture, and post-synthesis grafting. In all cases the Ti-loading was 1.5 - 1.8 wt%. Selectivity towards the epoxide was always 100%. Table 41.1 summarizes the results comparing Ti-TUD-1 and Ti-MCM-41 for cyclohexene epoxidation (15). For the direct incorporation, Ti-TUD-1 is five times more active than Ti-MCM-41, even though they have equivalent surface area. However, the grafted MCM-41 is also more active than its as-synthesized counterpart.

Table 41.1: Cyclohexene epoxidation using Ti-TUD-1 and Ti-MCM-41.

Catalyst	S_{BET} (m^2/g)	D (Å)	TOF (hr^{-1})
Ti-TUD-1	870	50	20
Ti grafted TUD-1	560	100	28
Ti-MCM-41	920	30	4
Ti grafted MCM-41	1020	30	23

Possible explanations for the higher activity of the Ti-TUD-1 are its three-dimensional pore structure and the distribution of the active Ti sites. The TUD-1 synthesis employs a bifunctional templating agent – triethanolamine - that is more selective to Ti than to Si. This results in most of the Ti being located within or at the edge of the organic phase. After calcination, the Ti in TUD-1 is preferentially dispersed on the pore wall surfaces, while the Ti in MCM-41 is expected to be homogeneously dispersed within the silica. If post-synthesis grafting is used, Ti is obviously present only at the wall surface for both mesoporous materials, thus reducing the difference between the two.

Other TUD-1 catalysts proven for selective oxidation (16) include Au/Ti-Si-TUD-1 for converting propylene to propylene oxide (96% selectivity at 3.5% conversion; see also (17), Ag/Ti-Si-TUD-1 for oxidizing ethylene to ethylene oxide (29% selectivity at 19.8% conversion), and Cr-Si-TUD-1 for cyclohexene to cyclohexene epoxide (94% selectivity at 46% conversion).

Fe-Si-TUD-1 for Friedel-Crafts Alkylation

Friedel-Crafts alkylations are among the most important reactions in organic synthesis. Solid acid catalysts have advantages in ease of product recovery, reduced waste streams, and reduction in corrosion and toxicity. In the past, people have used (pillared) clays (18), heteropolyacids (19) and zeolites (20) for Friedel-Craft alkylations, with mixed success. Problems included poor catalyst stability and low activity. Benzylation of benzene using benzyl chloride is interesting for the preparation of substitutes of polychlorobenzene in the application of dielectrics. The performance of Si-TUD-1 with different heteroatoms (Fe, Ga, Sn and Ti) was evaluated, and different levels of Fe inside Si-TUD-1 (denoted Fe_1, Fe_2, Fe_5 and Fe_{10}) were evaluated (21). The synthesis procedure of these materials was described in detail elsewhere (22).

Reactions have been carried out using 0.1 g of catalyst in an oil-bath-heated magnetically stirred round bottom flask equipped with a reflux condenser. The vessel was heated for 2 hrs at 120°C in vacuum, followed by cooling down to reaction temperature of 60°C and flowing dried nitrogen through the reaction flask to avoid the effect of moisture. Then 10 ml of molecular sieve dried benzene and 1.0 g of benzyl chloride were added. Liquid samples were taken at regular intervals and analyzed by gas chromatography. Tests to determine leaching were carried out by filtering off the catalyst at reaction temperature, and continuing to monitor the reaction using the filtrate only. The solid catalyst residue underwent quantitative elemental analysis using neutron activation analysis (INAA). For all TUD-1 samples, selectivity towards diphenyl methane was 100%. For Fe_{10}, 100% conversion was reached within 90 seconds.

Table 41.2 shows the results of the Friedel-Crafts benzylation of benzene at 60°C over several TUD-1 based materials, and compared with bulk Fe_2O_3. Comparison between different metal-containing Si-TUD-1 catalysts with Si/M ratio 50 indicates that Fe is the most active metal, as conversion reaches 100% within 180 minutes.

The order of catalytic activity was Fe > Ga > Sn > Ti, which is the same order as the standard reduction potential $E^0_{Mn+/M}$ for these metals. This illustrates that redox properties rather than acid properties are responsible for the activity. Comparison of the activation energies between the different Fe-Si-TUD-1 samples was carried out by conducting the reaction at temperatures between 40° and 80°C. For Fe_1, Fe_2, Fe_5 and Fe_{10} the activation energy was 47, 85, 182 and 216 kJ/mol, respectively. The large difference in activation energies between these samples may tentatively be explained by the increasing presence and activity of Fe-oxide nanoparticles besides the always present isolated Fe^{3+} sites.

Table 41.2: Conversion of benzyl chloride at 60°C over different TUD-1 samples.

Sample	Si/M ratio	Conversion (%)	reaction time (min.)
Si-TUD-1	∞	0	240
Fe_2O_3	0	5	240
Fe_1-Si-TUD-1 (Fe_1)	113	86	240
Fe_2-Si-TUD-1 (Fe_2)	54	100	180
Fe_5-Si-TUD-1 (Fe_5)	21	100	10
Fe_{10}-Si-TUD-1 (Fe_{10})	10	100	<1.5
Ga-Si-TUD-1	51	65	240
Sn-Si-TUD-1	50	16	240
Ti-Si-TUD-1	50	4	240

The leaching experiments showed that almost 33% of the Fe atoms were removed from the framework. However, the reaction did not proceed after removal of the solid catalyst, indicating that although leaching takes place, the reaction is truly catalyzed heterogeneously. The calculated catalytic activity of the Fe_{10}-Si-TUD-1 was 35, which is much higher than the activities of other mesoporous (Fe-MCM-41(23); TOF≈1) or microporous (Fe-ZSM-5 (18); TOF≈3) materials, where reactions were carried out under comparable conditions.

All in all, while an improved metals incorporation method may need to be developed in order to minimize leaching, Fe-Si-TUD-1 is a very promising candidate for heterogeneous Friedel-Crafts alkylations.

Pt/Pd TUD-1: Hydrogenation of Aromatics

The high-surface-area TUD-1 can serve as an anchor for many catalysts. Si- or Al-Si-TUD-1 (24,25) can be used as a support for various noble metals (Pt, PtPd, Ir, etc.). This will provide catalysts suitable for the hydrogenation of olefins and aromatics. In the refining industry, one use is the hydrogenation of polynuclear aromatics ("PNAs") in diesel fuel, which can lower the fuel's toxic properties. Also, jet fuel has an aromatics constraint, designed to lessen smoke formation. Cracked stocks (e.g., coker or visbreaker liquids) generally have undesirable olefins (especially α-olefins) that also need to be saturated prior to final processing.

Table 41.3 shows a performance comparison of Pt/Pd TUD-1 with a commercial Pt/Pd catalyst (26). The feedstock is a typical straight run gasoil ("SRGO"), a distillate precursor to diesel fuel. Under identical test conditions, the TUD-1 catalyst achieved 75% aromatics saturation versus 50% for the same volume of commercial catalyst. This superior result is particularly interesting because the TUD-1 catalyst had a much lower density than the commercial material, so that less catalyst by weight was required in the reactor.

Table 41.3: Aromatics hydrogenation of Pt/Pd TUD-1 compared to a commercial Pt/Pd catalyst.

	Feed	Commercial Pt/Pd Catalyst	Pt/Pd Si-TUD-1 Catalyst
Aromatics, vol%	21.2	10.1 (≈50% saturation)	5.1 (≈75% saturation)
Specific Gravity	0.8344	0.8241	0.8220
Cetane Index (D976)	44.7	46.7	47.8

Co-Si-TUD-1 for Selective Oxidation of Cyclohexane

A recent study (13,27) describes the use of Co-Si-TUD-1 for the liquid-phase oxidation of cyclohexane. Several other metals were tested as well. TBHP (tert-butyl hydroperoxide) was used as an oxidant and the reactions were carried out at 70°C. Oxidation of cyclohexane was carried out using 20 ml of a mixture of cyclohexane, 35mol% TBHP and 1 g of chlorobenzene as internal standard, in combination with the catalyst (0.1 mmol of active metal; pretreated overnight at 180°C). Identification of the products was carried out using GC-MS. The concentration of carboxylic side products was determined by GC analysis from separate samples after conversion into the respective methyl esters. Evolution and consumption of molecular oxygen was monitored volumetrically with an attached gas burette. All mass balances were 92% or better.

Table 41.4: Catalytic performance of different M-Si-TUD-1 samples for cyclohexane oxidation at 70°C.

Sample	Time (h)	Si/M ratio	Cyclohexane Conversion (%)	TBHP Conversion (%)	Mono-oxygenated selectivity (%)	Ketone/ Alcohol ratio
Ti-1	16	100	4.7	30.8	75.4	1.5
V-1	16	100	<2	36.5	-	-
Cr-1	18	100	14.1	97.5	74.2	86.5
Fe-1	16	100	3.1	19.5	83.4	3.3
Co-1	18	100	10.3	49.5	90.5	6.7
Cu-1	23	100	13.6	84.4	80.8	6.1
Co-2	18	50	8.8	41.1	88.6	5.8
Co-5	18	20	5.3	21.5	91	4.9
Co-10	24	10	5.8	25.9	88.7	5.1

Results of the cyclohexane oxidation tests are shown in Table 41.4. Mono-oxygenated products are cyclohexanone, cyclohexanol and cyclohexyl hydroperoxide. Cu and Cr were very active, but subsequent tests showed considerable leaching for both metals, whereas Co-Si-TUD-1 did not show any leaching. Tests with different Co loadings indicate that the lowest Co concentration has the best conversion and ketone selectivity. Isolated cobalt species are most efficient for the conversion of cyclohexane, as agglomeration of Co reduces accessibility of individual Co atoms and thus the number of active sites. It was also observed that the catalyst first converts cyclohexane to cyclohexanol, which is then further converted to cyclohexanone. A reusability test showed adequate recycling properties.

Thus, it was shown that Co-Si-TUD-1 is a highly active catalyst for the oxidation of cyclohexane, has a high selectivity for mono-oxygenated species and cyclohexanone, and can be used repeatedly.

Zeolites in TUD-1 for Benzene Alkylation and n-Hexane Cracking

Currently, benzene alkylation to produce ethylbenzene and cumene is routinely carried out using zeolites. We performed a study comparing a zeolite Y embedded in TUD-1 to a commercial zeolite Y for ethylbenzene synthesis. Two different particle diameters (0.3 and 1.3 mm) were used for each catalyst. In Figure 41.7, the first-order rate constants were plotted versus particle diameter, which is analogous to a linear plot of effectiveness factor versus Thiele modulus. In this way, the rate constants were fitted for both catalysts.

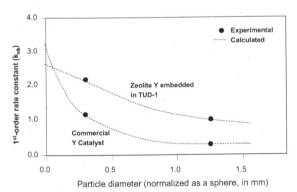

Figure 41.7: Ethylbenzene alkylation activity vs. catalyst particle diameter.

As can be seen in the graph, the Y/TUD-1 catalyst was twice as active as the commercial Y catalyst. This is primarily due to its very high calculated diffusivity of 131×10^{-6} cm^2/sec, which is over 10 times the diffusivity calculated for commercial zeolite Y, 11×10^{-6} cm^2/sec. Extrapolation of the curve to zero particle size shows that the commercial Y zeolite is in fact intrinsically more active than the Y zeolite embedded in the TUD-1. If the Y zeolite in TUD-1 had been optimized for this reaction like the commercial Y catalyst, one should expect an even greater boost in performance.

More recently, using the same methodology as described above, a commercially available alumina-bound zeolite Beta (80 wt %) was tested for ethylbenzene synthesis (28). A first-order rate constant of 0.29 cm^3 /g-sec was obtained. A 16% zeolite Beta in TUD-1 catalyst also had a rate constant of 0.30 cm^3

/g-sec for the same reaction. These results indicate that: (a) the integrity of the zeolite crystals in the mesoporous catalyst support is maintained during synthesis; (b) the microporous zeolite Beta in the mesoporous support was still accessible after the catalyst finishing; and (c) the mesopores of the support facilitate mass transfer in aromatic alkylation reactions.

Another example of performance enhancement using a zeolite/TUD-1 catalyst is shown in n-hexane cracking using a series of zeolite-Beta-embedded TUD-1 catalysts (29): 20, 40 and 60 wt% zeolite Beta in Al-Si-TUD-1 (Si/Al = 150). These are compared to pure zeolite Beta, and to a physical mixture of 40% zeolite Beta and 60% Al-Si-TUD-1. These catalysts were tested in a fixed bed reactor, at atmospheric pressure, with constant residence time at 538°C. The pseudo-first-order rate constants are shown in Figure 41.8. Note that the zeolite-loaded catalysts were clearly superior to both the pure zeolite Beta catalyst and the zeolite-TUD-1 physical mixture. Again, this is evidence that catalyst performance benefits from a hierarchical pore structure such as zeolite embedded in TUD-1.

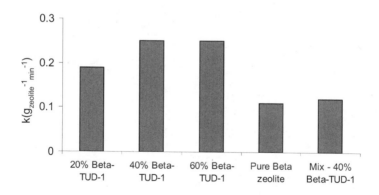

Figure 41.8: Pseudo-first-order rate constants based on the mass of zeolite for n-hexane cracking at 538°C over zeolite Beta-TUD-1 catalysts.

Other references (30) describe the synthesis and performance benefits of zeolites embedded in TUD-1. This concept has been reported in the literature with other mesoporous materials, but the three-dimensional nature of TUD-1 should make these embedded zeolite materials of special benefit.

Various other Applications of TUD-1

A recent patent describes the synthesis and catalytic use of Al-containing TUD-1 materials. Some of the reactions demonstrated include hydrogenation of mesitylene (Pt as active metal) and dehydration of 1-phenyl-ethanol to styrene. Several other conceptual reactions were also described, amongst others the Diels-Alder reaction of crotonaldehyde and dicyclopentadiene and the amination of phenol with ammonia.

Another recent patent (22) and related patent application (31) cover incorporation and use of many active metals into Si-TUD-1. Some active materials were incorporated simultaneously (e.g., NiW, NiMo, and Ga/Zn/Sn). The various catalysts have been used for many organic reactions [TUD-1 variants are shown in brackets]: Alkylation of naphthalene with 1-hexadecene [Al-Si]; Friedel-Crafts benzylation of benzene [Fe-Si, Ga-Si, Sn-Si and Ti-Si, see application 2 above]; oligomerization of 1-decene [Al-Si]; selective oxidation of ethylbenzene to acetophenone [Cr-Si, Mo-Si]; and selective oxidation of cyclohexanol to cyclohexanone [Mo-Si]. A dehydrogenation process (32) has been described using an immobilized pincer catalyst on a TUD-1 substrate. Previously these catalysts were homogeneous, which often caused problems in separation and recycle. Several other reactions were described, including acylation, hydrogenation, and ammoxidation.

Another patent application (28) describes the use of zeolite/TUD-1 with optionally a metal function for a variety of reactions. In an example, as-synthesized MCM-22 / TUD-1 was tested for acylation of 2-methoxynaphthalene with acetic anhydride to 2-acetyl-6-methoxynaphthalene at 240°C. After reaction for six hours, conversion of 2-methoxynaphthalene reached 56% with 100% selectivity to 2-acetyl-6-methoxynaphthalene. Other zeolite catalysts were similarly tested, but none were nearly as effective.

There are a host of other, non-catalytic applications of TUD-1 possible, for example in the area of sorption. In fact, the successful application of TUD-1 as a controlled release drug delivery system was recently demonstrated (33).

Concluding Remarks

TUD-1 is one of the few mesoporous materials having three-dimensional, intersecting pores as well as high surface area. As such, its catalytic potential appears quite promising. Since it can be made with many chemical versions, it has proven to be quite versatile. This article describes some of the catalytic applications for TUD-1. Since there are nearly 20 known chemical variants of TUD-1, these clearly are not all possible applications. We have focused on major hydrocarbon processes, but one can easily envision many reactions to form fine chemicals and pharmaceuticals. In refining, the areas of attention should be the upgrading of high molecular weight streams (e.g., resid and lubes). We expect that, with continued effort, many new uses will be discovered for TUD-1 and other ultra-large pore catalysts.

Acknowledgements

The technical leadership by Professors J.C. Jansen and Th. Maschmeyer is greatly appreciated. They played a visionary role in the early discovery, synthesis, and characterization of the TUD-1 family.

References

1. C.T. Kresge et al., *US 5,098,684* (1992); *US 5,102,643* (1992), and *US 5,198,203* (1993); J.S. Beck et al., *US 5,145,816* (1992).
2. F-S Xiao, *Topics in Catalysis* (2005), 35(1-2), 9-24; X.S. Zhao et al., *Ind. & Eng. Chem. Res.*(1996), **35(7)**, 2075-2090; H-P Lin et al., J. Chin. Chem. Soc.(Taipei) (1999), **46(3)**, 495-507.
3. W.J. Roth, *US 6,096,288* (1998).
4. J.C. Vartuli et al*., Stud. Surf. Sci. Catal.* **84** (1994), 53-60.
5. P.T. Tanev, T.J. Pinnavaia, *Science* (1995), 267(5199), 865-7.
6. D. Zhao et al., *Science* (1998), 279(5350), 548-552.
7. P.J. Angevine et al., *J. Ind. Eng. & Chem.* (Centenary issue), accepted for publication.
8. Z. Shan et al., *US 6,358,486* (2002).
9. F. M. Dautzenberg, *13th International Congress on Catalysis*, 2004, Paris.
10. Z. Shan et al., *International Symposium on Silica,* 2000, France.
11. Z. Shan et al., *US 7,211,238* (2007).
12. Z. Shan et al., *NCCC2000*, the Netherlands.
13. Z. Shan et al., *NCCCIII, 2002*, the Netherlands.
14. Z. Shan et al., *US 6,762,143* (2004).
15. Z. Shan et al., *Mesoporous and Mesoporous Materials* **48** (2001) 181-187; Z. Shan et al., *Chem. Eur. J.* **7(7)** (2001), 1437-1443; Z. Shan et al., *US 6,906,208* (2005).
16. Z. Shan et al., *US 6,762,143* (2004).
17. J. Lu et al*., J. Catal.* **250** (2007), 350-59.
18. B. Choudary et al., *Appl. Catal. A*, **149** (1997), 257.
19. Yusuke et al., *Appl. Catal. A*, **132** (1995), 127.
20. A. Singh, D. Bhattacharya, *Catal. Lett.*, **32** (1995), 327.
21. M.S. Hamdy et al., *Catal. Today*, **100** (2005), 255-60.
22. Z. Shan et al., *US 6,930,219* (2005).
23. Cao et al., *Stud. Surf. Sci. Catal.* **117** (1998), 461.
24. B. Ramachandran et al., *US Pat. Appl. 20060009665* (filed 7/8/04).
25. B. Ramachandran et al., *US Pat. Appl. 20060009666* (filed 1/10/05).
26. Z. Shan et al., *US Pat. Appl. 20040138051* (filed 10/22/03).
27. R. Anand et al*., Catal. Lett.*, **3-4** (2004), 113.
28. Z. Shan et al., *US Pat. Appl. 20060128555* (filed 2/8/06).
29. P. Waller et al., *Chemistry-A Eur. J.* **10(20)** (2004), 4970-4976.
30. Z. Shan et al., *US 6,814,950* (2004); *US 6,930,217* (2005) ; *US 7,084,087* (2006).
31. Z. Shan et al., *US Pat. Appl. 20030188991* (filed 12/6/02).
32. C.Y.Yeh et al., *US Pat. Appl. 20040181104* (filed 3/2/04).
33. T. Heikkila et al., *Int. J. Pharmaceutics* **331** (2007) 133-38.

42.   Bifunctional Catalysis of Alkene Isomerization and Its Applications

Douglas B. Grotjahn, Casey Larsen, Gulin Erdogan, Jeffery Gustafson, Abhinandini Sharma and Reji Nair

Department of Chemistry and Biochemistry, San Diego State University, 5500 Campanile Drive, San Diego, CA 92182-1030

grotjahn@chemistry.sdsu.edu

Abstract

Our group studies bifunctional catalysts created using phosphines which contain pendant bases in the vicinity of the metal center. Previously we have reported highly active catalysts for anti-Markovnikov alkyne hydration using this concept (*J. Am. Chem. Soc.* **2004**, *126*, 12232 and *Chem. Eur. J.* **2005**, *11*, 7146). Here we describe related catalyst **1**, which is very active for alkene isomerization. At the 2 mol % level, **1** isomerizes 1-pentene exclusively to (*E*)-2-pentene within 15 min at room temperature. More significantly, 2 mol % catalyst isomerizes 4-penten-1-ol to pentanal, a considerably more challenging task, within 1 h at 70°C. Control experiments with nonheterocyclic ligands show that the heterocycle accelerates these reactions by 10,000 times in the pentenol case. Very high selectivity for (*E*) isomers of unbranched alkenes is seen, a general phenomenon which can be explained by a mechanism involving alkene coordination followed by selective deprotonation at an allylic position by the pendant base. Here we expand on our initial communication (*J. Am. Chem. Soc.* **2007**, *129*, 9592), in which we showed that movement of an alkene double bond over 30 positions is possible. Thioether, amide, ester, ether and carboxylic acid functional groups are tolerated. Isomerizations of dienes show promising results. The high selectivity of the catalyst along with its activity presage interesting applications in organic synthesis.

Introduction

With the renaissance in alkene chemistry engendered by the rising versatility of olefin metathesis in both fine chemical and commodity production, new methods for alkene isomerization are of increasing interest and importance. Alkene isomerization can be performed using Brønsted-Lowry acid or base catalysis (1). However, these reactions are limited to substrates which tolerate carbanionic or carbocation intermediates, and are susceptible to undesired side reactions.

Transition metal catalysts offer a way to isomerize alkenes under neutral conditions. There are numerous catalysts capable of moving a double bond over a few positions (2-7). When it comes to more extensive movement, however, the apparent record for alkene isomerization is 20 positions on the hydrocarbon $CH_3(CH_2)_{19}CH=CH(CH_2)_{19}CH_3$ by *stoichiometric* amounts of the reagent

$Cp_2Zr(H)(Cl)$ (8). The apparent record for *catalyzed* double bond movement is on 9-decene-1-ol to decanal (nine positions) using $Fe_3(CO)_{12}$ (9). However, 30 mol % was required, which means that nearly a mole of metal was used per mole of alkenol. Herein we expand upon our initial report (10) of a very active catalyst (1) which has been shown to move a double bond over 30 positions. Catalyst 1 appears to have an intriguing and useful mode of action, in which the pendant base ligand performs proton transfer on coordinated alkene and π-allyl intermediates in a stereoselective fashion.

Experimental Section

The preparation of 1 and the detailed conditions used for alkene isomerization have been described elsewhere (10). Complex 1 will also soon be commercially available from Strem Chemicals. Reagents, solvents, and catalysts were measured or weighed out and combined in a nitrogen-filled glovebox. The reactions were conducted in J. Young resealable NMR tubes and monitored by ^1H and $^{31}P\{^1H\}$ NMR spectroscopy. As described elsewhere (10), an inert internal standard, $(Me_3Si)_4C$, was used along with short ^1H pulsewidths and long delay times between pulses in order to assure accurate integrations.

An example of a reaction from which product was isolated is conversion of 7 to (*E*)-8, Table 1, entry 3: in the glove box, internal standard $(Me_3Si)_4C$ and 7 (168.9 mg, 0.50 mmol) were combined with acetone-d_6 (~700 µL), and an initial NMR spectrum was acquired. Back in the glovebox, 1 (6.3 mg, 0.01 mmol, 2 mol %) was added, followed by enough acetone-d_6 to reach a final volume of 1.0 mL. The reaction was allowed to proceed at ambient temperature for 4 h, at which point the mixture was purified by flash chromatography over silica (6 in x ¾ in) using a gradient of 100% petroleum ether to 5% diethyl ether in petroleum ether. After concentration by rotary evaporation, white crystals were isolated in 78% yield (131.9 mg), which included approximately 3% of the starting material. For the 8 isolated: ^1H NMR (400 MHz, acetone-d_6) δ 7.81-7.87 (m, 2H), 7.43 (d, $J = 8.4$, 2H), 5.48-5.57 (m, 1H), 5.34-5.48 (m, 1H), 3.83 (~t, $J \approx 7.2$, 2H, AA' of AA'BB'), 2.43 (s, 3H), 2.34-2.43 (m, 2H), 1.63 (qd, $J = 1.2$, 6.4, 2H), 1.33 ppm (s, 9H). ^{13}C NMR (125.71 MHz, acetone-d_6) δ 151.78, 145.21, 138.98, 130.27, 128.85, 128.46, 128.17, 84.38, 47.52, 34.26, 28.10, 21.56, 18.23 ppm. For the product isolated: ^{13}C NMR (125.71 MHz, CDCl$_3$) 150.98, 143.98, 137.67, 129.19, 128.12, 127.90, 126.89, 83.98, 46.88, 33.45, 27.90, 21.59, 17.99 ppm. Elemental analysis calculated for $C_{17}H_{25}NO_4S$ (339.45): C, 60.15; H, 7.42; N, 4.13. Found: C, 60.53; H, 7.62; N, 4.59.

Results and Discussion

The catalyst reported by Grotjahn and Lev (11-13) for alkyne hydration (2) is capable of isomerizing alkenes, but very slowly. Because we knew that the rate of alkyne hydration was unchanged in the presence of excess phosphine ligand, we thought that like alkyne hydration, alkene isomerization would require loss of acetonitrile ligand (14) and alkene binding. Subsequent deprotonation at an allylic position would make an η^1-allyl intermediate which when reprotonated at the other end would promote alkene isomerization. We thought that a promising approach toward increasing catalyst reactivity would be to reduce the number of phosphines on the metal center, potentially allowing for η^3-allyl binding. We note that it was not obvious that this approach would work, because Kirchner et al. (15) reported that although $[CpRu(PR_3)(CH_3CN)_2]^+$ **(3a-3c)** was an improved catalyst for allylic alcohol isomerization relative to $CpRu(PR_3)_2Cl$, it failed in the case of 3-buten-1-ol or alkenes devoid of alcohol functionality.

Ligand screening experiments were conducted on the alkenes 1-pentene and pent-4-en-1-ol, because such substrates were inert to **3a-3c** (15). Pentene lacks any polar or protic group and pentenol contains the alkene and OH separated by 3 carbons. The preliminary studies involved phosphines with both imidazol-2-yl and pyrid-2-yl substituents on P as well as *t*-Bu, *i*-Pr, Ph, and Me groups (16). From the screening, complex **1** derived from the phosphine ligand **4** (17) was identified as the most capable (in terms of both reaction rate and final yield) of promoting isomerization of both 1-pentene and pent-4-en-1-ol.

To make **1**, $[CpRu(CH_3CN)_3]^+PF_6^-$ in acetone was allowed to react with **4**, forming a mixture of **3d** and chelate **1** (10). This mixture could be used but for purposes of catalyst characterization and further testing was converted to > 90% pure **1** by repeated coevaporation of added acetone with the acetonitrile liberated. Complex **1** showed diagnostic NMR data consistent with a chiral structure and four unique, diastereotopic methyl groups.

The isomerization of pent-4-en-1-ol to pentanal was optimized with respect to solvent, catalyst amount, and reaction temperature. As seen by looking at Table 42.1, entries 6a and 6b, at the 2 mol % level, heating at 70°C was advised to achieve a reasonable reaction rate, whereas using 5 mol %, reactions could be completed at 25°C within 1 d. Acetone emerged as the solvent of choice, though dichloromethane could also be used. The effects of a substrate hydroxyl group were apparent by noting that under the optimized conditions, isomerization of 1-pentene to (E)-2-pentene required only minutes at room temperature using 2 mol % catalyst (Table 42.1, entry 1).

Table 42.1. Scope and limitations using **1**[a]

Entry	Reactant	Product	1 (%)	Temp (°C)	Time	Yield (%)
1	1-pentene	(E)-2-pentene	2	25	15 min	95
2	5	(E, E)-6	2	25	40 min	96
3	7	(E)-8	2	25	4 h	75[b]
4	9	10	2	25	26 h	70:30[c]
5	11	12	2	25	4 d	61[d]
6a	15a	16a	2	70	1 h	95
6b	15a	16a	5	25	1 d	95
7	15b	16b	2	70	1 h	97
8	13	(E)-14	5	70	4 h	90
9	15c	16c	5	70	4 h	84
10	15d	16d	5	70	4 h	97
11	15e	16e	20	70	3.6 d	91[b]
12	15f	16f	30	70	3 d	81[b]
13	17	(E)-18	5	70	22 h	73
14	19	(E)-20	5	25	2 h	86
15	21	(E)-22	2	70	2 h	96

[a]Acetone-d_6 solvent; yields determined by ^{1}H NMR and internal standard unless otherwise indicated. [b]Isolated. [c]Equilibrium ratio of **9** and **10**, also reached from either side within 2-4 h using 2 mol% **2** at 70°C. [d]Unreacted **11** (29%) and an unidentified isomer (10%) also present.

Table 42.1 shows results on a variety of substrates, which are generally unoptimized, but give an idea of the scope and limitations of the catalyst. In general, full isomerization of substrates with polar or potentially coordinating groups (OH, amide, thioether, carboxylic acid) required heating at 70°C, whereas partial isomerization of these compounds or full isomerization of nonpolar substrates could be done at 25°C. Most dramatically, alkenol isomerization over 9, 22 and even 30 positions could be achieved (entries 9-12). Thioether and carboxylic acid functionalities were tolerated (entries 13 and 14). In the former case, traces of intermediate species (probably complexes of **17** or **18** or isomers) could be detected by ^{1}H and ^{31}P NMR, a subject of ongoing investigation. Considering that the proposed mechanism (see below) involves deprotonation of coordinated alkene by the catalyst, it is remarkable that the carboxylic acid functional group in **19** does not slow down isomerization any more than does an alcohol OH group (entries 14 and 6b).

The catalyst exhibits a high selectivity for formation of (*E*)-isomers, which is of both synthetic and mechanistic interest. The silyl ether **13** (entry 8) gave enol ether (*E*)-**14** in which only traces (< 2%) of (*Z*)-isomer are detected in 500 HMz ^{1}H NMR spectra. Remarkably, diallyl ether (entry 2) gave (*E, E*)-**4** in high yield without Claisen rearrangement of the intermediate allyl propenyl ether (18-20) and without any detectable (*Z, E*)-isomer, providing rapid entry under neutral conditions to a little-explored class of compounds which have typically been made as mixtures using strong base (21). Because this compound has two reacting double bonds, our inability to detect the (*Z, E*)-isomer is especially significant, because there would be twice the chance of forming this species.

The mechanism shown in Scheme 42.1 accounts for all the facts above, and three more: [1] complex **3e**, lacking a pendant base, is 330 times slower at isomerizing 1-pentene than is **1**, and moreover is 10,000 times slower at isomerizing pent-4-en-1-ol; [2] bubbling ethylene (ca. 25 equiv) into an acetone solution of **1** leads to formation of free acetonitrile and an ethylene complex similar to **23**, rather than a chelate-opened species; [3] bubbling propene into an acetone solution of **1** and D_2O leads to deuteration of the terminal positions of propene, but not to the internal position. Therefore, it is reasonable that the first step of the catalytic cycle is exchange of acetonitrile for alkene, forming alkene complex **23**. The basic nitrogen of the imidazole could then deprotonate the coordinated alkene at an allylic position, forming **24** with an anionic allyl and a protonated heterocyclic ligand, which could then return the proton to either end of the allyl moiety and promote isomerization to **25** and finally alkene product. An alternative mechanism involving metal-hydride formation and alkene insertion / β-hydride elimination is considered highly unlikely, because of complete lack of H/D exchange at the internal position of propene even after extended reaction times. The regioselectivity of an alkene insertion would have to be extremely high to produce these results.

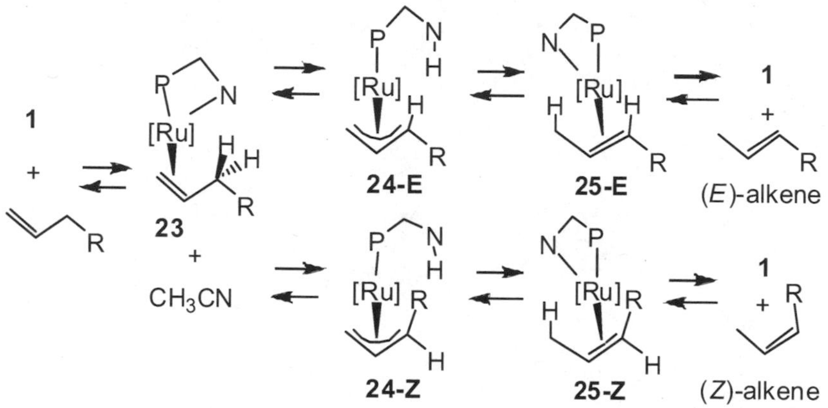

Scheme 42.1. Proposed Mechanism.

The high (*E*)-selectivity can also be explained by Scheme 42.1. Comparing proposed intermediates **24-E** and **24-Z**, we note that for steric reasons, the isomer **24-E** with the R group in the syn-position should be lower in energy (22). Therefore it is reasonable to propose that either the intermediate allyl complex **24-Z** leading to (*Z*)-alkene or the prior transition state (between **23-Z** and **24-Z**) is sufficiently high in energy to leave the system shuttling between terminal and (*E*)-alkene(s), creating mixtures consisting of almost exclusively these latter isomers, to the exclusion of (*Z*)-alkene. Indeed, when terminal alkenes are isomerized to internal (*E*)-alkenes, long after completion of the reaction (>> 10 times the reaction times shown in Table 42.1) some (*Z*)-isomer begins to build up. The mechanistic proposal in Scheme 42.1 helps rationalize the unusual ability of **1** to convert geraniol to (*E*)-iso-geraniol rather than to the aldehyde, because formation of the enol isomer would require the catalyst to form an allyl intermediate like **24-Z**. Finally, we note that since isomerization of the longer substrates presumably occurs by a random walk of the double bond up and down the chain *en route* to the most stable isomer, the observed movement of the double bond is actually just a lower estimate of the number of times the catalyst acts on substrate.

The high selectivity of the catalyst in forming (*E*)-alkenes can be used in interesting ways (eq. 1). For example, in acetone-d_6 solution, within 15 min at room temperature allyl alcohol is converted to nearly pure enol (*E*)-**26**. Under these mild conditions, the product slowly isomerizes to the more stable aldehyde tautomer. We know of one other report of rapid enol formation from allyl alcohol, using a Rh catalyst (23). However, the reported Rh catalyst gave approximately equal amounts of (*E*) and (*Z*) enol, whereas **1** gives 30-to-1 selectivity. Preliminary results on **15a** at room temperature suggest that here too one can detect an enol intermediate, but at 70 °C this intermediate disappears fast enough so that only **15a**, aldehyde product **16a**, and intermediate alkene isomers other than the enol are seen. Preliminary results also indicate that if CF_3CO_2H is present in the reaction shown in eq. 1, only the aldehyde product is seen, presumably because the protic acid speeds up tautomerization of **26** to propanal.

$$\text{HO}\diagup\diagup \xrightarrow[\text{acetone-}d_6]{\textbf{1} (5\ \%),\ 25\ ^\circ\text{C},\ <15\ \text{min}} \text{HO}\diagup\diagup\diagdown\ (E)\text{-}\mathbf{26}\ +\ \text{HO}\diagup\diagdown\ (Z)\text{-}\mathbf{26} \qquad (\text{eq. 1})$$

97 : 3 ratio

25 °C, hours ↓

$$\underset{\text{H}}{\overset{\text{O}}{\diagdown}}\diagup\diagdown$$

Another example of the usefulness of high (*E*) selectivity is shown in eq. 2, where without optimization, 1,5-hexadiene (commercially available at ca. $1 per gram) is converted to (*E,E*)-2,4-hexadiene (which costs close to $100 per gram). Further analysis of the mixture will be needed to identify the composition of the 12% unaccounted for in eq. 2, but the 88% yield of product in this simple procedure is promising.

$$\diagup\diagdown\diagup\diagdown\ \xrightarrow[\text{acetone-}d_6]{\substack{\textbf{1} (2\ \%),\\ 25\ ^\circ\text{C},\ 4.5\ \text{h}\\ \text{followed by}\\ 70\ ^\circ\text{C},\ 20\ \text{h}}}\ \diagup\diagdown\diagup\diagdown\ \qquad (\text{eq. 2})$$

88%

Finally, we are examining the role of branching on selectivity. It is useful to compare four substrates (Scheme 42.2), the first two of which (**9** and **15b**) were presented in Table 42.1. In the case of **9**, we saw no evidence for the formation of **27**, even after extended reaction times. This and related results on geraniol would be consistent with an energetic barrier to removal of the required proton at position α of **9** when coordinated to the catalyst. The fact that isomerization of **9** *does* occur at all (leading to some **10**) verifies that the catalyst can act on the trisubstituted alkene.

Secondary alcohol **15b** presumably is isomerized to enol **28**, but at temperatures of 70°C the latter is quickly converted to the most stable keto form (**16b**), the observed product. Interestingly, silylation of the alcohol (**29**) allows for a mild and complete isomerization to (*E*)-**30** at room temperature, confirming that an alcohol OH group somehow slows the first isomerization step. However, the alcohol

OH must *help* subsequent isomerization, because (*E*)-**30** is stable in the presence of **1**, even after 2 d at 70°C. To test whether the bulk of the silyl ether protecting group was responsible for this reactivity difference, compound **32** was allowed to isomerize, forming a high yield of (*E*)-**33** at room temperature. In this case, heating at 70°C for 15 h does seem to lead to traces of what is tentatively identified as **34**, but the second isomerization is clearly much slower than the first. Further studies are planned to elucidate the role of a hydroxyl substituent in alkene isomerization catalyzed by **1**.

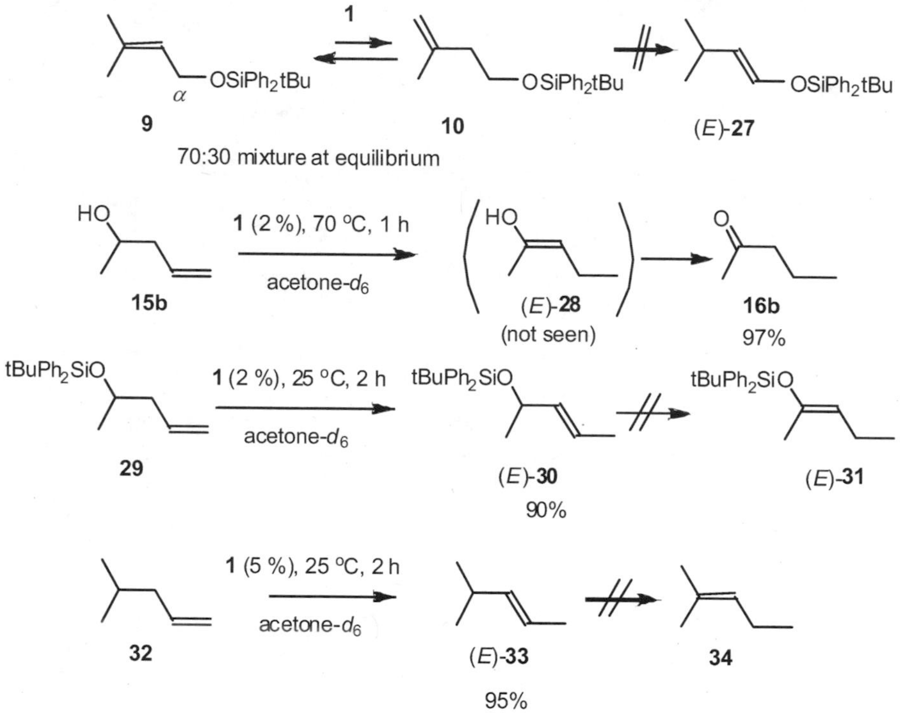

Scheme 42.2. Comparing Branched Substrates.

Conclusions

In summary, complex **1** is shown to be an active catalyst for alkene isomerization, with high selectivity for formation and reactions of (*E*)-alkenes. A variety of functional groups, including alcohols, amides, thioethers and even carboxylic acids are tolerated, but typically require slightly elevated temperatures (70°C) for complete reaction. One outstanding characteristic of **1** is that it enables movement of double bonds over many positions along an unbranched chain. As for the mechanism, evidence to date is wholly consistent with alkene binding and activation by the metal and allylic deprotonation-reprotonation involving the pendant imidazole base. The application of this novel bifunctional catalyst to alkene isomerization reactions of synthetic interest is the subject of ongoing investigations, as is the extension of bifunctional catalysis to other organic transformations.

Acknowledgements

We thank the NSF for continuing support under CHE 0415783 and 0719575, and Dr. LeRoy Lafferty for assisting with NMR experiments.

References

1. C. A. Brown, *Synthesis*, 754 (1978).
2. W. A. Herrmann and M. Prinz, in *Applied Homogeneous Catalysis with Organometallic Compounds (2nd Edition)*, (eds. B. Cornils and W. A. Herrmann), Wiley-VCH, Weinheim, Germany, 2002, p. 1119.
3. E.-I. Negishi, in *Handbook of Organopalladium Chemistry for Organic Synthesis*, (ed. E.-i. Negishi), John Wiley & Sons, Hoboken, 2002, p. 2783.
4. R. Uma, C. Crevisy and R. Gree, *Chem. Rev.* **103**, 27 (2003).
5. S. Otsuka and K. Tani, in *Transition Metals for Organic Synthesis (2nd Edition)*, (eds. M. Beller and C. Bolm), Wiley-VCH, Weinheim, Germany, 2004, p. 199.
6. H. Suzuki and T. Takao, in *Ruthenium in Organic Synthesis*, (ed. S.-I. Murahashi), Wiley-VCH, Weinheim, Germany, 2004, p. 309.
7. N. Kuznik and S. Krompiec, *Coord. Chem. Rev.* **251**, 222 (2007).
8. T. Gibson and L. Tulich, *J. Org. Chem.* **46**, 1821 (1981).
9. N. Iranpoor and E. Mottaghinejad, *J. Organomet. Chem.* **423**, 399 (1992).
10. D. B. Grotjahn, C. R. Larsen, J. L. Gustafson, R. Nair and A. Sharma, *J. Am. Chem. Soc.* **129**, 9592 (2007).
11. D. B. Grotjahn and D. A. Lev, *J. Am. Chem. Soc.* **126**, 12232 (2004).
12. D. B. Grotjahn and D. A. Lev, *Catal. Org. React.* **104**, 227 (2005).
13. D. A. Lev and D. B. Grotjahn, *Catal. Org. React.* **104**, 237 (2005).
14. E. Rüba, W. Simanko, K. Mauthner, K. M. Soldouzi, C. Slugovc, K. Mereiter, R. Schmid and K. Kirchner, *Organometallics* **18**, 3843 (1999).
15. C. Slugovc, E. Rueba, R. Schmid and K. Kirchner, *Organometallics* **18**, 4230 (1999).
16. For examples of ligands we have made and tested here, see D. B. Grotjahn, C. D. Incarvito and A. L. Rheingold, *Angew. Chem., Int. Ed..* **40**, 3884 (2001). D. B. Grotjahn and D. A. Lev, *J. Am. Chem. Soc.* **126**, 12232 (2004).
17. D. B. Grotjahn, Y. Gong, L. N. Zakharov, J. A. Golen and A. L. Rheingold, *J. Am. Chem. Soc.* **128**, 438 (2006).
18. B. M. Trost and T. Zhang, *Org. Lett.* **8**, 6007 (2006).
19. B. Schmidt, *J. Mol. Catal. A: Chemical* **254**, 53 (2006).
20. S. G. Nelson, C. J. Bungard and K. Wang, *J. Am. Chem. Soc.* **125**, 13000 (2003).
21. E. Taskinen and R. Virtanen, *J. Org. Chem.* **42**, 1443 (1977).
22. J. W. Faller, M. E. Thomsen and M. J. Mattina, *J. Am. Chem. Soc.* **93**, 2642 (1971).
23. S. H. Bergens and B. Bosnich, *J. Am. Chem. Soc.* **113**, 958 (1991).

43. Investigation of Deactivation of Co-Salen Catalysts in the Hydrolytic Kinetic Resolution of Epichlorohydrin

Surbhi Jain[1], Xiaolai Zheng[2], Krishnan Venkatasubbaiah[3], Christopher W. Jones[3], Marcus Weck[4] and Robert J. Davis[1]

[1]*Chemical Engineering Department, University of Virginia, Charlottesville, VA 22904*
[2]*School of Chemistry and Biochemistry,* [3]*School of Chemical & Biomolecular Engineering, Georgia Institute of Technology, Atlanta, GA 30332*
[4]*Molecular Design Institute and Department of Chemistry, New York University, New York, NY 10003*

rjd4f@virginia.edu

Abstract

The possible modes of deactivation of Jacobsen's Co-salen catalyst during the hydrolytic kinetic resolution (HKR) of epichlorohydrin were investigated by combining recycling studies with UV-Vis and X-ray absorption spectroscopy. In addition, electrospray ionization mass spectrometry was used to check for deactivation by dimer formation. Recycling studies without catalyst regeneration showed substantial deactivation after multiple runs; however, no reduction of the original Co(III) salen to Co(II) salen was observed spectroscopically. Furthermore, the mass spectrum of a deactivated catalyst did not show significant formation of dimer. Since the loss in activity was not caused by Co reduction or catalyst dimerization, the effect of different Co-salen counterions (acetate, tosylate, chloride, or iodide) on the HKR reaction was explored. Kinetic results suggested that the rate of addition of the Co-salen counterions to epoxide during the HKR reaction was related to catalyst deactivation. Moreover, the rate of deactivation was influenced by the exposure time of the catalyst to the reactants. An oligo(cyclooctene) supported Co-OAc salen catalyst, which was 25 times more active than the monomeric Co-salen catalyst, could be recycled at shorter reaction times with negligible deactivation. Preliminary experiments with silica-supported Co-salen catalysts indicated leaching of Co species is another possible mode of deactivation.

Introduction

The hydrolytic kinetic resolution (HKR) of terminal epoxides using Co-salen catalysts provides a convenient route to the synthesis of enantioenriched chiral compounds by selectively converting one enantiomer of the racemic mixture (with a maximum 50% yield and 100% ee) (1-3). The use of water as the nucleophile makes this reaction straightforward to perform at a relatively low cost. The homogeneous Co(III) salen catalyst developed by Jacobsen's group has been shown to provide high ee's for a variety of terminal epoxides with only 0.2-2 mol% catalyst loading in solution (4, 5).

The proposed mechanism for the HKR of terminal epoxides involves the cooperation of two Co-salen complexes, leading to the squared dependence of rate on catalyst concentration (6, 7). Because of this bimetallic mechanism of the HKR reaction, researchers have developed multimeric Co-salen catalysts that exhibit activities 1-2 orders of magnitude greater than that of the monomeric Co-salen catalyst (8-17). For example, the Co-salen complexes bound to polymeric or dendrimeric frameworks as well as the dimers and oligomers of Co-salen complexes synthesized by Jacobsen's group have revealed similar high enantioselectivity but substantially enhanced reactivity relative to the monomeric Co-salen analog (8-14). Song et al. (15, 16) have also synthesized poly-salen Co(III) salen complexes that exhibited an enantiomeric excess up to 98% in the HKR of terminal epoxides. Zheng et al. (17) have synthesized poly(styrene) supported Co-salen catalysts and obtained high activity and selectivity in the HKR of epichlorohydrin.

Recently, we have prepared an oligo(cyclooctene) supported Co-salen catalyst (**A**), with the idea that the higher local concentration of Co-salen species in the macrocyclic framework would enhance the reactivity and enantioselectivity in the HKR reaction (18).

(**A**)   (n=2-1 0)

Although high activities and enantioselectivities can be obtained from these multimeric homogeneous as well as supported catalysts, most of them are difficult to synthesize and cannot be recycled unless they are regenerated after the HKR reaction. The general speculation about the catalyst deactivation during the HKR reaction has been reported to be the reduction of active Co(III) salen complex to an inactive Co(II) salen complex (19-21). However, the mechanism of the proposed Co reduction is not clear. In this work, we investigated the deactivation of Co-salen catalyst during the HKR of epichlorohydrin by performing the recycling of homogeneous Co-salen catalyst without catalyst regeneration and following the chemical state of the catalyst by UV-Vis and X-ray absorption spectroscopy (22). The effect of dimer formation on deactivation was probed by electrospray ionization mass spectrometry. The influence of counterion on the rate of epichlorohydrin HKR has also been evaluated. In addition, we compared the activity and recyclability of the oligo(cyclooctene) supported Co-OAc salen catalyst (**A**) to that of the standard molecular Co-salen catalyst. Finally, an insoluble SBA-15 silica-supported Co-salen catalyst (**B**) was synthesized and evaluated in the HKR reaction under conditions of continuous flow.

(**B**)

Experimental Section

Catalyst preparation. Various R,R-Jacobsen's Co-salen catalysts were prepared with four different counterions, namely acetate, tosylate, chloride and iodide. The detailed preparation procedure for all these catalysts can be found in a previous paper (22). Catalyst (**A**) was prepared via ring-expanding olefin metathesis of a monocyclooct-4-en-1-yl functionalized salen ligand and the corresponding Co(II) salen complex in the presence of Grubbs catalyst (18). For the preparation of catalyst (**B**), a solution of (2.0 g) of compound (**C**) in dry toluene was added to a solution of SBA-15 (2.0 g) in dry toluene in a nitrogen dry box. The solution was stirred under reflux conditions for 24 h, filtered and washed with toluene and hexanes. The red solid obtained was soxhlet extracted with dichloromethane for 12 h, and the resulting solid was dried under vacuum at 50°C, overnight (Co loading 1.92%, IR (KBr): ν (cm^{-1}) = 2952, 2863, 1610, 1521, 1434, 1388). The dried solid was then oxidized in the presence of acetic acid (270 eq.) and dichloromethane to yield catalyst (**B**).

(**C**)

Hydrolytic Kinetic Resolution (HKR) of epichlorohydrin. The HKR reaction was performed by the standard procedure as reported by us earlier (17, 22). After the completion of the HKR reaction, all of the reaction products were removed by evacuation (epoxide was removed at room temperature (~300 K) and diol was removed at a temperature of 323-329 K). The recovered catalyst was then recycled up to three times in the HKR reaction. For flow experiments, a mixture of racemic epichlorohydrin (600 mmol), water (0.7 eq., 7.56 ml) and chlorobenzene (7.2 ml) in isopropyl alcohol (600 mmol) as the co-solvent was pumped across a 12 cm long stainless steel fixed bed reactor containing SBA-15 Co-OAc salen catalyst (**B**) bed (~297 mg) via syringe pump at a flow rate of 35 µl/min. Approximately 10 cm of the reactor inlet was filled with glass beads and a 2 µm stainless steel frit was installed at the outlet of the reactor. Reaction products were analyzed by gas chromatography using ChiralDex GTA capillary column and an FID detector.

Spectroscopic evaluation of the catalysts. The UV-Vis spectra of the Jacobsen Co-salen catalysts were collected in the transmission mode on a CARY-3E UV-Vis spectrophotometer by dissolving the catalysts in epichlorohydrin. The Co K-edge (7709 eV) X-ray absorption near edge structure, XANES, of Jacobsen's Co-salen catalyst was collected during the HKR reaction at beamline X10-C at National Synchrotron Light Source (NSLS), Brookhaven National Lab, Upton, NY.

Electrospray Ionization - Mass Spectrometry (ESI-MS). The Jacobsen's Co-salen catalysts dissolved in dichloromethane were pumped to the mass spectrometer system after dilution with methanol at a flow rate of 50 µl min^{-1} and 600 scans were collected in 1 min.

Results and Discussion

Recycling studies on Jacobsen's Co-OAc salen catalyst (without regeneration). The HKR of racemic epichlorohydrin (10 mmol) was performed in the presence of 0.5 mol% R,R-Jacobsen's Co-OAc salen catalyst and water (0.7 eq), producing S-epichlorohydrin and R-3-chloro 1,2-propane diol as the major products (eq. 1).

$$\text{(1)}$$

(+/-) epichlorohydrin   R,R-Co-OAc salen   S-epichlorohydrin   R-3-chloro 1,2-propane diol

We monitored the percent conversion of epichlorohydrin and enantiomeric excess of the recovered S-epichlorohydrin with time by using GC-FID. Approximately 54% conversion and >99% ee were obtained in about 4 h reaction time. After 4 h, the epichlorohydrin was removed under vacuum at room temperature and diol was removed at a temperature of 329 K. The recovered catalyst was further treated in the HKR of racemic mixture of fresh epichlorohydrin. In the second run, we observed a decrease in the conversion and ee compared to the fresh catalyst. The Co-salen was again recovered after the second run by removing all the products under vacuum and recycled two more times. With each subsequent HKR reaction, the conversion and ee were found to decrease with time (22). Table 43.1 summarizes the initial rates and ee's determined from the four runs without intermediate catalyst regeneration. Interestingly, the initial catalyst activity was resumed when the catalyst was regenerated with acetic acid prior to recycle.

Kim et al. (19) also observed that the ee of recovered epichlorohydrin was reduced to 17% in the second hydrolysis reaction with Jacobsen's Co-OAc salen catalyst, if the catalyst was not regenerated with acetic acid in air. Although they attributed the loss of enantioselectivity to the reduction of Co(III) to Co(II) salen complex after the HKR reaction, no spectroscopic evidence was provided. Therefore, we probed the catalyst by UV-Vis and XANES spectroscopy before and after the HKR reaction.

Table 43.1. Initial rates and ee's of the HKR of epichlorohydrin with Jacobsen's Co-OAc salen catalyst without intermediate catalyst regeneration[a]

Runs	Initial rate $(mol\ ECH)\ (mol\ cat)^{-1}\ (min)^{-1}$	% ee after 60 min
1	2.98	75.3
2	1.41	47.5
3	0.41	17.2
4	0.28	8.8

[a]Reaction conditions: 10 mmol racemic epichlorohydrin (ECH), 0.7 eq. water, 0.5 mol% R,R-Jacobsen's Co-OAc salen catalyst, 120 μl chlorobenzene, ~300 K reaction temperature.

UV-Vis and XANES spectroscopic studies on Jacobsen's Co-salen catalyst. The UV-Vis and XANES spectra of Jacobsen's Co-salen complexes were recorded before and after the HKR reaction to investigate the possible catalytic reduction during the HKR reaction as stated by Kim et al. (19). In the UV-Vis measurements, two strong absorption bands at ~ 360 and 415 nm were observed for the inactive Jacobsen's Co (II) salen complex, whereas a weak band at ~ 408 nm was present in the active Jacobsen's Co(III)-OAc salen complex (22). The Jacobsen's Co-salen complex recovered after one run of the HKR reaction of epichlorohydrin did not show any feature at ~360 nm, indicating no reduction after the HKR reaction. The catalyst recovered after the fourth run had only a very small feature at ~360 nm, indicating only slight reduction of the metal center after multiple reactions. Although Jacobsen's Co(III)-OAc salen catalyst exhibited substantial deactivation on recycling without intermediate acetic acid regeneration in the HKR of epichlorohydrin, very little of the Co(III) reduced to Co(II).

To complement results from UV-Vis spectroscopy, the in-situ Co XANES were measured during the HKR of epichlorohydrin with Jacobsen's Co-OAc salen catalyst. The XANES of Jacobsen's Co(II) and Co(III)-OAc salen catalysts were recorded in the transmission mode. A pre-edge peak was observed at about 6.6 eV followed by a peak at ~17.7 eV above the edge energy in Co(II) salen catalyst, whereas a small pre-edge peak was at about 2.1 eV followed by a peak at ~19.2 eV above the edge energy in Co(III) salen catalyst (22). It should be noted that the HKR of epichlorohydrin in the X-ray absorption experiment was conducted in the presence of isopropyl alcohol as a co-solvent. The co-solvent was needed to make the reaction mixture a single-phase and allowed the HKR to be performed without agitation. Under these static conditions, about 50% conversion and >99% ee were obtained in about six and a half hours. The in-situ XANES were collected every 22 min of the reaction with Jacobsen's Co-OAc salen catalyst under the similar static HKR conditions. In every spectrum, a pre-edge peak at about 2.1 eV above the Co metal edge was observed, and the overall XANES was characteristic of the Co(III) salen catalyst. These results were consistent with the UV-Vis results presented above, which indicated negligible reduction of the Co(III) to Co(II) during the HKR of epichlorohydrin.

Effect of dimer formation on deactivation. Another possible mode of deactivation is formation of inactive Co dimers or oligomers. To test for these species, we examined the ESI-mass spectrum of fresh and deactivated Co-salen catalysts in dichloromethane solvent (22). The major peak in the mass spectrum occurred at m/z of 603.5 for both Jacobsen's Co(II) and Co(III)-OAc salen catalysts, whereas much smaller peaks were observed in the m/z range of 1207 to 1251. The major feature at 603.5 corresponds to the parent peak of Jacobsen's Co(II) salen catalyst (formula weight = 603.76) and the minor peaks (1207 to 1251) are attributed to dimers in the solution or formed in the ESI-MS. The ESI-MS spectrum of the deactivated Co-salen catalyst, which was recovered after 12 h HKR reaction with epichlorohydrin, was similar to that of Co(II) and Co(III)-OAc salen. Evidently, only a small amount of dimer species was formed during the HKR reaction. However, the mass spectrum of a fresh Co(III)-OAc salen catalyst diluted in dichloromethane for 24 h showed substantial formation of dimer. The activity and selectivity of HKR of epichlorohydrin with the dimerized catalyst recovered after 24 h exposure to dichloromethane were similar to those observed with a fresh Co-OAc salen catalyst. Therefore, we concluded that catalyst dimerization cannot account for the observed deactivation.

Influence of counterion on catalyst deactivation in the HKR reaction. Recent studies by the Jacobsen group (7) indicate a significant role of the counterion on the observed activity of the Co-salen catalyst, which can be rationalized by the elementary steps involved in the reaction. The well-recognized cooperative bimetallic mechanism of the HKR of epoxides involves one Co-salen complex bound to OH and a second Co-salen complex containing a non-specific counterion such as acetate. Nielsen et al. (7) suggested that a maximum rate of epoxide ring opening is achieved by a mixed Co-salen catalyst system in which 50% of the catalyst has an OH counterion. However, the rate is thought to be low when the HKR is carried out with a pure Co-OH salen catalyst. They also suggested that the rate of counterion addition determines the kinetics of HKR (7). The stronger nucleophilic counterions such as Cl⁻ add to the epoxide so rapidly that almost all of the HKR reaction is carried out with less active Co(OH)-salen. In support of that idea, we detected the mono-acetylated side product (**D**) by the GC-MS analysis of epichlorohydrin HKR reaction products. Therefore, the addition of acetate counterion to epichlorohydrin to give a less active Co-OH salen species may account for the observed deactivation in this system.

$$Cl \diagdown \diagup \underset{\displaystyle \text{OH}}{\diagup} \diagdown \diagup \text{OAc} \qquad \textbf{(D)}$$

This hypothesis was tested by carrying out a kinetic study of the HKR of epichlorohydrin using Jacobsen Co(III)-salen catalyst with four different counterions, namely, acetate (OAc), tosylate (OTs), chloride (Cl) and iodide (I) (22). Approximately, 0.5 mol% loading of all the catalysts was used to perform the HKR of epichlorohydrin. As shown in Table 43.2, the 1ˢᵗ run initial rates with Co-OAc and Co-OTs salen catalysts were similar and slightly below those with Co-Cl and Co-I salen. Nevertheless, all of the catalysts were quite active initially. After conducting the HKR for 12 h reaction time, the catalysts were recycled by evaporating all the reaction products under vacuum and high temperature.

Table 43.2. Effect of counterions on the deactivation of Jacobsen's Co-salen catalyst in the HKR of epichlorohydrin[a]

Catalyst	Initial rate (mol ECH) (mol cat)$^{-1}$ (min)$^{-1}$ 1st run	Initial rate (mol ECH) (mol cat)$^{-1}$ (min)$^{-1}$ 2nd run
Co-OTs salen	2.85	2.45
Co-OAc salen	2.98	0.33
Co-Cl salen	4.18	0.16
Co-I salen	4.20	0.06

[a]Reaction conditions: 10 mmol epichlorohydrin (ECH), 120 µl chlorobenzene, 126 µl water, 0.5 mol% R,R-Jacobsen Co(III) salen catalyst, ~300 K reaction temperature.

The initial rates in the 2nd runs decreased in the following order: Co-OTs > Co-OAc > Co-Cl > Co-I (see Table 43.2), which is consistent with the idea of deactivation by counterion addition to epoxide. As reported by Nielsen et al. (7), the rate of counterion addition to epoxide determines the kinetics of HKR. We suspect that the iodide and chloride counterions undergo such rapid addition to epoxide that most of the HKR reaction in the 2nd run occurred with the much less active Co-OH catalyst formed in situ. In contrast, the weakly nucleophilic tosylate counterion reacted more slowly with the epoxide and was therefore more stable.

Pretreatment tests. Since exposure of the Co-salen catalyst to the HKR reaction mixture deactivated the catalyst, the active R,R-Jacobsen Co-OAc salen catalyst was pretreated with different reagents used in the HKR reaction for a period of 12 h at room temperature followed by a subsequent recovery of the catalyst by removal of the reagents under vacuum. Then, the HKR reaction was performed at room temperature. The regular HKR reaction (no pretreatment) with 10 mmol of racemic epichlorohydrin produced 30% conversion and 42% ee in 20 min of reaction time. A 12 h pretreatment with all the HKR reagents severely deactivated the catalyst, producing 5% conversion and 5% ee in 20 min of reaction time (22). Upon exposing a fresh catalyst to R/S-epichlorohydrin for a period of 12 h, the subsequent HKR reaction exhibited low conversions and ee's in 20 min. We conclude that the epichlorohydrin binds to the Co-OAc salen and extracts the acetate counterion during pretreatment forming monoacetylated side product and less active Co-OH salen species upon hydrolysis leading to catalyst deactivation. The catalyst was then pretreated with the major and minor HKR reaction products R/S-3-chloro 1,2-propane diol, respectively, for a period of 12 h. Almost similar conversions and ee's were obtained after diol pretreatments as that obtained with the fresh Co-OAc salen catalyst without any pretreatment. As reported by Nielsen et al. (7), the product diol does not bind to the catalyst, and hence does not cause deactivation. Lastly, a 12 h pretreatment with a mixture of R-3-chloro 1,2-propane diol and water also did not show significant deactivation. Therefore, water alone did not lead to the formation of Co-OH salen species.

Oligo(cyclooctene) Co-OAc salen catalyst vs. molecular Co-OAc salen catalyst. The oligo(cyclooctene) Co-salen catalyst (**A**) was expected to be highly active and enantioselective in the HKR reaction because of the flexible oligomer backbone and the high local concentration of Co-salen species resulting from the macrocyclic framework (18). Catalyst (**A**) was examined for its catalytic efficiency in the HKR of epichlorohydrin. At about 0.02 mol% catalyst loading (based on Co) of the oligomeric catalyst, similar conversion and ee as those with 0.5 mol% of Jacobsen's Co-OAc salen catalyst were obtained, which corresponds to an activity enhancement of 25 at our conditions. Interestingly, the oligomeric catalyst was more selective than Jacobsen's catalyst, producing >99% ee at 50% conversion in 90 min vs 92% ee at 50% conversion in 120 min for the Jacobsen's Co-OAc salen catalyst. As suggested previously (18, 22), this high activity and selectivity of the oligomeric catalyst can be attributed to an increase in local molarity of the cobalt centers in contrast to the high dilution of the molecular catalyst.

The oligo(cyclooctene) Co-OAc salen catalyst was then recycled without intermediate regeneration with acetic acid. As seen in Table 43.3, the initial rate of the oligomeric catalyst that was recycled after short reaction time (105 min) did not decrease significantly. However, the initial rate decreased substantially when the oligomeric catalyst was recycled after 12 h of the HKR reaction (run 2b).

Table 43.3. Effect of reaction time on catalyst deactivation[a]

Run[b]	Initial rate (mol ECH) (mol cat)$^{-1}$ (min)$^{-1}$
1	71
2a	64
2b	2

[a]Reaction conditions: 10 mmol epichlorohydrin (ECH), 120 μl chlorobenzene, 126 μl water, 0.02 mol% R,R-Oligo(cyclooctene) Co-OAc salen catalyst, ~300 K reaction temperature.
[b]Run 2a was performed with a catalyst recovered after 105 min and run 2b was performed with a catalyst recovered after 12 h.

This result supports the idea that counterion addition to the epoxide also occurred with this material. The longer the exposure time of the catalyst to the HKR reagents (mainly epichlorohydrin), the greater is the extent of deactivation. Nevertheless, this oligomeric catalyst is certainly more effective than the monomeric catalyst as it can be used for shorter times to achieve comparable conversion and ee, thus reducing the influence of the reaction time on catalyst deactivation.

Toward HKR in a continuous reactor. Although multimeric homogeneous soluble Co-salen catalysts exhibit high activity and selectivity in the HKR reaction, they cannot be utilized in a fixed-bed reactor without some form of immobilization. Thus, it is highly desirable to synthesize an insoluble supported Co-salen catalyst that can be used in a continuous reactor under HKR conditions. We have recently prepared an SBA-15 supported Co-salen catalyst (**B**) that has been tested in a fixed-bed reactor carrying out the HKR of epichlorohydrin in the presence of isopropyl alcohol as the co-solvent. Under our conditions of flow and catalyst loading, approximately 22% conversion and 25% ee were obtained after running for about 1 h. Subsequently, the conversion and the ee decreased with time on-stream. The elemental analysis of the effluent solution showed that some of the Co had leached from the silica support during the reaction. An attempt to regenerate the remaining anchored Co-salen catalyst by in situ treatment with acetic acid was unsuccessful. In fact, treatment with acetic acid caused additional Co leaching since the effluent from the reactor was highly colored during this procedure and elemental analysis revealed Co in the solution. Currently, we are working on the synthesis of more stable insoluble supported Co-salen catalysts that could be utilized in the HKR of epichlorohydrin in a fixed bed reactor.

Conclusions

Recycling Jacobsen's Co-salen catalyst without intermediate regeneration revealed catalyst deactivation with each run; however, almost negligible reduction of Co(III) to Co(II) species was observed spectroscopically during the HKR reaction. Results from ESI-MS showed that catalyst dimerization did not account for deactivation. The highest level of deactivation with Co-salen catalysts with various counterions was observed for Co-I salen, presumably because of the fast addition of iodide to epoxide and rapid formation of Co-OH salen species. Results from various catalyst pretreatments showed that long time exposure to R/S-epichlorohydrin caused significant deactivation of the catalyst. An oligo(cyclooctene) supported Co-OAc salen catalyst was about 25 times more active than a molecular Co-OAc salen catalyst and also deactivated after prolonged reaction times. However, the high activity of the multimeric catalyst allowed for multiple runs without significant deactivation as long as the reaction time was short. An insoluble SBA-15 supported Co-OAc salen catalyst has also been tested in a continuous reactor. However, the catalyst released Co into the reactant solution, which is another mode of deactivation of these catalysts.

Acknowledgements

We thank the U.S. Department of Energy, Basic Energy Sciences, for financial support through Catalysis Science Grant/Contract Nos. DE-FG02-03ER15459 and DE-FG02-03ER15460. We also thank Prof. Fernandez and his student Wei Qi in the Department of Chemical Engineering, University of Virginia, for assistance with the ESI-MS experiments. Use of the National Synchrotron Light Source, Brookhaven National Laboratory, was supported by the U.S. Department of Energy, Office of Science, Office of Basic Energy Sciences, under Contract No. DE-AC02-98CH10886.

References

1. S. E. Schaus, B. D. Brandes, J. F. Larrow, M. Tokunaga, K. B. Hansen, A. E. Gould, M. E. Furrow and E. N. Jacobsen, *J. Am. Chem. Soc.*, **124**, 1307 (2002).
2. M. E. Furrow, S. E. Schaus and E. N. Jacobsen, *J. Org. Chem.*, **63**, 6776 (1998).
3. J. F. Larrow, K. E. Hemberger, S. Jasmin, H. Kabir and P. Morel, *Tetrahedron-Asymmetry*, **14**, 3589 (2003).
4. P. S. Savle, M. J. Lamoreaux, J. F. Berry and R. D. Gandour, *Tetrahedron-Asymmetry*, **9**, 1843 (1998).
5. J. M. Ready and E. N. Jacobsen, *J. Am. Chem. Soc.*, **121**, 6086 (1999).
6. E. N. Jacobsen, *Acc. Chem. Res.*, **33**, 421 (2000).
7. L. P. C. Nielsen, C. P. Stevenson, D. G. Blackmond and E. N. Jacobsen, *J. Am. Chem. Soc.*, **126**, 1360 (2004).
8. R. G. Konsler, J. Karl and E. N. Jacobsen, *J. Am. Chem. Soc.*, **120**, 10780 (1998).
9. J. M. Ready and E. N. Jacobsen, *Angew. Chem. Intl. Ed.*, **41**, 1374 (2002).
10. J. M. Ready and E. N. Jacobsen, *J. Am. Chem. Soc.*, **123**, 2687 (2001).
11. D. E. White and E. N. Jacobsen, *Tetrahedron-Asymmetry*, **14**, 3633 (2003).
12. D. A. Annis and E. N. Jacobsen, *J. Am. Chem. Soc.*, **121**, 4147 (1999).
13. S. Peukert and E. N. Jacobsen, *Organic Letters*, **1**, 1245 (1999).
14. R. Breinbauer and E. N. Jacobsen, *Angew. Chem. Intl. Ed.*, **39**, 3604 (2000).
15. Y. M. Song, H. L. Chen, X. Q. Hu, C. M. Bai and Z. Zheng, *Tetrahedron Letters*, **44**, 7081 (2003).
16. Y. Song, X. Y., H. Chen, C. Bai, X. Hu and Z. Zheng, *Tetrahedron Letters,* **43**, 6625 (2002).
17. X. Zheng, C. W. Jones and M. Weck, *Chemistry-a European Journal*, **12**, 576 (2006).
18. X. Zheng, C. W. Jones and M. Weck, *J. Am. Chem. Soc.*, **129**, 1105 (2007).
19. G. J. Kim, H. Lee and S. J. Kim, *Tetrahedron Letters*, **44**, 5005 (2003).
20. G. J. Kim, D. W. Park and Y. S. Tak, *Catalysis Letters*, **65**, 127 (2000).
21. M. Kwon and G. J. Kim, *Catalysis Today*, **87**, 145 (2003).
22. S. Jain, X. Zheng, C. W. Jones, M. Weck and R. J. Davis, *Inorg. Chem.*, **46**, 8887 (2007).

Supporting Information

Original spectra and results from pretreatment tests can be found in the following paper: Jain, Surbhi; Zheng, Xiaolai; Jones, Christopher W.; Weck, Marcus; Davis, Robert J. **Inorganic Chemistry**, 2007, 46(21), 8887-8896.

44. A Comparison of the Reaction Mechanism for the Gas-Phase Methylation of Phenol with Methanol Catalyzed by Acid and by Basic Catalysts

Nicola Ballarini, Fabrizio Cavani, Stefania Guidetti, Luca Maselli, Ambra Montaletti, Sauro Passeri and Sara Rovinetti

Dipartimento di Chimica Industriale e dei Materiali, Università di Bologna, Viale Risorgimento 4, 40136 Bologna, Italy. INSTM, Research Unit of Bologna

fabrizio.cavani@unibo.it

Abstract

This chapter compares the reaction of gas-phase methylation of phenol with methanol in basic and in acid catalysis, with the aim of investigating how the transformations occurring on methanol affect the catalytic performance and the reaction mechanism. It is proposed that with the basic catalyst, Mg/Fe/O, the true alkylating agent is formaldehyde, obtained by dehydrogenation of methanol. Formaldehyde reacts with phenol to yield salicyl alcohol, which rapidly dehydrogenates to salicyladehyde. The latter was isolated in tests made by feeding directly a formalin/phenol aqueous solution. Salicylaldehyde then transforms to *o*-cresol, the main product of the basic-catalyzed methylation of phenol, likely by means of an intramolecular H-transfer with formaldehyde. With an acid catalyst, H-mordenite, the main products were anisole and cresols; moreover, methanol was transformed to alkylaromatics.

Introduction

The methylation of phenol and phenol derivatives has been widely investigated in the literature, and several papers report about the effect of the catalyst characteristics on the nature and distribution of phenolic products [1]. The reaction can be carried out with acid or with basic catalysts; for instance, when the methylation is aimed at the production of *o*-cresol or of 2,6-xylenol, methanol is the alkylating agent and catalysts possessing basic characteristics are used. The latter exhibit the following characteristics [2]: (i) a very high regio-selectivity in *C*-methylation, since the ortho/para-methylation ratio is in all cases largely higher than 2, and (ii) a high chemo-selectivity, since the *O/C*-selectivity ratio, a function of the basic strength of catalysts, is in general, very low.

One major problem of the industrial process of phenol methylation is the low yield with respect to methanol, due to its decomposition; consequently, a large excess of methanol is usually fed in order to reach an acceptable per-pass conversion of phenol. This aspect, however, is often forgotten in scientific literature, and only few papers take into consideration the methanol decomposition [3]. Information is very scarce on (a) the nature of transformations occurring on methanol, (b) how the latter are affected by the catalyst type, and (c) how they affect the mechanism of the reaction. In this communication, we report about an investigation of the reaction mechanism of phenol methylation with methanol, using either an acid catalyst (a H-mordenite) or a catalyst with basic/dehydrogenating properties (Mg/Fe/O).

Experimental

Mg/Fe/O catalyst was prepared by precipitation from an aqueous solution containing the corresponding metal nitrates. The solid was then dried at 110°C overnight, and calcined at 450°C for 8h in air. The acid catalyst was a commercial H-mordenite, having an atomic Si/Al ratio equal to 20, shaped in 1/16" extrudates (binder alumina). This sample was supplied by Süd-Chemie AG. Catalytic tests were carried out by vaporization of a methanol/phenol liquid mixture (methanol/phenol molar ratio 10/1) in a N_2 stream. Overall gas residence time was 2.68 s. Total pressure was atmospheric. Other tests were carried out by feeding an aqueous phenol/formaldehyde solution in a N_2 stream. The aqueous formalin solution was prepared containing the minimal amount of methanol.

Results and Discussion

Figures 44.1 and 44.2 report the performance in the gas-phase phenol methylation of the H-mordenite and of the Mg/Fe/O catalyst, respectively. The differences between the two catalysts concerned both the transformations occurring on methanol and the type of phenolic products obtained. The H-mordenite was very active; at 350°C the conversion of phenol was 80%. A further increase of temperature led to a decrease of conversion. This can be attributed to a progressive deactivation of the catalyst, due to

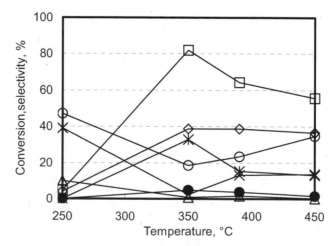

Figure 44.1. Conversion of phenol (□), selectivity to *o*-cresol (○), 2,6-xylenol (◇), *p*-cresol (△), anisole (✗), 2,4-xylenol (●) and polyalkylated phenols (✱) as

functions of temperature. Catalyst H-mordenite. Feed gas composition (molar fractions): methanol 0.108, phenol 0.011, nitrogen 0.881.

the formation of condensed aromatics (coke precursors) by transformation of methanol [4]. Indeed, the spent catalyst was coked. The main products at low temperature were *o*-cresol and anisole, but also 2,6-xylenol and *p*-cresol formed in non-negligible amounts. The increase of temperature led to a lower selectivity to all mono-alkylated compounds, in favor of the formation of xylenols and polyalkylated compounds. However, above 350°C the decrease of phenol conversion caused an increase of the selectivity to *o*-cresol and anisole.

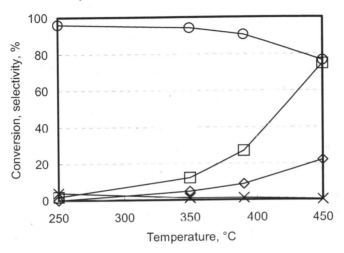

Figure 44.2. Conversion of phenol (□), selectivity to *o*-cresol (○), 2,6-xylenol (◇), *p*-cresol (△) and anisole (✗) as functions of temperature. Catalyst: Mg/Fe/O. Feed gas composition as in Figure 44.1.

With the Mg/Fe/O catalyst, the main product at low temperature was *o*-cresol, but its selectivity decreased in favor of 2,6-xylenol when the reaction temperature was increased. Noteworthy is the very high chemo- and regio-selectivity of the catalyst, typical of systems possessing basic/dehydrogenating properties [2], and the absence of any deactivation phenomena.

Another relevant difference between the acid and the basic catalyst concerned the side reactions occurring on methanol. With the H-mordenite (Figure 44.3), two different classes of compounds formed: (i) alkylbenzenes, ranging from toluene to hexamethylbenzene, and (ii) light decomposition products. Alkylaromatics were the prevailing compounds in the temperature range comprised between 250°C and 350°C; in this range the formation of light decomposition products was nil (only methane formed). At above 390°C, the formation of alkylaromatics decreased considerably and the main product was methane, with also small amounts of CO, CO_2 and H_2. With the Mg/Fe/O catalyst, a great fraction of methanol dehydrogenated to formaldehyde, which then underwent transformation to methylformate and to decomposition products, i.e., CO, CO_2, CH_4 and H_2 (Figure 44.4).

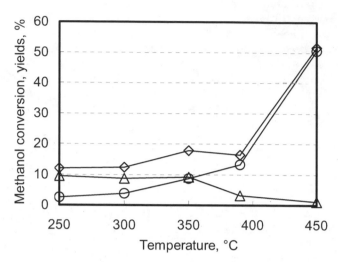

Figure 44.3. Conversion of methanol ($\Diamond$), yields of alkylaromatics ($\triangle$) and yields to light compounds ($CO + CO_2 + CH_4$) ($\bigcirc$) as functions of temperature. Feed composition (molar fractions): methanol 0.12, nitrogen 0.88. Catalyst H-mordenite.

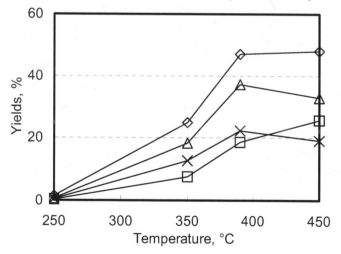

Figure 44.4. Yields to CO ($\square$), CO_2 ($\triangle$), CH_4 ($\times$) and H_2 ($\Diamond$) as functions of the temperature. Feed composition as in Figure 44.3. Catalyst Mg/Fe/O.

The considerable methanol decomposition indicates that with the catalyst having basic/dehydrogenating properties, the true alkylating agent is not methanol, but formaldehyde. This is supported by the experimental evidence that the conversion of phenol began when the conversion of methanol to light compounds also started. Due to the presence of the carbonyl moiety, formaldehyde is more electrophylic than methanol, and hence is able to attack the aromatic ring even in the absence of an acid-catalyzed activation of the molecule. Moreover, the deprotonation of phenol and the generation of the phenolate species further activate the aromatic substrate for methylation.

In order to confirm this hypothesis, tests were made by feeding a formalin solution, with the H-mordenite (Figure 44.5) and the Mg/Fe/O catalyst (Figure 44.6). In the former case, the results obtained with formaldehyde were different from those obtained with methanol (Figure 44.1); the prevailing products in fact were *o*-cresol and 2,6-xylenol, with small amount of anisole, *p*-cresol and of polyalkylated phenols. At low temperature, also 2-hydroxybenzaldehyde (salicylaldehyde) formed, with selectivity of 30%. Polyalkylbenzenes, which were obtained in large amounts from methanol, formed with negligible yield from formaldehyde. Furthermore, no short-term deactivation phenomenon was observed, while the latter was evident when starting from methanol (Figure 44.1). With the basic catalyst the phenolic products were *o*-cresol and 2,6-xylenol, with also very small amount of trimethylphenols, and with no formation of anisole at all. In this case, the distribution of products was similar to that one obtained from methanol (Figure 44.2). The only difference concerned the formation of salicylaldehyde; its selectivity was close to 30% at 250°C, and then decreased down to zero at 350°C, with a corresponding increase in the selectivity to *o*-cresol. Salicylaldehyde did not form when methanol was the reactant for phenol methylation (Figure 44.2).

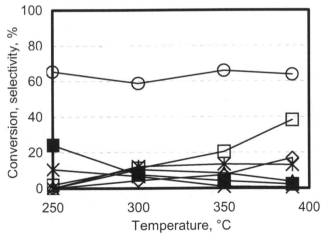

Figure 44.5. Conversion of phenol (□), molar selectivity to anisole (✕), *o*-cresol (○), *p*-cresol (△), 2,6-xylenol (◇), salicylaldehyde (■) and polyalkylated phenols (✱) as functions of temperature. Catalyst: H-mordenite. Feed composition: N_2 89.3%, formaldehyde 1.7%, phenol 0.46%, methanol 0.03% and water 8.5%.

Tests were carried out with the Mg/Fe/O catalyst, with the formalin/phenol feed and with variation of the residence time, in order to obtain information on the reaction intermediates. Very low residence times (e.g., smaller than 0.1 s) and a low temperature (250°C) were used in order to isolate the intermediates, which were supposed to be extremely reactive. Salicylaldehyde was found to be a primary product, since its selectivity extrapolated to nil conversion was higher than zero;

however, its selectivity then rapidly declined when the contact time was increased. The decrease of selectivity to the aldehyde corresponded to an increase of that to *o*-cresol; this supports the hypothesis that the latter compound forms by consecutive transformation of salicylaldehyde, possibly by means of an intramolecular H-transfer with formaldehyde.

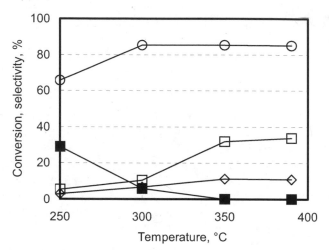

Figure 44.6. Conversion of phenol (□), molar selectivity to *o*-cresol (O), xylenols (◇) and salicylaldehyde (■) as functions of temperature. Catalyst: Mg/Fe/O. Feed composition as in Figure 44.5.

The results obtained indicate that in the reaction between phenol and methanol, formaldehyde is the true methylating agent when basic catalysts are used. This indicates that the type of transformation occurring with methanol is the factor that mainly differentiates performances in phenol methylation when catalyzed by either basic or acid catalysts. The catalyst plays its role in the generation of the methylating species; the nature of the latter then determines the type of phenolic products obtained.

Acknowledgements

INSTM is acknowledged for the grant of Luca Maselli and Sauro Passeri.

References

1. N. Ballarini, F. Cavani, L. Maselli, A. Montaletti and S. Passeri, *J. Catal*. **251**, 423 (2007).
2. R.F. Parton, J.M. Jacobs, H. van Ooteghem and P.A. Jacobs, *Stud. Surf. Sci. Catal*. **46**, 211 (1989).
3. S. Sato, K. Koizumi and F. Nozaki, *J. Catal*. **178**, 264 (1998).
4. I.M. Dahl and S. Kolboe, *J. Catal*. **149**, 458 (1994).

45. In Situ ATR Study of Photocatalytic Dehydrogenation of Alcohols on Au and Pd Catalysts

Duane D. Miller and Steven S.C. Chuang

Chemical and Biomolecular Engineering, The University of Akron, Akron, OH 44325-3906

schuang@uakron.edu

Abstract

Photocatalytic dehydrogenation of 2-propanol to acetone over Au/TiO_2 exhibited a higher activity for the generation of photogenerated electron than the photocatalytic inactive Pd/TiO_2 and Pd/Al_2O_3 at 298 K.

Introduction

Catalytic dehydrogenation of alcohol is an important process for the production of aldehyde and ketone (1). The majority of these dehydrogenation processes occur at the liquid-metal interface. The liquid phase catalytic reaction presents a challenge for identifying reaction intermediates and reaction pathways due to the strong overlapping infrared absorption of the solvent molecules. The objective of this study is to explore the feasibility of photocatalytic alcohol dehydrogenation.

In this paper we examined isopropyl alcohol dehydrogenation on Au/TiO_2, Pd/Al_2O_3, and Pd/TiO_2 under UV-irradiation using attenuated total reflection infrared spectroscopy (ATR-IR). Dehydrogenation was carried out in the presence and absence of UV-irradiation as well as in the presence of hydrogen. Hydrogen was used to limit further conversion of aldehyde to acid. The ATR-IR technique allows monitoring changes in the concentration of adsorbed species in the liquid environment of a UV-initiated reaction (2, 3). Au/TiO_2 was used for its unique activity and selectivity of partial oxidation which may exhibit unique activity for oxidative dehydrogenation. Pd/Al_2O_3 and Pd/TiO_2 were used for their activities in alcohol dehydrogenation at 323 K (4, 5).

Experimental Section

1 wt% Pd/TiO_2 and 1 wt% Pd/Al_2O_3 catalysts were prepared by depositing Pd onto Degussa P25 TiO_2 and Al_2O_3 (STREM) using the impregnation method. An aqueous solution of $Pd(NO_3)_2$ (Aldrich) with 2 g of TiO_2 or Al_2O_3 was stirred at room temperature for 30 min, heated to 373 K at a rate of 10 K/min, and held until complete evaporation of the solvent. The resulting catalyst was further dried in air at 353 K for 24 h and then calcined in air at 673 K for 30 min.

1 wt% Au/TiO$_2$ was prepared by depositing Au onto TiO$_2$ by the homogeneous deposition-precipitation method. A 200 ml of aqueous solution for the deposition-precipitation was prepared from 0.15 g HAuCl$_4$ (Aldrich), 1 g TiO$_2$, and 2.65 g urea and then vigorously stirred at 353 K for 8 h. The resulting solid catalyst was separated by centrifugation and then washed with 353 K deionized water. This procedure was repeated five times to remove residual Cl ions as well as any non-interacting Au particles. The Au/TiO$_2$ catalyst was further dried in air at 373 K for 24 h.

The experimental apparatus consists of a gas flow system with a four-port valve, a multi-reflection Attenuated Total Reflection (ATR) accessory (Pike Technologies), and a custom reactor manifold mounted to the ATR top plate, shown in Fig. 45.1. The ATR reactor manifold consists of (i) a CaF$_2$ window for UV irradiation, (ii) an inlet and outlet port, and (iii) an injection port for the liquid phase reactant.

Thin-film was prepared from a slurry of catalyst powder which was prepared from 10 mg catalyst in 5 ml of 2-propanol. The catalyst slurry was sonicated for 30 min. and allowed to sit stagnant overnight. Before preparing the films, the slurry was sonicated for 15 min., 20 drops (0.1 ml) were added onto a ZnSe trough plate internal reflection element (022-2010-45, Pike Technologies). The solvent was allowed to evaporate, the procedure was repeated a total of five times. After drying in air at room temperature, the catalyst thin-film was ready for 2-propanol dehydrogenation studies.

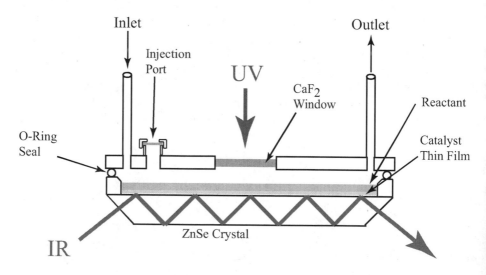

Figure 45.1   ATR reactor system.

Results and Discussion

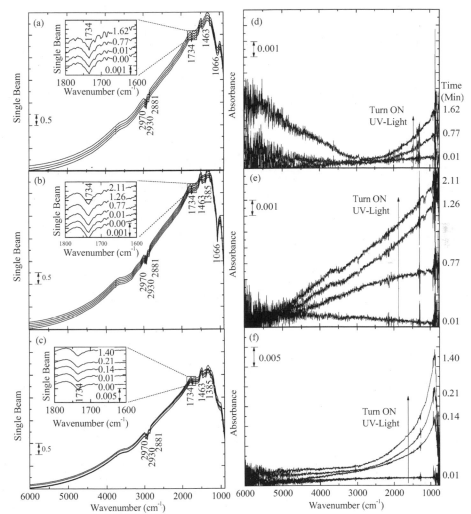

Figure 45.2 ATR single beam spectrum of a catalyst thin-film on ZnSe crystal under argon upon UV-irradiation (a) Pd/TiO_2 catalyst, (b) Pd/Al_2O_3 catalyst, and (c) Au/TiO_2 catalyst; difference spectra reveals the background shift on (d) Pd/TiO_2 catalyst, (e) Pd/Al_2O_3 catalyst, and (f) Au/TiO_2 catalyst.

Irradiating a radiating catalyst thin-film with UV-radiation causes the excitation of trapped electrons producing adsorption in the mid-IR region (6, 7). Irradiating a catalyst thin-film with UV-radiation causes the excitation of trapped electrons producing adsorption in the mid-IR region (6, 7). Figures 45.2(a), (b), and (c) show the development of changes in the single beam IR spectra for Pd/TiO_2, Pd/Al_2O_3, and Au/TiO_2 catalysts, respectively, during UV-irradiation. Residual 2-propanol from the preparation of the catalyst thin film is converted to acetone

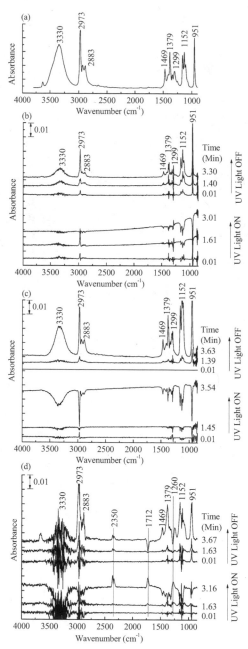

exhibiting the C=O band at 1734 cm⁻¹, shown in the single beam spectra in Fig. 45.2(a), (b), and (c). To highlight the change in IR absorption of the catalyst thin film due to UV-irradiation, the difference spectra, obtained by subtracting the single beam spectrum at 0.00 min before UV-irradiation from those after UV-irradiation at times after 0.01 min, are shown in Fig. 45.2 (d), (e), and (f). The shape of the broad absorption varies with each catalyst suggests strong dependence of IR adsorption behavior on the catalyst composition. Fig 45.2(d) shows there is an inflection point around 2700 cm⁻¹ on Pd/TiO₂ catalyst, indicating IR adsorption at lower wavenumbers results from Pd metal and the higher wavenumbers results from the TiO₂ support. Fig. 45.2(e) presents the absence of the IR adsorption at high wavenumber, 3500-6000 cm⁻¹ on the Pd/Al₂O₃ catalyst, indicating the lack of the reduced Pd metal on the Al₂O₃ support as compared to the interaction with the semiconductor TiO₂.

Fig. 45.2(f) shows the background shift occurs around 4000 cm⁻¹ on the Au/TiO₂ catalyst. The intensity of the Au/TiO₂ adsorption is four times greater than for the Pd/TiO₂ and Pd/Al₂O₃ catalysts. The results suggest that Au/TiO₂ could be a more effective photo-catalyst than Pd catalysts due to its ability to produce more photogenerated electrons.

Figure 45.3 ATR difference spectrum of catalyst thin-film on ZnSe crystal in contact with 2-propanol; (a) 2-propanol, (b) Pd/TiO₂ catalyst, (c) Pd/Al₂O₃ catalyst, and (d) Au/TiO₂ catalyst. The difference spectra were obtained by subtracting the single beam spectra at time 0.00 min.

Injecting 0.5 ml 2-propanol onto the catalyst thin-film produced the characteristic bands (8) at (i) υ_{OH} band at $3330cm^{-1}$, (ii) υ_{C-H} stretching bands at 2973 and 2883 cm^{-1}, (iii) δ_{a-CH3} bending mode at 1469 cm^{-1}, (iv) δ_{s-CH3} bending mode at 1379 cm^{-1}, (v) δ_{OH} band at 1299 cm^{-1}, (vi) υ_{C-C} band at 1152 cm^{-1}, and (vii) r_{CH3} at 951 cm^{-1}, shown in Fig. 45.3(a).

Figures 45.3(b), (c), and (d) show the difference spectra for 2-propanol dehydrogenation; UV-irradiation caused the disappearance of the characteristic bands for 2-propanol on all of the catalysts. The disappearance of the 2-propanol bands on the Au/TiO_2 catalyst correspond to the emergence of C=O band at 1712 cm^{-1} confirms the dehydrogenation of 2-propanol to acetone. The Pd/TiO_2 and Pd/Al_2O_3 catalysts in Fig. 45.3 (b) and (c) were not active for dehydrogenation by the missing C=O band. Turning off UV-irradiation causes the reappearance of the characteristic bands for 2-propanol on both Pd catalysts indicating Pd catalysts were not active for the reaction. In contrast, Au/TiO_2 catalyst, acetone is hydrogenated to 2-propanol upon removal of UV-irradiation.

Conclusions

The Au/TiO_2 catalyst shows activity for photocatalytic dehydrogenation of 2-propanol at 298 K. The activity of Au/TiO_2 is attributed to its unique capability for producing photogenerated electrons evidenced by the featureless IR adsorption during UV-irradiation.

Acknowledgements

This work was partially supported by the Ohio Board of Regents (Grant R4552-OBR).

References

1. Weissermel, K., *Industrial Organic Chemistry*. 3rd ed. 1997, Weinheim, Germany: VCH.
2. Buergi, T., *J. Catal.,* **229** (1) 55-63., (2005).
3. Buergi, T., R. Wirz and A. Baiker, *J. Phys. Chem. B*, **107** (28) 6774-6781, (2003).
4. Grunwaldt, J.-D., M. Caravati and A. Baiker, *J. Phys. Chem. B*, **110** (51) 25586-25589, (2006).
5. Keresszegi, C., et al., *J. Catal.*, **234**(1) 64-75, (2005).
6. Z. Yu and S.S.C. Chuang, *J. Catal.*, **246**(1) 118-126, (2007).
7. S.H. Szczepankiewicz, J.A. Moss and M.R. Hoffmann, *J. Phys. Chem. B*, **106** (11) 2922-2927, (2002).
8. Rossi, P.F. et al., *Langmuir*, **3** (1) 52-8, (1987).

46. Novel Hydride Transfer Catalysis for Carbohydrate Conversions

Johnathan E. Holladay, Heather M. Brown, Aaron M. Appel and Z. Conrad Zhang

Pacific Northwest National Laboratory, Institute for Interfacial Catalysis, P.O. Box 999, Richland, WA 99352

Conrad.Zhang@pnl.gov

Abstract

5-(Hydroxymethyl)furfural, (HMF), an important versatile sugar derivative has been synthesized from glucose using catalytic amounts of $CrCl_2$ in 1-ethyl-3-methylimidizolium chloride. Glycerol and glyceraldehyde were tested as sugar model compounds. Glycerol is unreactive and does not interfere with glucose conversion. Glyceraldehyde is reactive and does interfere with glucose conversion in competitive experiments. $MnCl_2$ or $FeCl_2$ catalyze dehydration of glyceraldehyde dimer to form compound I, a cyclic hemiacetal with an exocyclic double bond. Upon aqueous work-up I forms pyruvaldehyde. $CrCl_2$ or VCl_3 further catalyze a hydride transfer of I to form lactide. Upon aqueous work-up lactide is converted to lactic acid.

Introduction

HMF is an important versatile sugar derivative and is a key intermediate between bio-based carbohydrate chemistry and petroleum based industrial organic chemistry (1, 2). The most common feedstock for HMF is fructose and reactions are carried out in water-based solvent systems using acid catalysis (3,4). HMF is unstable in water at low pH and breaks down to form levulinic acid and formic acid, resulting in an expensive HMF recovery process. In strongly polar organic co-solvents, such as dimethylsulfoxide (DMSO), levulinic acid formation is reduced and HMF yields are improved (5).

Recently several publications have examined replacing aqueous solvents with ionic liquids. Since simple and complex sugars are soluble in many imidazolium halides, water is not required as a co-solvent and degradation of HMF is minimal. Lansalot-Matras et al. reported on the dehydration of fructose in imidazolium ionic liquids using acid catalyst (6). Moreau et al. reported that 1-H-3-methylimidazolium chloride has sufficient acidity to operate without added acid (7). And we reported that a 0.5 wt% loading (6 mole% compared to substrate) of many metal halides in 1-ethyl-3-methylimidazolium chloride ([EMIM]Cl) result in catalytically active materials particularly useful for dehydration reactions (8).

When starting with fructose, we obtained high HMF yields using a number of metal halides (batch reaction, 3 h, 80°C). Metals halides successfully tested included: $CrCl_2$, $CrCl_3$, $FeCl_2$, $FeCl_3$, $CuCl$, $CuCl_2$, VCl_3, $MoCl_3$, $PdCl_2$, $PtCl_2$, $PtCl_4$, $RuCl_3$, and $RhCl_3$. The product mixtures are very clean; yields of levulinic acid and angelicalactones are less than 0.08%. Interestingly, $AlCl_3$ is effective at 0.5 wt%, but the widely studied $AlCl_3$-[EMIM]Cl systems having mole ratios between 0.5 and 2 (9) do not result in high HMF yield (8).

Very different results are obtained when starting with glucose. For this feedstock, only chromium chloride is effective. Other metal halides are either inactive (for example, $MnCl_2$ or $FeCl_2$) or lead to uncharacterized "heavies" (for example, $CuCl_2$ or VCl_3). HMF is stable under reaction conditions which suggest that "heavies" are not HMF degradation products but rather arise through separate pathways from glucose (8). Although debated, the mechanism for HMF formation likely requires that glucose be converted to fructose (Figure 46.1). Whether the

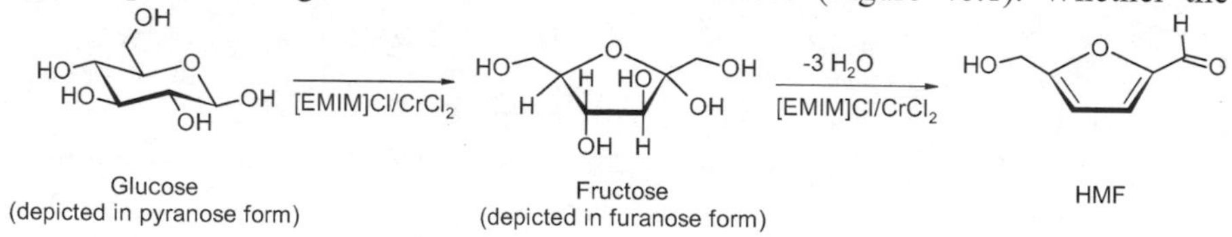

Glucose
(depicted in pyranose form)

Fructose
(depicted in furanose form)

HMF

Figure 46.1. Conversion of glucose to HMF via fructose.

mechanism goes through fructose or a common intermediate between fructose and glucose (10-13), the conversion involves a formal hydride transfer facilitated by $CrCl_2$. In this paper, we investigate the catalytic mechanism involving simple model molecules, using glycerol to simulate the polyhydroxy portion of glucose and using glyceraldehyde dimer to simulate the semiacetal port.

Experimental Section

Glucose (99%) was supplied by Acros. Glycerol (99.9%) was supplied by Fischer and glyceraldehyde dimer (99%), $CrCl_2$, $CuCl_2$, $FeCl_2$, $MnCl_2$ and VCl_3 were supplied by Sigma-Aldrich. 1-Ethyl-3-methylimidazolium chloride (99%) was supplied by Solvent-Innovation.

In a standard experiment [EMIM]Cl (500 mg) was loaded into 24 vials (15.5 mm O.D. x 50 mm). $CrCl_2$, $CuCl_2$, $FeCl_2$, $MnCl_2$, or VCl_3 (approximately 6 mol% with respect to reactant) were added individually to four vials per metal halide (accounting for 20 vials), the remaining four vials did not contain metal chloride and were used as controls. The vials were sealed and installed into a Symyx high throughput batch reactor that had the capability for heating and orbital shaking. The reactor was heated to 150°C and shaken at 700 rpm for 20 minutes. After the reactor was cooled to room temperature, 50 mg of reactant(s) was added to each vial. The vials were sealed and installed back into the high throughput reactor. The reactor was heated to 100°C and shaken at 700 rpm. After 3 h reaction, the reactor was cooled to room temperature and 2.0 ml of water was added to each vial. The vials were sealed and re-installed into the high throughput reactor. The reactor was shaken at 700 rpm for 10 minutes at room temperature. A single liquid layer was formed and the liquid products were analyzed by HPLC. Analyses were obtained using a Bio-Rad Aminex HPX-87H column with refractive index detection. ^{1}H and ^{13}C liquid NMR spectra were obtained on samples stripped of water and re-diluted in CD$_3$CN using a Varian Infinity CMX 500-MHz NMR spectrometer. Spectra were accumulated using a 5 µs pulse width at 300 K. Chemical shifts were measured on the δ scale (ppm) relative to tetramethylsilane.

Results and Discussion

Our hypothesis is that sugar-metal coordination is responsible for the catalysis. It has been well established that imidazolium chloride readily forms adduct with many metal chlorides, MeCl$_x$, resulting in new ionic liquids with a [MeCl$_{x+1}$]$^-$ anion and with reduced melting points (14). However, it remains unclear how the new anion adduct, as the catalyst, interacts with a molecule like glucose. As a first step we chose to examine glycerol and glyceraldehyde as compounds that model the polyhydroxy and the hemiacetal portions of glucose in an effort to improve our understanding of sugar-metal coordination and the resultant impact on chemistry (Figure 46.2).

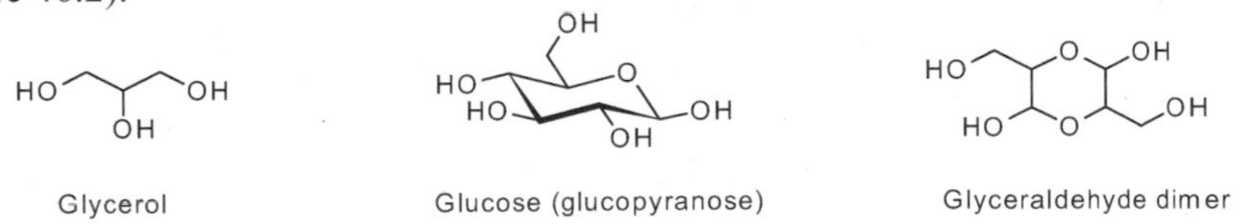

| Glycerol | Glucose (glucopyranose) | Glyceraldehyde dimer |

Figure 46.2. Glycerol, glucopyranose and glyceraldehyde dimmer.

Employing high-throughput tools a series of experiments were run using glycerol or glyceraldehyde as model compounds for glucose. Each test was run in duplicate and showed good agreement. The averaged results are presented in this paper. In Experiment A model compounds were tested in the presence of glucose in 1:1 or 1:2 molar ratios (model compound to sugar). In these tests the model compounds compete with sugar as ligands for the metal halide catalysts. Glucose

Table 46.1. Experiment A, Glucose Results (see Figure 46.3).

Catalyst	Model Compound	Model Compound to Glucose Ratio (mole ratio)	Glucose Conversion (mol%)	HMF Yield (mol%)	Other (~mol%)
None	None	-	6.1	0.0	0.0
CrCl2	None		93.9	69.0	8.6
CrCl2	Glycerol	1:1	93.3	71.1	6.5
CrCl2	Glycerol	1:2	94.7	70.9	8.2
CrCl2	Glyceraldehyde	1:1	60.9	17.5	10.7
CrCl2	Glyceraldehyde	1:2	61.7	22.4	11.4
VCl3	Glyceraldehyde	1:2	71.1	10.2	10.2
MnCl2	Glyceraldehyde	1:2	16.1	0.0	8.7
CuCl2	Glyceraldehyde	1:2	65.5	11.0	7.9
FeCl2	Glyceraldehyde	1:2	18.1	0.2	5.7

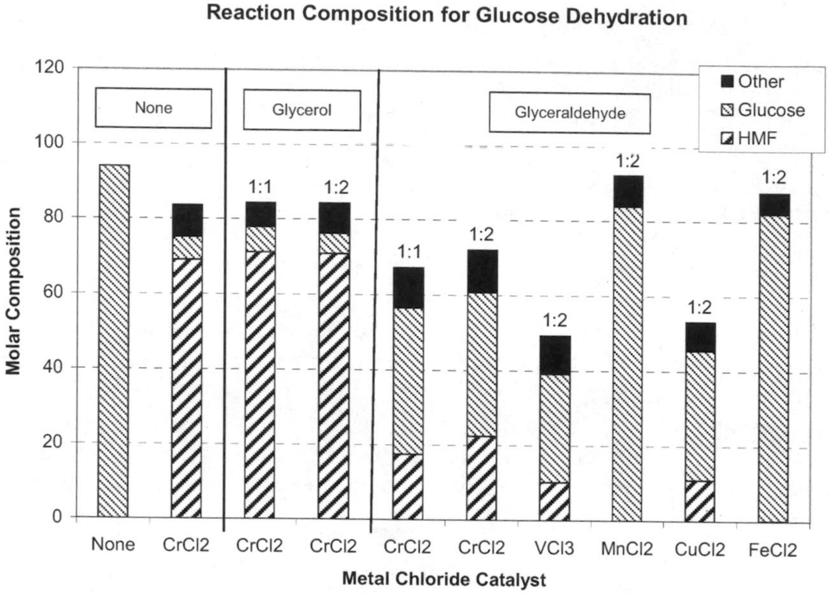

Figure 46.3. Experiment A, molar composition of products of glucose conversion after 3h, 100°C reaction (includes control - with no model compounds, and competitive reactions with 1:1 and 1:2 model compound to glucose molar ratios).

reactivity is summarized in Table 46.1 and Figure 46.3. Glyceraldehyde reactivity is summarized in Table 46.2.

Glycerol had negligible impact on the catalysis. Glucose conversion and HMF yield were essentially the same as the control experiment which did not contain glycerol. Glycerol was stable under the reaction conditions and was recovered in high yields. There is no evidence that glycerol is able to compete with glucose as a ligand for metal binding and we conclude that the only effect glycerol has is that of a diluent.

The addition of glyceraldehyde dimer had a significant impact on the catalysis of glucose. $CrCl_2$ is the preferred catalyst and resulted in a 94% conversion of glucose with a 70% yield of HMF. In the presence of glyceraldehyde the conversion decreased to about 60% and the yield of HMF fell to 20%. This means that HMF selectivity also decreased (selectivity = yield/conversion). The loss of selectivity was primarily due to formation of heavies via intermolecular condensation reactions. The heavies were not characterized.

The other four metal halides, VCl_3, $MnCl_2$, $CuCl_2$, and $FeCl_2$, were not effective converting glucose to HMF. Glucose conversion was low (16-18%) when $MnCl_2$ and $FeCl_2$ were used and the recovery of the starting material was greater than 80% indicating the stability level of glucose in the presence of glyceraldehyde under reaction conditions. Glucose conversion was high (65-71%) when VCl_3 and $CuCl_2$

were used as catalyst but the HMF yield was low (10-11%). Similar results were obtained in reactions that did not contain glyceraldehyde. We examined the stability of HMF under reaction conditions (no glyceraldehyde) and obtained 86% recovery with VCl₃ and 85% with CuCl₂. This suggests that a substantial amount of the glucose conversion occurs through a second mechanism which subsequently leads to heavies rather than HMF formation followed by degradation.

Glyceraldehyde dimer was consumed during the reaction which complicates the mechanistic picture. Glyceraldehyde conversion and product yields are shown in Table 46.2. Major products include pyruvaldehyde and—surprisingly—lactic acid (See Figure 46.4 for chemical structures). Lactic acid is the primary product from $CrCl_2$ and VCl_3 and pyruvaldehyde is the primary product from $MnCl_2$ and $FeCl_2$. Both products are made with $CuCl_2$. Molar recoveries ranged from 41% to 76% suggesting that intermolecular condensation reactions with glucose also occur.

Table 46.2. Molar Yield of Glyceraldehyde Products from Experiment A

Catalyst	Lactic Acid Yield (mol%)	Pyruvaldehyde Yield (mol%)
VCl3	41.0	-
MnCl2	-	44.0
CuCl2	27.0	18.0
CrCl2	73.0	1.5
FeCl2	-	76.0

We ran a second set of experiments to examine the reactivity of glyceraldehyde without interference from glucose. This set of studies is referred to as Experiment B. The results of the experiment are shown in Table 46.3 and Figure 46.5. The results match what was observed in Experiment A. The molar balance of around 80% indicates about 20% formation of condensation heavies not observed by HPLC.

| Glyceraldehyde | Lactic acid | Pyruvaldehyde | 1,3-Dihydroxyacetone |

Figure 46.4. Glyceraldehyde and its products.

Table 46.3. Experiment B Results (Figure 46.5).

Catalyst	Glyceraldehyde Conversion	Lactic Acid Yield	Pyruvaldehyde Yield	1,3-Dihydroxyacetone Yield	Other
VCl3	100.0	68.5	4.9	0.0	5.8
MnCl2	87.5	0.0	44.0	4.1	13.8
CuCl2	100.0	43.1	37.5	0.0	7.4
CrCl2	100.0	76.5	0.0	0.0	6.6
FeCl2	97.9	0.0	49.9	0.0	20.3

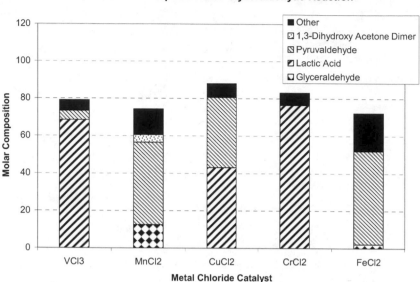

Figure 46.5. Experiment B Results, Molar composition after 3 h at 100°C (glucose not present during run).

Glyceraldehyde gave a clean NMR spectrum in [EMIM]Cl showing three resonances consistent with the dimer species, with the furthest downfield signal at 92.9 ppm for the hemiacetal carbon. The product solution from the CrCl₂ sample show three resonances at δ = 177.0, 66.7 and 20.6 ppm consistent with lactic acid. The product solution from the runs where pyruvaldehyde was identified by HPLC also show three new resonances, δ = 207.9, 91.4 and 24.0 ppm. The downfield resonance at 91 ppm indicates that pyruvaldehyde has strong hydrate character (Figure 46.6).

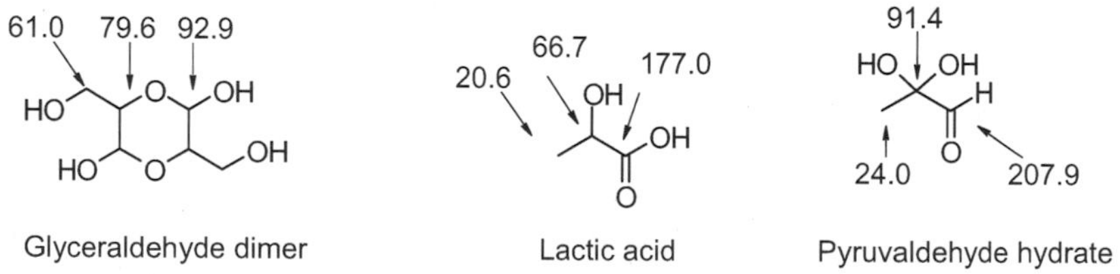

61.0   79.6   92.9   66.7   177.0   91.4

Glyceraldehyde dimer   Lactic acid   Pyruvaldehyde hydrate

Figure 46.6. Major products identified by ^{13}C NMR with proposed chemical shift assignments.

Based on the HPLC and NMR results we can propose the mechanism shown in Figure 46.7. Initial dehydration of glyceraldehyde dimer gives compound **I** with an exocyclic double bond. Although exocyclic double bonds are thermodynamically higher in energy than endocyclic double bonds, MnCl₂ and FeCl₂ are not able to catalyze hydride transfer and no further reaction occurs until aqueous work-up when the hemiacetal is disrupted to give **II** and rapid tautomerization results in pyruvaldehyde. When $CrCl_2$ or VCl_3 are used as the catalyst, the double bond is isomerized to form compound **III** which rapidly tautomerizes to lactide and upon aqueous work-up gives lactic acid.

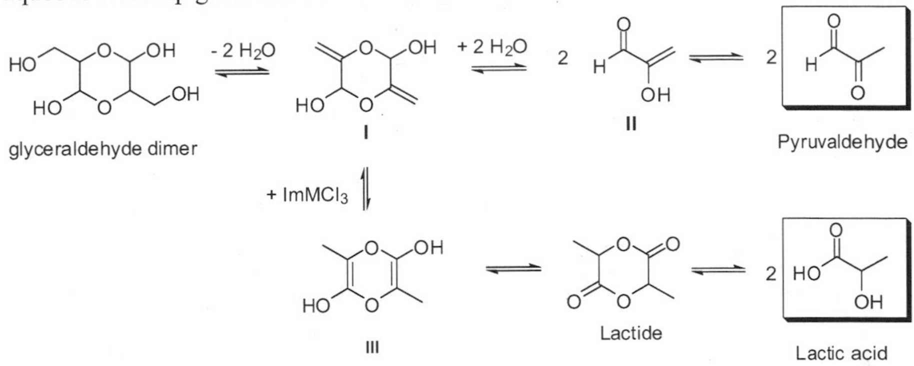

Figure 46.7. Reactions of glyceraldehyde dimer in ionic liquids with metal halides.

At this time we do not have a firm understanding of how $CrCl_2$ and VCl_3 catalyze the double bond isomerization and why other metal chlorides are less effective. We propose that $CrCl_3^-$ or VCl_4^- anion plays a role in hydride transfer, facilitating double bond isomerization. $CuCl_3^-$ is less effective and both lactic acid and pyruvaldehyde are formed. $FeCl_3^-$ and $MnCl_3^-$ anions are ineffective in the transformation and only pyruvaldehyde is formed. The fact that only a small amount of 1,3-dihydroxyacetone is formed is consistent with the NMR observation that the compounds exist as hemiacetal dimers in ionic liquids and not as monomers. Otherwise 1,3-dihydroxyacetone would be expected as a major product (16).

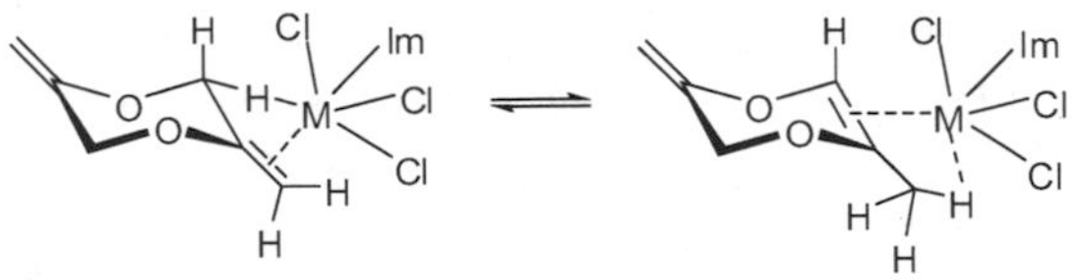

Conclusions

Metal halides in imidazolium ionic liquids offer unique environments able to facilitate dehydration reactions. Under such conditions certain metal halides are able to catalyze formal hydride transfer reactions that otherwise do not occur in the ionic liquid media. We have now discovered two systems in which this transformation has been observed. The initial system involves the conversion of glucose to fructose followed by dehydration; the second system involves the dehydration of glycedraldehyde dimer followed by isomerization to lactide. $CrCl_3^-$ anion is the only catalyst that has been effective for both systems. VCl_3^- is effective for the glyceraldehyde dimer system but not for glucose.

The lack of glycerol conversion and no suppression to glucose conversion in the presence of glycerol suggests that the interaction of the polyol groups with the catalyst is weak. Glyceraldehyde, a smaller molecule than glucose with a hemiacetal group, favorably competes with glucose and is catalyzed by the metal chloride catalysts, resulting in a decreased glucose conversion.

Acknowledgements

This work was supported by the Laboratory Directed Research and Development Program at the Pacific Northwest National Laboratory (PNNL), a multiprogram national laboratory operated by Battelle for the U.S. Department of Energy under Contract DE-AC06-76RL01830. Part of the research described in this paper was performed at the Environmental Molecular Science Laboratory, a national scientific user facility located at PNNL.

References

1. T. A. Werpy, G. Peterson, A. Aden, J. Bozell, J. E. Holladay, J. F. White and A. Manheim, "Top Value Added Chemicals from Biomass" United States Department of Energy report number DOE/GO-102004-1992.
2. B. Kamm, M. Kamm, M. Schmidt, T. Hirth and M. Schulze, in "Biorefineries-Industrial Processes and Products", (Eds. B. Kamm, P. R. Gruber, and M. Kamm), WILEY-VCH Verlag GmbH & Co. KGaA, Weinheim, 2006, Vol. 2.
3. F. S. Asghari and H. Yoshida, *Ind. Eng. Chem. Res.* **45**, 2163 (2006).
4. B. F. M, Kuster, *STARCH-STARKE* **42**, 314 (1990).
5. Y. R. Leshkov, J. N. Chheda and J. A. Dumesic, *Science* **312**, 1933 (2006).
6. C. Lansalot-Matras and C. Moreau, *Catal. Commun.* **4**, 517 (2003).
7. C. Moreau, A. Finiels and L. Vanoye, *J. Mol. Catal. A: Chem.* **253**, 165 (2006).
8. H. Zhao, J. E. Holladay, H. M. Brown and Z. C. Zhang, *Science* **316**, 1597 (2007).
9. K. M. Dieter, C. J. Dymek, Jr., N. E. Heimer, J. W. Rovang and J. S. Wilkes, *J. Am. Chem. Soc.* **110**, 2722 (1988).
10. M. J. Antal, W. S. L. Mok and G. N. Richards, *Carbohydr. Res.* **199**, 91 (1990).
11. Z. Srokol, A. G. Bouche, A. V. Estrik, R. C. J. Strik, T. Maschmeyer and J. A. Peters, *Carbohydr. Res.* **339**, 1717 (2004).
12. B. M. Kabyemela, T. Adschiri, R. M. Malaluan and K. Arai, *Ind. Eng. Chem. Res.* **38**, 2888 (1999).
13. B. M. Kabyemela, T. Adschiri, R. M. Malaluan and K. Arai, *Ind. Eng. Chem. Res.* **36**, 1552 (1997).
14. Z. C. Zhang, *Advance in Catalysis*, **49**, 153 (2006).
15. G. L. Lookhart and M. S. Feather, *Carbohydr. Res.* **60**, 259 (1978).
16. B. M. Kabyemela, T. Adschiri, R. Malaluan and K. Arai, *Ind. Eng. Chem. Res.* **36**, 2025, (1997).

47. A Reasonable Parallel Screening System for Catalytic Multiphase Processes

T. Salmi, D. Murzin, Kari Eränen, P. Mäki-Arvela, J.Wärnå, N. Kumar, J. Villegas and K. Arve

Åbo Akademi, Process Chemistry Centre, Laboratory of Industrial Chemistry, FI-20500 Åbo/Turku, Finland

Kari.Eranen@abo.fi

Abstract

A reasonable throughput screening equipment consisting of six parallel reactor tubes was constructed. The system operates continuously and can be used for screening of various catalysts, different particle sizes and temperatures. Gas, gas-solid and gas-solid-liquid applications are possible. The screening equipment is coupled to gas chromatographic-mass spectrometric analysis. The construction principles, the equipment as well as the application of the equipment is demonstrated with three-phase catalytic systems.

Introduction

High throughput screening is one of the hot topics in heterogeneous catalysis. Advanced experimental techniques have been developed to screen and develop solid catalysts for gas-phase systems. However, for catalytic three-phase systems, rapid screening has got much less attention [1-6]. Three-phase catalysis is applied in numerous industrial processes, from synthesis of fine chemicals to refining of crude oil.

Batchwise operating three-phase reactors are frequently used in the production of fine and specialty chemicals, such as ingredients in drugs, perfumes and alimentary products. Large-scale chemical industry, on the other hand, is often used with continuous reactors. As we developed a parallel screening system for catalytic three-phase processes, the first decision concerned the operation mode: batchwise or continuous. We decided for a continuous reactor system. Batchwise operated parallel slurry reactors are commercially available, but it is in many cases difficult to reveal catalyst deactivation from batch experiments. In addition, investigation of the effect of catalyst particle size on the overall activity and product distribution is easier in a continuous device.

Equipment

A parallel screening system consisting of six equivalent tube reactors was constructed (10 mm diameter, 50 mm length). The reactor system operates in liquid downflow mode; the liquid phase is fed into the reactor by an HPLC pump through capillary tubes to ensure equal flow rates in the tubes. Gases are introduced through four mass flow controllers to enable a wide range of flow rates. Each reactor is surrounded by a heating block (aluminium), so that six different reaction temperatures can be screened in parallel. High pressures (up to 90 bar) and temperatures (873 K) can be used in the system. The device is not limited to catalytic three-phase processes, but liquid-phase and gas-phase systems can be investigated as well. Liquid-phase samples can be vaporized in a heated vaporizer-mixer system. On-line gas samples are taken through an eight-port selection valve and six-port injection valve through a gas chromatograph mass spectrometer system. Liquid samples are withdrawn off-line from each reactor outlet. The whole system is placed in an oven to prevent condensation and crystallization.

The reactor system works nicely and two model systems were studied in detail; catalytic hydrogenation of citral to citronellal and citronellol on Ni (application in perfumery industry) and ring opening of decalin on supported Ir and Pt catalysts (application in oil refining to get better diesel oil). Both systems represent very complex parallel-consecutive reaction schemes. Various temperatures, catalyst particle sizes and flow rates were thoroughly screened.

Figure 47.1. The parallel screening equipment.

The inner diameters of the reactor tubes were 10 mm and the catalysts were placed into the reactors by using glass wool and metal nets as supports for the catalyst beds. The liquid was pumped into the system by using an HPLC pump. An equal liquid-phase distribution between the reactors was obtained by using calibrated capillary tubes located upstream of each reactor. For introduction of gases, the system is equipped with four mass flow controllers with different flow ranges to be able to utilise a broad range of flow rates and mixtures. The main gas flow was divided into six equal flows by using the same type of calibrated capillaries as for the distribution of the liquid. In addition, a backpressure controller, located downstream of the reactors, was used to maintain system pressure. The reactors were surrounded by separate aluminium blocks, which were heated by cartridge heaters. The temperature of each reactor was controlled separately by using temperature controllers and the temperature was recorded inside the catalyst bed by using internal thermocouples.

Online gas samples can be analysed by a GC/MS equipped with an eight-port selection valve and a six-port injection valve. To maintain the system pressure during gas sampling, the sampling line was equipped with a separate backpressure controller. Liquid samples can be withdrawn from the outlet of each reactor by high pressure valves for off-line analysis by GC or GC/MS. During three phase reactions, the liquid passing through each reactor is collected in separate gas-liquid separators cooled by a cooling thermostat. For gas-solid reactions, it is possible to vaporise a liquid reactant by a heated vaporiser/mixer. The vaporiser/mixer, capillaries, reactors and selection valves are all housed in an oven to prevent condensation.

Demonstration case

To evaluate the performance of the reactor system, the catalytic hydrogenation of citral to citronellal and citronellol in ethanol was used as a sample reaction. The reaction scheme is displayed below.

| Citral | Citronellal | Citronellol | 3,7-Dimethyloctanol |

The reaction was performed on a nickel/silica catalyst (0.5 wt. % nickel). Typically, 0.1 g of the catalyst (particle size 63-150 μm) diluted with 0.9 g pure silica was used for the tests. A bed of 0.3 g pure silica was placed above this diluted catalyst bed to spread the liquid over the entire catalyst cross-section. The catalysts reduced in situ with hydrogen at 400°C for 90 min. and cooled down to the reaction temperature. Ethanol was flushed prior to the experiment with a hydrogen flow to remove dissolved oxygen and this flow was maintained during the reaction. Citral (0.02 M) was injected into the solvent and the reaction commenced. The liquid feed rate was between 0.6–1.2 ml/min and the hydrogen gas flow rate was 100 ml/min through each reactor. The temperature interval investigated was 25–65°C and the total pressure was 6.1 bar. To test the influence of the particle size on the internal mass transfer, two separate experiments with catalyst particles in the range of 63–90 μm as well as 150–200 μm were performed. In this case the bed was diluted with 0.9 g silica of the particle sizes between 63–150 μm. Gas chromatographic analysis was applied.

The bed void fraction and the Reynolds number were determined with the experimental procedures reported in literature [7]. In preliminary experiments, citral hydrogenation was investigated in six parallel reactors under identical reaction conditions, i.e., at 25°C and 6.1 bar hydrogen with the residence time of 156 s. The aim was to investigate, whether the same rates and conversions can be achieved for different reactor tubes. The initial conversions determined after 15–20 min varied from 38–44% demonstrating that the results were in a good accordance in different reactor tubes. The overall rate, defined as

$$\frac{(c_0 - c)\dot{V}_L}{m_{cat}} = \frac{1}{\tau} \int_0^{\tau_{TOT}} r dt \tag{1}$$

where c_0 is the initial reactant concentration, c the concentration at the residence time τ_{TOT}, r is the reaction rate, $\dot{V}_L$ the liquid flow rate and m_{cat} the catalyst mass.

Effect of residence time and particle size

The effect of residence time was investigated by using two liquid flow rates, 0.6 ml/min and 1.2 ml/min. These flow rates corresponded to the residence times of 156 s and 80 s, respectively. The results are displayed in Figure 47.2. In the former case a fresh, prereduced catalyst was used, whereas in the latter case the very same catalyst was applied during citral hydrogenation by using a dublicate liquid flow rate. Between these experiments, the catalyst was reduced at 400°C for 90 min with flowing hydrogen. Utilization of the same catalyst in two consecutive experiments with different liquid flow rates would not allow to separate completely the effect of catalyst deactivation from the effect of the residence time. This procedure was applied, however, since the catalytic performance deviated only slightly within the first 40 min time-on-stream, thereafter the same conversion levels were achieved. The ratio between the initial rates obtained with the residence times of 156 s and of 80 s was about 1.95 indicating that the effect of catalyst deactivation was minor. The conversion after 90 min time-on-stream was, however, 4.3 times higher for the former case compared to the latter residence time, which might suggest the catalyst deactivation was more prominent than expected, as the catalyst was in contact with the double amount of citral.

The effect of catalyst particle size was investigated by two different catalyst particle size fractions: 63–93 μm and 150–250 μm, respectively. The effect of the particle size is very clear as demonstrated by Figure 47.2. The overall hydrogenation rate was for smaller particles 0.17 mol/min/g_{Ni} while it was 0.06 mol/min/g_{Ni}, for the larger particles, showing the presence of diffusion limitation. This kind of studies can be used to determine the effectiveness factors. The conversion levels after 70 min time-on-stream were 21% and 3%, respectively, for these two cases.

Parallel screening of temperature effects

Citral hydrogenation was carried out in parallel reactor tubes at six different temperatures to screen the temperature effect. The initial hydrogenation rates increased with increasing temperature until 45°C, thereafter the rates were about the same due to extensive catalyst deactivation, which was more prominent at higher temperatures, especially above 60°C. The results were well reproducible.

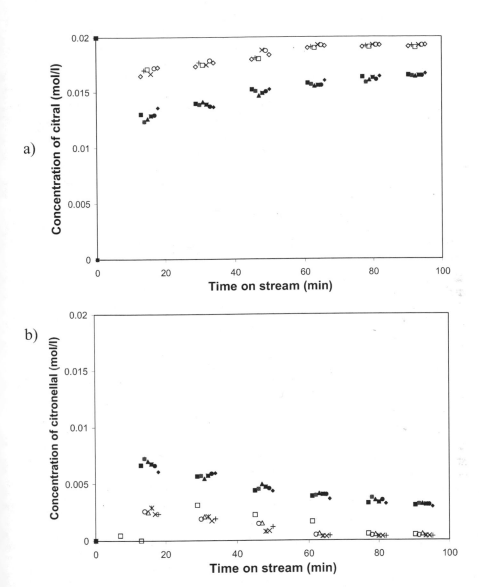

Figure 47.2. (a) Effect of residence time 156 s, fresh catalyst (solid symbol); 80 s, catalyst used once (open symbol) and (b) effect of catalyst particle size in citral hydrogenation at 25°C, 6.1 bar total pressure, residence time 156 s, solvent ethanol, 0.1 g catalyst Ni/SiO$_2$, initial citral concentration 0.02 M.

Conclusions

In this work it has been demonstrated that parallel fixed bed reactor system facilitates experimentation of heterogeneous catalysts in three-phase systems. The residence times and Reynolds numbers were determined. The Reynolds numbers were very small suggesting that the absence of external mass transfer limitations could not be totally excluded, which is well known for three-phase fixed-bed reactors. The results from hydrogenation experiments with two different particle sizes revealed the presence of diffusion limitations when using larger particles. The reproducibility of the reactor system was confirmed by achieving the same kinetic curves under identical reaction conditions. The effect of residence time was studied in two consecutive experiments with the same catalyst. The results showed that initial conversion was proportional to the residence times. The decrease of conversion levels revealed, however, that the catalyst deactivation was more prominent with the higher liquid flow rate. The reactor system is a strong tool in investigating the catalyst performance and durability, as well as diffusion effects inside the particles.

Acknowledgement

This work is part of activities at the Åbo Akademi Process Chemistry Centre of Excellence Programmes (2000–2011) by the Academy of Finland.

References

1. S. Möhmel, S. Engelschalt, M. Baerns and D. Wolf, DE 10304217, 2004.
2. J. A. Moulijn, J. P9rez-RamPreż, R. J. Berger, G. Hamminga, G. Mul and F. Kaptejn, Catal. Today 81 (2003) 457.
3. A. Hagemeyer, P. Strasser and A. F. Volpe, Jr, (Eds), High-throughput screening in chemical catalysis (Wiley-VCH, Weinheim, 2004).
4. E. G. Derouane, V. Parmon, F. Lemos and F. R. Ribeiro, Principles and methods for accelerated catalyst design and testing, Series II : Mathematics, Physics and Chemistry, vol. 69, (Kluwer Academic Publishers, Dordrecht, 2002).
5. U. Rodemerck and M. Baerns, in Basic Principles in Applied Catalysis, M. Baerns, ed.,(Springer-Verlag, Berlin, 2004) p.261.
6. S. Van der Beken, E. Dejaegere, K. A. Tehrani, J. S. Paul, P. A. Jacobs, G. V. Baron and J. F. M. Denayer, J. Catal. 235 (2005) 128; S. Van der Beken, E. Dejaegere, H. Verelst, G. V. Baron and J. F. M. Denayer, Int. J. Chem. Reactor Engng. 3 (2005) A23.
7. R. Lange, M. Schubert, T. Bauer, Ind. Eng. Chem. Res. 44 (2002) 6504.

48. "Green" Acylation of Aromatic Sulfonamides in Heterogeneous Catalysis

Simona M. Coman[1], Cristina Stere[1], Jamal El Haskouri[2], Daniel Beltrán[2], Pedro Amorós[2] and Vasile I. Parvulescu[1]

[1]University of Bucharest, Faculty of Chemistry, Department of Chemical Technology and Catalysis, Bdul Regina Elisabeta 4-12, 030016 Bucharest, Romania
[2] Institut de Ciencia dels Material, Universitat de Valencia, Valencia E46071, Spain

v_parvulescu@chem.unibuc.ro

Introduction

Increasingly stringent environmental legislations have generated a pressing need for cleaner methods of chemical manufacture, for instance, technologies that reduce, or preferably eliminate the generation of waste, and avoid the use of toxic and/or hazardous reagents [1]. In this context, some of the major goals are to develop new sustainable chemical products, to increase process efficiency, and to simplify process work-up so as to facilitate the recovery and reuse of the catalyst or reagents, and hence minimize waste.

In manufacturing chemistry many of the least efficient processes are those commonly used by fine chemicals companies [2]. An illustrating example is given by acylation reactions which are carried out in very environmentally unfriendly conditions, using homogeneous catalysts [3-5], or stoichiometric inorganic acids [4, 6], which generate large amounts of corrosive and toxic waste products. They also often necessitate the presence of environmentally unfriendly solvents like dichloromethane [7].

Particularly, some newly developed drugs, which incorporate the N-acyl sulfonamide moiety [8-10], are synthesized from the parent sulfonamides, by their coupling with acid chlorides or carboxylic anhydrides in basic conditions [11-15]. Unfortunately all these methods lead to substantial waste products. Less common reports mentioning this transformation under acidic conditions (Brønsted or Lewis acids) do not systematically examine the purpose and limitations of the reaction [16].

In order to overcome these problems, attention was focused on the use of heterogeneous catalysis. We have found that functionalized solid materials, e.g., ionic liquids or tin triflates immobilized into mesoporous materials, can be used in N-acylation reactions as environmentally friendly replacements for traditional homogeneous acids which are useful but environmentally unacceptable catalysts [17, 18]. They had comparable activity to homogeneous reagents but can offer greater stability, safer and easier handling and can be recycled with little loss in activity. Therefore, these materials represent a significant break-through in green chemistry.

Here we report on an efficient and greener procedure for the synthesis of N-acyl sulfonamides using anhydrides and organic acids as acylating agents, and a large family of functionalized solid materials: $AlCl_3/MCM-41$, $La(OTf)_3/SiO_2$, $Sn(OTf)_4/MCM-41$, $Sn(OTf)_4/UMV-7$, silica-embedded Nafion (SAC-13), CF_3COOH/SiO_2, and HSO_3F/SiO_2 as acid heterogeneous catalysts.

Experimental Section

Catalysts Preparation

Immobilized Lewis Acids: *Supported Aluminum Chloride (AlCl₃/MCM):* $AlCl_3$-grafted onto MCM-41 was prepared by reacting anhydrous $AlCl_3$ with Si-MCM-41, as follows: a mixture of 10 g of Si-MCM-41 and 250 ml of dry CCl_4 was refluxed for 1.5 h, while bubbling moisture-free N_2 (30 cm^3/min) through the mixture. Then 2.6 g (19 mmol) of anhydrous $AlCl_3$ was added, and the resulting mixture was refluxed under N_2 flow for 3 h. After reaction, the obtained AlCl₃/MCM suspension was filtered out, washed with CCl_4 and dried under vacuum at 80°C.

Supported Lanthanum Triflate (LaOTf-SiO₂): The sol-gel synthesis was carried out in inert atmosphere (Ar), using an adapted route in which the silica sol was obtained by acid hydrolysis of a solution of tetraethoxyorthosilicate (TEOS). Water was added to the acidic solution in a TEOS:H_2O molar ratio of 1:10 and the mixture was refluxed at 70°C for 2 h. After cooling the silica sol solution at rt, the lanthanum triflate was added under vigorous stirring as an ethanolic solution. Hexadecyltrimethylammonium bromide (HDTMABr) was then introduced and the gelation was carried out at 90°C for two days in a Teflon cylinder within an autoclave. The resulting gel was dried under vacuum, at rt for 24 h, and then at 100°C for 6 h. Resulted sample has 15 wt% $La(OTf)_3$.

Supported Tin Triflate Catalysts (SnOTf-MCM-41) and (SnOTf-UMV-7): Tin triflate $(Sn(OTf)_4)$ based catalysts with unimodal and bimodal pore system were prepared in a two-step synthesis in which the triflic acid (HOTf) was incorporated to previously synthesized mesoporous tin-containing silicas. The Sn incorporation inside the pore walls was carried out through the Atrane method. SnOTf-UMV-7 catalysts were constructed by aggregation of nanometric mesoporous particles defining a hierarchic textural-type additional pore system. Following these procedures catalysts with different Si/Sn ratios were prepared. They varied for SnOTf-MCM-41 in the range Si/Sn = 21.8 to 50.8, while for SnOTf-UVM-7 the Si/Sn ratio was 18.4. The detailed preparation procedure is given elsewhere [18].

Immobilized Brønsted Acids: *Supported Fluorosulphuric Acid (FSO₃H/SiO₂) and Trifluoroacetic Acid (CF₃COOH/SiO₂):* 1.6 mmoles of (3-aminopropyl) trimethoxysilane was dissolved in 100 mL of chloroform, and 1 g of silica gel was added. The slurry was heated under reflux for 24 h. The slurry was then cooled at rt, and the solid was separated by filtration, washed with 50 ml of pentane, 100 ml of acetonitrile, and 100 ml of diethyl ether before being dried under vacuum. The functionalized silica was then impregnated with an equimolecular amount of FSO_3H or CF_3COOH in ethyl alcohol (1 g of silica for 2 ml of acidic solution). After impregnation the samples were dried at 60°C for 12 h.

Nafion (13 wt%) embedded silica (SAC13) was purchased from Aldrich (S_{sp} = 400 m^2/g; D_p=10 nm; V_p=0.6 mL/g; stable to 200°C, 0.15 meq/g, H_0 ≥ -12, 96% H_2SO_4).

Catalysts Characterization: Catalysts were characterized by nitrogen adsorption-desorption isotherms, XRD, XPS, TEM, and FT-IR. The concentration and the strength of the acid sites were determined using a combination of NH_3-chemisorption and FTIR. Detailed procedures are given elsewhere [18, 19].

Catalytic Tests: In a typical procedure, 1.25 mmols of sulfonamide (benzenesulfonamide, *p*-nitrobenzenesulfonamide, *p*-methoxybenzenesulfonamide) were dissolved in 4 mL of acetonitrile in a glass vessel of 10 mL. To this mixture, 3.75 mmols of acylating agent (acetic anhydride, maleic anhydride, *p*-methoxybenzoic acid, acetic acid) and 15 mg of catalyst were added (acylating agent : substrate ratio = 3:1). All reactions were carried out at 80°C for 18 h under vigorous stirring. After the reaction was stopped, the catalyst was filtered off and the product was separated from solvent under vacuum, at 80°C, as a white crystalline solid. The product was re-dissolved in the HPLC eluent and analyzed, in the followed conditions: column: GROM-SIL 80 ODS-2 FE; eluent: MeOH:H_2O = 40 : 60; flow rate: 0.5 mL/min; wavelength: 235 nm; volume sample: 15 µL. The products were characterized by HPLC-MS and ^{1}H-NMR techniques.

Results and Discussion

All the prepared samples show a very high surface area ranging from 357 to 872 m^2/g combined with an average pore size diameter in the mesoporous (21-31 Å) or macroporous region (100-366 Å) (Table 48.1). Characterization of the superficial acidity performed by recording Py-FTIR or NH_3-FTIR (LaOTf/SiO$_2$) spectra showed that all catalysts exhibit either Lewis (AlCl$_3$/MCM, LaOTf/SiO$_2$, SnOTf-MCM and SnOTf-UVM-7) or Brønsted (CF$_3$COOH/SiO$_2$ and FSO$_3$H/SiO$_2$) acid sites. The Brønsted acidity of the last two samples was accounted for by the residual acidity left after the neutralization of a weak base (-Si–(CH$_2$)$_3$–NH$_2$) with a strong acid (CF$_3$COOH, pKa = 0.3) or a super acid (FSO$_3$H, H_0 = -15.07).

Besides different types and strengths of acidity, the catalysts were characterized by a different density of acidic sites (expressed as number of centers/m^2 of the surface, Table 48.1). While AlCl$_3$/MCM showed the highest density of acid centers, the SnOTf -MCM samples were the opposite, having the lowest density (17 – 23 x 10^{16}).

Table 48.1. The textural and chemical properties of the catalysts

Catalyst	The nature of acidity	Amount of active phase (mmol/g)	S_{BET} (m²/g)	D_p (Å)	Number of centers /m² *
AlCl₃/MCM	Lewis	1.90	490	24.0	230 x 10¹⁶
LaOTf/SiO₂	Lewis	0.26	357	28.0	40 x 10¹⁶
SnOTf(22)-MCM	Lewis	0.28	725	25.0	23 x 10¹⁶
SnOTf(47)-MCM	Lewis	0.23	878	26.0	16 x 10¹⁶
SnOTf(51)-MCM	Lewis	0.23	812	31.0	17 x 10¹⁶
SnOTf(18)-UVM-7	Lewis	0.65	646	21.0 & 366.0	60 x 10¹⁶
CF₃COOH/SiO₂	Brønsted	0.80	460	26.0	100 x 10¹⁶
FSO₃H/SiO₂	Brønsted	0.90	470	25.0	110 x 10¹⁶
SAC-13	Brønsted	0.15	400	100.0	22 x 10¹⁶

*- calculated from columns 3 and 4

Initial catalytic tests, using acetic anhydride as acylation agent, indicated that both immobilized Lewis and Brønsted acid catalysts were active in N-acylation of aromatic sulfonamides (see Table 48.2). With a few exceptions the reaction gave high yields and worked for all tested substrates. It has to be specified that in all cases the selectivity to mono N-acylsulfonamide was 100%. As Table 48.2 shows, the highest yields to N-acylsulfonamides were obtained in the presence of SAC-13 irrespective of the sulfonamide structure, followed by FSO₃H/SiO₂, and tin triflate based samples. In opposition, AlCl₃/MCM gave the lowest yields. The CF₃COOH/SiO₂ catalyst, with a similar density of acidic sites as FSO₃H/SiO₂ (see Table 48.1) but considerably weaker, gave only modest yields to N-acylsulfonamide. On the other hand, taking into account the different

Table 48.2. Catalytic performances in the N-acylation of sulfonamides with acetic anhydride

	Substrate					
	C₆H₅SO₂NH₂		(MeO)C₆H₄SO₂NH₂		(NO₂)C₆H₄SO₂Nh₂	
Catalyst	Yield (%)	TOF (h⁻¹)	Yield (%)	TOF (h⁻¹)	Yield (%)	TOF (h⁻¹)
AlCl₃/MCM	20.1	0.5	14.3	0.3	19.6	0.5
LaOTf/SiO₂	50.6	9.0	32.5	5.8	49.3	8.8
SnOTf(22)-MCM	95.0	15.7	52.7	8.7	94.4	15.6
SnOTf(47)-MCM	99.0	20.0	80.7	16.3	99.0	19.9
SnOTf(51)-MCM	97.0	19.5	64.6	13.0	96.8	19.4
SnOTf(18)-UVM-7	96.0	6.8	67.2	4.8	95.0	6.7
CF₃COOH/SiO₂	45.7	2.6	22.6	1.3	29.7	1.7
FSO₃H/SiO₂	100	5.3	99.4	5.1	100	5.1
SAC-13	100	30.8	99.5	30.7	100	30.8

Conditions: substrate (1.25 mmols), acetic anhydride (3.75 mmols), catalyst: (15 mg), solvent (4 mL of acetonitrile), reaction temperature (80°C), reaction time (18 h). TOF (h⁻¹) = [mmols of transformed substrate] / [mmols of active phase x time]

density of the catalytic active sites, the situation is different. In agreement with the turnover frequency, TOF, (h^{-1}) values, SAC-13 seems to be the most active among the tested catalysts, followed by tin triflate based catalysts and LaOTf/SiO$_2$. The electronic effect exerted by the aromatic ring of the sulfonamide only had a little contribution on the efficiency of the process while the steric one was more important.

During the immobilization of AlCl$_3$ one or two mole equivalent of HCl were released forming mostly –OAlCl$_2$ and –O$_2$AlCl units on the surface. The loss in activity on going from AlCl$_3$ to –OAlCl$_2$ or –O$_2$AlCl is partially compensated by the existent isolated Al centers (Scheme 48.1), rather than by the dimeric Al$_2$Cl$_6$ which predominates in many homogeneous aluminum chloride catalyzed reactions [20]. Drago and co-workers [21] proposed that the active sites may be formed by the reaction of AlCl$_3$ with a silanol to form Si-O-AlCl$_2$ and HCl. Coordinate covalent bonding of a second silanol unit to the aluminum yields an acidic bridging hydroxyl analogous to the Brønsted site in zeolites (Si-O-AlCl$_2$-O(H)-Si). However, the poor activity of this sample in the acylation reactions might prove that –OAlCl$_2$ and –O$_2$AlCl units, with a low activity, predominate on the solid surface.

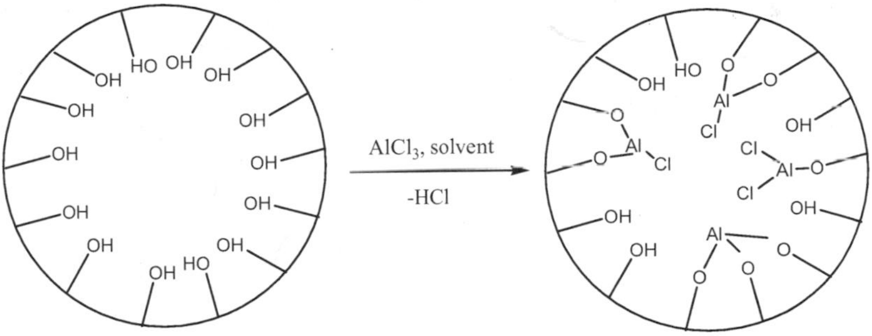

Scheme 48.1.

Usually in heterogeneous acid catalysis, isolated species are much more active than oligo- and polymeric species. Using conventional co-hydrolysis methods it is hard to achieve a good dispersion of monomeric species in the silica network. For example, the tin (IV) hydrolysis rate is much higher than that of silicon, leading to low chemical homogeneity and even phase segregations. Atrane route to synthesis of Sn-MCM-41 is an effective solution because it favors harmonic heterometallic hydrolysis-condensation processes. With this preparation method, a high concentration of well-dispersed Sn species into the framework of the mesoporous silica is attainable (Scheme 48.2A) [18].

Silica sol–gel immobilized La(OTf)$_3$ (Scheme 48.2B) previously used in the acylation of a series of alcohols and activated aromatic compounds using acetic anhydride as acylating agent, showed a poor activity compared with other various silica sol–gel immobilized triflate derivatives: (tert-butyl-dimethylsilyl-trifluoromethane-sulfonate (BDMST), or triflic acid (HOTf)). Acylation at the aromatic ring occurred over the BDMST and HOTf catalysts, while the La(OTf)$_3$ catalysts only led to *O*-acetylated products [22]. Such behavior is characteristic of catalysts with middle or low acidity. Our results are in agreement with these previous results, the low reactivity to the acylation of sulfonamides being a result of this low acidity.

A wide variety of organically modified mesoporous silicas can be made via sol-gel methodology. In this way materials with high loadings of functional groups including –NH$_2$, -SH and –CN can be directly prepared by using commercially available organosilanes [20]. This methodology is generally superior to the treatment of a pre-existing solid with a silane since the sol-gel route generally gives higher loadings, and the thermal stabilities are usually significantly better, presumably due to a higher proportion of multiple bonded groups. Subsequent modification of these groups can lead to a wide range of functionalities or to a certain functionality (e.g., acidic or basic) but with different strengths, as in the case of FSO$_3$H/SiO$_2$ and CF$_3$COOH/SiO$_2$ (Scheme 48.3).

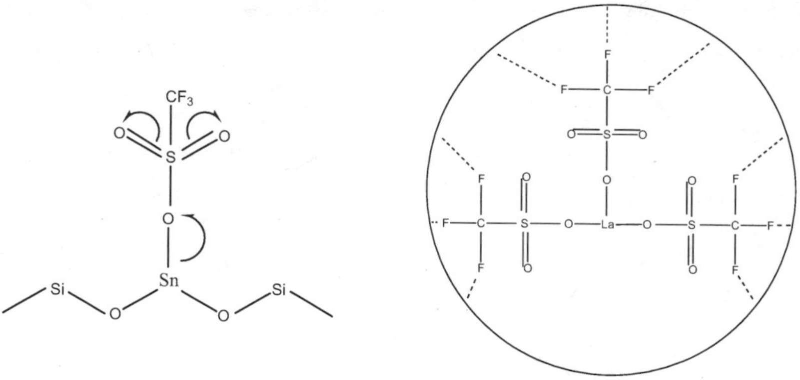

A.   B.

Scheme 48.2.

Catalytic results are well correlated with the acid strength of the active species irrespective of their nature (Lewis or Brønsted). On the other hand, there is no clear correlation between the density of the active sites and the catalytic performances. While the FSO$_3$H/SiO$_2$ catalyst is very active (yields: 99.5 -100%, Table 48.2), AlCl$_3$/MCM shows only moderate yields (14.3-20.1%) to N-acylsulfonamide, even if both samples exhibit a similar density (25 x 10^{16}, Table 48.1).

Obviously, the reaction rate of nucleophilic substrates with acylium cations increases as the induced electronic density on C4 increases, thus the N basicity also increases. In our experiments the reaction rate followed the sequence: benzenesulfonamide   >   *p*-nitrobenzenesulfonamide   >   *p*-methoxybenzenesulfonamide.

In some cases the effect of the nature of aromatic hydrogen substituent has been also observed (Table 48.2). These results contrast with those obtained using typical homogeneous Brønsted acid catalysts (e.g., sulphuric) for the acylation of the same substrate with acetic anhydride, under the same experimental conditions, where the yields (98%, 95%, 91% for R=NO$_2$, H, OMe, respectively) do not significantly depend on the nucleophile's substituent nature [23]. These data underline the contribution of the heterogeneous catalyst.

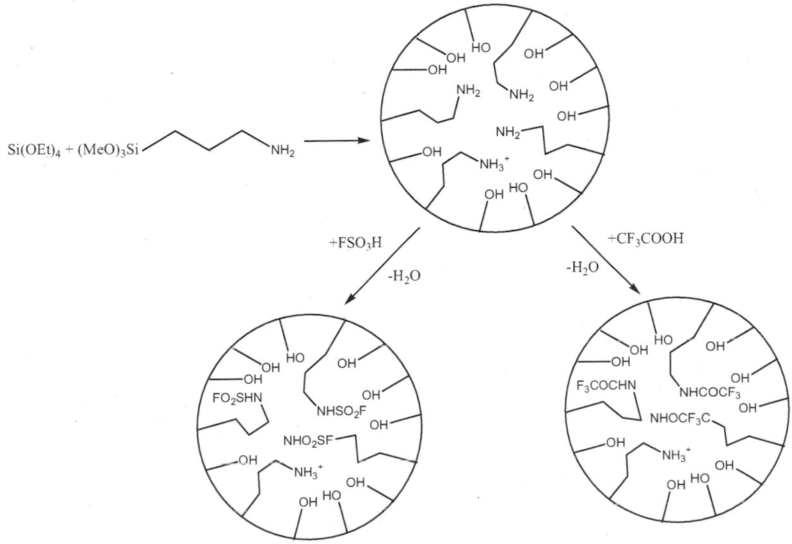

Scheme 48.3.

In order to extend the scope of the reaction, and with the aim of designing a greener approach to the above set of reactions, we preformed the acylation of the same substrates with different acylation agents, such as maleic anhydride, *p*-methoxybenzoic acid and acetic acid (Scheme 48.4). Table 48.3 shows the results for acylation of benzenesulfonamide.

Obviously, in the presence of a Lewis acid, the acylating agent forms an adduct via the carbonyl oxygen atoms. As a result, the carbonyl bond is strongly polarized, and the carbon atom becomes sufficiently electrophillic to interact with the substrate so that the reaction may occur. The formation of these adducts is limited by their reactivity. It is known that an increased stability of the carbonyl group induces a lower reactivity. On this basis, the carboxylic acids are more stable than the corresponding acid chlorides or carboxylic anhydrides, and therefore are less reactive. To activate these acylating agents there are two options: a high reaction temperature and/or the use of a catalyst with a strong acidity, even superacidity.

As we expected, the nature of the acylating agents has a great influence on the N-acylsulfonamide yields, the highest yields being obtained with acetic anhydride irrespective of the catalyst's nature. The acylation with acetic acid only took place with samples characterized by a very strong acidity, such as SAC-13 and FSO_3H/SiO_2 catalysts (Table 48.3).

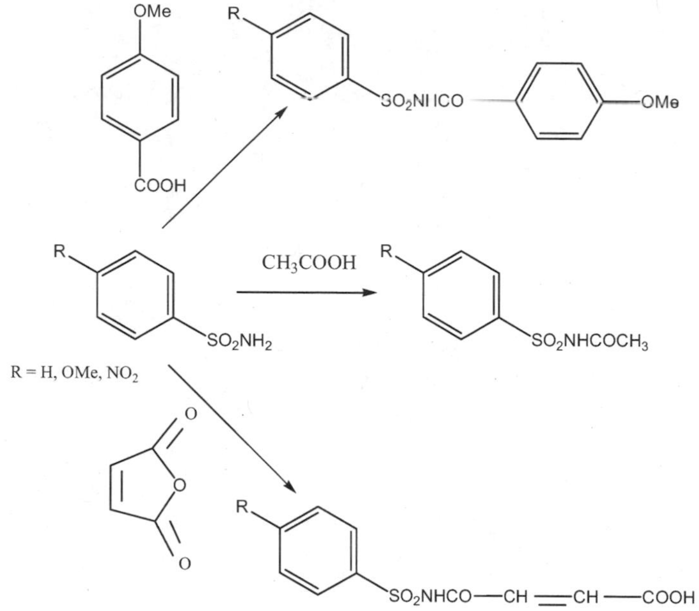

Scheme 48.4.

Table 48.3. The acylation of benzenesulfonamide with different acylating agents

Catalyst	Acylating agent					
	Maleic anhydride		*p*-methoxybenzoic acid		Acetic acid	
	Yield (%)	TOF (h^{-1})	Yield (%)	TOF (h^{-1})	Yield (%)	TOF (h^{-1})
$AlCl_3/MCM$	89.2	2.2	66.0	1.6	0	0
$LaOTf-SiO_2$	89.6	15.9	66.4	11.8	0	0
CF_3COOH/SiO_2	87.5	5.1	64.3	3.7	0	0
FSO_3H/SiO_2	84.6	4.3	80.2	4.1	37.0	1.9
SAC-13	83.8	40.6	71.0	21.9	62.3	19.2

Conditions: substrate (1.25 mmols), acylating agent (3.75 mmols), catalyst: (15 mg), solvent (4 mL of acetonitrile), reaction temperature (80°C), reaction time (18 h)

Results obtained in the acylation of aromatic sulfonamides with acetic acid, in the presence of SnOTf based catalysts are presented in Table 48.4. The rate of the sulfonamide acylation follows the sequence: benzenesulfonamide > *p*-nitrobenzenesulfonamide > *p*-methoxybenzenesulfonamide, and is very sensitive towards the nature of the aromatic hydrogen substituent (the selectivity in acylated *p*-methoxybenzenesulfonamide did not exceed 7% irrespective of the catalyst nature; this corresponds to an approximate relative yield benzenesulfonamide: *p*-nitrobenzenesulfonamide : *p*-methoxybenzenesulfon-amide of 10:4:1).

Table 48.4. The N-acyl sulfonamides yield and TOF as a function of the substrate and catalyst nature

Catalyst	Substrate	Yield (%)	TOF (h^{-1})
SnOTf(22)-MCM	benzenesulfonamide	54.5	8.9
SnOTf(22)-MCM	*p*-nitrobenzenesulfonamide	17.4	2.8
SnOTf(47)-MCM	benzenesulfonamide	45.2	9.1
SnOTf(47)-MCM	*p*-nitrobenzenesulfonamide	11.4	2.9
SnOTf(51)-MCM	benzenesulfonamide	48.6	9.6
SnOTf(51)-MCM	*p*-nitrobenzenesulfonamide	11.7	2.3
SnOTf(18)-UVM	benzenesulfonamide	65.5	4.6
SnOTf(18)-UVM	*p*-nitrobenzenesulfonamide	24.2	1.7

Conditions: substrate (1.25 mmols), acetic acid (3.75 mmols), catalyst: (15 mg), solvent (4 mL of acetonitrile), reaction temperature (80°C), reaction time (18 h)

The best yield (65.5%, Table 48.4) to N-acylbenzenesulfonamide was obtained in the presence of the SnOTf(18)-UVM-7 catalyst, very close to the yield obtained using SAC-13 (62.3%, Table 48.3). TOF values given in the same Table 48.4 are in line with the model of the active sites proposed for these catalysts (Scheme 48.3A, [18]). They proved that the catalytic activity was associated with highly dispersed tin species, and this implicitly means tin triflate species. Accordingly, the values of TOF are pretty similar irrespective of the Sn content. The similarity of the yields obtained in this case with those obtained on a superacidic SAC-13 sample, can be a suggestion that these samples behave like superacidic catalysts. The superacid active sites of the catalyst would be the anchored $Sn(OTf)_x$ species and the silica network acts as the dehydrating agent.

Conclusion

The large dimensions of the catalyst pores allow the substrate to diffuse into the pores system and to react with the active sites anchored on their surface. While the yields to N-acylsulfonamides depend on the parent sulfonamide structure, the acylating agent reactivity, and acid strength of the catalysts, the selectivity to N-acylsulfonamide was 100% irrespective of all these parameters. The direct acylation with acetic acid might be evidence of the catalyst's superacidity. Beside the commercial SAC-13, FSO_3H/SiO_2 and SnOTf based catalysts also seem to exhibit superacidic properties. All these catalysts were active in acylation of sulfonamides with acetic acid (yields of 37.0 and 62.3%, respectively).

References

1. R. A. Sheldon, Pure Appl. Chem., 72 (2000) 1233.
2. M. Lancaster, in "Green Chemistry: an introductory text", The Royal Society of Chemistry Publisher, 2002.
3. R.I. Zhdanov and S.R. Zhenadavora, Synthesis (1975) 222.

4. G.A. Olah, Friedel–Crafts and Related Reaction, vol. III, Interscience Publisher, New York, London, Sydney, 1964, p. 1606.
5. S. Lee and J.H. Park, J. Mol. Catal. A: Chem. 194 (2003) 49.
6. M. Spagnol, L. Gilbert and D. Alby, in: J.R. Desmurs, S. Rattoy (Eds.), The Roots of Organic Development, Elsevier, Amsterdam, 1996, p. 29.
7. R. Gosh, S. Maiti and A. Chakraborty, Tetrahedron Lett. 46 (2005) 147.
8. T. Hasegawa and H. Yamamoto, Bull. Chem. Soc. Jpn., 73 (2000) 423.
9. M. G. Banwell, C. F. Crasto, C. J. Easton, A. K. Forrest, T. Karoli, D. R. March, L. Mensah, M. R. Nairn, P. J. O'Hanlon, M. D. Oldham and W. Yue, Bioorg. Med. Chem. Lett., 10 (2000) 2263.
10. Y. Wang, D. L. Soper, M. J. Dirr, M. A. DeLong, B. De and J. A. Wos, Chem. Pharm. Bull., 48 (2000) 1332.
11. K. Kondo, E. Sekimoto, J. Nakao and Y. Murakami, Tetrahedron, 56 (2000) 5843.
12. K. Kondo, E. Sekimoto, K. Miki and Y. Murakami, J. Chem. Soc., Perkin Trans., 1 (1998) 2973.
13. N. Ishizuka, K.-I. Matsumura, K. Hayashi, K. Sakai and T. Yamamori, Synthesis, 6 (2000) 784.
14. N. Ishizuka and K. Matsumura, Japanese Patent JP-10045705, 1998
15. T. Inoe, O. Myahara, A. Takahashi and Y. Nakamura, Japanese Patent JP-08198840, 1996.
16. Y. Morisawa, M. Kataoka, H. Negahori, T. Sakamoto, N. Kitano and K. Kusano, J. Med. Chem., 23 (1980) 1376.
17. S. M. Coman, M. Florea, V. I. Parvulescu, V. David, A. Medvedovici, D. De Vos, P. A. Jacobs, G. Poncelet and P. Grange, J. Catal. 249 (2007) 359.
18. S. M. Coman, G. Pop, C. Stere, V. I. Parvulescu, J. El Haskouri, D. Beltrán and P. Amorós, J. Catal. 251 (2007) 388.
19. A.N. Parvulescu, B.C. Gagea, V. Parvulescu, V.I. Parvulescu, G. Poncelet and P. Grange, Catal. Today 73 (2002) 177.
20. J.H. Clark, D.J. Macquarrie and K. Wilson, Stud. Surf. Sci. Catal., 129 (2000) 251.
21. T. Xu, N. Kob, R. S. Drago, J. B. Nicholas and J. F. Haw, J. Am. Chem. Soc., 119 (1997) 12231.
22. A.N. Parvulescu, B.C. Gagea, G. Poncelet and V.I. Parvulescu, Appl. Catal. A: General 301 (2006) 133.
23. M.T. Martin, F. Roschangar and F. F. Eaddy, Tetrahedron Letters, 44 (2003) 5461.

49.   Synthesis of α, β-Unsaturated Ketones

Lina M. Gonzalez[1], Aida L. Villa[1], Consuelo Montes de C.[1] and Alexander B. Sorokin[2]

[1]Environmental Catalysis Research Group, Universidad de Antioquia-CENIVAM, AA 1226, Colombia
[2]IRCELYON-CNRS, 69626 Villeurbanne-Cedex, France

lgonzale@udea.edu.co

Abstract

The activity of the **FePcCl$_{16}$-S**/tert-butyl hydroperoxide (TBHP) catalytic system was studied under mild reaction conditions for the synthesis of three α,β-unsaturated ketones: 2-cyclohexen-1-one, carvone and verbenone by allylic oxidation of cyclohexene, limonene, and α-pinene, respectively. Substrate conversions were higher than 80% and ketone yields decreased in the following order: cyclohexen-1-one (47%), verbenone (22%), and carvone (12%). The large amount of oxidized sites of monoterpenes, especially limonene, may be the reason for the lower ketone yield obtained with this substrate. Additional tests suggested that molecular oxygen can act as co-oxidant and alcohol oxidation is an intermediate step in ketone formation.

Introduction

α, β-unsaturated ketones are important raw materials in organic synthesis owing to their highly reactive carbonyl group. In particular, ketones obtained by allylic oxidation of cyclohexene (**1**), limonene (**2**), and α-pinene (**3**), (see equations 1 – 3) as 2-cyclohexen-1-one (**4**), carvone (**5**), and verbenone (**6**) are useful raw materials in pharmaceutical, cosmetic and flavor industries (1-3).

Since the allylic oxidation is a process that involves free radicals where the hydrogen abstraction is the dominant reaction, the double bond and also other derivatives can be epoxidized (4). Several procedures for oxidizing allylically activated hydrogen compounds are known, but unsatisfactory yields, tedious workups and/or expensive or environmentally unacceptable reagents are required (5, 6). Various homogeneous and heterogeneous catalytic procedures have been proposed in the literature for replacing typical allylic oxidation methods for the synthesis of **4**, **5**, and **6**. Ketone yields above 60% have been reported for homogeneous catalytic systems using catalysts based on Mn, Co, and Pd complexes (7-9). Notwithstanding, less successful results have been obtained with heterogeneous systems, because of the leaching of active species under reaction conditions (4).

(1)

(1)   (4)

(2)

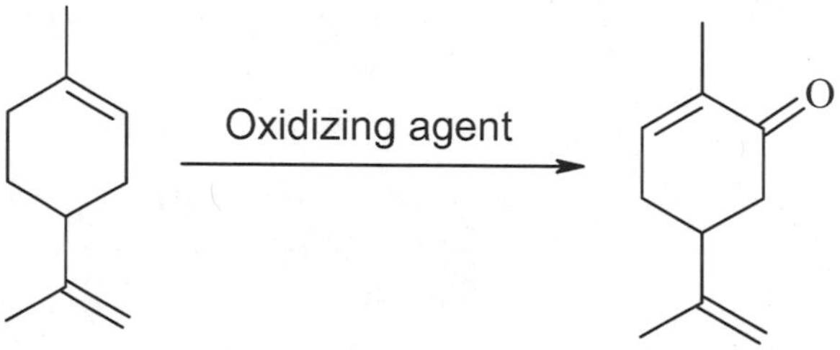

(2)   (5)

(3)

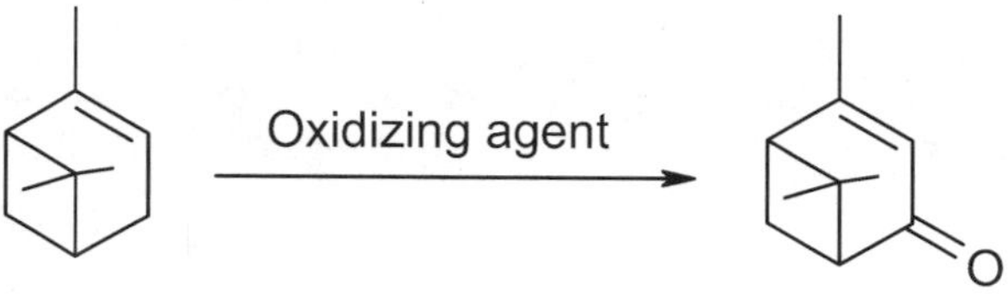

(3)   (6)

The heterogeneous catalytic system iron phthalocyanine (**7**) immobilized on silica and tert-butyl hydroperoxide, TBHP, has been proposed for allylic oxidation reactions (10). This catalytic system has shown good activity in the oxidation of 2,3,6-trimethylphenol for the production of 1,4-trimethylbenzoquinone (yield > 80%), a vitamin E precursor (11), and in the oxidation of alkynes and propargylic alcohols to α,β-acetylenic ketones (yields > 60%) (12). A 43% yield of 2-cyclohexen-1-one was obtained (10) over the μ-oxo dimeric form of iron tetrasulfophthalocyanine (**7a**) immobilized on silica using TBHP as oxidant and CH₃CN as solvent; however, the catalyst deactivated under reaction conditions.

Since iron phthalocyanine complexes can be activated and stabilized by a chlorine ring substitution (12), the activity of iron hexadecachlorophthalocyanine (**7b**) immobilized on silica was examined for the synthesis of **4**, **5**, and **6** with TBHP as oxidant.

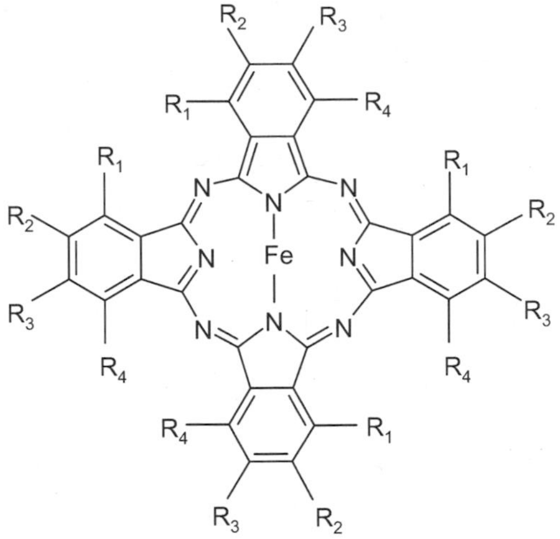

(a) R_1, R_2, R_4 = H, R_3 = SO$_2$

(b) R_1, R_2, R_3, R_4 = Cl

(7)

Experimental Section

All reagents and solvents were from commercial sources (Aldrich, J.T. Baker, Merck, Fluka, Mallinckrodt) and used as received. The complex **7b** was synthesized following a modification of the method reported by Metz and co-workers (13).

For the synthesis of **7b** a mixture of urea (11.7 g, 190 mmol), tetrachlorophthalic anhydride (16 g, 56 mmol), FeCl$_2$.4H$_2$O (2.8 g, 14.1 mmol) and ammonium heptamolibdate tetrahydrate (0.1 g, 0.081 mmol) was well grounded. The finely pulverized mixture was added to nitrobenzene (50 mL). Then, the temperature of the mixture was increased to 180–190°C within 30 min, and it was maintained until complete ammonia evolution (~4 h). The viscous olive green suspension was cooled to 90°C, diluted with ethanol (50 mL), and the solid recovered by filtration. The solid was copiously washed with boiling water until a clear filtrate was obtained. For complex purification it was firstly boiled in 1% HCl aqueous solution (100 mL), then filtered and washed with water until pH=7. After that, the solid was dispersed in 1% NaOH aqueous solution (100 mL), boiled for 5 min, filtered and washed with water until pH=7. Finally, the product was dissolved in 95% H$_2$SO$_4$ (50 mL), obtaining a dark violet solution. The green solid recovered by precipitation in crushed ice (300 mL) was water washed until pH=7, and then with dichloromethane (10 mL) and acetone (10 mL). The resulting solid was dried overnight at 80°C under vacuum.

Before the immobilization of **7b** on silica, the support was modified with 3-aminopropyltriethoxysilane (3-APTES) (14). For complex immobilization 0.06 g of **7b** was dissolved in pyridine (20 mL), and the mixture was stirred for 7 h at room temperature. A suspension of 1 g of amino-modified silica in pyridine (10 mL) was stirred for 5 min in a round bottom flask; then, the complex-pyridine mixture was added dropwise under flowing argon. After the addition of the complex, the new suspension was stirred at room temperature for 1 h and at reflux temperature for 24 h. After cooling, the green solid obtained was washed with acetone until a colorless filtrate was observed. Finally, the catalyst was dried under vacuum at 80°C for 15 h. The immobilized catalyst was coded **FePcCl$_{16}$-S**. The immobilized **FePcCl$_{16}$-S** complex was characterized by diffuse reflectance UV-vis analysis in a Perkin-Elmer Precisely/Lambda 35 spectrophotometer without previous preparation. The iron content was determined by ICP-MS method.

The oxidation reactions were performed in a 25 mL round bottom flask. In a typical reaction the catalyst (0.5 % Fe mol) was added to 0.125 M olefin solution in acetone; then dry TBHP (3.5 M in CH$_2$Cl$_2$ or in PhCl) was added in one step, and the reaction mixture was stirred and heated in an oil bath at 40°C for 7 h. For the allylic oxidation of cyclohexene with isotopically labelled oxygen (^{18}O$_2$) the following procedure was carried out: the suspension of the catalyst (0.5% Fe mol) in cyclohexene (4 mL, 0.125 M) was frozen and the air in the reactor was evacuated and replaced by an oxygen (21% mol) – argon (79% mol) mixture. Then, the suspension was allowed to warm at room temperature and 1.3 mmol of degasified TBHP was added to the solution and the reaction mixture was stirred at 40°C for 3 h.

The products were identified by comparing the retention times of the reaction products with commercial compounds, and by GC-MS analysis in a Hewlett-Packard 5973/6890 GC equipped with an electron impact ionization at 70 eV detector and a cross-linked 5% PH ME siloxane (0.25 mm coating) capillary column. The reaction products were separated from the catalyst with filter syringes and analyzed in an Agilent 4890D and a Varian 3400 GC equipped with a flame ionization detector, and CP-Sil 8CB (30 m x 0.53 mm x 1.5 μm) and DB-1 (50 m x 0.52 mm x 1.2 μm) columns, respectively. Decane was used as an internal standard. The catalyst was thoroughly washed after reaction with acetonitrile, acetone and water, and dried overnight under vacuum at 40°C.

Results and Discussion

Catalyst characterization. The green color of the **FePcCl$_{16}$- S** (21.4 μmol Fe/g) catalyst is attributed to the $\pi - \pi^*$ ligand transitions of the C-N bonds in the phthalocyanine ring (15). Also the green color can be associated with the presence of monomeric complex species on the support (14). The catalyst exhibited four UV-vis bands at 365, 434, 620, and 668 nm. The band around 365 nm can be assigned to $\pi - \pi$ transition of the C = C double bond (16), which suggests that the complex remained intact during the immobilization process. The bands at 434 and 668 nm confirm the presence of monomeric species (14, 17). Furthermore, the UV-vis band at 620 nm suggests the presence of some μ-oxo dimeric species in the catalyst (18).

Catalytic activity. Figure 49.1 shows the catalytic activity of **FePcCl$_{16}$-S** for the allylic oxidation of **1**, **2**, and **3**. High olefin conversion was observed; the highest ketone yield was obtained with cyclohexene. The lower ketone yields

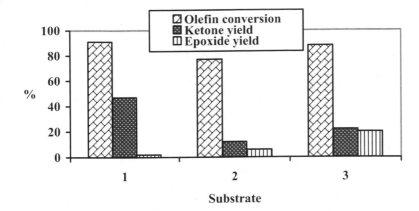

Figure 49.1. Catalytic allylic oxidation of **1**, **2** and **3** with **FePcCl₁₆-S**. Reaction conditions: 0.125 M olefin in acetone, 0.5 % mol **FePcCl₁₆-S**, 1.3 mmol of dry TBHP, 40°C, 3 h.

obtained from tested monoterpenes can be associated with the larger amount of oxidized sites present in these substrates, especially limonene (19, 20).

The formation of ROO* and RO* radicals, and M^{IV}=O species is expected when phthalocyanines and porphyrins are used as catalysts (21, 22). The formation of the epoxide, Figure 49.1, may be associated with the attack of metal oxo species (Fe = O) to the double bond (23). For α-pinene this attack is possibly favored by its rigid structure that causes an orbital overlapping, making the allylic hydrogen abstraction difficult (24).

It has been reported that molecular oxygen plays an important role in the allylic oxidation of olefins with TBHP (25, 26). Rothenberg and coworkers (25) proposed the formation of an alcoxy radical via one-electron transfer to hydroperoxide, Equation 4, as the initiation step of the allylic oxidation of cyclohexene in the presence of molecular oxygen. Then, the alcoxy radical abstracts an allylic hydrogen from the cyclohexene molecule, Equation 5. The allylic radical (**8**) formed reacts with molecular oxygen to yield 2-cyclohexenyl hydroperoxide (**10**), Equations 6 and 7, with the radical (**9**) as intermediate. The 2-cyclohexen-1-ol (**11**) and cyclohexene oxide (**12**) are formed from the decomposition of **10** (27, 28), Equation 8, and then the alcohol is re-oxidized by M=O species to the corresponding ketone 2-cyclohexen-1-one (29), equation 9.

The role of oxygen on the allylic oxidation of cyclohexene over the **FePcCl₁₆-S**/TBHP catalytic system was determined by using $^{18}O_2$ labelled oxygen. Since more than 70% of the main cyclohexene oxidation products, **4**, **11**, and **12**, had labelled oxygen, we can assure that molecular oxygen acts as co-oxidant. However, under the reaction conditions the over-oxidation of **4** seems to be unavoidable. Labelled 2, 3- epoxy-1-cyclohexanone (**13**), 2-cyclohexen-1, 4-dione (**14**), and 4-hydroxy-2-cyclohexen-1-one (**15**) were detected as reaction products.

$$t\text{-BuOOH} + M^n \longrightarrow t\text{-BuO*} + M^{n+1} + OH^- \qquad (4)$$

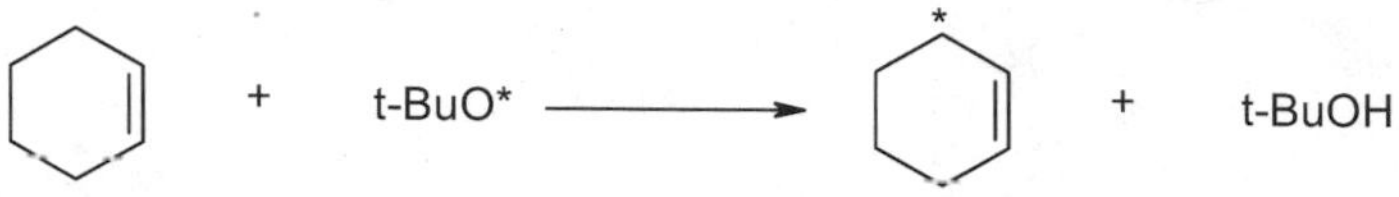

(1) + t-BuO* ⟶ (8) + t-BuOH   (5)

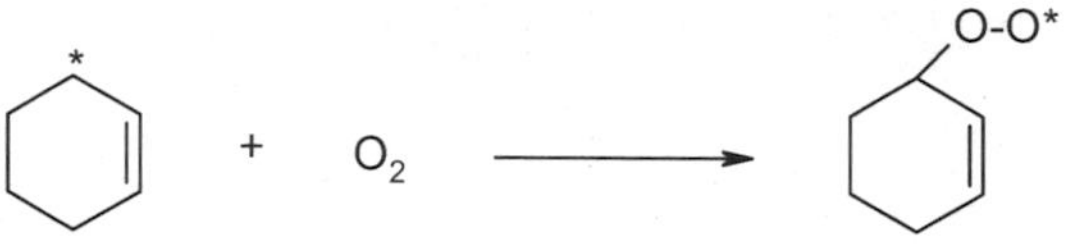

(8) + O_2 ⟶ (9)   (6)

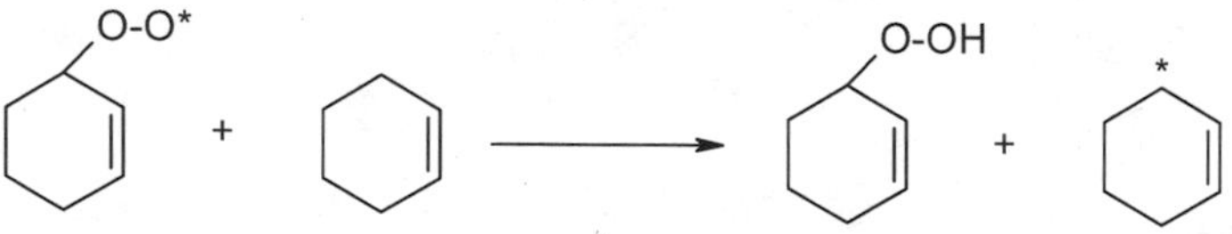

(9) + (1) ⟶ (10) + (8)   (7)

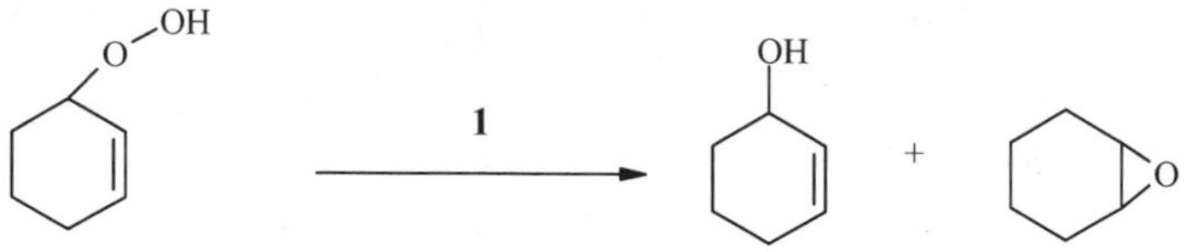

(10) $\xrightarrow{\ 1\ }$ (11) + (12)   (8)

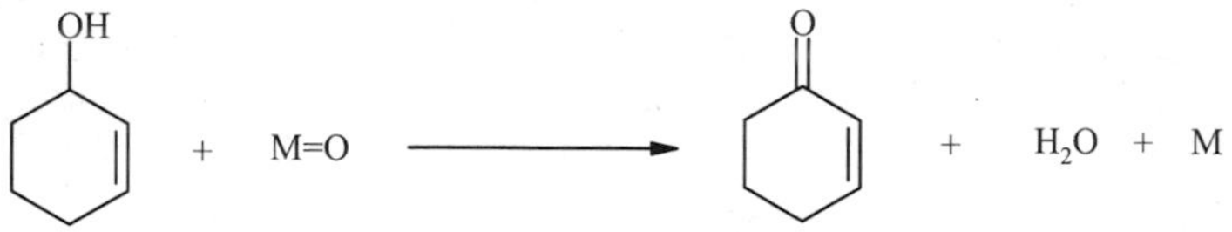

(11) + M=O ⟶ (4) + H_2O + M   (9)

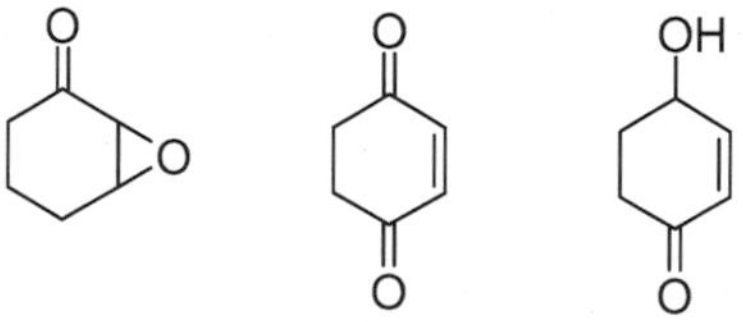

(13)   (14)   (15)

The low allylic alcohols (≤ 6%) yield suggests that the alcohol is re-oxidized to the corresponding ketone, probably by $Fe^{IV}=O$ species (M = O) (29). Alcohol oxidation tests (Figure 49.2) confirm that the formation of alcohol is an intermediate step in the synthesis of the corresponding ketone, equations 9 – 11.

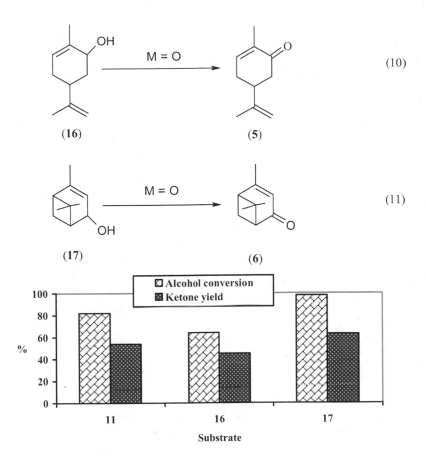

$$(10)$$

$$(11)$$

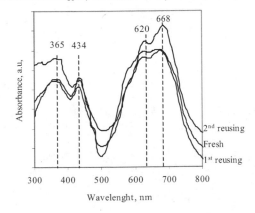

Figure 49.2. Catalytic oxidation of the allylic alcohols **11**, **16** and **17** with **FePcCl₁₆-S** to the corresponding α,β-unsaturated ketones. Reaction conditions: 0.125 M allylic alcohol in acetone, 0.5 % mol **FePcCl₁₆-S**, 1.3 mmol of dry TBHP, 40°C, 3 h.

Figure 49.3. UV-vis spectra of **FePcCl₁₆-S** before and after cyclohexene oxidation.

Stability tests of the **FePcCl$_{16}$-S** catalyst confirms that this material is stable under the reaction conditions, as it was previously reported (17). Neither leaching of the complex nor significant changes in the UV-vis spectra of the catalyst after reaction (Figure 49.3) were detected. The catalyst can be reused at least three times without a significant activity loss.

Conclusions

The oxidation of cyclohexene, limonene and α–pinene to their corresponding α, β-unsaturated ketones was carried out with the **FePcCl$_{16}$-S**/tert-butyl hydroperoxide (TBHP) catalytic system. The reactivity and the structure of the molecule are two key factors in the allylic ketone formation. A preferential attack of the active oxidizing species on the allylic C-H bond instead of on the C=C bond was observed in the oxidation of cyclohexene; the formation of the α,β-unsaturated ketone, 2-cyclohexen-1-one, was favored. The yields to carvone and verbenone were low, probably because of the large amount of oxidized sites in limonene and α-pinene. In the oxidation of α-pinene the similar yield of the epoxide and the α,β-unsaturated ketone can be associated with an allylic hydrogen overlapping. The allylic oxidation of cyclohexene in the presence of $^{18}O_2$ confirm that molecular oxygen participates in the formation of the main products, together with metal oxo species, M=O. Alcohol oxidation tests suggest that the alcohol is formed first and then it is re-oxidized to the corresponding ketone.

Acknowledgements

The authors acknowledge financial support of Colciencias through CENIVAM Center of Excellence RC No.432, Universidad de Antioquia (Colombia) and CNRS (France). L.M.G. acknowledges a doctoral fellowship from Colciencias.

References

1. A. Sakthivel, S.E. Dapurkar and P. Selvam, *Appl. Catal. A*, **246**, 283 (2003).
2. P.A. Wender and T.P. Mucciaro, *J. Am. Chem. Soc.,* **114**, 5878 (1992).
3. P.N. Davey, C.P. Newman, W.A. Thiam and C.-L. Tse, US Patent 6500989 B1, (2002).
4. E.F. Murphy, T. Mallat and A. Baiker, *Catal. Today*, **57**, 115 (2000).
5. T. Joseph, D.P. Sawant, C.S. Gonipath and S.B. Halligudi, *J. Mol. Catal. A*, **184**, 289 (2002).
6. J.M. Thomas and R. Raja, *Annu. Rev. Mater. Res.*, **35**, 315 (2005).
7. K. Kasuga, K. Tsubai, M. Handa, T. Sugimori and K. Sagabe, *Inorg. Chem. Commun.*, **2**, 507 (1999).
8. M. Lajunen and A.M.P. Koskinen, *Tetrahedron Lett.*, **35**, 4461 (1994).
9. A.D. Silva, M.L. Patitucci, H.R. Bizzo, E. D′Elia and O.A.C. Antunes, *Catal. Commun.*, **3**, 435 (2002).
10. L.M. González, A.L. Villa de P., C. Montes de C. and A.B. Sorokin, *Tetrahedron Lett.* **47**, 6465 (2006).

11. A.B. Sorokin, S. Mangematin and C. Pergrale, *J. Mol. Catal. A*, **182–183**, 267 (2002).
12. C. Pérollier and A.B Sorokin, *Chem. Commun.* **14**, 1548 (2002).
13. J. Metz, O. Schneider and M. Hanack, *Inorg. Chem.*, **23**, 1065 (1984).
14. A.B. Sorokin and A. Tuel, *New J. Chem.* **23**, 473 (1999).
15. K. Balkus Jr., A. Gabrielov and S. Bell, *Inorg. Chem.* **33**, 67 (1994).
16. M.M. El-Nahass, K.F. Abd-El-Rahman and A.A.A Darwish, *Mater. Chem. Phys.,* **92**, 185 (2005).
17. M.N. Golovin, P. Seymour, K. Jayaraj, Y. Fu and A.B.P. Lever, *Inorg. Chem.,* **29**, 1719 (1990).
18. C. Pergrale and A.B. Sorokin, *C.R. Acad. Sci., Ser. IIc: Chim.* **3**, 803 (2000).
19. B.A. Allal, L. El Firdoussi, S. Allaoud, A. Karim, Y. Castanet and A. Mortreux, *J. Mol. Catal. A*, **200**, 177 (2003).
20. M.R. Maróstica Jr. and G.M. Pastore, *Quimica Nova*, **30**, 382 (2007).
21. A.B. Sorokin and A. Tuel, *Catal. Today,* **57**, 45 (2000).
22. N. Sehlotho and T. Nyokong, *J. Mol. Catal. A,* **209**, 51 (2004).
23. J. Poltowicz, E.M. Serwicka, E. Bastardo-Gonzalez, W. Jones and R. Mokaya, *Appl. Catal. A*, **218**, 211 (2001).
24. G. Rothenberg, Y. Yatziv and Y. Sasson, *Tetrahedron*, **54**, 593 (1998).
25. G. Rothenberg, H. Weiner and Y. Sasson, *J. Mol. Catal. A*, **136**, 253 (1998).
26. M.K. Lajunen, M. Myllyskoski and J. Asikkala, *J. Mol. Catal. A,* **198**, 223 (2003).
27. H. Tohma, T. Maegawa, S. Takizawa and Y. Kita, *Adv. Synth. Catal.*, **344**, 328 (2002).
28. G. Yin, M. Buchalova, A.M. Danby, C.M. Perkins, D. Kitto, J.D. Carter, W.M. Scheper and D.H. Busch, *Inorg. Chem.*, **45**, 3467 (2006).
29. N.I. Kuznetsova, L.I. Kuznetsova, N.V. Kirillova, L.G. Detusheva, V.A. Likholobov, M.I. Khramov and J.E. Ansel, *Kinet. Catal.*, **46**, 206 (2005).

50. Reaction Progress Kinetic Analysis: A Powerful Methodology for Streamlining Mechanistic Analysis of Complex Organic Reactions

Natalia Zotova[2], Fernando Valera[2] and Donna G. Blackmond[1,2]

[1]Department of Chemistry

[2]Department of Chemical Engineering and Chemical Technology

Imperial College, London SW7 2AZ United Kingdom

d.blackmond@imperial.ac.uk

Abstract

The investigation of the driving forces and robustness of proline-catalyzed aldol reaction is performed by utilizing the methodology of reaction progress kinetic analysis.

Introduction

The past several decades have witnessed tremendous growth in the application of catalytic routes to the synthesis of complex organic molecules, driven by academic and industrial discoveries of efficient, selective catalysts for a wide variety of liquid- and multi-phase organic transformations. These developments coincide with the beginning of major efforts in the pharmaceutical industry toward streamlining the costs of manufacture and waste disposal in an ever more economically competitive and ecologically aware market. Complex liquid and multi-phase reactions carried out in batch or semi-batch mode make up the bulk of pharmaceutical reaction steps, including as prominent examples homogeneously catalyzed C-C and C-N coupling reactions and heterogeneous catalytic hydrogenations. Development of methodologies to obtain a rapid and comprehensive understanding of the driving forces of a reaction, as well as its robustness, are important goals in pharmaceutical research. These goals have brought an increased focus on the role of kinetic studies of these reactions.

One of the most important developments in pharmaceutical process research of the past several decades has been the introduction of bench-scale experimental tools for efficient and accurate in situ monitoring of chemical reactions. Such tools include reaction calorimetry, which measures a property directly proportional to reaction rate (a "differential" method), and spectroscopic methods such as FTIR spectroscopy, which measures a property proportional to concentration, or the integral of reaction rate. Monitoring the temporal profile of a reaction's progress can provide invaluable information to help accelerate process development and scale-up, can help to probe proposed reaction mechanisms for further catalyst design and optimization, and can play an important role in the demonstration of good manufacturing practice for regulatory approval of a drug process.

Interestingly, even with significant advances in experimental and computational tools, kinetic analysis in the 21st century is carried out largely by the same methods used nearly a century ago. These methods focus primarily on acquiring initial rate data under batch reaction conditions. The data obtained from a series of such experiments are then plotted in a linearized version of the rate equation in order to extract kinetic parameters from the slope and intercepts of the resulting straight line. Reactions involving two or more different substrate concentrations are problematic for this approach, however, and typically must be analyzed in a sequence of experiments where one substrate's concentration is held constant while probing the concentration dependence of the other. This classical kinetic methodology is time-consuming, involving a large number of repetitive experiments, and in most cases the reactions are necessarily carried out under concentration conditions far removed from those of practical operation. Further, methodologies based on initial rate data provide no information about critical issues such as catalyst activation or deactivation and product inhibition.

The purpose of this work is to highlight the methodology we term "reaction progress kinetic analysis,"[1] which streamlines kinetic studies by exploiting the extensive data sets available from accurate in situ monitoring of the entire reaction progress under practical operating conditions, where two or more concentration variables are changing simultaneously. This methodology involves the graphical manipulation of a critical minimum set of carefully designed experiments; the key to this graphical approach lies in defining a reaction parameter termed the *excess*, or [*e*], which is related to the stoichiometry of the reaction under study. We have developed two different experimental protocols: (a) *"different excess"* experiments; and (b) *"same excess"* experiments. Employing these protocols, reaction progress kinetic analysis can provide a rapid and comprehensive understanding of the kinetics of a steady-state catalytic cycle in a fraction of the time and number of experiments that would be required in a classical kinetic approach featuring initial rate measurements.

Kinetic Studies Using In Situ Tools

Our experimental technique of choice in many cases is reaction calorimetry.[2] This technique relies on the accurate measurement of the heat evolved or consumed when chemical transformations occur. Consider a catalytic reaction proceeding in the absence of side reactions or other thermal effects. The energy characteristic of the transformation – the heat of reaction, ΔH_{rxn} – is manifested each time a substrate molecule is converted to a product molecule. This thermodynamic quantity serves as the proportionality constant between the heat evolved and the reaction rate (eq. 1). The heat evolved at any given time during the reaction may be divided by the total heat evolved when all the molecules have been converted to give the fractional heat evolution (eq. 2). When the reaction under study is the predominant source of heat flow, the fractional heat evolution at any point in time is identical to the fraction conversion of the limiting substrate. Fraction conversion is then related to the concentration of the limiting substrate via eq. (3).

First and foremost in any kinetic study using reaction calorimetry, we must confirm the validity of the method for the system under study by showing that eq. (2) holds. Comparing the temporal fraction conversion obtained from the heat flow measurement with that measured by an independently verified measurement technique, such as chromatographic sample analysis or FTIR or NMR spectroscopy, confirms the use of the calorimetric method. This is shown in Figure 50.1 for the proline-mediated intermolecular aldol reaction shown in Scheme 50.1.[3,4] Sample analysis is typically the most common method of validating in situ reaction progress data; while the number of sample data points available in a sampling study alone is rarely sufficient to carry out the graphical manipulations of reaction progress kinetic analysis. The comparison of Figure 50.1 provides assurance that the data provided by the in situ method is valid for this treatment.

$$q = \Delta H_{rxn} \cdot \left(\frac{reaction}{volume} \right) \cdot rate \tag{1}$$

$$fraction\ conversion = f = f_{final} \cdot \frac{\int_0^t q(t)\,dt}{\int_0^{t(final)} q(t)\,dt} \tag{2}$$

$$\left[substrate \right] = \left[substrate \right]_0 \cdot (1 - f) \tag{3}$$

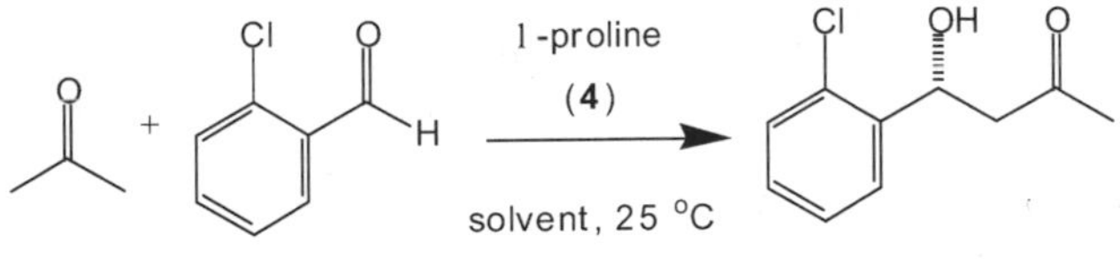

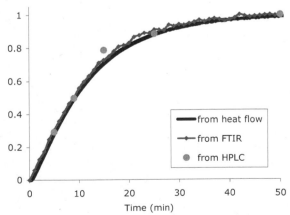

Scheme 50.1. Proline-mediated aldol reaction.

Figure 50.1. Comparison of conversion vs. time for the reaction of Scheme 50.1 using HPLC sampling of product concentration to in situ monitoring by FTIR spectroscopy and reaction calorimetry.

It is important to point out that the methodology described in this work is not limited to studies using reaction calorimetry but may be employed using any in situ experimental technique that provides accurate (rate, concentration) data pairs. Once we have established the validity of our experimental technique for following reaction rate, we turn to our experimental protocols for rapid kinetic analysis. We have introduced the terminology "graphical rate equation" to describe plots constructed from graphical manipulations of kinetic data that relate reaction rate behaviour to the reaction's concentration dependences. In the next sections, we describe this graphical approach using "different excess" and "same excess" experiments.

A full kinetic study of the proline-mediated aldol reaction based on a detailed catalytic reaction mechanism will be published separately.

"Different Excess" Experiments

We first consider the simple example of an uncatalyzed Diels-Alder reaction shown in Scheme 50.2 in order to demonstrate the use of "different excess" experiments. The Diels-Alder reaction is known to exhibit second overall order kinetics, as shown in eq. (4). We demonstrate with this known case how reaction progress kinetic analysis may be used to extract the reaction orders in both substrate concentrations, [5] and [6].[1b]

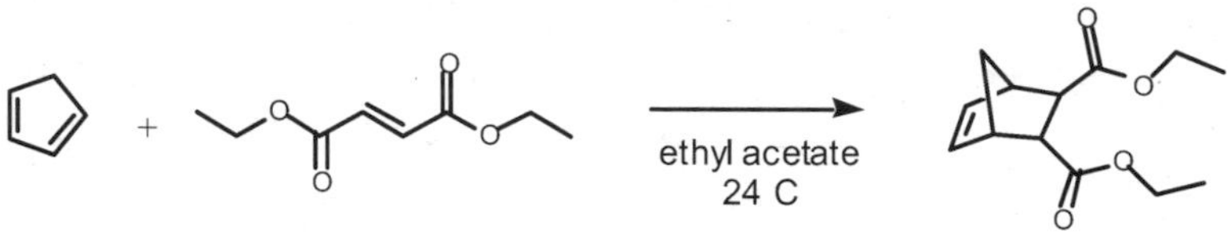

5   **6**   **7**

Scheme 50.2. Diels-Alder reaction.

$$rate = k \cdot [5] \cdot [6] \tag{4}$$

Figure 50.2a shows reaction heat flow as a function of time for two reactions of Scheme 50.2 carried out in sequence. In this protocol, an aliquot of substrate **5** at 0.2 M is injected into a reaction vial containing substrate **6** at 0.61 M, resulting in the first heat flow curve of Figure 50.2a. When this curve returns to the baseline, signalling the end of the reaction, a second identical aliquot of **5** is injected into the vial, and the second heat flow curve in Figure 50.2a is observed. The difference between the initial concentrations of substrates **5** and **6** equals 0.41 M in the first reaction and 0.21 M in the second reaction of the sequence. We term this *difference* between the initial concentrations of the two substrates the *excess*, or [e], with units of molarity, eq. (5). Making use of the reaction stoichiometry, which tells us that for each molecule of consumed diene **5**, one molecule of dienophile **6** is also consumed, we see that the quantity [e] allows us to relate the two substrates' concentrations at any point during the reaction, eq. (6).

$$[e] = [6]_0 - [5]_0 \tag{5}$$

$$[6] = [e] + [5] \tag{6}$$

The excess [*e*] does not change over the course of the reaction, provided that the density does not change appreciably. The excess [*e*] can be large, small, positive or negative. However, under practical conditions of practical synthetic experiments, [*e*] is usually small, and both concentrations [**5**] and [**6**] change over the course of the reaction in the manner dictated by eq. (6). Thus at any point in time we know the values of both even by measuring only one.

The temporal reaction heat flow data may be graphically manipulated to reveal the overall second order dependence in a quantitative manner. Reaction heat flow is converted to reaction rate using eq. (1), and the concentration of the limiting substrate **5** may be calculated according to eq. (3). From these calculations we may construct the plot in Figure 50.2b of reaction rate vs. [**5**]. The reaction is known to be first order in both [**5**] and [**6**]: these plots reveal the curvature typical of overall second order kinetics.

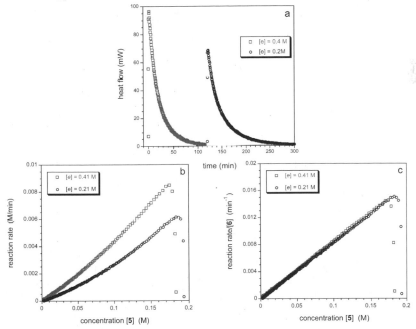

Figure 50.2. Reaction progress data for a one-pot, two reaction sequence for the reaction of Scheme 50.2. (a) reaction heat flow vs. time; (b) reaction rate vs. [**5**]; (c) reaction rate/[**6**] vs. [**5**].

A simple procedure to "normalize" the value of rate by concentration is shown in eq. (7), in which both sides of the rate eq. (4) have been divided by the substrate [**6**]. Eq. (7) gives the formula for a straight line if the function on the left side of the equation is plotted on the y-axis and the quantity [**5**] on the x-axis. Thus the data in Figure 50.2b are replotted as this new "graphical rate equation" in Figure 50.2c.

$$\textit{normalized rate} = \frac{rate}{[\mathbf{6}]} = k \cdot [\mathbf{5}] \tag{7}$$

Two key features of reaction progress kinetic analysis are revealed in the plots in Figure 50.2c. First, we see that the curvature observed when plotting the data as in Figure 50.2b disappears, and we obtain a straight line with slope = k. This confirms that the reaction is first order in [**5**], the quantity plotted on the x-axis. Second, we now see that the two curves from two reactions carried out at two different [e] values fall on top of another – i.e., they "overlay." The significance of this may be understood looking back at eq. (6): any two reactions with different values of [e] have *different* concentrations of **6** at any given concentration of **5**. Thus the "overlay" in Figure 50.2c shows that the quantity (rate/[**6**]) is not a function of [**6**]; this can be true only if the reaction exhibits first order kinetics in the "normalized" substrate [**6**].

This simple example of a non-catalytic reaction demonstrates how a reaction rate law may be comprehensively defined in two substrates by just *two* reaction progress experiments employing two *different* values of excess [e]. A classical kinetics approach using initial rate measurements would require perhaps a dozen separate initial rate or pseudo-zero-order experiments to obtain the same information.

Many catalytic reactions do not, however, exhibit simple integer orders in substrate concentrations. In such cases, our "graphical rate equation" derived by normalizing the rate by the concentration of one substrate will not result in the "overlay" we saw for the Diels-Alder reaction. This is illustrated in Figure 50.3a, which shows data from three "different excess" experiments of the proline-mediated aldol reaction in Scheme 50.1 plotted as the normalized rate function (rate/[**1**]) vs. [**2**].[5] The fact that the three curves do *not* overlay tells us that the reaction does *not* exhibit simple first order kinetics in the normalized substrate concentration, [**1**]. The true kinetic order in [**1**] may be a complex function as is characteristic of catalytic reactions and could require detailed kinetic modelling and further mechanistic analysis to obtain a full mechanistic understanding. However, the approach we have outlined in this work allows us to assess the magnitude of the driving force attributable to [**1**] and [**2**] via graphical manipulation by assuming a power-law form for the rate law as in eq. (8), where the catalyst driving force has been lumped into k'. The normalization of rate is now carried out by dividing both sides of eq. (8) by [**1**]^x to give eq. (9).

$$rate = k' \cdot [\mathbf{1}]^x \cdot [\mathbf{2}]^y \tag{8}$$

$$\frac{rate}{[\mathbf{1}]^x} = k' \cdot [\mathbf{2}]^y \tag{9}$$

The next task is to find a value for x that causes the three experimental data sets to exhibit "overlay." As shown in Figure 50.3b, overlay is found for $x \approx 0.5$. We also note that at this value of x, the curves become close to straight lines, meaning that $y \approx 1$. Thus a reasonable description of the reaction's concentration driving forces is given by eq. (8) with $x \approx 0.5$ and $y \approx 1$.

We may compare our graphical result with the result obtained from solving for x and y by nonlinear regression fitting of the experimental rate curves to the power law form of eq. (8). Carrying this out using the Excel Solver tool and the program Solvstat,[6] we obtain $x = 0.46 \pm 0.002$ and $y = 0.90 \pm 0.001$. This reasonable agreement between our graphical methodology and the more mathematically rigorous approach lends support to the concept that kinetic information from experimental data sets may be extracted from graphical as well as mathematical manipulations.

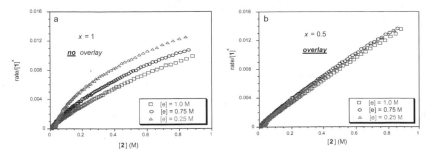

Figure 50.3. Comparison of rate/$[1]^x$ vs. [aldehyde] **2** for three reactions of Scheme 50.1 carried out at different excess [e] as shown in the Figure legends. $[2]_0 = 1$ M, $[4] = 0.1$ M in each case. (a) $x = 1$; (b) $x − 0.5$.

The values of $x = 0.5$ and $y = 1$ for the kinetic orders in acetone [**1**] and aldehyde [**2**] are not true kinetic orders for this reaction. Rather, these values represent the power-law "compromise" for a catalytic reaction with a more complex catalytic rate law that corresponds to the proposed steady-state catalytic cycle shown in Scheme 50.3.[7] In the generally accepted mechanism for the intermolecular direct aldol reaction, proline reacts with the ketone substrate to form an enamine, which then attacks the aldehyde substrate.[4] A reaction exhibiting saturation kinetics in [**1**] and rate-limiting addition of [**2**] can show "apparent" power law kinetics with both x and y exhibiting orders between zero and one.

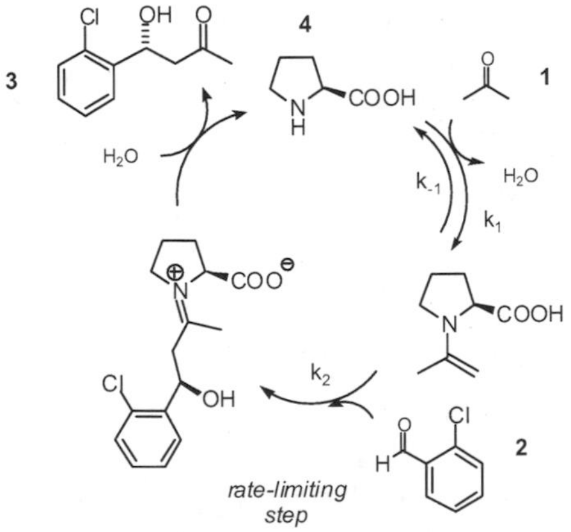

$$rate = \frac{k_1 \cdot k_2 \cdot [1] \cdot [2] \cdot [4]}{k_{-1} \cdot [H_2O] + k_1 \cdot [1] + k_2 \cdot [2]}$$

Scheme 50.3. Proposed mechanism for the proline-mediated aldol reaction of Scheme 50.1.

"Same Excess" Experiments

We now describe a second set of experiments that are critical to reaction progress kinetic analysis when it is carried out on catalytic systems. Determining reaction orders in [substrate] from rate measurements in catalytic reactions implicitly requires that the active catalyst concentration remains unchanged during the kinetic measurement.[8] Catalyst activation and deactivation are common phenomena that can change the active catalyst concentration, both in homogeneous and in heterogeneous catalytic reactions. Making certain that such effects do not complicate the analysis is critical in kinetic studies of catalytic reactions. Initial rate methods are often employed in the hope of avoiding the need to address the problem of catalyst deactivation. Extensive spectroscopic studies are sometimes carried out to try to observe changes in the concentration of catalytic intermediates, but a case must be made to show that such experiments do in fact probe *active* catalyst species.

Reaction progress kinetic analysis offers a reliable alternative method to assess the stability of the active catalyst concentration, again based on our concept of excess [e]. In contrast to our "different excess" experiments described above, now we carry out a set of experiments at the *same* value of excess [e]. We consider again the proline-mediated aldol reaction shown in Scheme 50.1. Under reaction conditions, the proline catalyst can undergo side reactions with aldehydes to form inactive cyclic species called oxazolidinones, effectively decreasing the active catalyst concentration. It has recently been shown that addition of small amounts of water to the reaction mixture can eliminate this catalyst deactivation.[9] Reaction progress kinetic analysis of experiments carried out at the ***same excess*** [e] can be used to confirm the deactivation of proline in the absence of added water as well to demonstrate that the proline concentration remains constant when water is present.

Table 50.1 shows two sets of conditions for carrying out the proline-mediated aldol reaction in Scheme 1, employing the same catalyst concentration but different initial concentrations of the two reactants chosen such that the value of the excess, [e], is the same in the two experiments. Note that more conventionally reported parameters – for example, the number of equivalents of acetone and the catalyst mol% – are ***not*** the same for the two experiments.

Table 50.1. Comparison of the initial conditions of two reactions carried out at the same excess [e] and their concentrations during the reaction.

Component:	Acetone **1**	Aldehyde **2**	[e]	Product **3**
Exp. 1: Initial conditions	2.5 M	0.5 M	2 M	0 M
Exp. 2: Initial conditions	**2.25 M**	**0.25 M**	2 M	0 M
Exp. 1: 50% conversion	**2.25 M**	**0.25 M**	2 M	0. 25 M

What is the significance of this choice of conditions? Table 50.1 shows that the initial conditions of Experiment 2 are identical to the concentrations of reactants found in Experiment 1 at the point when it reaches 50% conversion (Table 50.1 bolded entries). In fact, from this point onwards the two experiments exhibit identical [**1**] at any given [**2**] throughout the course of both reactions. The only differences between the two experiments are (1) the catalyst has performed more turnovers in experiment 1; and (2) the reaction vial has a greater concentration of product in experiment 1.

This "same [*e*]" experimental protocol leads to a graphical "overlay" plot that yields valuable kinetic information: if the two experiments described in Table 50.1 are plotted together as reaction rate vs. [**2**], the two curves will fall on top of one another ("overlay") over the range of [**2**] common to both *only* if the rate is not significantly influenced by changes in the overall catalyst concentration within the cycle, including catalyst activation, deactivation or product inhibition. Overlay in "same excess" plots, therefore, may be used to confirm catalyst robustness or identify problems such as catalyst deactivation or product inhibition.

"Same excess" experiments for the aldol reaction of Scheme 50.1 carried out in the absence of water and in the presence of water are shown in Figure 50.4a and 50.4b, respectively. Figure 50.4a shows that the plots do *not* overlay in the absence of water. By contrast, the "overlay" in these "same [*e*]" experiments in Figure 50.4b means that the total concentration of active catalyst within the cycle is *constant* and is the same in these two experiments where water is present. Once conditions are found for obtaining constant catalyst concentration during reactions such as the aldol reaction of Scheme 50.1, we may proceed to our "different excess" experiments to probe substrate concentration dependences, in full confidence that the complicating influence of a changing active catalyst concentration is absent.

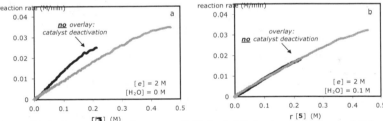

Figure 50.4. Comparison of rate vs. [aldehyde] **2** for two reactions of Scheme 50.1 carried out at the same excess [*e*] with initial conditions as given in Table 50.1 (gray dots: Exp. 1; black dots: Exp. 2). (a) no water added; (b) 0.1 M water added to the reaction.

To summarize what we have learned from our "different excess" and "same excess" experimental protocols and graphical rate equations:
Different Excess:
- **"Overlay" between curves** plotted as rate/[substrate 1]^x vs. [substrate 2] reveals that the reaction exhibits first order kinetics in concentration of the "normalized" substrate 1.
- **The shape of the curve** reveals the reaction order in the concentration of the substrate plotted as the x-axis variable; for example, first order kinetics are revealed when the overlaid curves give a straight line.

Same Excess:
- **"Overlay" between curves** plotted as rate vs. [substrate] reveals that the plots manifest the intrinsic reaction kinetics; rate is not influenced

by catalyst deactivation, catalyst activation, or inhibition or acceleration by product or substrate.

Summary

We showed in this work that by applying the methodology of reaction progress kinetic analysis to data acquired by accurate and continuous reaction monitoring of complex organic reactions, we are able both to provide a quantitative assessment of the reaction orders in two separate substrate concentrations as well as to delineate conditions under which the stability of the catalyst and the robustness of the process is insured. Carrying out this kinetic analysis at the outset of a mechanistic investigation provides a framework for further work aimed at seeking a molecular-level understanding of the nature of the species within the catalytic cycle. In order that any independent mechanistic proposal would be considered convincing, it must be consistent with the global kinetic analysis.

Acknowledgements

Funding from the EPSRC, AstraZeneca, Pfizer and Mitsubishi Pharma is gratefully acknowledged. Stimulating discussions with our co-workers, including Dr. Suju P. Mathew, Dr. Hiroshi Iwamura, Dr. Martin Klussmann, and Dr. Jinu S. Mathew, are gratefully acknowledged.

References

1. a) Blackmond, D.G., *Angew. Chemie Int. Ed.*, **2004**, *44*, 4302; b) Mathew, J. S.; Klussmann, M.; Iwamura, H.; Valera, F.; Futran, A.; Emanuelsson, E. A. C.; Blackmond, D. G. *J. Org. Chem.* **2006**, *71*, 4711.
2. a) Beezer, A. E. *Thermochimica Acta* **2001**, *380*, 205-208; b) Mathot, V. B. F. *Thermochimica Acta* **2000**, *355*, 1-33; c) Kemp, R. B.; Lamprecht, I. *Thermochimica Acta* **2000**, *348*, 1-17; d) Ferguson, H. F.; Frurip, D. J.; Pastor, A. J.; Peerey, K. M.; Whiting, L. F. *Thermochimica Acta* **2000**, *363*, 1; (e) Regenass, W. *J. Thermal Anal.* **1997**, *49* 1661.
3. List, B., Lerner, R.A., Barbas, C.F. III, J. Am. Chem. Soc., **2000**, *122*, 2395.
4. a) J. Seayad, B. List, *Org. Biomol. Chem.* **2005**, *3*, 719; b) B. List, *Acc. Chem. Res.* **2004**, *37*, 548; c) Notz, F. Tanaka, C. F. Barbas, III, *Acc. Chem. Res.* **2004**, *37*, 580; d) P. I. Dalko, L. Moisan, *Angew. Chem. Int. Ed.* **2004**, *43*, 5138.
5. Zotova, N., Iwamura, H., Mathew, S.P., Blackmond, D.G., in: *Trends in Process Chemistry*, ed. By T. Braish and K. Gadasamatti, CRC Press: New York, in press.
6. E.J. Billo, *Excel for Chemists*, Wiley-VCH, New York, **2001**.
7. Blackmond D. G., Schultz T., Mathew J.S., Loew C., Pfaltz A., *Synlett* **2006**, 3135.
8. Nyberg, A. I.; Usano, A.; Pihko, P. M. *Synlett* **2004**, 1891.

51. Dinuclear Copper(II) Complexes for Regioselective Glycoside Oxidation

Susanne Striegler

Department of Chemistry and Biochemistry, Auburn University, 179 Chemistry Building, Auburn, AL 36849

Susanne.Striegler@auburn.edu

Abstract

The combination of the ability of a dinuclear copper(II) complex to discriminate between sugars and its capacity to oxidize primary alcohols is a topic of our ongoing investigations. Towards this end, we studied the oxidation of methyl-α-D-glucopyranoside into the corresponding glucopyranosyl-6-aldehyde using oxygen in air and the developed metal complex. Preliminary data reveal catalytic transformation in aqueous alkaline acetonitrile. The yields obtained, however, are very low, most likely due to the high reactivity of the aldehyde in alkaline solution. Current efforts to develop an assay to capture and stabilize the sugar-6-aldehydes by the addition of formaldehyde or ethylamine as additives are discussed.

Introduction

Obesity has reached epidemic proportions in most industrialized countries over the last two decades. Many serious diseases, such as stroke, diabetes mellitus and other chronic conditions, are directly related to being overweight (1, 2). The number of adults who are classified as *extremely obese* tallies in the millions for the United States alone (3). The number of children that are most likely obese adults for the rest of their lives has tripled in the same time period (3).

This has led to an interest in unnatural sugars as food components to fight obesity. Major emphasis has been given to the preparation of 5-C-(hydroxymethyl)hexoses (**1**) that resist metabolism by mammalian enzymes and by intestinal anaerobic bacteria (4). The second hydroxymethyl group at C-5 in **1** not only prevents degradation of the hexose framework, but also inhibits cleavage of the glycosidic linkages in simple glycosides and in disaccharides containing this unit (4, 5). An aldol-type condensation of hexose-6-aldehydes (**2**) and formaldehyde in alkaline solution yields an intermediate **3** (step i, eq. 1) that subsequently undergoes a spontaneous Cannizzaro reduction with excess formaldehyde to afford **1** (step ii, eq. 1).

Low-fat food prepared from unnatural sugars **1** will have a zero calorie value and might contribute to battling obesity. The availability of **2** as precursor compound for the synthesis of **1**, however, is a current bottleneck. Sugar substrates

(1)

that are unreactive to the enzyme galactose oxidase are therefore only accessible by elaborate chemical methods (6-12).

Oxidation of alcohol groups in carbohydrates is conventionally achieved after protection of the remaining other hydroxyl groups using stoichiometric amounts of two-electron oxidants, such as iodosylbenzene (13), amine N-oxides (8, 14, 15), peroxides (16), or transition metal complexes (17). The reagents can be used in catalytic amounts when reoxidation by molecular oxygen, sodium hypochlorite or other oxidants are provided (8). Unselective oxidation of both primary and secondary hydroxyl groups, overoxidation of the primary alcohols to carboxylic acids (8, 17) or limited catalyst stability during the reaction (12) reduce the applicability of these methods for the oxidation of hexopyranoses into hexose-6-aldehydes. A catalytic system consisting of bipyridine, copper(II) bromide and 2,2,6,6-tetramethylpiperidine-N-oxyl (TEMPO) has been recently demonstrated to prevent overoxidation during aerobic oxidation of benzyl alcohol in aqueous acetonitrile (18, 19).

Based on these results, a novel method for the synthesis of hexose-6-aldehydes from natural carbohydrate sources has been developed by us. Preliminary results for the regioselective oxidation of the primary alcohol group in glycosides by the dinuclear copper(II) complex *N*, *N*-bis[(2-pyridylmethyl)-1,3-diaminopropan-2-olato] (μ-acetato) dicopper(II) perchlorate (Cu$_2$(bpdpo), (**4**) are described below.

Experimental Section

Instrumentation. ^{1}H and ^{13}C NMR spectra were recorded on a Bruker AV 400 spectrometer (400.2 MHz for proton and 100.6 MHz for carbon) at 310 K. Chemical shifts (δ) are expressed in ppm; coupling constants (*J*) in Hz. Deuterated DMSO and/or water were used as solvent; chemical shift values are reported relative to residual signals (DMSO: δ = 2.50 for ^{1}H and δ = 39.5 for ^{13}C). ESI-MS data were obtained on a VG Trio-2000 Fisons Instruments Mass Spectrometer with VG MassLynx software, Vers. 2.00 in CH$_3$CN/H$_2$O at 60°C. Isothermal titration calorimetry (ITC) experiments were conducted on a VP isothermal titration calorimeter from Microcal® at 30°C.

Reagents. Dinuclear copper(II) complex **4** was prepared as described earlier (20). Nanopure water was obtained from an EASYpure® II water system by Barnstead (18.2 MΩ/cm). HPLC grade acetonitrile (99.93%) was purchased from

Aldrich and used for all oxidation experiments. TEMPO, sodium hydroxide, sodium sulfide, formaldehyde, ethylamine and methyl-α-D-glucopyranoside were reagent grade or higher and were used as received from Sigma and Aldrich.

Oxidation procedure. Typically, a 50 mL round bottom flask was charged with the dinuclear copper(II) complex **4** (30 mg, 4.57 mmol) dissolved in 9 mL of alkaline aqueous acetonitrile ($CH_3CN/H_2O = 2/1$), with an aqueous layer of 18 mM NaOH. The solutions turned from dark blue to yellow-green after vigorous stirring at ambient temperature for 45 min. Subsequently, 200 μL aliquots of a TEMPO stock solution in acetonitrile were added to adjust the total TEMPO concentration in the final reaction mixtures to 18 mM. The solutions were stirred for another 30 min to allow formation of the catalytically active species (see below, eq. 4) in situ. Darkening of the yellow-green solutions to dark green was observed. Methyl-α-D-glucopyranoside was then added, and the total volume of the solutions adjusted with the solvent mixture to a final volume of 10 mL yielding final substrate concentrations of 450 mM. Sample aliquots of the solution were diluted with 10 μL of a 1 M aqueous Na_2S solution and 100 μL acetonitrile to decompose the catalyst centrifuged and filtered to remove precipitated CuS. The supernatant was evaporated and the residue dissolved in DMSO-d6 and/or D_2O for NMR analysis. For control experiments, the reaction was repeated in the absence of glycoside or of dinuclear copper(II) complex **4**, respectively. In a second set of experiments, all supernatants were treated with an excess of formaldehyde solution or ethylamine, respectively, prior to evaporation. The residue was prepared for NMR analysis as described above.

Results and Discussion

Catalytically active species derived from 4. Spectrophotometric titration of the backbone ligand of the sugar discriminating dinuclear copper(II) complex *N, N*-bis[(2-pyridylmethyl)-1,3-diaminopropan-2-olato] (μ-acetato) dicopper(II) perchlorate (Cu_2(bpdpo), **4**) in the presence of two equivalents of copper(II) ions with sodium hydroxide indicates successive replacement of the bridging acetate anion bound in the solid state with two hydroxyl ions and two water molecules in alkaline aqueous solution (eqs. 2 and 3) (20-22). Two species, $[Cu_2(L_{-H})(OH)]^{2+}$ (**4a**) and $[Cu_2(L_{-H})(OH)_2]^+$ (**4b**), are thus observed in a pH-dependent equilibrium (20).

The addition of TEMPO in acetonitrile to a solution of **4b** in alkaline water leads in situ to formation of the catalytically active dinuclear copper(II) species **4c** (eq. 4) as determined by isothermal titration calorimetry (23). Studies on a model system employing benzyl alcohol further revealed a linear dependence of the oxidation rate on the TEMPO co-catalyst concentration when the molar ratio of **4** : TEMPO is below 2 : 5 (23). The same model system also showed linear dependence

of the oxidation rate on the sodium hydroxide concentration. Using less than equimolar amounts of base compared to the amount of copper(II) ions in the metal complex core of **4** leads to incomplete substrate oxidation indicating that formation of the catalytically active species **4c** requires at least one hydroxyl ion per copper(II) ion.

Catalytic oxidation of methyl-α-D-glucopyranoside. Subsequently, the catalytic aerobic oxidation of methyl-α-D-glucopyranoside (**5**) to the corresponding sugar 6-aldehyde (**6**) with **4c** (5 mol %) was studied *qualitatively* by ^{1}H and ^{13}C NMR spectroscopy (eq. 5).

The ^{1}H NMR spectrum of this mixture, measured in DMSO-d6, revealed a new signal at 8.23 ppm (in DMSO-d6). This signal is assigned to the methine proton at C-6 of **6** and corresponds to the hydrated form of the sugar-aldehyde. A carbonyl signal of hydrated sugar-carbaldehydes has been reported with a chemical shift of 88 ppm in D$_2$O (4, 24, 25). A carbonyl signal above 200 ppm is usually not observed in the presence of traces of water (4). Indeed, a new ^{13}C resonance signal at 87 ppm (**6**, in D$_2$O) is observed that correlates to the carbon C-6 in hydrated **6**. The catalyst backbone ligand, TEMPO and the carbohydrate **5** themselves do not show any signal between 90 and 80 ppm.

Control experiments do not provide evidence for oxidation of the secondary alcohol groups in the glycoside or for degradation of the ligand backbone. A similar regioselectivity was also observed in a benzyl alcohol/1-phenylethanol model system that showed no proof for the oxidation of the secondary alcohol by formation of acetophenone (18, 23, 26).

Furthermore, there is no proof for over-oxidation of the primary alcohol of **5** in a carboxylate or for cleavage of the glycosidic bond and release of methanol under the conditions applied. However, additional signals in the ^{13}C NMR spectrum of the reaction mixture are observed between 80 and 85 ppm, which are ascribed to side products formed from **6** by aldol condensations in alkaline solution.

(6)   excess HCHO / base → (7)   (6)

(7)   excess HCHO → (8)   + HCOO⁻   (7)

Further indirect evidence for the oxidation of the primary alcohol in **5** and the formation of glycoside **6** during the course of the reaction was obtained by electrospray mass spectrometry. Towards this end, excess formaldehyde was added to the reaction mixture after the oxidation of **5** into **6**, and the resulting solution stirred for an additional 30 min at ambient temperature to form the instable intermediate 7 (eq 6). The unnatural sugar 5-hydroxymethyl-α-methylglucoside (**8**) is spontaneously derived from 7 at ambient temperature via a Cannizzaro-like reaction in the presence of excess formaldehyde (eq. 7).

Our attempts to isolate larger amounts of **6** from the solution have failed so far. The mass peak of the sodium adduct of **8** ($M_{8–Na^+} = 246.8$), however, strongly supports the oxidation of **5** into **6** as a prerequisite for the subsequent formation of **8** from 7 as outlined (eqs. 6 and 7). The addition of ethylamine to capture the reactive aldehyde **6** as imine did not result in recognized reaction products.

Conclusions

A regioselective oxidation of an unprotected glycoside with catalytic amounts of a sugar discriminating dinuclear copper(II) species and oxygen from air has been achieved. Experimental evidence for the oxidation reaction was obtained by NMR spectroscopy and the observation of reaction products by ESI mass spectrometry. Stabilization of the highly reactive sugar-6-aldehyde **6** was obtained by condensation with formaldehyde yielding unnatural hydroxymethylglycoside **8**. Further studies will focus on a putative stereoselective interaction of the sugar-discriminating complex with 4,5-unsaturated glycosides during oxidation of their primary alcohol group.

Acknowledgements

Partial support of this work by a grant in the Competitive Research Program, Auburn University, is gratefully acknowledged.

References

1. S. C. Woods and R. J. Seeley, *Int. J. Obes.*, **26 (Suppl. 4)**, S8 (2002).
2. F. X. Pi-Sunyer, *Obesity surgery*, **12 (Suppl. 1)**, 6S (2003).
3. I. Sadaf Farroqi, *Best Pract. Res. Clin. Endocrinol. Metabol.*, **19(3)**, 359 (2005).
4. A. W. Mazur and G. D. Hiler, *J. Org. Chem.*, **62 (13)**, 4471 (1997).
5. A. W. Mazur, M. J. Mohlenkamp, G. Hiler, II, T. D. Wilkins and R. L. Van Tassell, *J. Agric. Food Chem.*, **41 (11)**, 1925 (1993).
6. T. T. Tidwell, *Org. React. (N.Y.)*, **39**, 297 (1990).
7. J. Muzart, *Tetrahedron*, **59 (31)**, 5789 (2003).
8. S. Trombotto, E. Violet-Courtens, L. Cottier and Y. Queneau, *Top. Catal.*, **27 (1-4)**, 31 (2004).
9. L. M. Mirica, X. Ottenwaelder and T. D. P. Stack, *Chem. Rev.*, **104 (2)**, 1013 (2004).
10. P. Chaudhuri, M. Hess, U. Florke and K. Wieghardt, *Angew. Chem., Int. Ed. Engl.*, **37(16)**, 2217 (1998).
11. P. Chaudhuri, M. Hess, J. Mueller, K. Hildenbrand, E. Bill, T. Weyhermueller and K. Wieghardt, *J. Am. Chem. Soc.*, **121 (41)**, 9599 (1999).
12. M. J. Schultz, S. S. Hamilton, D. R. Jensen and M. S. Sigman, *J. Org. Chem.*, **70(9)**, 3343 (2005).
13. R. A. Sheldon and I. W. C. E. Arends, *Adv. Synth. Catal.*, **346 (9 + 10)**, 1051 (2004).
14. P. L. Bragd, H. van Bekkum and A. C. Besemer, *Top. Catal.*, **27 (1-4)**, 49 (2004).
15. C. Annunziatini, M. F. Gerini, O. Lanzalunga and M. Lucarini, *J. Org. Chem.*, **69 (10)**, 3431 (2004).
16. S. S. Stahl, *Angew. Chem., Int. Ed. Engl.*, **43 (26)**, 3400 (2004).

17. M. Besson and P. Gallezot, *Catal. Today*, **57 (1-2)**, 127 (2000).
18. P. Gamez, I. W. C. E. Arends, R. A. Sheldon and J. Reedijk, *Advanced Synthesis & Catalysis*, **346 (7)**, 805 (2004).
19. R. A. Sheldon and I. W. C. E. Arends, *Catalysis by Metal Complexes*, **26**, 123 (2003).
20. S. Striegler and M. Dittel, *J. Am. Chem. Soc.*, **125 (38)**, 11518 (2003).
21. H. Gampp, M. Maeder, C. J. Meyer and A. D. Zuberbühler, *Talanta*, **33 (12)**, 943 (1986).
22. R. A. Binstead, B. Jung and A. D. Zuberbühler, SPECFIT/32™ Global Analysis System, Vers. 3.0 (2000), Marlborough, MA, USA, Spectrum Software Associates.
23. S. Striegler, *Tetrahedron*, **62 (39)**, 9109 (2006).
24. L. Sun, T. Bulter, M. Alcalde, I. P. Petrounia and F. H. Arnold, *ChemBioChem*, **3 (8)**, 781 (2002).
25. R. Schoevaart and A. P. G. Kieboom, *Carbohydr. Res.*, **334**, 1 (2001).
26. P. Gamez, I. Arends, J. Reedijk and R. A. Sheldon, *Chem. Commun.*, **19**, 2414 (2003).

52. Photocatalytic Oxidation of Alcohols on Pt/TiO$_2$ and Nafion-Coated TiO$_2$

Felipe Guzman, Zhiqiang Yu and Steven S.C. Chuang

Department of Chemical and Biomolecular Engineering, The University of Akron, 200 E Buchtel Commons, Akron, OH 44325

chuang@uakron.edu

Abstract

Photocatalytic oxidation of ethanol on Pt/TiO$_2$ and Nafion coated TiO$_2$ catalysts were studied using in situ infrared IR techniques. Infrared studies show that the reaction produced acetaldehyde, acetic acid, acetate, formic acid, formate, and CO$_2$/H$_2$O. Modification of the TiO$_2$ catalyst by Pt and Nafion slowed down the oxidation reaction through site blocking. Incorporation of Pt was found to favor formation of formate (HCOO$^-$), indicating Pt decreases the rate of oxidation of formate more than that of its formation.

Introduction

Photocatalytic oxidation is a novel approach for the selective synthesis of aldehyde and acid from alcohol because the synthesis reaction can take place at mild conditions. These reactions are characterized by the transfer of light-induced charge carriers (i.e., photogenerated electron and hole pairs) to the electron donors and acceptors adsorbed on the semiconductor catalyst surface (1-4). Infrared (IR) spectroscopy is a useful technique for determining the dynamic behavior of adsorbed species and photogenerated electrons (5-7).

The objective of this study is to investigate the dynamic behavior of IR-observable species and their relation to photogenerated electrons during photocatalytic reactions of ethanol on the TiO$_2$-based catalysts at 30°C and 1 atm. Nafion and Pt are used to modifying catalytic properties of TiO$_2$. Diffuse Reflectance Infrared Fourier Transform Spectroscopy (DRIFTS) was used to study the photocatalytic oxidation of ethanol vapor while the Attenuated Total Reflectance (ATR) technique was employed to investigate the ethanol reaction in liquid phase. The DRIFTS and ATR results of the reaction on Nafion-TiO$_2$ catalysts were compared to evaluate the effect of the reaction environment on the photocatalytic oxidation.

Experimental Section

A 0.5 wt% Pt/TiO$_2$ catalyst was prepared by the photodeposition method. A 100 mg TiO$_2$ (Aeroxide® P25S Degussa, ~50 m^2/g, approximately 70% anatase and 30% rutile) were suspended in 30 mL aqueous solution with 1.4 mg chloroplatinic acid hexahydrate in a rectangular quartz reactor. The suspending solution was ultrasonicated for 20 min and then deoxygenized by N_2-bubbling for 30 min before UV illumination. A 350 W Xe mercury lamp (Oriel 6286) with full UV-visible light range was used for initiating Pt deposition onto TiO_2 in the suspending solution for 24 hours under continuous magnetic stirring. The resulting catalyst was centrifuged, washed 5 times with deionized water, and dried in a vacuum oven at 373 K. The Nafion modified TiO_2 (Nafion-TiO_2) catalysts were prepared by weighting 0.5 g of a Nafion solution (EC-NS-05, ElectroChem Inc.) containing 5 wt % Nafion dissolved in a mixture of aliphatic alcohols, and mixing sufficient amounts TiO_2 in order to obtain 1:1 and 2:1 Nafion:TiO_2 weight ratio slurries.

For the DRIFTS study, the Nafion-TiO_2 slurries were sonicated for 2 hours, dried at ambient conditions for 5 hr, and ground with a pestle and mortar until a fine powder catalyst was formed. 30 mg of the resulting catalysts were placed on top of 80 mg of inert CaF_2 powder (325 mesh, Alfa Aesar) in a DRIFTS cell's sample holder. The sample holder was enclosed by a dome with two IR transparent ZnSe windows and a third CaF_2 window for UV illumination. For the ATR study, the Nafion-TiO_2 slurries, which were sonicated for two hours, were cast directly on the surface of the ATR ZnSe crystal to form a continuous solid film. The films were enclosed with a stainless steel cover equipped with a CaF_2 window for UV illumination.

The photocatalytic catalytic reaction of ethanol was carried out on the Pt/TiO_2 and Nafion-TiO_2 catalysts in a DRIFTS cell (Harrick Scientific) and an HATR cell (Horizontal Attenuated Total Reflectance, Pike Technologies) at 303 K and 1 atm using a 350 W Xe mercury lamp (Oriel 6286) with a light condenser (Oriel 77800). The quantity of ethanol adsorbed on the catalyst was controlled by introducing specific amounts of O_2 (Praxair, 99.999%) saturated with ethanol (10.3% mol ethanol on O_2) at 298 K to the IR cells. The IR spectra were collected by an FTIR (DigiLab FTS 4000) bench with 20 co-added scan at a resolution of 4 cm^{-1}.

Results and Discussion

Figure 52.1 shows infrared (IR) single beam spectra of Pt/TiO_2 as well as adsorbed ethanol on Pt/TiO_2 during UV illumination. Adsorption of ethanol on the Pt/TiO_2 catalyst formed molecularly adsorbed ethanol ($CH_3CH_2OH_{ad}$), as evidenced by the presence of the characteristic C-H stretching vibrations at 2971, 2931 and 2869 cm^{-1}; the CH_2 bending vibrations at 1450 and 1380 cm^{-1}, and the COH bending at 1274 cm^{-1}. Ethanol adsorbed on part of the OH sites, causing the intensity of OH at 3731 cm^{-1} and 3675 cm^{-1} to decrease.

Upon UV light illumination, the photocatalytic reactions were initiated at the surface of the Pt/TiO_2 catalyst, resulting in the formation of CO_2, H_2O, and intermediate species. Because of the overlap of the bands of $CH_3CH_2OH_{ad}$ (adsorbed ethanol) with the bands of the intermediate species, the intensity variations in the 1300-1750 cm^{-1} region can be revealed through the difference spectra obtained by subtracting the spectrum at 0 min (i.e., before the reaction) from the subsequent spectra. For example, the difference spectrum (0-2 min) in Figure 52.2 was obtained by subtracting the spectrum at 0 minutes from the spectrum at 2 min in Figure 52.1.

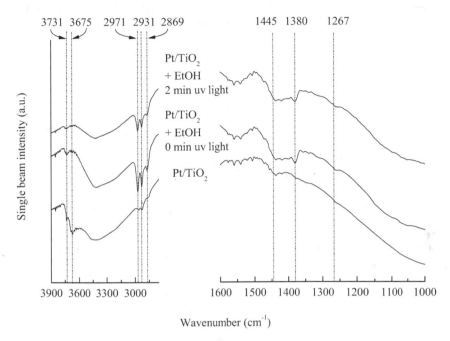

Figure 52.1. Single beam IR spectra of the Pt/TiO$_2$ catalyst on the DRIFT cell before and after introducing ethanol, and during UV illumination.

Formation of products and intermediate species, as well as disappearance of reactants during the photocatalytic reactions can be discerned by the evolution of positive (i.e., concave shape) bands and negative (i.e., convex shape) bands, respectively.

Figure 52.2 presents the different spectra obtained during the photocatalytic reactions of ethanol adsorbed on the surface of thePt/TiO$_2$ catalyst. Figure 52.2 shows the rapid formation of adsorbed acetaldehyde (CH$_3$CHO$_{ad}$) at 1721 cm^{-1}; adsorbed formic acid (HCOOH$_{ad}$) at 1689 cm^{-1}; adsorbed acetic acid (CH$_3$COOH$_{ad}$) at 1645 cm^{-1}; adsorbed formate (HCOO$^-_{ad}$) at 1587 and 1350 cm^{-1}; adsorbed acetate (CH$_3$COO$^-_{ad}$) at 1542, 1446, and 1423 cm^{-1}; and CO$_2$ at 2350 cm^{-1} upon UV illumination. Subsequent difference spectra show that the intensity of HCOOH$_{ad}$ gradually decreases after reaching a maximum, but the intensity of HCOO$^-$, CH$_3$COO$^-$, and CH$_3$CHO$_{ad}$ increases progressively. After 20 min of reaction, CH$_3$CHO$_{ad}$ exhibited a prominent band at 1721 cm^{-1}.

The adsorbed acetaldehyde (CH$_3$CHO$_{ad}$), adsorbed acetate (CH$_3$COO$^-_{ad}$) and adsorbed formate (HCOO$^-_{ad}$) relative IR intensity during the first 20 min of photocatalytic oxidation are presented in Figure 52.3 for the Pt/TiO$_2$ catalyst and TiO$_2$ catalyst (Aeroxide$^®$ P25S Degussa). Results of TiO$_2$ are included for comparison. Detailed results of photocatalytic oxidation of ethanol over TiO_2 have been reported elsewhere. It is interesting to note that the formation of acetate led that of formate which further led that of adsorbed acetaldehyde. The formation of adsorbed acetaldehyde showed an upward trend.

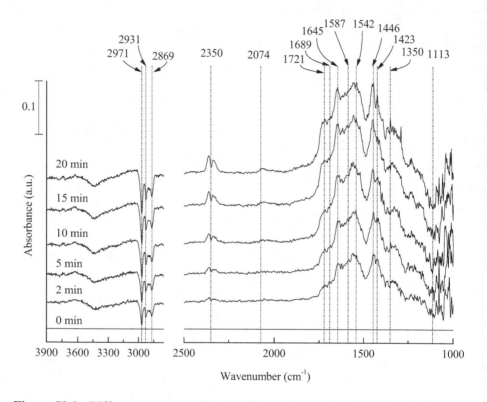

Figure 52.2. Difference spectra of the Pt/TiO_2 catalyst on the DRIFTS cell during UV illumination.

Figure 52.4 presents the formation of CO_2 during the photocatalytic oxidation of ethanol on Pt/TiO_2, TiO_2, and 1:1 Nafion-TiO_2 catalysts using the DRIFTS cell. The TiO_2 catalyst is significantly more active towards the oxidation of ethanol to CO_2 than the Pt/TiO_2 catalyst and the 1:1 Nafion-TiO_2 catalyst, suggesting that modification of TiO_2 by incorporation of Pt and Nafion could block the catalyst active sites for the oxidation of ethanol to CO_2.

The observations of CH_3COO^-, CH_3CHO_{ad}, and $HCOO^-$ confirm that the photocatalytic oxidation of ethanol follow a reaction pathway consistent with our previous study (8), as shown in the following scheme.

Adsorbed ethanol ($CH_3CH_2OH_{ad}$) can be converted to CH_3CHO_{ad} by hydrogen abstraction, as shown in path A. $CH_3CH_2OH_{ad}$ may react with h^+ to produce I_{tran1}, I_{tran2}, and I_{tran3}, which represent transient intermediates with lifetimes that could be too short and concentrations that could be too small to be detected by

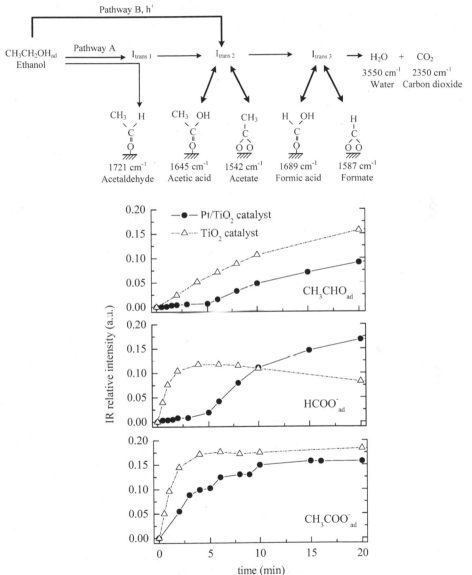

Figure 52.3. Adsorbed acetaldehyde (CH_3CHO_{ad}), adsorbed acetate ($CH_3COO^-_{ad}$), and adsorbed formate ($HCOO^-_{ad}$) IR relative intensities during the first 20 min of photocatalytic oxidation for the Pt/TiO_2 and TiO_2 catalysts. IR intensities were obtained by measuring the peak height of bands corresponding to intermediate species directly from the difference spectra presented in Figure 52.2.

our IR spectrometer. Formation of $CH_3COO^-_{ad}$ led that of other oxygenated species, indicating that most of the $CH_3COO^-_{ad}$ was produced from pathway B. As the reaction proceeds, the decline in the C-H stretching was accompanied by the increasing CH_3CHO_{ad}, CO_2 and H_2O. Incorporation of Pt to the TiO_2 catalyst may increase the rate of formation of formate $HCOO^-$ either by facilitating oxidation of I_{tran1}, or by slowing conversion of formate to CO_2 and H_2O, as it is suggested in Figure 52.3. Modification of the TiO_2 catalyst with Nafion blocked the sites for ethanol oxidation, producing significantly smaller amounts of intermediates, CO_2, and H_2O.

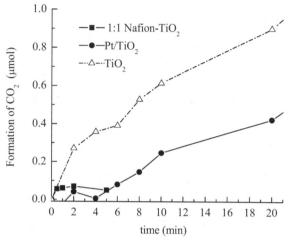

Figure 52.4. CO_2 Formation during the photocatalytic oxidation of ethanol on the Pt/TiO_2, TiO_2, and 1:1 Nafion-TiO_2, catalyst surface at 303 K.

Figure 52.5 presents the difference spectra during the first 20 min of photocatalytic oxidation on a 1:1 Nafion-TiO_2 weight ratio catalyst in the DRIFTS cell. The spectra clearly show the rapid rising of the IR background from 3000 cm^{-1}, corresponding to the generation of photogenerated electrons (8), was accompanied by the formation of small amounts of CO_2 and intermediate species including adsorbed formic acid, at 1691 cm^{-1} and adsorbed formate at 1573 and 1355 cm^{-1}.

Figure 52.6 shows infrared (IR) single beam spectra of ethanol collected using the ATR cell coated with thin Nafion-TiO_2 films corresponding to (a) 2:1 and (b) 1:1 Nafion-TiO_2 catalysts at 303 K. The spectra present the dry Nafion-TiO_2 films before and after injecting 5 ml gas phase ethanol, as well as during the subsequent UV illumination. Table 52.1 summarizes the IR band assignments for Nafion, adsorbed ethanol and intermediate species formed during the photocatalytic reaction (9-15). Figure 52.6 shows Nafion exhibited the CF_2 asymmetric vibration at 1148 cm^{-1}, SO_3H asymmetric vibration at 1440 cm^{-1}, SO_3^- symmetric vibration at 1058 cm^{-1}, C-C vibration at 1318 cm^{-1}, and the C-O-C vibrations at 983 and 965 cm^{-1}, in agreement with previously reported results (16). Adsorption of ethanol on the Nafion-TiO_2 formed molecularly adsorbed ethanol ($CH_3CH_2OH_{ad}$), as evidenced by the presence of the characteristic C-H stretching vibrations at 2971, 2931 and 2869 cm^{-1}, the C-O vibration at 1052 cm^{-1}, the C-H$_2$ vibration at 1380 cm^{-1} as well as the COH vibration at 1274 cm^{-1} (8). The intensity of CH$_3$CH$_2$OH$_{ad}$ vibration bands in the 2:1 Nafion- TiO$_2$ film were significantly weaker than those obtained from the 1:1 Nafion-TiO$_2$ film, as evidenced by comparison of Figure 52.6a and Figure 52.6b.

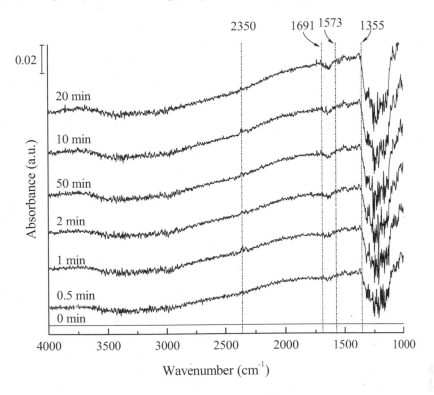

Figure 52.5. Difference spectra of a 1:1 Nafion:TiO$_2$ weight ratio catalyst on the DRIFTS cell during UV illumination. Difference spectra was obtained by subtracting the spectrum at 0 min (i.e., before the reaction starts) from the subsequent spectra.

Figure 52.7 presents the difference spectra obtained during the photocatalytic reactions of ethanol adsorbed on the surface of the 2:1 and 1:1 Nafion-TiO$_2$ films in the ATR cell. Figure 52.7(a) shows that the illumination of UV light onto the 2:1 Nafion-TiO$_2$ film resulted in a rise in IR background between 3000 and 1000 cm^{-1}, indicating the production of photogenerated electrons (8). The absence of variation in the IR bands of adsorbed ethanol indicates the ethanol reaction did not occur. In contrast, Figure 52.7(b) shows the rapid consumption of CH$_3$CH$_2$OH$_{ad}$ marked by the decrease of the C-H vibration bands at 2971, 2931 and 2872 cm^{-1} as well as the OH stretching at 1274 cm^{-1}.

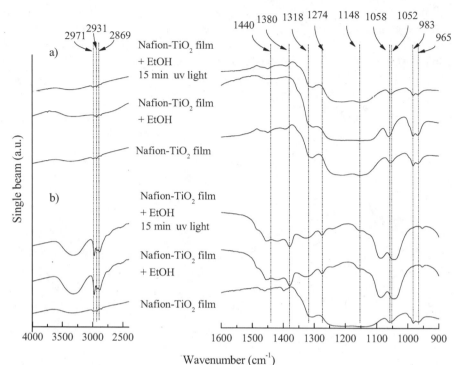

Figure 52.6. Single beam IR spectra of the Nafion-TiO$_2$ film coated on the ATR crystal before and after introducing gas phase ethanol, and during UV illumination for (a) 2:1 and (b) 1:1 Nafion-TiO$_2$ films.

Table 52.1. Band Assignments and Their Vibration modes.

Species	Bands (cm^{-1}) and modes
C$_2$H$_5$OH$_{ad}$	ν_{as}(CH$_3$)/2971; ν_{as}(CH$_2$)/2931; ν_s(CH$_3$)/2872, 2869; δ_{as}(CH$_2$)/1450; δ_s(CH$_2$)/1380; δ(OH)/1274; ν(C-O)/1052
CH$_3$CHO$_{ad}$	N(C=O)/1721
HCOOH$_{ad}$	N(C=O)/1691, 1689
HCOO$^-_{ad}$	ν_{as}(COO)/1587, 1573; ν_s(COO)/1350, 1355
CH$_3$COOH$_{ad}$	N(C=O)/1645
CH$_3$COO$^-_{ad}$	ν_{as}(COO)/1542; ν_s(COO)/1446, 1423
Nafion	ν_{as}(CF$_2$)/1148; ν_{as}(SO$_3$H)/1440; ν_s(SO$_3^-$)/1058; ν(C-C)/1318; ν(C-O-C)/983, 9650
Isolated -OH	ν(OH)/3731, 3675
CO$_2$	ν_{as}(C=O)/2350

The relative intensities corresponding to the intermediate species formed during the photocatalytic oxidation reaction on the 1:1 Nafion: TiO_2 catalyst using the DRIFTS and ATR cells, presented in Figure 52.5 and Figure 52.7 (b), respectively, evidenced the higher sensitivity of the DRFITS technique to monitor the course of the reaction.

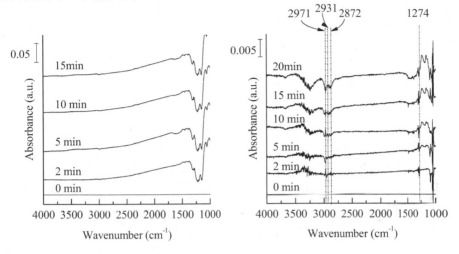

Figure 52.7. Difference spectra of the Nafion-TiO_2 film coated on the ATR crystal during UV illumination for (a) 2:1 and (b) 1:1 Nafion: TiO_2 weight ratio films.

Conclusions

The photocatalytic oxidation of alcohols constitutes a novel approach for the synthesis of aldehydes and acid from alcohols. Modification of TiO_2 catalyst with Pt and Nafion could block the catalyst active sites for the oxidation of ethanol to CO_2. Incorporation of Pt resulted in enhanced selectivity towards formate ($HCOO^-_{ad}$). Blocking of active sites by Nafion resulted in formation of significantly smaller amounts of intermediate species, CO_2 and H_2O, and accumulation of photogenerated electrons. The IR experimental technique has been extended to Attenuated Total Reflectance (ATR), enabling the study of liquid phase photocatalytic systems.

Acknowledgements

This work was partially supported by the Ohio Board of Regents (Grant R4552-OBR).

References

1. Kamat, P. V., *Chem. Rev.* 1993, 93, (1), 267-300.
2. Fox, M. A.; Dulay, M. T., *Chem. Rev.* 1993, 93, (1), 341-357.
3. Hoffmann, M. R.; Martin, S. T.; Choi, W.; Bahnemann, D. W., *Chem. Rev.* 1995, 95, (1), 69-96.
4. Linsebigler, A. L.; Lu, G.; Yates, J. T., Jr., *Chem. Rev.* 1995, 95, (3), 735-758.

5. Szczepankiewicz, S. H.; Colussi, A. J.; Hoffmann, M. R., *J. Phys. Chem. B* 2000, 104, (42), 9842-9850.
6. Yamakata, A.; Ishibashi, T.-a.; Onishi, H., *J. Mol. Catal. A: Chem.* 2003, 199, (1-2), 85-94.
7. Szczepankiewicz, S. H.; Moss, J. A.; Hoffmann, M. R., *J. Phys. Chem. B* 2002, 106, (31), 7654-7658.
8. Yu, Z.; Chuang, S. S. C., *J. Catal.* 2007, 246, (1), 118-126.
9. Wu, W.-C.; Chuang, C.-C.; Lin, J.-L., *J. Phys. Chem. B* 2000, 104, (36), 8719-8724.
10. Liao, L.-F.; Wu, W.-C.; Chen, C.-Y.; Lin, J.-L., *J. Phys. Chem. B* 2001, 105, (32), 7678-7685.
11. Coronado, J. M.; Kataoka, S.; Tejedor-Tejedor, I.; Anderson, M. A., *J. Catal.* 2003, 219, (1), 219-230.
12. El-Maazawi, M.; Finken, A. N.; Nair, A. B.; Grassian, V. H., *J. Catal.* 2000, 191, (1), 138-146.
13. http:// webbook.nist.gov/chemistry/
14. Arana, J.; Garriga Cabo, C.; Dona-Rodriguez, J. M.; Gonzalez-Diaz, O.; Herrera-Melian, J. A.; Perez-Pena, J., *Appl. Surf. Sci.* 2004, 239, (1), 60-71.
15. Colthup, N. B.; Daly, L. H.; Wiberley, S. E., *Introduction to Infrared and Raman Spectroscopy*. 3rd ed.; Academic Press: Boston, 1990.
16. Tsai, C. E.; Hwang, B. J, *Fuel cells*. 2007, 5, 408-416.

53. Enhanced Catalytic Oxidation Activity of a Dinuclear Catalyst Using Water as a Co-Solvent

Moses G. Gichinga and Susanne Striegler

Department of Chemistry and Biochemistry, Auburn University, 179 Chemistry Building, Auburn, AL 36849

Susanne.Striegler@auburn.edu

Abstract

The development of catalysts for the efficient oxidation of catechol and its derivatives in water is topic of ongoing work in this laboratory. Towards this end, polyethylene glycol side-chains were incorporated in a pentadentate salen ligand to enhance the water solubility of the complexes derived thereof. A dinuclear copper(II) complex is found to catalyze the oxidation of 3,5-di-tert.-butylcatechol into 3,5-di-tert-butyl-o-benzoquinone more than twice as fast in aqueous organic solution as in purely organic solvents (k_{cat}/k_{non}= 140,000). Preliminary data are discussed.

Introduction

The oxidation of alcohols is an important reaction in organic chemistry. While this transformation is traditionally performed in organic solvents, the use of aqueous organic solutions has just recently become a field of intense study (1-6). The effect of water on transition metal-catalyzed reactions, however, remains widely unexplored as most of these reactions require dry organic solvents to avoid decomposition of the transition metal catalyst, of water sensitive reagents, and/or intermediates by a nucleophilic attack of water (1). *Comparative* studies focusing on the effect of water as a co-solvent on the catalyst and the proceedings of a reaction are therefore rare (7).

In our ongoing efforts to develop oxidation catalysts that are functional in water as environmentally benign solvent, we synthesized a water-soluble pentadentate salen ligand with polyethylene glycol side chains (8). After coordination of copper(II) ions to the salen ligand, a dinuclear copper(II) complex is obtained that is soluble in water, methanol and mixtures of both solvents. The aerobic oxidation of 3,5-di-*tert*.-butylcatechol (DTBC) into 3,5-di-*tert*.-butylquinone (DTBQ) was used as a model reaction to determine the catalytically active species and initial data on its catalytic activity in 80% methanol.

Experimental Section

Instrumentation. UV-Vis spectra were recorded at 30°C over a range of 200-900 nm on a Varian Cary 50 with WinUV Analysis Suite software, Vers. 3.0, using Suprasil®

standard cells (200-2000 nm) of 1 cm thickness and 4.5 ml volume for the determination of the distribution of species. Disposable 1.5 ml semi-micro Brandtech UV-cuvettes (220-900 nm) of 10 mm light path with caps were used for the oxidation studies. Samples for elemental analysis were sent to Atlantic Microlab Inc., Atlanta, GA.

Reagents. Dinuclear copper(II) complex **1** was prepared as described (9); 3,5-di-*tert.*-butylcatechol (3,5-DTBC, **8**) was purchased from Aldrich and recrystallized twice from *n*-pentane under argon before use. Nanopure water was obtained from an EASYpure® II water system by Barnstead (18.2 MΩ/cm). HPLC grade methanol (99.93%) was purchased from Aldrich and used for all oxidation experiments. All other reagents were reagent grade or higher, and were used as received from commercial suppliers.

Oxidation procedure. Typically, a 2 mM catalyst stock solution was prepared by mixing a solution of 8 mg (40 μmol) copper(II) acetate monohydrate and 12.5 g (20 μmol) of ligand **5** and adjusting the total volume to 10 ml. A 0.125 M DTBC stock solution was prepared by dissolving 222 mg of **8** in 8 ml of MeOH. Aliquots of this DTBC stock solution (120-400 μl) were pipetted into the cuvettes, and 780-500 μl of MeOH were added and thermostated at 30°C. Subsequently, aliquots of the thermostated catalyst stock solution (100 μl) were pipetted into these substrate solutions resulting in a total substrate concentration of 15 mM to 50 mM **8** in methanol and a total 0.2 mM catalyst concentration. The molar absorbance of **9** at 420nm length was determined to be 1670 M^{-1} cm^{-1} by using a commercial sample. The self-oxidation of **8** into **9** was determined without catalyst under otherwise identical conditions. The concentration of substrate **8** was varied from 50-400 μM M at a 12 μM apparent catalyst concentration for the experiments in 80% MeOH.

Results and Discussion

Synthesis of complex 1. The pentadentate salen catalyst **1** was synthesized as described (9). In short, the tosylated 2-[2-(2-methoxyethoxy)-ethoxy]-ethanol **2** (10) was reacted with 2,4-dihydroxybenzaldehyde **3** to yield 4-alkoxy salicylaldehyde **4** after chromatographic purification (eq. 1). Subsequent condensation of **4** with 1,3-diaminopropanol yielded water-soluble salen ligand **5** in sufficient purity and 89% yield (11). The formation of an azomethine bond is indicated by a shift of the ^{13}C NMR signal for the carbonyl carbon from 194.4 ppm in aldehyde **4** to 166.4 ppm for the imino carbon in **5**. The pentadentate ligand **5** was then treated with copper(II) acetate in methanol to obtain the dinuclear copper(II) complex **1** as a green solid (eq. 2) (11).

(eq. 1)

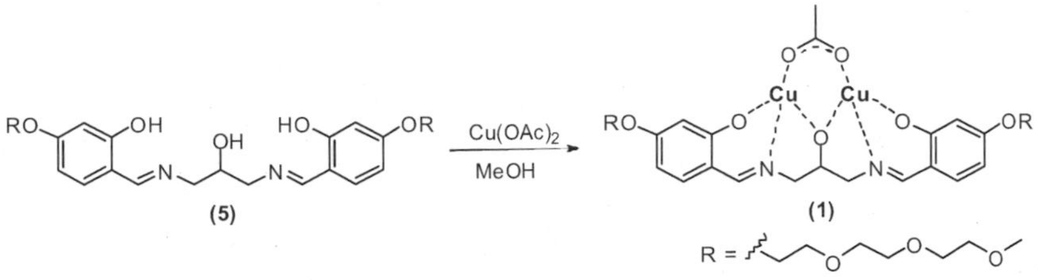

(eq. 2)

Determination of the catalytically active species derived from 1 in solution. Spectrophotometric titration of the backbone ligand **5** with copper(II) acetate in methanol revealed formation of a dinuclear copper(II) complex species $Cu_2L_{-3H}(OAc)$ above a 1 : 2 molar ratio. A mononuclear copper(II) species CuL_{-2H} (**6**) dominates at a 1 : 1 molar ratio of **5** and copper(II) acetate. Control experiments for the assignment of putative structures based on the obtained spectroscopic data included a UV/Vis spectroscopic titration of **5** with anhydrous sodium acetate in the presence of copper(II) chloride and revealed that acetate is necessary for the formation of a copper (II) complex in methanol. The composition of **1** in methanol is the same as determined by elemental analysis for the solid state.

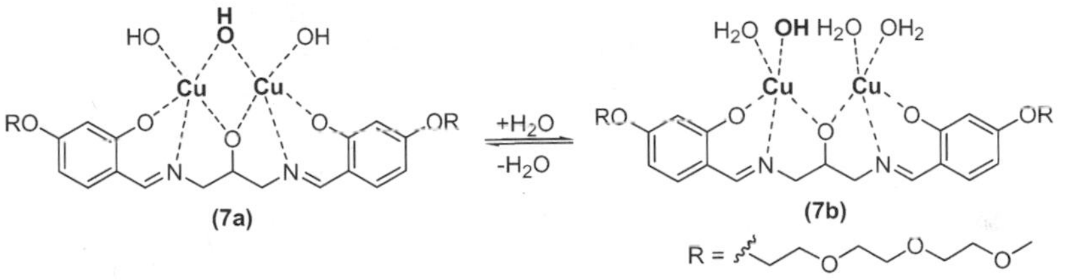

(eq. 3)

The composition of a dinuclear complex derived from ligand **5** and copper(II) acetate is drastically altered in *aqueous* methanol (MeOH/H$_2$O = 80/20). The bridging acetate anion in **1** is exchanged by water and hydroxyl ions under these conditions as concluded from spectrophotometric and calorimetric titration experiments. The resulting dinuclear copper(II) species $Cu_2L_{-3H}(OH)$ (**7**) is consequently described with a μ^1-coordinating or η^2-bridging hydroxyl group in equilibrium structures (eq. 3). The amount of **7** formed at a 2 : 1 molar ratio of copper(II) acetate and **5** is diminished in 80% MeOH to 59% of the theoretical value due to the presence of mononuclear complex **6**.

Catalytic oxidation of 3,5-di-tert.-butylcatechol. The catalytic oxidation of 3,5-di-*tert.*-butylcatechol, DTBC (**8**), to 3,5-di-*tert.*-butylquinone, DTBQ (**9**), was chosen as a model reaction to test our hypothesis and establish the catalytic abilities of **1** and **7** in organic and aqueous organic solution (eq. 4).

(eq. 4)

The reaction was monitored by UV/Vis spectroscopy by following the product formation at 420 nm. The initial rates were used for analysis of the catalyzed oxidation of **8** into **9** that follows Michaelis-Menten kinetics. Control experiments show a linear increase of the reaction rates with the catalyst concentration at constant substrate concentration.

The first set of experiments was conducted in methanol. The substrate concentration was varied from 15 to 50 mM at a 200 μM concentration of **1** for the determination of kinetic parameters for the transformation of **8** into **9**. The catalytic rate constant k_{cat} was determined to be 0.04 min^{-1} and the Michael constant K_m was determined to be 40 mM at 30°C. The rate constant is comparable to those reported for other dinuclear Cu(II) complexes with a comparable Cu···Cu distance of 3.5 Å, but about one magnitude lower than those observed for complexes with a shorter intermetallic distances (12-14), e.g. 2.9 Å (k_{cat} = 0.21 min^{-1}) (12) or 3.075 Å (k_{cat} = 0.32 min^{-1} (13). The rate constant k_{non} for the spontaneous (uncatalyzed) oxidation of **8** into **9** was determined to be 6×10^{-7} min^{-1} and corresponds to the oxidation without catalyst under otherwise identical conditions. The rate acceleration (k_{cat}/k_{non}) deduced from these values is 60,000-fold.

The transformation of **8** into **9** was then monitored in 80% aqueous MeOH for substrate concentrations between 0.05 to 0.4 mM, and 12 μM of apparent concentration of **7**. Unbuffered nanopure water was always used, as the addition of base accelerates the uncatalyzed oxidation of **8** by air significantly. The catalytic rate constant k_{cat} in 80% aqueous MeOH was determined to be 0.13 min^{-1}. The Michaelis–Menten constant K_m was determined to be 0.07 mM, which refers to a higher affinity of the substrate to the metal complex in aqueous methanol than in pure methanol. The rate constant for the spontaneous reaction k_{non} was determined to be 1×10^{-6} min^{-1} in 80% aqueous MeOH. The transformation of **8** into **9** is 140,000-fold accelerated over background under these conditions, and is thus more than twice as fast as accelerated than the reaction in pure methanol.

Conclusions

A dinuclear salen complex was investigated as catalyst for the aerobic oxidation of 3,5-di-*tert*.-butylcatechol into 3,5-di-*tert*.-butylquinone in organic and aqueous organic solution. The actual catalyst composition varies in both solvent systems. Formation of a mononuclear species competes with formation of a dinuclear copper(II) catalyst. The aerobic oxidation of **8** into **9** is 140,000-fold accelerated over background in aqueous methanol, and is about twice as fast as the same reaction in pure methanol.

References

1. H. C. Hailes, *Org. Process Res. Dev.*, **11 (1)**, 114 (2007).
2. Y. Jung and R. A. Marcus, *J. Am. Chem. Soc*, **129 (17)**, 5492 (2007).
3. S. Otto and J. B. F. N. Engberts, *Org. Biomol. Chem.*, **1 (16)**, 2809 (2003).
4. D. C. Rideout and R. Breslow, *J. Am. Chem. Soc*, **102 (26)**, 7816 (1980).
5. R. Breslow and U. Maitra, *Tetrahedron Lett.*, **25 (12)**, 1239 (1984).
6. R. Breslow, U. Maitra and D. Rideout, *Tetrahedron Lett.*, **24 (18)**, 1901 (1983).
7. R. N. Butler, W. J. Cunningham, A. G. Coyne and L. A. Burke, *J. Am. Chem. Soc*, **126 (38)**, 11923 (2004).
8. M. G. Gichinga, D. T. Guilford and S. Striegler *unpublished results*.
9. M. G. Gichinga and S. Striegler, *J. Am. Chem. Soc*, **submitted**.
10. O. Kocian, K. W. Chiu, R. Demeure, B. Gallez, C. J. Jones and J. R. Thornback, *J. Chem. Soc. Perkin Trans. 1*, **(5)**, 527 (1994).
11. G. D. Fallon, K. S. Murray, W. Mazurek and M. J. O'Connor, *Inorg. Chim. Acta*, **96 (2)**, L53 (1985).
12. N. A. Rey, A. Neves, A. J. Bortoluzzi, C. T. Pich and H. Terenzi, *Inorg. Chem.*, **46 (2)**, 348 (2007).
13. R. A. Peralta, A. Neves, A. J. Bortoluzzi, A. dos Anjos, F. R. Xavier, B. Szpoganicz, H. Terenzi, M. C. B. de Oliveira, E. Castellano, G. R. Friedermann, A. S. Mangrich and M. A. Novak, *J. Inorg. Biochem.*, **100 (5-6)**, 992 (2006).
14. C. Belle, C. Beguin, I. Gautier-Luneau, S. Hamman, C. Philouze, J. L. Pierre, F. Thomas, S. Torelli, E. Saint-Aman and M. Bonin, *Inorg. Chem.*, **41 (3)**, 479 (2002).

54. Effect of the Oxidizing Agent and Catalyst Chirality on the Diastereoselective Epoxidation of R-(+)-Limonene

Jairo A. Cubillos, Juliana Reyes, Aída L. Villa and Consuelo Montes de C

Environmental Catalysis Research Group, Universidad de Antioquia-CENIVAM, AA 1226, Medellín, Colombia

jacubil@gmail.com

Abstract

Enantiomerically pure limonene oxides are synthetically useful for the preparation of fragrances, perfumes and food additives. Although these compounds are commercially available and relatively inexpensive, they are marketed as 1:1 mixtures of the *cis*- and *trans*-epoxides. To date, the best known chemical method to asymmetrically epoxidize unfunctionalized olefins is the Jacobsen's epoxidation, in which a chiral Mn (III) salen complex is employed as catalyst. Since R-(+)-limonene is a pure enantiomerically compound, the asymmetric epoxidation of this substrate offers the possibility of using an achiral catalyst which is less expensive than its chiral counterpart. In this work, we report on the effect of the oxidizing agent and catalyst chirality on the diastereoselective epoxidation of R-(+)-limonene. Asymmetric induction was governed by the substrate when in situ prepared oxidizing agents were used (dimethyldioxirane and peroxyacid), whereas the asymmetric induction was governed by the chiral catalyst with commercially available oxidizing agents (NaOCl/4-PPNO and m-CPBA/4-NMO). It was found that the use of dimethyldioxirane (DMD) as oxidizing agent, remarkably improves the stability toward oxidative degradation of the chiral catalyst.

Introduction

Enantiomerically pure limonene oxides (**2**) and (**3**) are important building blocks for obtaining fine chemicals (1). Usually, epoxidation of R-(+)-limonene (**1**) results in the formation of a 1:1 mixture of **2** and **3** (2, 3). On the other hand, physical methods for separation of these diastereomers generally involve fractional distillation or chromatography and are expensive and not trivial. As the diastereomerically pure oxides **2** and **3** are useful and their separation is tough, development of an asymmetric synthesis method of these epoxides from **1** is of significant importance. To date, the best known chemical method to prepare these compounds is the Jacobsen's epoxidation, in which a chiral Mn (III) salen complex (**4a**) is employed as catalyst (4, 5). Although the asymmetric epoxidation of **1** offers the possibility of using an achiral catalyst (**4b**), which is less expensive than its chiral counterpart, there are no comparative studies between these catalyst types.

(1)

(4a)   (4b)

It is generally accepted that oxo-Mn (V) is the catalytically active species; however, the reaction mechanism is currently under intensive debate since it strongly depends on the oxygen donor and the reaction conditions (6). On the other hand, low catalytic productivities are obtained due to oxidative degradation of the salen ligand (7).

Herein, we report on the effect of the oxidizing agent and catalyst chirality on the diastereoselective epoxidation of R-(+)-limonene. Four oxidizing agents were used together with **4a** and **4b**. The oxidizing agents explored were: dimethyldioxirane (DMD) prepared in situ from acetone and KHSO$_5$ (Oxone®), O$_2$/pivalaldehyde which generates an in situ peroxy acid, a commercially available sodium hypochlorite/4-phenylpyridine N-oxide (NaOCl/4-PPNO) and meta-chloroperoxybenzoic acid/4-methylmorpholine N-oxide (m-CPBA/4-NMO). Results indicate that when the oxidizing agent was prepared in situ (DMD and peroxy acid), the chiral center of **1** might facilitate the formation of diastereomer **2**. On the other hand, when the commercially available oxidizing agents (NaOCl/4-PPNO and m-CPBA/4-NMO) were used, the chiral center of **4a** appears to favor the formation of diastereomer **2**. Catalyst **4a** was shown to be very stable under the reaction conditions of this study when DMD was used as oxidizing agent.

Experimental Section

The reagents and solvents used for synthesizing the catalysts and for the catalytic experiments were purchased from Aldrich, Carlo Erba, J.T. Baker, and Merck, and used as received. The salen ligands were prepared by the standard procedure refluxing ethanolic solutions of the corresponding diamine ((R,R)-1,2 diammoniumcyclohexane mono-(+)-tartrate salt (Aldrich, 99% pure) to obtain the chiral ligand or 1,2-diaminoethane (Merck, 99% pure) to obtain the achiral ligand) and 3,5-di-tert-butylsalicylaldehyde (Aldrich, 99% pure) in a 1:2 molar ratio. Finally, the ligands were complexed with manganese acetate (Aldrich, 98% pure) yielding **4a** and **4b,** respectively (8). The purity of **4a** and **4b** was checked by FT-IR and DR UV–VIS. The FT-IR spectra (KBr pellets) were recorded on a Nicolet Avatar 330 spectrophotometer while DR UV–VIS electronic spectra were recorded using a Lamda 4B Perkin Elmer spectrophotometer.

The complexes **4a** and **4b** were used as catalysts for the liquid phase diastereoselective epoxidation of R-(+)-limonene (Aldrich, 98% pure). In this study we examined four different oxidizing agents: dimethyldioxirane (DMD) prepared in situ from $KHSO_5$ (Oxone®, Aldrich) and acetone (Aldrich, 99.5% pure), a peroxyacid prepared in situ from oxygen and pivalaldehyde (Aldrich, 96% pure), a commercially available sodium hypochlorite/4-phenylpyridine N-oxide (aqueous NaOCl, Carlo Erba 13%/4-PPNO, Aldrich 98% pure) and metha-chloroperbenzoic acid/4-methylmorpholine N-oxide (m-CPBA, Aldrich 77% pure/4-NMO, Aldrich 97% pure). Catalytic activity experiments were performed following established procedures (9-12). In general, **1** together with the solvent and catalyst were first introduced into a round-bottom flask when DMD, NaOCl/4-PPNO and m-CPBA/4-NMO were used as the primary oxidants. After the addition of the oxidant, the reaction was started with vigorous stirring at atmospheric pressure and room temperature. A stainless steel Parr reactor at 450 psi was used when O_2/pivalaldehyde was the oxidant. In the case of DMD, an aqueous solution of Oxone® was slowly added, while the reaction mixture was kept at a pH between 8.0-8.5 by means of a $NaHCO_3$ aqueous solution (5 wt %) (9). The reaction products and the catalyst were recovered by dichloromethane extraction in the experiments performed under aqueous-organic biphasic conditions (DMD and NaOCl/4-PPNO). In all cases, the reaction products were separated from the catalyst by vacuum distillation at 160°C and 0.08 MPa. Blank reactions were carried out the same way as the standard reactions with the exception that no catalyst was added. R-(+)-limonene conversion and total epoxide selectivity were quantified by gas chromatography (GC) using a Varian 3400 GC equipped with a capillary column DB-1 (50 m x 0.52 mm x 1.2 mm) and a FID detector. Diastereomeric excesses (*de*) were quantified using an Agilent 7890 A GC Chromatograph equipped with a capillary column, β-Dex (60 m x 0.25 mm x 0.25 mm), and a FID detector.

Results and Discussion

Catalyst characterization. FT-IR and DR UV–VIS spectra of complexes **4a** and **4b** were in good agreement with the expected chemical structure of the Mn(III) salen complexes. The FT-IR spectra showed the vibration of the azomethine group (C=N) of the salen ligands around 1630 cm^{-1} (13). This band was shifted to a lower frequency, indicating that the nitrogen atom of the C=N group is coordinated to the manganese ion. Furthermore, the characteristic band of these Mn(III) salen complexes around 1540 cm^{-1} was obtained (13). This band is ascribed to the N–Mn and O-Mn stretching vibrations (13). DR UV–Vis spectra of the free ligand exhibited two absorption bands in the 250 and 323 nm regions. These bands are attributed to π-π* transitions. The first one belongs to the benzene ring and the second one to the C=N groups (14). For the complexes **4a** and **4b**, the C=C π-π* transition is shifted to a longer wavelength due to metal coordination, confirming the formation of Mn(III) salen complexes (14).

      Catalytic activity. Table 54.1 shows the catalytic activity of **4a** and **4b** on the diastereoselective epoxidation of R-(+)-limonene using DMD, O₂/pivalaldehyde, NaOCl/4-PPNO and m-CPBA/4-NMO. As can be observed in Table 54.1, in the absence of catalyst the oxidation took place with the oxidizing agents used. However, diastereomeric excesses (*de*) with the oxidizing agents prepared in situ, were greater than those obtained when the commercially available oxidizing agents were used (entries 1, 4, 7 and 10). These results suggest that the in situ preparation of the oxidizing agent promotes the chiral induction of **1**. When commercially available oxidizing agents were used, conversions of **1** were improved in the presence of catalysts in comparison to the un-catalyzed reactions. On the other hand, the selectivity and *de* were remarkably improved when using catalyst (except for m-CPBA/4-NMO, entries 11 and 12), indicating the positive catalytic effect of **4a** and **4b** on the reaction under study. Interestingly, very similar diastereomeric excesses were obtained using the in situ prepared oxidizing agents with the catalysts (entries 2-3 and 5-6), indicating that the chiral center of **1** might favor the formation of diastereomer **2**. Very recently, Gusmão and colleagues (15) reported that the chirality of **1** plays an important role in the stereochemical formation of new chiral centers because the synthesis occurs with double asymmetric induction. However, they did not report the use of an achiral homologue of the Jacobsen's catalyst. In contrast,

Table 54.1. Catalytic Activity of **4a** and **4b** Using Different Oxidizing Agents.

Entry	Oxidizing agent	Catalyst	Conversion (%)	Selectivity to(2)+(3)	de[e] (%)
1	DMD[a]	None	53	34	23
2		**4a**	55	100	56
3		**4b**	53	100	50
4	O₂/pivaladehyde[b]	None	97	30	29
5		**4a**	83	90	38
6		**4b**	76	88	38
7	NaOCl/4-PPNO[c]	None	34	13	0
8		**4a**	72	77	65
9		**4b**	79	76	38
10	m-CPBA/4-NMO[d]	None	5	100	1
11		**4a**	12	88	79
12		**4b**	17	84	39

[a]**1**/Oxone (KHSO₅)/**4a** or **4b**= 5/4 (8)/0.05 mmol/mmol (mmol)/mmol; acetone= 4 mL; reaction time= 35 min. [b]**1**/pivalaldehyde/**4a** or **4b**= 3.7/9.2/0.05 mmo/mmol/mmol; toluene= 18.5 mL; O₂ (450 psi); reaction time= 24 h. [c]**1**/NaOCl (pH= 11.2)/4-PPNO/**4a** or **4b**= 1/3/0.3/0.02(mmol/mmol/mmol/mmol); dichloromethane= 5 mL; reaction time= 24 h. [d](**1**)/m-CPBA/4-NMO/**4a** or **4b**= 0.5/0.4/1/0.02/0.005 (mmol/mmol/mmol/mmol); dichloromethane= 3 mL; reaction time = 30 min..[e]*de*= (**2-3**)/(**2+3**)×100 %.

when the commercially available oxidizing agents were used, higher diastereomeric excesses with **4a** than **4b** (entries 8-9 and 11-12) were achieved. In this case, the chiral center of **4a** appears to favor the formation of diastereomer **2**. The fact that the nature of the oxygen donor affects diastereoselectivity suggests that besides the oxo-Mn(V) species, at least one other active oxidant species must be involved in this catalytic process (16).

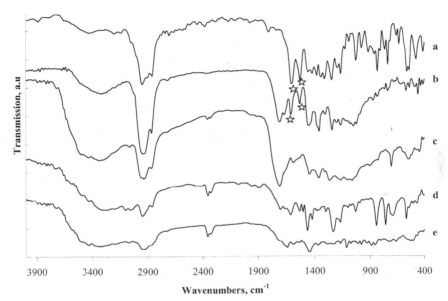

Figure 54.1. FT-IR spectra of catalyst (**4a**) used once on the diastereoselective epoxidation of (**1**) using different oxidants and recovered by reduced pressure: (a.) fresh catalyst. (b.) DMD. (c.) O₂/pivaladehyde. (d.) NaOCl/4-PPNO. (e.) m-CPBA/4-NMO.

Finally, we investigated the stability of (**4a**) to oxidative degradation. Figure 54.1 shows the FT-IR spectra of the catalyst samples (**4a**) used with each oxidizing agent recovered under reduced pressure. It can be seen that (**4a**) kept the main bands associated with the Mn(III) salen complexes, only when DMD was used as the oxidizing agent. A control experiment, where fresh (**4a**) was recovered by reduced pressure, did not show changes in the chemical structure. Therefore, under the reaction conditions accomplished with DMD, the stability of the catalyst towards the oxidative degradation was improved.

Conclusions

Chiral and achiral Jacobsen's catalysts exhibit similar diatereomeric excesses during the diastereoselective epoxidation of R-(+)-limonene using in situ prepared oxidizing agents. Therefore, the chiral center of the substrate appears to govern the chiral induction. In contrast, the chirality of the Jacobsen's catalyst appears to be responsible for the chiral induction when commercially available oxidants were used.

Furthermore, the chemical structure of the Jacobsen's catalyst is stable in the presence of DMD as oxidant.

Acknowledgements

The authors acknowledge COLCIENCIAS-Universidad de Antioquia financial support to this work through CENIVAM Center of Excellence RC No. 432.

References

1. E.L. Grimm, J. Methot and M. Shamji, *Pure. Appl. Chem.,* **75**, 231 (2002).
2. P.C. Andrews, M. Blair, B. H. Fraser, P. C. Junk, M. Massi and K. L. Tuck, *Tetrahedron: Asymmetry*, **17**, 2833 (2006).
3. C.B. Woitiski, Y.N. Kozlov, D. Mandelli, G.V. Nizova, U. Schuchardt and G.B. Shul'pin, *J. Mol. Catal A. Chem.*, **222**, 103 (2004).
4. C. T. Dalton, K. M. Ryan, V. M. Wall, C. Bousquet and D.G. Gilheany, *Top. Catal.*, **5**, 75 (1998).
5. M.D.T. Gomes and O.A.C. Antunes, *Catal. Lett.*, **38**, 133 (1996).
6. K.P. Bryliakov, D.E. Babushkin and E.P. Talsi, *J. Mol. Catal A. Chem.,* **158**, 19 (2002).
7. C. Baleizão and H. García, *Chem. Rev.,* 106, 3987 (2006).
8. J.F. Larrow and E. N. Jacobsen, *J. Org. Chem.*, **59**, 1939 (1994).
9. J. Cubillos and W.F. Hoelderich, *Rev. Fac. Ing. Univ. Antioquia.* **41**, 31 (2007).
10. S. Bhattacharjee and J.A. Anderson, *Catal. Lett.*, **95**, 119 (2004).
11. L. Deng and E.N. Jacobsen, *J. Org. Chem.*, **57**, 4320 (1992).
12. M. Palucki, G.J. McCormick and E.N. Jacobsen, *Tetrahedron Lett.*, **36**, 5457 (1995).
13. A. Doménech, P. Formentin, H. García and M.J. Sabater, *Eur. J. Inorg. Chem.*, 1339 (2000).
14. B. Bahramian, V. Mirkhani, M. Moghadam and A.H. Amin, *Appl. Catal. A. Gen.*, **315**, 52 (2006).
15. L.D. Pinto, J. Dupont, R.F. de Souza and K. Bernardo-Gusmão, *Catal. Commun.*, **9**, 135 (2008).
16. W. Adam, C. Mock-Knoblauch, C.R. Saha-Moeller and M. Herderich, *J. Am. Chem. Soc.*, **112**, 9685 (2000).

55. The Dynamic Nature of Vanadyl Pyrophosphate, Catalyst for the Selective Oxidation of *n*-Butane to Maleic Anhydride

Nicola Ballarini[1], Fabrizio Cavani[1], Elisa Degli Esposti[1], Davide De Santi[1], Silvia Luciani[1], Ferruccio Trifirò[1], Carlotta Cortelli[2], Roberto Leanza[2], Gianluca Mazzoni[2], Angelika Brückner[3] and Elisabeth Bordes-Richard[4]

[1]*Dipartimento di Chimica Industriale e dei Materiali, Università di Bologna, Viale Risorgimento 4, 40136 Bologna, Italy. INSTM, Research Unit of Bologna: a Partner of NoE Idecat, FP6 of the EU*
[2] *Polynt SpA, Via E. Fermi 51, 24020 Scanzorosciate (BG), Italy*
[3] *Leibniz-Institut für Katalyse an der Universität Rostock e. V., Branch Berlin, P.O. Box 96 11 56, D-12474 Berlin, Germany: a Partner of NoE Idecat, FP6 of EU*
[4]*Unité de Catalyse et de Chimie du Solide, UMR CNRS 8181, Université des Sciences et Technologies de Lille, 59655 Villeneuve d'Ascq Cedex, France: a Partner of NoE Idecat, FP6 of the EU*

fabrizio.cavani@unibo.it

Abstract

The reactivity of vanadyl pyrophosphate $(VO)_2P_2O_7$, catalyst for *n*-butane oxidation to maleic anhydride, was investigated under steady and unsteady conditions, in order to obtain information on the status of the active surface in reaction conditions. Specific treatments of hydrolysis and oxidation were applied in order to modify the characteristics of the surface layer of the catalyst, and then the unsteady catalytic performance was followed along with the reaction time, until the steady original behavior was restored. It was found that the transformations occurring on the vanadyl pyrophosphate surface depend on the catalyst characteristics (i.e., on the P/V atomic ratio) and on the reaction conditions.

Introduction

The industrial catalyst for *n*-butane oxidation to maleic anhydride (MA) is a vanadium/phosphorus mixed oxide, in which bulk vanadyl pyrophosphate (VPP) $(VO)_2P_2O_7$ is the main component. The nature of the active surface in VPP has been studied by several authors, often with the use of in situ techniques (1-3). While in all cases bulk VPP is assumed to constitute the core of the active phase, the different hypotheses concern the nature of the first atomic layers that are in direct contact with the gas phase. Either the development of surface amorphous layers, which play a direct role in the reaction, is invoked (4), or the participation of specific planes contributing to the reaction pattern is assumed (2,5), the redox process occurring reversibly between VPP and $VOPO_4$.

In the present work, the transient reactivity and the changes of the surface characteristics of an equilibrated VPP in response to modifications of the gas-phase composition have been investigated. As the P/V atomic ratio is one of the most important factors affecting the catalytic performance of the VPP (6), two catalysts differing in P/V ratio were studied. Data obtained were used to draw a model about the nature of the surface active layer, and on how the latter is modified in function of the reaction conditions.

Experimental Section

Catalysts were prepared by decomposition of $VOHPO_4.0.5H_2O$, the precursor of VPP. The precursor was synthesized by the "organic procedure" by refluxing V_2O_5 and H_3PO_4 in isobutanol. The precipitate finally obtained was thermally treated according to the following procedure: (a) drying at 120°C for 12 h, in static air; (b) pre-calcination step in flowing air, with temperature gradient from room temperature up to 300°C, followed by isothermal step at 300°C in air for 6 h; (c) calcination step in flowing N_2, with temperature gradient from 300°C to 550°C, and final isothermal step at the latter temperature for 6 h. Two samples were prepared, having P/V atomic ratio (as determined in the dried catalyst by means of X-Ray Fluorescence and Scanning Electron Microscopy-EDX) equaling 1.00 ± 0.01 and 1.06 ± 0.02. The catalytic tests were carried out in a laboratory glass continuous-flow reactor, loading 0.8 g of catalyst. The feed stream contained 1.8% *n*-butane, 18% O_2, and remainder N_2. Products were analysed on-line by means of a gas chromatograph equipped with a HP-1 column for the separation of C_4 hydrocarbons, formaldehyde, acetic acid, acrylic acid, and maleic anhydride. A Carbosieve SII column was used for the analysis of O_2, CO and CO_2. The steady state was considered after an equilibration period of 100 hours reaction time.

Results and Discussion

The catalytic behaviors in *n*-butane oxidation at steady state of the two catalysts differing by P/V atomic ratio were considerably different. This is shown in Figure 55.1, which compares the conversion of *n*-butane and the selectivity to MA as a function of the reaction temperature for the two catalysts investigated. The differences were relevant at intermediate reaction temperature (between 340 and 400°C), whereas they were minimal at T < 340°C and at T > 400°C. The catalyst P/V 1.06 (containing an excess P with respect to the stoichiometric ratio), showed the expected increase of conversion and decrease of selectivity to MA when the reaction temperature was increased. On the contrary, the catalyst P/V 1.00 with the stoichiometric ratio showed an anomalous behavior, with a maximum in conversion and a corresponding minimum of selectivity to MA at 360-380°C. These phenomena were quite reproducible and reversible, as they could be repeated several times by temperature cycling between 320 and 450°C. They can be attributed to changes in the nature of the active surface layer, possibly because of reversible oxidation and hydrolysis (7) phenomena occurring on VPP by exposure to the reactive atmosphere.

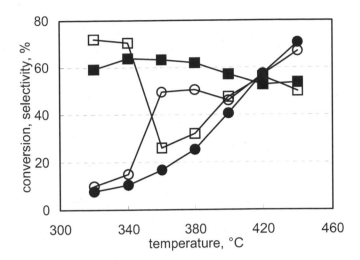

Figure 55.1. Conversion of *n*-butane (○●) and selectivity to MA (□■) as a function of the reaction temperature for catalyst P/V 1.00 (open symbols) and P/V 1.06 (full symbols).

In order to reproduce the transformations occurring at the surface of VPP, the catalyst performance was studied in forced-unsteady regime, by applying conditions different from those underwent by the catalyst under reaction. Specifically, we chose three representative temperatures: 320°C, 380°C and 440°C, and at each temperature either oxidizing treatments (feed of air), or hydrolyzing treatments (feed of 4% steam in He), or both, were applied. Then, the reaction mixture was fed again, and the variation of catalytic performance that occurred with time-on-stream was examined during the transition from the forced regime to the steady one.

For example, Figure 55.2 compares the unsteady performance of the two equilibrated catalysts after an oxidizing treatment with air at 380°C, while Figure 55.3 shows the behavior after a hydrolyzing treatment with steam at 380°C. Both treatments were applied on the two catalysts after they had been exposed to the reaction atmosphere at 380°C. It is shown that the response of the two catalysts to the two treatments was quite different. At 380°C, catalyst P/V 1.00 did not show any change of catalytic performance after the treatment with air, while the latter modified the characteristics of catalyst P/V 1.06. After the oxidizing treatment that catalyst was slightly less active but clearly more selective than it was before. After ~1 h, the original steady performance was restored. This means that the treatment in air did not modify the catalytically active surface of the equilibrated catalyst P/V 1.00, whereas it did alter the surface of the catalyst P/V 1.06. It can be inferred that P/V 1.00 was already in an oxidized state during reaction, or that it was insensitive to any oxidizing treatment. On the other hand, the catalyst P/V 1.06 was not in its higher oxidized state under reaction conditions; this stable active surface was less selective to MA and almost as active as a more oxidized active surface.

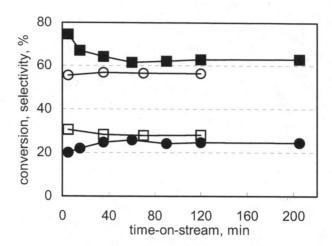

Figure 55.2. Effect of time-on-stream on *n*-butane conversion (○●) and selectivity to MA (□■) of catalysts P/V 1.00 (open symbols) and P/V 1.06 (full symbols) at 380°C after treatment of the equilibrated catalysts with air at 380°C for 1 h.

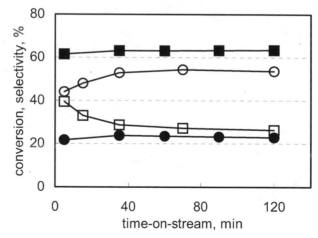

Figure 55.3. Effect of time-on-stream on *n*-butane conversion (○●) and selectivity to MA (□■) of catalysts P/V 1.00 (open symbols) and P/V 1.06 (full symbols) at 380°C after treatment of catalysts with 4% H_2O in He at 380°C for 1 h.

The opposite behavior was observed after the treatment of the two catalysts in the steam-containing stream, at 380°C. The catalyst P/V 1.06 did not show any change of catalytic performance, whereas in the case of P/V 1.00 the treatment rendered the catalyst less active but more selective than the sample equilibrated in the reactive atmosphere at 380°C. This means that with P/V 1.00, the active layer is not fully hydrolyzed under reaction conditions, and that a hydrolyzed surface is more selective than the active surface of the equilibrated P/V 1.00 catalyst. On the contrary, the active surface of catalyst P/V 1.06 either was already hydrolyzed under reaction conditions, and hence it was not modified by a hydrolyzing treatment. An alternative is that it was fully stable towards hydrolysis.

Figure 55.4 compares the Raman spectra of the two samples; spectra were recorded at 380°C in a 15% O_2/N_2 stream, on equilibrated catalysts downloaded after reaction. Catalyst P/V 1.06 was not oxidized in the air stream, whereas in the case of catalyst P/V 1.00 bands typical of a V^{5+} phosphate, α_I-VOPO$_4$, appeared in the spectrum. These bands were not present in the spectrum of the equilibrated catalyst recorded at room temperature. Indeed, the spectra of the two equilibrated catalysts were quite similar when recorded at room temperature. This result confirms that the surface of catalyst P/V 1.06 is less oxidizable than that of catalyst P/V 1.00. Therefore, the latter is likely more oxidized than the former one under reaction conditions. A treatment in a more oxidant atmosphere than the reactive *n*-butane/air feed modifies the surface of catalyst P/V 1.06, and leads to the unsteady behavior shown in Figure 55.1. The same treatment did not alter the surface of the equilibrated catalyst P/V 1.00 that was already in an oxidized state under reaction conditions.

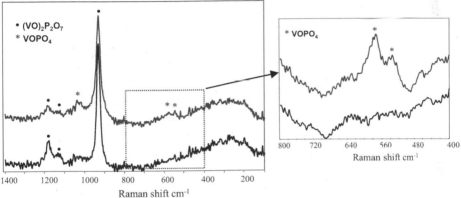

Figure 55.4. Raman spectra recorded in situ after exposure of catalysts P/V 1.00 (top) and P/V 1.06 (bottom) to a 15% O_2/N_2 stream at 380°C for 30 min.

In situ UV-Vis Diffuse Reflectance Spectroscopy was performed under reactive atmosphere (*n*-butane/oxygen). These experiments confirmed that submitting the catalyst to the reaction mixture favors the development of a more oxidized active surface, and that the extent of transformation depends on the reaction temperature and on the catalyst P/V ratio. For instance, catalyst P/V 1.06 was less oxidized than catalyst P/V 1.00 at a temperature lower than 340°C. X-ray Photoelectron spectra of catalysts recorded after reaction at 380°C confirmed that catalyst P/V 1.00 was considerably more oxidized (average oxidation state for surface V: 4.23) than the P/V 1.06 catalyst (average oxidation state: 4.03).

The results of the unsteady-state reactivity tests and of the catalysts characterization allow us to propose a model for the active layer of VPP under reaction conditions, illustrated in Figure 55.5. In this model, the surface is in dynamic equilibrium with the gas phase, and its nature is a function of both reaction conditions (temperature, gas-phase composition) and catalyst P/V ratio. The surface may be closer to either VPP, or to $VOPO_4$, or to a mixture of (poly)phosphoric acid and V_xO_y (in which V may be present as V^{5+} and as V^{4+}). The oxidation/reduction and hydration/dehydration equilibria which are set up oblige the surface to adapt itself, their composition moving from one to another inside the composition area when the reaction temperature is changed as depicted in Figure 55.5. Each zone exhibits its own catalytic performance, and the catalyst P/V ratio addresses the changes of the surface composition either through the intermediate formation of an oxidized, non-hydrolized surface (for catalyst P/V 1.00), or through the formation of a hydrolyzed, non-oxidized surface (for catalyst P/V 1.06). The former corresponds to a very active but poorly selective surface at temperatures between 340 and 400°C. At high temperature (i.e., 440°C) the active surface of both catalysts is hydrolyzed and oxidized, leading therefore to the same catalytic behavior (Figure 55.1).

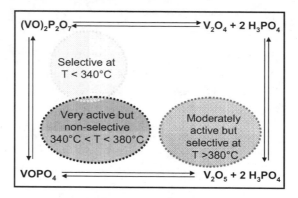

Figure 55.5. A picture summarising the catalytic performance of different active surfaces in VPP catalyst.

References

1. H. Bluhm, M. Hävecker, E. Kleimenov, A. Knop-Gericke, A. Liskowski, R. Schlögl and D.S. Su, *Topics Catal.* **23**, 99 (2003).
2. I.E. Wachs, J.M. Jehng, G. Deo, B.M. Weckhuysen, V.V. Guliants and J.B. Benziger, *Catal. Today* **32**, 47 (1996).
3. D. Wang, H.H. Kung, M.A. Barteau, *Appl. Catal. A* **201** (2000) 203.
4. A. Brückner, B. Kubias, B. Lücke, *Catal. Today* **32**, 215 (1996).
5. P.A. Agaskar, L. DeCaul and R.K. Grasselli, *Catal. Lett.* **23**, 339 (1994).
6. F. Cavani, G. Centi and F. Trifirò, *Appl. Catal.* **15**, 151 (1985).
7. Z.Y. Xue and G.L. Schrader, *J. Phys. Chem. B* **103**, 9459 (1999).

56. A Five-Week Teaching Laboratory at Wabash College Introduces Students to Independent, Original Research in the Field of Asymmetric Organocatalysis

Paul R. LePlae, Andrew C. Oehler and Colin M. Ridenour

Department of Chemistry, Wabash College, 301 W. Wabash Avenue, Crawfordsville, IN 47933

leplaep@wabash.edu

Abstract

A five-week teaching laboratory in which students synthesize and evaluate enantiomerically pure organocatalysts for the asymmetric epoxidation of two activated alkenes is described. Students prepared Gringard reagents derived from substituted aryl bromides for reaction with Boc-L-proline methyl ester to yield N-protected diarylprolinols. Removal of the Boc protecting group under thermal conditions followed by crystallization of the corresponding tosylate salts provided the conjugate acid of the desired catalyst in a pure form. Catalyzed asymmetric epoxidations of *trans*-chalcone and (*E*)-4-phenyl-3-buten-2-one with *tert*-butyl peroxide were used to evaluate catalyst performance. Experimental results and the inherent challenges and pedagogical value of providing undergraduate students the opportunity to participate in original research are described.

Introduction

Wabash College is a private, independent, four-year liberal arts college for men, granting Bachelor of Arts degrees. The college is located in Crawfordsville, Indiana, a community of 14,000 located 45 miles northwest of Indianapolis and 150 miles southeast of Chicago. Wabash's 900 male students come from 34 states and about two dozen foreign countries. Approximately 75 percent of Wabash students enter graduate or professional school within five years of graduating from Wabash.

Each year 45–70 students begin a rigorous organic chemistry two-term sequence in which the average grade earned by those who complete the course is B-. After studying the chemistry of amines during the second term, two lectures concerning organocatalysis are presented. A subsequent literature assignment requires students to choose three publications in the field of organocatalysis and summarize them. The lectures and literature assignment prepare the students for a five-week, original research laboratory in the field of organocatalysis.

The five-week laboratory is based on the report by Lattanzi[1] that describes the asymmetric expoxidation of unsaturated ketones utilizing (S)-α,αdiphenylprolinol as an asymmetric organocatalyst (Figure 56.1). The five-week laboratory was designed to extend this research by investigating the effect of replacing simple phenyl groups with electronically varied substituted phenyl groups (Figure 56.2).

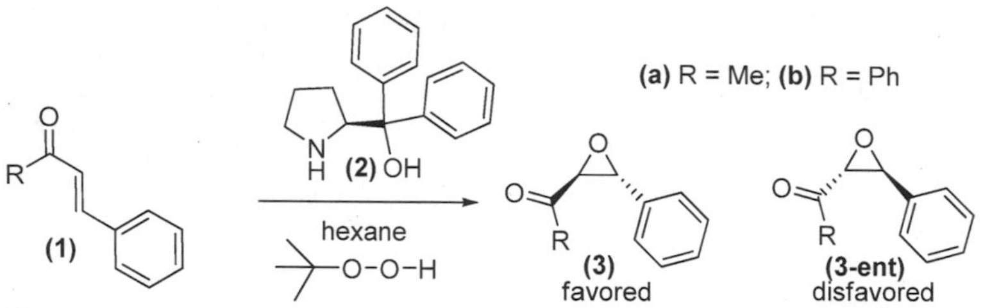

Figure 56.1: Enantioselective epoxidation reported by Lattanzi.[1]

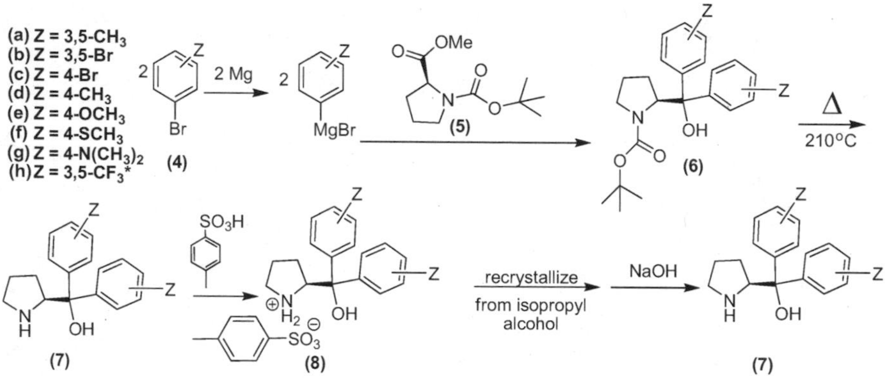

Figure 56.2: Synthetic sequence followed by students for the synthesis of catalysts for evaluation. *Bis(trifluoromethyl) derivative (**6h**) was prepared by the procedure of Leazer et al.[2]

The pedagogical purpose of the laboratory exercise was to provide students with the full research experience. Students were given the opportunity to apply their classroom and laboratory knowledge in a real research project. Previous laboratories introduced students to various purification techniques including recrystallization, thin-layer chromatography, and flash chromatography. Students were trained in the acquisition and interpretation of proton and carbon nuclear magnetic resonance (NMR) spectroscopy. Students had also completed a two-week laboratory that involved the preparation and use of phenyl Gringard reagents. This five-week laboratory was aimed to convince students that they were capable of performing original research.

One week before the beginning of the laboratory, students were provided with Lattanzi's paper and informed that they would engage in original research by preparing catalysts that utilized various aryl groups in place of the phenyl groups present in the original report. Students were provided fairly detailed procedures to perform the Gringard chemistry and the thermally induced deprotection.

Recrystallizations of the Gringard addition product as the toluenesulfonate salt were less detailed, providing students the opportunity to optimize their purification procedures. Figure 56.3 shows the synthesis of the starting L-proline derivative used in the laboratory.

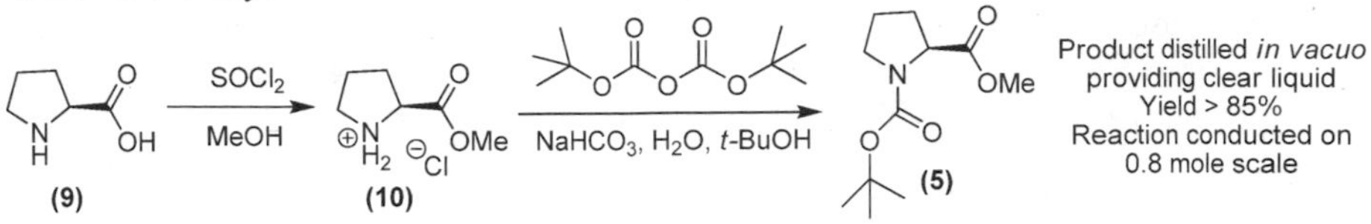

Figure 56.3: The preparation of Boc-protected L-proline methyl ester.

The Gringard addition was performed in groups of four, the deprotection and recrystallization were performed in groups of two, and the epoxidations were performed individually. Students were required to spend additional time in the laboratory outside of their traditional three-hour laboratory period. The average student committed five hours per week to the completion of this project. In the course of this work, students were exposed to failed reactions and unsuccessful recrystallizations.

Experimental Section

General Information
All the reactions were carried out under air unless otherwise specified. THF was of anhydrous grade, hexane was of HPLC grade, and isopropanol was of histological grade. Reactions were monitored by thin layer chromatography (TLC) on Merck silica gel plates (0.25 mm) and visualized by $KMnO_4$ staining procedures. Flash chromatography was performed on Merck silica gel (60, particle size: 0.040-0.063 mm). ^{1}H NMR spectra were recorded in $CDCl_3$ solutions on a Jeol 400 spectrometer (400 MHz) at room temperature. The following abbreviations were used to explain the multiplicities: s=singlet, d=doublet, dd=doublet of doublets, m=multiplet.

L-(*tert*-Butyloxycarbonyl)proline Methyl Ester (5) A solution of 92.0 g (.80 mol) of L-proline and 550 mL of methanol were combined under Argon and 64.4 mL (.88 mol) of thionyl chloride was added dropwise over 5 hours. After stirring over 20 hr, the solvent was removed *in vacuo* to afford 145 g of a yellow oil which was then dissolved in 600 mL of tert-butyl alcohol and 600 mL of water; then 336 g (4.0 mol) sodium bicarbonate was added and the mixture was stirred (mechanically) 10 min. A solution of 170 g (0.78 mol) of di-*tert*-butyl dicarbonate in 200 mL of *tert*-butyl alcohol was added dropwise over 2 hr. After stirring for 18 h, the solution was filtered and the bulk of the filtrate was evaporated. The remaining solution was diluted with 1.0 L of ether and washed with 1 N HCl (2 x 500 mL), 500 mL of saturated sodium dicarbonate, and 500 mL of brine. The organic layer was dried ($MgSO_4$) and concentrated *in vacuo*. Further drying on a vacuum pump, followed by vacuum distillation gave 156.8 g (0.684 mol 85.5%) of (5) as a colorless liquid :NMR ($CDCl_3$, 400 MHz) δ 4.32 (dd, J = 8,4 Hz, 0.4 H), 4.22 (dd, *J* = 8.5 Hz, 0.6 H), 3.72 (s, 3 H), 3.3-3.6 (m, 2.6 H), 2.1-2.3 (m, 1.2 H), 1.8-2.0 (m, 3.5 H), 1.46 (s, 3.7 H), 1.41 (s, 6.6 H).

General procedure for the production of Boc-protected α,α-diaryl-L-prolinols (6). To 100 mL of a 0.42 M solution of ArMgBr in THF (0.042 mol) at 0°C was added *N*-(*t*- BOC)-L-proline methyl ester **(5)** (3.0 mL, 0.014 mol) dropwise via syringe over a five-minute period. The solution was stirred for at least 4 h at 25°C and then cooled to 0°C. After slow addition of 3 mL of water, the solution was slowly warmed to rt with stirring. The mixture was decanted and the solid was washed with 100 mL of ethyl ether. The organics were pooled and washed with brine, dried with sodium sulfate and evaporated to yield the final compound.

General procedure for the deprotection of Boc-protected α,α-diaryl-L-prolinols. Boc-protected α,α-diaryl-L-prolinol **(6)** was placed in a sand bath under a weak vacuum source (approx. 15 mmHg) and heated to 210°C. Ten minutes after the formation of bubbles had ceased. The reaction was cooled to room temperature and ^{1}H-NMR was used to determine the extent of reaction. If the Boc group was still detectable by ^{1}H-NMR, then the reaction was reheated to 210°C for an additional twenty minutes and allowed to cool and the extent of reaction was reevaluated by ^{1}H-NMR. This procedure was repeated until the Boc group could not be detected by ^{1}H-NMR.

General procedure for the enantioselective epoxidation of α,β-enones. To a solution of catalyst **(9)** (0.3 mmol) and *activated alkene* **(1)** (1.0 mmol) in hexane (6 mL) was added TBHP (5-6 M decane solution (0.8 mL, 4.0 mmol) at room temperature (23°C) and the homogenous reaction was maintained at room temperature for approximately one week. The crude reaction mixture was then purified by flash chromatography on silica gel (petroleum ether/ diethyl ether 90/10) to provide the epoxy ketone, which was subjected to chiral HPLC for ee analysis.

trans-(3R,4S)-**Epoxy-4-phenylbutan-2-one (3a).** ^{1}H-NMR (400 MHz) δ 2.20 (s, 3H), 3.51 (d, *J* 1.96 Hz, 1H), 4.03 (d, *J* 1.96 Hz, 1H), 7.17-7.47 (m, 5H). The absolute configuration was determined by chiral HPLC using Chiralcel AD column and compared with the literature data.[1]

trans-(2R,3S)-**Epoxy-1,3-diphenylpropan-1-one (3b).** ^{1}H-NMR (400 MHz) δ 4.09 (d, *J* 1.86 Hz, 1H), 4.30 (d, *J* 1.86 Hz, 1H), 7.37-7.67 (m, 8H), 7.98-8.05 (m, 2H). The absolute configuration was determined by chiral HPLC using Chiralcel OD column and compared with the literature data.[1]

Results and Discussion

True to the nature of actual research, not all groups were able to prepare and purify their catalyst. Students assigned to the 4-dimethylamino aryl group were not able to isolate their protected prolinol presumably due to aniline coordination of the magnesium salts. Three groups discovered that bromine containing aryl groups rendered their catalyst insoluble in hexane, the optimal solvent for enantioselectivity of the conjugate addition reaction. Another group of students obtained spectral evidence that supports the notion that thioanisole-containing catalysts were susceptible to oxidation. Nevertheless, approximately 60% of the students were able to successfully carryout asymmetric epoxidations and isolate their desired epoxide.

Due to time constraints, *ee* values were not determined by the students. These values were determined by the authors during the following term utilizing chiral HPLC. Enantiomeric excess for reactions conducted in hexane at room temperature ranged from 50% to 75%, consistent with Lattanzi's original report. This result suggests that the nature of aryl substitution has little to no beneficial effect on the asymmetric induction by the catalyst under the conditions tested. Interestingly, the 3,5-trifluoromethyl groups rendered the catalyst inactive at room temperature. Higher temperatures resulted in epoxidation with diminished ee. Control experiments indicated that the catalyst was necessary to induce epoxidation at the elevated temperature. We suspect that the electron withdrawing nature of these groups attenuates the pyrrolidine nitrogen's basicity. We are currently investigating this hypothesis. Some students demonstrated that reactions conducted at higher temperatures or in more polar solvents resulted in diminished chiral discrimination.

Two months after the completion of the laboratory, seven students presented their results in poster format at the Wabash Celebration of Student Research, an annual event highlighting the research efforts of Wabash students.

Challenges and Student Feedback

Performing actual research in a teaching laboratory presents several challenges to the students as well as the instructor. In order for students to complete this project, they were required to schedule additional laboratory time outside of the traditional laboratory period. In order to provide all students equal opportunity to complete their research, the instructor spent significant additional time in the laboratory. Additionally, as the research proceeded, procedures were modified and this led to frustration amongst some of the students. Representative student comments at the conclusion of the laboratory are provided below.

"Several steps of the reaction required extra heating or extra chemicals added to dissolve product into solution. Despite the setbacks, though, I feel like I genuinely learned organic chemistry."

"Overall, I thought this lab was a lot of fun and provided insight to what chemists do in real labs. I think it is important for the class to participate in current studies of organic chemistry. I thought that this was one of the best labs of the semester."

"I enjoyed the idea of doing original research, but I really thought the extra time in the lab was over the top, especially so late in the semester."

"It has been one of the most exciting labs I have worked on throughout my [science] career at Wabash and I hope to have some further experience with this sort of thing in my college experience."

"The five-week lab had many pros and fewer cons. This was my favorite [organic] lab."

"This lab reaffirmed my desire to be involved in medicine and not research. Lessons abound from undertaking original research: success, failure, trial and error. These lessons are simply indispensable and should be rated among the top experiences a student can undergo."

"Compared to my other laboratory experiences, I would say that this lab was by far the most difficult and time consuming, but probably the most educational. I never fully appreciated the effort that it took to come up with one's own procedure when the prescribed procedure did not pan out."

Conclusions

An original research laboratory project involving the synthesis and evaluation of organocatalysts for the asymmetric epoxidation of activated alkenes was carried out within an undergraduate organic chemistry teaching laboratory. Although demanding and time consuming for both instructor and students, the laboratory provided students with a true research experience.

Acknowledgements

We thank the Department of Chemistry at Wabash College for financial support. P.L. thanks the students who participated in this research.

References

1. Lattanzi, A. *Org. Lett.* (2005) **7**, 2579-2582.
2. Leazer, J. L., Jr.; Cvetovich, R.; Tsay, F.-R.; Dolling, U.; Vickery, T.; Bachert, D. *J. Org. Chem.*, (2003) **68**, 3695-3698.

57. 2008 Murray Raney Award Lecture: The Scientific Design of Activated Nickel Catalysts for Chemical Industry

Daniel J. Ostgard

Evonik Industries GmbH, Rodenbacher Chausse 4, 63457 Hanau, Germany

dan.ostgard@evonik.com

Abstract

Murray Raney discovered in 1927 that he could generate a hydrogen-rich activated Ni catalyst by leaching most of the Al out of a Ni/Al alloy with alkaline solutions (1). This catalyst is typically referred to as Raney® Ni, but it is also known as skeletal or activated Ni and its resulting spongy pore structure (2) has been accredited with affording it a high surface area (3,4). It is a highly active catalyst whose robust chemical and mechanical properties make it ideal for industrial hydrogenation, isomerization, dehydrogenation and reductive alkylation reactions. Although it has been widely used, the mechanism of its activation (leaching) and the optimization of its resulting properties still need to be studied and improved. This work has elucidated the mechanism of Ni/Al leaching for the Ni_2Al_3 phase and correlated it to the properties of the alloy so that improvements could be made to the ensuing catalyst. These catalysts were also modified via the decomposition of carbonaceous residues on their surface to provide steric Ni ensemble control for the enhanced selective hydrogenation of nitriles to primary amines.

Introduction

Contrary to common misconceptions, the types of activated Ni catalysts are as numerous as their applications. The desired properties of the catalyst will be determined by the type, medium and conditions of the reaction to be performed and its design may utilize different Ni-to-Al ratios, alloying techniques, Al leaching methods and modifiers. The alloy's Ni-to-Al ratio, cooling rate and cooling medium determine the structures, sizes and amounts of the Ni_2Al_3, $NiAl_3$ and Al-$NiAl_3$-eutectic (a.k.a, Al-eutectic) domains in the alloy. Although the two Al-richer phases are easily leached to more active surfaces (5) with more accessible pores (6), they also sinter (7-9) and deactivate faster at temperatures greater than ~110°C. Thus the amount and structures of the alloy's Ni_2Al_3 domains will determine the resulting catalyst's stability and steady state activity for most reactions. Consequently, the effect of the Ni_2Al_3 properties on its leachability with caustic will also impact the performance of the catalyst and this has yet to be systematically studied. If fact, there is no consensus in the literature concerning the mechanism of Ni/Al alloy leaching. Presnyakov et al. (10) suggested that the stable phases of the phase diagram form in the sequence of first Ni_2Al_3 then NiAl and finally Ni_3Al with the uninterrupted loss of Al during leaching. The formation of these stable intermediate nickel-rich phases already present in the phase diagram can be excluded, since no NiAl with CsCl-type structure or Ni_3Al with Cu_3Au structure were found during leaching (11). Furthermore, these phases are reported to be resistant to alkali leaching (12) and once formed, they would have remained in the nickel catalyst. Gros et al. (13) suggested that Ni/Al leaching with caustic involves the formation of intermediate phases such as disordered cubic-centered NiAl (50:50 at.%) and consecutively ordered domains of Ni_2Al-type structure. Reynaud (14) found that Ni_2Al is hexagonal instead of cubic and recent results have not confirmed its presence during the activation of the Ni_2Al_3 phase (15). Wang and colleagues (15) proposed that the Ni_2Al_3 phase does not go through a bcc intermediate during leaching before being transformed into the fcc structure. Instead, they suggest that the face-centered tetragonal Ni_3Al_2 phase is the only possible intermediate, if there even is one, and they also presented a series of phase transformations to fit their theory. Unfortunately the concepts of Wang do not agree with the extent of alloy restructuring observed by the XANES (x-ray adsorption of the near-edge spectroscopy) studies of Rothe et al. (16), where they also found that the structure of Ni/Al alloys approach that of fcc Ni as they lose Al. However, Rothe and colleagues were not able to describe neither the mechanism of this transformation nor the detailed structure of the resulting catalyst. To add to the confusion, not all researchers use the same types of alloys for their studies and rarely are there comparisons made between them. Some researchers have activated rapidly quenched (15) or single-phase alloys (17,18) to find that they are leached via an "advancing interface mechanism," in which is another name for the shrinking core model (19). On the other hand, when Choudhary and colleagues (20,21) activated their multiphase casted alloy, the shrinking core model no longer seemed to apply and the leaching process proceeded via a mathematic kinetic expression that is best described as a site nucleation model (20,21,22). Hence, both the mechanism of Ni/Al leaching and its kinetics need to be resolved so that the preferred alloy types and activation methods can be effectively chosen for the desired application.

Controlling the size of the catalyst's active site is another way to optimize it. This can be done by "fencing in" smaller more selective steric Ni ensembles (active sites with the desired number of contiguous metal atoms defined by their steric environment) with coke residues made by formaldehyde (H_2CO) decomposition (23). These steric Ni ensembles not only improve the catalyst's primary (1°) amine yield during nitrile hydrogenation (24,25), but they also enhance the chemo-selective reduction of nitriles over that of olefins to produce 1° fatty amines with high olefin retention (26,27). Moreover, this method proved to be very useful in determining the mechanism of fructose hydrogenation (28,29). Thus, we will also use steric Ni ensembles to further elucidate the mechanism of nitrile hydrogenation.

Experimental Section

This research was carried out in two parts. The first one explored the relationships between alloy structure, alloy activation and the resulting catalyst, and the second one looked at the use of sterically controlled Ni ensembles for improved reaction selectivity. Most of the results of the first part came from a joint project between Evonik Degussa and the Technical University of Darmstadt, and the resulting literature (11,20,22) can be consulted for additional details. This investigation looked at both single- and multiphase alloys. The casted single-phase $NiAl_3$ (42 wt.% Ni / 58 wt.% Al) and Ni_2Al_3 (60 wt.% Ni / 40 wt.% Al) casted powder alloys were bought from Goodfellow in Germany. The X-ray diffraction (XRD) analysis of the $NiAl_3$ powder found that a few of its particles had Ni_2Al_3 impurities. Thus, we had to embed these powders so that we could identify and measure only the single-phase particles. Each of these alloys was embedded in POLYFAST® (Struers) and polished to expose the particles. The samples were mounted at a well-defined position in a scanning electron microscope (SEM) outfitted with energy dispersive X-ray (EDX) so that the desired single-phase particles could be identified by their EDX-measured wt.% Al and their positions could be noted for future measurements. The embedded alloy was activated in a 20 wt.% NaOH aqueous solution at room temperature (rt) for a set time followed by washing with distilled water before it was analyzed by SEM to determine the wt.%Al of the identified single-phase particles. This procedure was repeated many times to record the loss of Al during activation over 240 min.

The multiphase alloys were made by melting the required metals in an induction oven over a set period of time to obtain a uniform melt (the inductive oven's magnetic field stirs the Ni in the alloy) followed by cooling it either as a casted ingot, a nitrogen atomized (NA) powder or as a water atomized (WA) powder. The activation of the 53 wt.% Ni / 47 wt.% Al casted alloy (CA) was studied both as a coarse chunk and a finely ground powder. The CA chunk was embedded in EPOFIX® (Struers), polished and analyzed at a well-defined position in the SEM to identify and locate the various phases. It was then leached in a 20 wt.% NaOH aqueous solution at rt for a set time and washed with distilled water before SEM analysis via backscattering electron imaging (BEI) and the EDX of the previously identified phases. This procedure was repeated many times to record the Al loss from the Ni_2Al_3, $NiAl_3$ and Al-eutectic phases over 120 min of activation and a BEI of the catalyst was also made after 570 min of leaching at rt. The powder multiphase alloys described in Table 57.2 were sieved between the particle sizes of 32 and 125 μm before activation. They were activated by uniformly adding 300 g of each of them over 9 min. to 2.6 l of a stirred 20 wt.% NaOH aqueous solution whose temperature at the end of this addition time briefly reached 110°C for a very short period of time before it was either maintained at 110°C or cooled to 75°C for the following 120 min of post-addition leaching (a.k.a., activation time). Catalyst samples were removed after 10, 20, 30, 40, 60, 90 and 120 minutes of post-addition leaching, washed twice with fresh 20 wt.% NaOH and washed to a pH of 9 with distilled water for storage. The samples were characterized by XRD, nitrobenzene hydrogenation, light microscope imaging and SEM/EDX. Another catalyst was made for TEM/EDX and XRD analysis by adding 2 g of a < 35 μm 53 wt.% Ni / 47 wt.% Al CA to 200 ml of a stirred 10 wt.% NaOH aqueous solution at rt for 45 min followed by washing twice with 20 wt.% NaOH, washing with water and storage in ethanol.

The characterization methods used here include SEM, high resolution transmission electron microscopy (HRTEM), EDX, XRD, light microscope imaging and temperature programmed oxidation (TPO). The SEM BEI and metal content determinations of the embedded and powder samples were performed with a DSM 962 apparatus from Zeiss outfitted with EDX. The HRTEM was done with a Philips CM2OUT microscope working at 200 kV and equipped with an EDX analysis system. The HRTEM samples were prepared by spreading a drop of the powder's ethanol suspension on a carbon foil covered copper microgrid that was then quickly transferred to the HRTEM's airlock to avoid oxidation. HRTEM analyses were made in the transmission, the diffraction and the selected area electron diffraction (SAED) modes. The XRD with Cu Kα_1 radiation was carried out on a Stoe STADI-P diffractometer in the Bragg angle 2θ range of 10–110°, and some of the diffraction patterns were analyzed by the Rietveld method (31) using the software FULLPROF. The phases were identified against the ICCD XRD standards and the Ni crystal size was calculated from the width of the Ni(111) peak at half the maximum intensity (FWHM) using the Scherrer equation (32) with correction for instrumental broadening. The light microscope imaging of the alloys were done with an Olympus BH2-UMA microscope and a Nikon OPTIPHOT-2 microscope. The volume percentages (vol.%) of the Ni_2Al_3, $NiAl_3$ and Al-eutectic phases and the interface densities (S_v) of the Ni_2Al_3 domains were determined via the computer analysis of the light microscope images according to Exner and Hougardy (33) with the Quantimet system from the company Leica. The S_v of the Ni_2Al_3 phases were calculated with eq. 1 (22), and the larger the average Ni_2Al_3 S_v of the alloy was, the smaller its Ni_2Al_3 domains were.

$$S_v = \frac{4}{\pi} \bullet \frac{\left(\text{Perimeter of the } Ni_2Al_3 \text{ phase}\right)}{\left(\text{Area of the } Ni_2Al_3 \text{ phase}\right)} \left[\mu m^{-1}\right] \qquad (1)$$

Before performing TPO, the wet catalyst was dried under N_2 flowing 10 l/h at 120°C for a period of 17 h before being slowly cooled to 20°C. Once the catalyst was stabilized at 20°C the pure N_2 was switched to a 4 vol.% O_2 in N_2 mixture flowing at 10 l/h while the catalyst's temperature was ramped to 800°C at the rate of 8°/min. The O_2 content of the gas mixture leaving the catalyst bed was measured by an Oxynos 100 paramagnetic detector and plotted with respect to either time or temperature. The plot of the vol.% O_2 with respect to time was used to determine the O_2 consumption of the catalyst.

All the reactions performed here used hallow shaft bubble inducing stirrers rotating at high enough speeds to eliminate any problems associated with the mass transfer of H_2. The atmospheric hydrogenation of nitrobenzene (NB) was carried out over 1.5 g of catalyst in 110 ml of a 9.1% NB ethanolic solution at 25°C in a baffled glass reactor stirred at 2000 rpm, and this was primarily used to characterize the single- and multiphase alloys. The hydrogenation of nitriles was used to show the enhanced selectivity of the catalysts modified by stirring them in an aqueous H_2CO solution at rt for 1 h (23-29). The level of modification is given in terms of mmol H_2CO per g of catalyst in Table 57.3, Table 57.4 and Table 57.5. Catalysts were also modified with an aqueous solution of sodium formate at 90°C over 1 h (23) at the levels of 1.66, 2.0 and 3.0 mmol sodium formate per g of catalyst. The other catalyst modifiers were carbon monoxide, acetone and acetaldehyde and they were used at the conditions and levels mentioned in Table 57.3. The hydrogenation of 32.5 g of benzonitrile (BN) in 514.2 g of methanol (MeOH) stirring at 2000 rpm was performed with 1.3 g of catalyst at 41 bar while ramping the reaction temperature from 25 to 105°C with a hold at 105°C. Depending on the catalyst, the reaction started once the temperature reached somewhere between 40 to 80°C and the hydrogenation time was calculated from the start of H_2 uptake to its end. The effect of ammonia (NH_3) on the hydrogenation of BN was studied by adding 10 ml of a 32 wt.% NH_3 aqueous solution to the reaction mixture. The hydrogenation of 40 g of Pynitrile (24) in 1475 ml MeOH with 300 g of NH_3 was accomplished with either 5 g of activated Ni (the acetone modified activated Ni required 6 g), 10 g of activated Co or 8 g of Ni/SiO_2 at 40 bar and 110°C over 5 h. The reduction of 40 g of pyridine-3-carbonitrile (P3C) in 2 l of MeOH with 31 g of NH_3 was done with 5.3 g of unmodified catalyst over 5 h. Surprisingly, the same amount of a catalyst modified with 0.908 mmol H_2CO per g of catalyst was able to hydrogenate P3C within 3 h under the same conditions. The reduction of 80 g of valeronitrile (VN) in 2 l of MeOH with 15 g of NH_3 was done with 20 g of unmodified catalyst over 5 h. Unexpectedly, the same amount of a catalyst treated to the level of 3.090 mmol H_2CO per g of catalyst was able to hydrogenate VN within 3 h under the same conditions. The hydrogenation of a 500 g tallow nitrile mixture (mostly C18 with some C16 and traces of C14 and C20) with an iodine value of ~51 over 1 g of either an untreated or a treated (1.3 mmol H_2CO per g of catalyst) catalyst was performed with 17 bar of NH_3 and 23 bar of H_2 at 140°C. Samples were taken at regular intervals and titrated to determine the total amine value (TAV - AOCS method Tf 1a-64) as an indication of nitrile conversion, the total amount of 2° and 3° amines (2/3 A - AOCS method Tf 2a-64) and the iodine value (IV - AOCS modified Wijs method Tg 1-64) for olefin retention. Further details can be seen in the literature (23,25-27).

Results and Discussion

Ni / Al Alloy Leaching with NaOH. The catalysts investigated here are activated by leaching Al out of a Ni/Al alloy with aqueous NaOH via eq. 2 to form reduced Ni° (the active site), sodium aluminate and H_2. The Al is not totally leached out and what remains in the catalyst has both structural and electronic influences on the performance of the catalyst. This reaction is exothermic and it generates a lot of H_2 that is mostly lost as a gas; however, a considerable amount of it also remains with the catalyst (34), thereby enhancing its hydrogenation and pyrophoric properties. The type of alloy, the concentration of NaOH, the NaOH-to-alloy ratio, the leaching temperature, the addition time of the alloy and the post-addition leaching time are some of the parameters that can be changed to modify the properties of the resulting catalyst. If this process is carried out in an excess of NaOH (see eq. 3), the sodium aluminate remains in solution so that it can be easily removed and sold for use in the paper industry, accelerated concrete solidification, alumina production, the manufacture of zeolites, water treatment and other applications.

$$Ni_xAl_3 + 3\,NaOH + 9\,H_2O \rightarrow X\,Ni^{\circ} + 3Na[Al(OH)_4] + 4.5\,H_2 \qquad (2)$$

$$Na[Al(OH)_4] + 2\,NaOH \rightarrow Na_3[Al(OH)_6] \qquad (3)$$

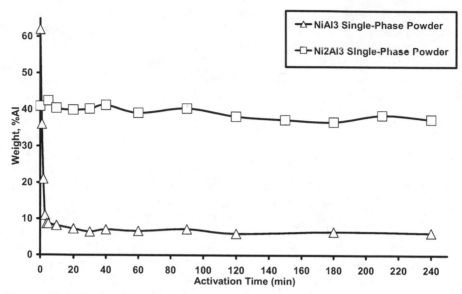

Figure 57.1. The leaching of casted single-phase powders with 20 wt.% NaOH at rt.

Leaching the catalyst under NaOH-deficient conditions will convert the sodium aluminate to $Al(OH)_3$ according to eq. 4 and this can react further to form Al_2O_3 (see eq. 5). This series of reactions will produce a white powder that is not only hard to remove from the catalyst, but it can also cause by-products and filtration problems during the catalyst's use. Moreover, some of this $Al(OH)_3$ and Al_2O_3 will adsorb onto the catalyst to block its surface and fill its pores leading to lower activity.

$$Na[Al(OH)_4] \rightarrow Al(OH)_3 + NaOH \qquad (4)$$

$$2\,Al(OH)_3 \rightarrow Al_2O_3 + 3\,H_2O \qquad (5)$$

In some cases, the catalyst can be leached in highly NaOH-deficient conditions so that most of the Al will precipitate out as $Al(OH)_3$ or Al_2O_3 and act as a catalyst support. According to Petró this will also help to make the catalyst non-pyrophoric (35).

The Leaching of Casted Single- and Multiphase Alloys. The Ni_2Al_3 single-phase powder casted alloy (CA) is practically inert to aqueous NaOH at rt (see Figure 57.1), while the Al is readily leached from of the $NiAl_3$ single-phase CA under the same conditions. It is expected that the Al-richer $NiAl_3$ phase activates much faster than Ni_2Al_3, but it is nonetheless interesting that the Al content dropped from 62 to 6.5 wt.% over 30 min with 20 wt.% NaOH at rt. Figure 57.2 displays the SEM BEI

Figure 57.2. The SEM BEI of the multiphase casted alloy (CA) block leached at rt with 20 wt.% NaOH for 2 min (a), 5 min (b) and 570 min (c).

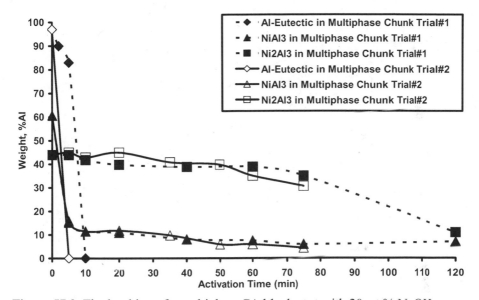

Figure 57.3. The leaching of a multiphase CA block at rt with 20 wt.% NaOH.

of the multiphase CA chunk after 2, 5 and 570 min of leaching at rt with 20 wt.% NaOH. After 2 min most of the Al-eutectic phase has been leached, but the $NiAl_3$ and Ni_2Al_3 phases still seem to be intact. After 5 min, all of the Al-eutectic has been leached and the $NiAl_3$ is starting to show some very deep cracks, while the Ni_2Al_3 is still rather pristine. The SEM BEI of the 570 min rt leached CA chunk demonstrates that both the $NiAl_3$ and Ni_2Al_3 phases are activated; however, it is much easier to recognize the initial structure of the Ni_2Al_3 phase than it is for the $NiAl_3$ that clearly has more cracks which are also deeper suggesting that this phase is far easier to sinter during leaching (even at rt) than Ni_2Al_3. Figure 57.3 displays the leaching profiles of the Al-eutectic, $NiAl_3$ and Ni_2Al_3 of this multiphase CA block at rt with 20 wt.% NaOH and they confirm the SEM BEI impressions. The Al-eutectic is immediately leached within 5 to 10 min. The $NiAl_3$ phase is quickly leached after or simultaneously as the Al-eutectic and it reaches a steady state of ~6 wt.% Al after ~ 50 min of activation. The Ni_2Al_3 only leaches slightly up to the activation time of 75 min, after which its Al content drops from ~35 to 8 wt.% Al over the next 45 min.

Even after 570 min of leaching, the Ni_2Al_3 has only been activated to 7.5 wt.% Al and it is unlikely that it would drop lower than that. Comparing the leaching behavior of the single-phase CA to that of the multiphase block indicates that the $NiAl_3$ phase activates very fast in both situations; however, the leachability of the Ni_2Al_3 phase is considerably higher for the multiphase block than it is for the isolated single-phase powder. The most obvious difference that seems to have enhanced the leachability of the Ni_2Al_3 phase in the multiphase alloy was the presence of the $NiAl_3$ phase. Timofeeva (36) and colleagues have also noted that the activation of Ni_2Al_3 is enhanced when it is in contact with the $NiAl_3$ phase and they attributed this to the nearby heat of reaction produced by leaching the $NiAl_3$ phase. They also suggested that the two phases form a $NiAl_3$-Ni_2Al_3 galvanic couple that should also accelerate the leaching of Ni_2Al_3; however, they did not present any data to support this claim.

Table 57.1. The Effects of Grinding on the Structure of a 53 wt.% Ni / 47 wt.% Al CA.

Form	Size	Vol.% Ni_2Al_3	S_v
Chunk	> 1 mm	60 ± 5	0.14
Powder	From 32 to 125 μm	80 ± 2	0.20

The Leaching of Powder Multiphase Alloys with Different Cooling Rates. The cooling rate of an alloy plays a significant role in determining its structure. The most common method is to cast it into an ingot and allow it to slowly cool. This may be accelerated somewhat by pouring water on it; however, the extent this influences the alloy's structure depends on if the melt was solidified before or after the addition of water. Even if it was already solidified, pouring water onto a hot Ni/Al alloy can oxidize some of the surface Al according to eq. 6 and the amount of Al_2O_3 this forms will depend on the temperature and the surface-to-volume ratio of the alloy solid. Once the CA is solidified, it needs to be ground into a fine powder of the desired particle size. Table 57.1 shows that the grinding process increases the amount of the Ni_2Al_3 phase while decreasing the average size of its domains as indicated by the 43% increase of its S_v. It was mentioned in the literature (37,38) that particles of multiphase alloys smaller than 60 μm are richer in Ni_2Al_3 due to the higher brittleness of this phase. This would only explain the shift towards a higher Ni_2Al_3 concentration if sieving the original powder resulted in the loss of a considerable amount of coarse particles. However, the yield of the sieved alloy was too high to explain this drastic increase in the Ni_2Al_3 vol.% and the only losses were the finer particles that went through the 32 μm sieve. Hence this increase in the vol.% Ni_2Al_3 is not simply due to the selection of a particle size richer in the Ni_2Al_3 phase, instead there must have been some conversion of the $NiAl_3$ phase to Ni_2Al_3 and Al-eutectic during grinding. It has been reported by Lei and colleagues (39,40) that the $NiAl_3$ can be converted to Ni_2Al_3 by treating the alloy in flowing H_2 at 500°C. It isn't obvious from their data if the phase conversion is caused by the presence of H_2, the high temperature or both; however, this does not preclude or conflict with the conversion of the $NiAl_3$ phase to Ni_2Al_3 that we have noticed during grinding. In fact, some alloys are made by the high energy grinding of the separate components (41,42) and even though this mechanical alloying method is not commercially viable due to the long batch times and high energy use, it does demonstrate the ability of alloys to change during grinding. Gostikin et al. (43) have found that normally casted and ground alloys can be also changed considerably by milling them at high energy for only 30 to 40 min resulting in higher catalytic activity after activation. Even the normal grinding processes are very energy intensive and it is known that Al can readily migrate in Ni/Al alloys (44), hence it should not be surprising that the ground alloy has a different phase content and size than what the original ingot had.

$$2\,Al + 3\,H_2O \rightarrow Al_2O_3 + 3\,H_2 \tag{6}$$

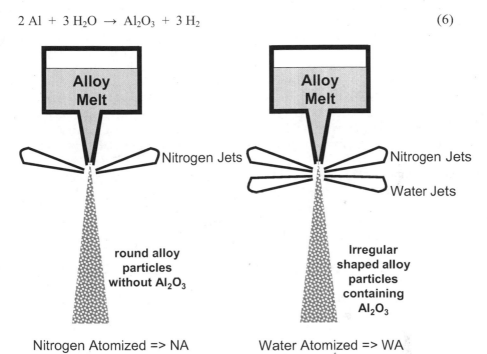

Figure 57.4. The making of nitrogen atomized (NA) and water atomized (WA) alloys.

Rapidly under cooling the alloy before a significant amount of solids are formed can extend the solubility limits of the alloy and create non-equilibrium crystalline or quasi-crystalline intermediate phases (45). This increases the homogeneity of the alloy and suppresses the formation of the extremely Al-rich eutectic phase leading to an alloy that is more corrosion resistant. In fact, the initial driving force for the development of rapidly cooled alloys was their high strength and excellent corrosion resistance. Unlike the slowly cooled casted alloys with their large bulky quasi-dendritic phase domains (see Figure 57.2), the quickly cooled ones form very fine dendritic phases that become even finer as the cooling rate increases. There is a strong correlation between the cooling rate of Al alloys from more than 10^9 to almost 10^{-3} K per s and the secondary dendritic arm spacing as shown in eq. 7 below, where K and n are alloy specific constants. The secondary dendritic arm spacing is used for this calculation because the formation of the primary dendritic arms is too fast and random to fit any type of mathematical relationship (45).

Secondary dendritic arm spacing $= K * (\text{cooling rate})^{-n}$ (7)

Rapidly cooled alloys can be made by many different methods; however, the most commercially viable ones involve atomizing the melt into fine alloy droplets before solidification to yield a finely structured alloy powder. Not only does the formation of droplets result in the desired powder form, but it also produces a favorable surface-to-volume ratio for very high and more uniform cooling rates. The selection of cooling medium is also important. Figure 57.4 displays the nitrogen atomizing (NA) and water atomizing (WA) techniques used here for the production of rapidly solidified alloys. The NA alloy is produced by breaking down a constant stream of the liquid Ni/Al alloy at 100 to 200°C above its solidification temperature (i.e., the metal superheat) with one or more high-velocity jets (or a ring arrangement) of N_2. The NA alloys are spherical and both their particle sizes and phase structures are controlled by the jet distance to the melt stream, jet pressure, nozzle geometry, velocity and mass flow rate of both gas and metal, metal superheat, angle of impingement for the cooling medium, metal surface tension and metal melting range. The typical cooling rates of NA alloys range from 10^2 to 10^3 K/s (45). The H_2O cooling medium for WA alloys is at a higher pressure, a faster velocity and a lower angle of impingement than N_2 for the NA ones, and WA also works better with a higher alloy superheat. The WA alloys have irregular particle shapes with rough surfaces and both their particle sizes and phase structures are determined by the same parameters as were mentioned above for the NA alloys. The usual cooling rates for WA alloys range from 10^2 to 10^4 K/s (45). Unlike with NA, the H_2O cooling medium of the WA alloys is not inert and it can react with the Ni/Al alloy according to eq. 6 to give Al_2O_3 and H_2. To avoid the development of an explosive atmosphere, WA is typically done in an inert atmosphere (e.g., N_2) and this may or may not be part of the jet system (Figure 57.4). The Al_2O_3 that is produced during this reaction remains on the alloy and it also influences the performance of the catalyst.

Figure 57.5 shows the particle structures of the alloys described in Table 57.2. As mentioned above, the NA alloys are spherical, the WA alloy is very irregular with thin brittle parts and the CA53 consists of block shaped particles with sharp edges. The phase with the darkest color is the Ni-rich Ni_2Al_3, the moderate gray color is $NiAl_3$ and the white phase is the Al-eutectic. The comparison of the NA50 to the NA40 demonstrates that the Ni_2Al_3 content decreases considerably and the size of the Ni_2Al_3 phases are also much smaller (as seen with the larger S_v) when the Al content of the alloy is increased from 50 to 60 wt.%. This is to be expected since the Al-richer $NiAl_3$ phase is more stable at higher Al-to-Ni ratios. Contrasting the WA50 to the NA50 indicates that the faster H_2O quenching leads to slightly more Ni_2Al_3 whose phases are plainly smaller than those of NA50. While both the Al-to-Ni ratio and the cooling rate strongly influence the size of the resulting Ni_2Al_3 phases, the Al-to-Ni ratio has a greater impact on the Ni_2Al_3 content. The data also unveils that the CA53 alloy is clearly different than the rapidly cooled ones in that it has the highest Ni_2Al_3 content and by far the largest Ni_2Al_3 phase domains.

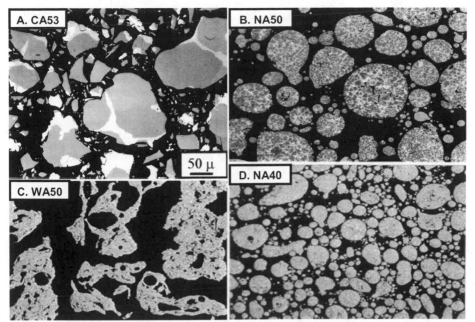

Figure 57.5. The light microscope images of the different multiphase alloys in Table 57.2.

Table 57.2. The Properties of the Multiphase Alloy Powders and the Resulting Catalysts.

Alloy	Cooling	wt.% Ni[a]	vol.% Ni$_2$Al$_3$[a]	S$_v$ of Ni$_2$Al$_3$[a]	Leaching Temp., °C	Ni$_2$Al$_3$ content after 120 min leaching[b]
CA53	Casted	53	80 ± 2	0.2	75	zero
					110	zero
NA40	N$_2$ Atomized	40	44 ± 5	4	75	zero
					110	zero
NA50	N$_2$ Atomized	50	55 ± 5	2	75	high
					110	low
WA50	H$_2$O Atomized	50	60 ± 5	3	75	very high
					110	very high

[a] The properties of the alloys containing only Ni and Al before leaching.
[b] The XRD determined Ni$_2$Al$_3$ content after leaching for 120 min.

Samples of the alloys in Figure 57.5 were activated for 120 min with 20 wt.% NaOH at either 75 or 110°C and Table 57.2 lists the residual Ni$_2$Al$_3$ contents of the resulting catalysts. Both CA53 and the NA40 were leached to zero Ni$_2$Al$_3$ content at 75 and 110°C. The NA50 contained a high level of Ni$_2$Al$_3$ after the 75°C activation and as expected, the residual Ni$_2$Al$_3$ content was much lower after leaching at 110°C. The WA50 was the most difficult to leach and it contained a very high level of the Ni$_2$Al$_3$ phase after both the 75 and the 110°C activation. The cooling rate of the alloy evidently determined the leachability of its Ni$_2$Al$_3$ phase.

Figure 57.6 displays the structures of these leached catalyst particles. The leached CA53 retained the structure of its original Ni_2Al_3 phase and although the lighter parts seem to be from the leached $NiAl_3$, this Al-richer phase was not able to exhibit the structural memory effect as was seen with Ni_2Al_3. The activated CA53 had large block shaped particles with visible cracks and sharp edges. The leached NA40 and NA50 also exhibited the Ni_2Al_3 structural memory effect and one does not see a great difference between their light microscope images due to their complex dendritic structure. The activated WA50 also displayed a Ni_2Al_3 structural memory effect and the very bright white spots in the middle of the former Ni_2Al_3 phases of the leached particles are most probably the remaining unleached Ni_2Al_3 residues. The central location of the remaining Ni_2Al_3 in its former domains strongly suggests that WA50 is activated by the shrinking core model. The Al_2O_3 domains resulting from the water cooling of the alloy (see eq. 6) would be too small to see with a light microscope. Another feature of the leached WA50 is the obvious frailness of the particles meaning that they would be more susceptible to mechanical attrition that can lead to more difficulties during filtration and the pumping of this catalyst's slurry.

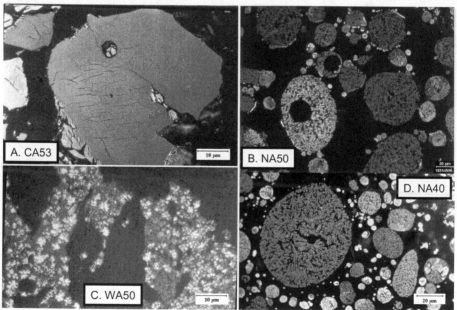

Figure 57.6. The light microscope images of the catalysts leached from the alloys of Figure 57.5 and Table 57.2 at 110°C.

Figure 57.7 shows the leaching profiles of the multiphase alloys from Table 57.2. Regardless of the alloy type, a vast majority of the Al that is leachable under the applied activation conditions is removed from the alloy after only 15 min of leaching. The alloy that leached the fastest and achieved the lowest wt.% Al levels at both 75 and 110°C was the NA40 due to it high $NiAl_3$ content. Notably, the CA53 with the highest wt.% Ni, the highest vol.% Ni_2Al_3, and the largest Ni_2Al_3 phase domains was much easier to leach than NA50 and WA50. The CA53 was almost as easily leached as the NA40. One would think that smaller Ni_2Al_3 phases with higher S_v should activate faster due to their increased contact to the readily leached $NiAl_3$, but this is not the case when comparing rapidly cooled alloys to slowly cooled ones of roughly the same Ni content. Figure 57.8A shows that the Al leachability in rapidly cooled alloys decreases as the Ni_2Al_3 content goes up and that the slowly cooled CA53 does not fit on this trend. Figure 57.8B clearly demonstrates that the relationship between the S_v of the Ni_2Al_3 phase and the leachability of its Al depends on how the size of the phase was controlled. If the phase size is controlled by the cooling rate of the alloy, then the leachability of the Al will decrease with decreasing Ni_2Al_3 phase sizes. If this phase size is controlled by increasing the Al content and consequently the $NiAl_3$ concentration, then the smaller phases are more leachable. These leachability relationships plainly indicate that the Al-leaching kinetics is very different for the rapidly cooled alloys in comparison to the casted one.

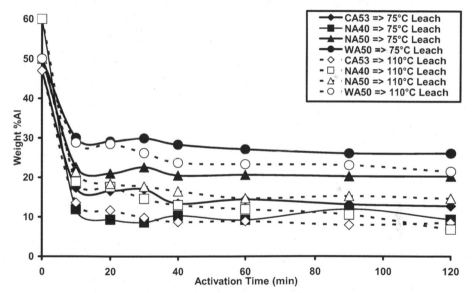

Figure 57.7. The leaching profiles of the multiphase powder alloys in Table 57.2.

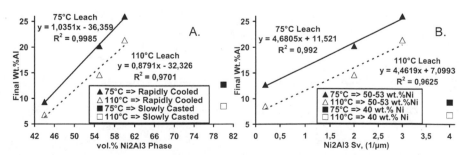

Figure 57.8. The correlations of the vol.% Ni_2Al_3 (A.) and the Ni_2Al_3 S_v (B.) to the final wt.% Al after 120 min of leaching in 20 wt.% NaOH.

Figure 57.9 shows the influence of the residual Al content of the catalysts from the alloys of Table 57.2 on their nitrobenzene (NB) hydrogenation activity leached at 75°C (Figure 57.9A) and 110°C (Figure 57.9B). When the catalysts are leached at 75°C, their NB activity increases linearly as the wt.% Al decreases down to the levels reached in Figure 57.9A due to the resulting formation of more Ni°. Practically all of the catalysts, independent of their Al-leachability, fall on the same NB activity to wt.% Al curve after leaching at 75°C. However, there are some small differences. The 75°C leached NA40 catalysts reached the lowest wt.% Al and they also had some of the highest activities. The CA53 catalysts were not leached to the same low wt.% Al levels at 75°C as the NA40 catalysts, but the ones that were leached the longest were able to obtain some of the highest NB activities. The CA53 catalyst that was leached to 13.4 wt.% Al over 40 min had a NB activity of 22.4 ml H_2/(min)(g of cat) and the one that was leached to 12.7 wt.% Al over 120 min was able to reach the NB activity of 33 ml H_2/(min)(g of cat). This increase in activity is not simply due to the loss of Al, but instead it appears to be from the restructuring of the CA53 catalyst to provide a more effective surface. The NB activities of the 75°C leached NA50 and WA50 catalysts predictably landed on the overall NB activity to wt.% Al curve and the lower leachability of their Al undoubtedly limited their activity.

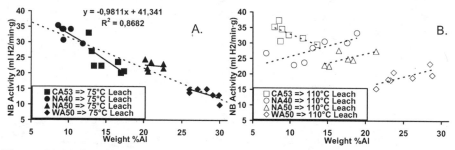

Figure 57.9. The effect of wt.% Al on the catalysts' nitrobenzene hydrogenation activity after being leached with 20 wt.% NaOH at 75°C (A) and 110°C (B).

The results after 110°C activation are considerably different than those after 75°C. The catalysts leached from NA40 were still able to reach the lowest Al contents at 110°C; however, the activity of these catalysts decreased as the wt.% Al went down. It is well known that the catalysts from $NiAl_3$-rich alloys are more prone to sinter (7-9) than ones from alloys with higher Ni_2Al_3 contents. Nonetheless, it is still surprising that the catalysts from NA40 displayed such sensitivity to sintering when the leaching temperature has only been increased from 75 to 110°C. The catalysts from CA53 were the only ones that did not sinter during their leaching at 110°C. In fact, these catalysts became more active as the Al content decreased and the catalysts leached from CA53 at 110°C were clearly more active than those that were leached at 75°C. This observation confirms that the catalysts from CA53 not only lose Al to generate more activated Ni surface, but they also undergo restructuring to produce a more active surface. This restructuring is unmistakably more effective at 110°C than 75°C for the enhancement of NB activity. The catalysts from NA50 and WA50 also lost activity as their Al contents decreased during the

110°C activation and this is also most likely due to sintering. It has never been reported in the literature before that rapidly cooled alloys with 55 to 60 vol.% Ni_2Al_3 sinter when activated at 110°C; however, it makes sense that the further extraction of Al from these practically leach resistant Ni_2Al_3 phases still remaining in the middle of their old domains will result in the sintering of the overall catalyst structure. In this respect, the leftover Ni_2Al_3 phases of the leached rapidly cooled alloys activate more like isolated Ni_2Al_3 than they do as Ni_2Al_3 in the presence of the $NiAl_3$.

The Restructuring of the Ni_2Al_3 Phase in Casted Alloys During Leaching. To best understand the restructuring of the Ni_2Al_3 phase in CA53 during leaching, we analyzed it after it has been activated to different levels of Al with TPO, HRTEM and XRD. Figure 57.10 shows the TPO of the CA53 after it has been activated at 75°C for 10 min to 17.4 wt.% Al and after 120 min at 75°C to 12.7 wt.% Al. Our experiences with TPO in the past have shown that the O_2 uptake from rt to ~140°C is mostly due to the oxidation of the hydrogen on the catalyst. Since this catalyst was dried 17 h at 120°C before measuring the TPO, most of the weakly adsorbed hydrogen has been desorbed and the O_2 uptake prior to ~140°C is more an indication of how much strongly adsorbed hydrogen is remaining on the catalyst. The O_2 uptake between ~140 to ~350°C is from the oxidation of surface Ni and Al, and the uptake after 350°C is from the oxidation of the bulk phase Ni and Al. The comparison of the 10 min leached catalyst to the one activated 120 min suggests that the longer leaching process increases the concentration of the strongly held H_2, while increasing the depth of activation. The catalyst's surface structure doesn't seem to change much and there may actually be a small drop in surface Ni and Al area during the longer leaching process due to the loss of Al and possibly the restructuring of the catalyst.

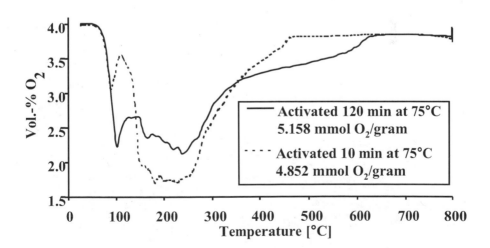

Figure 57.10. The TPO of the CA53 activated for 10 and 120 min at 75°C (22).

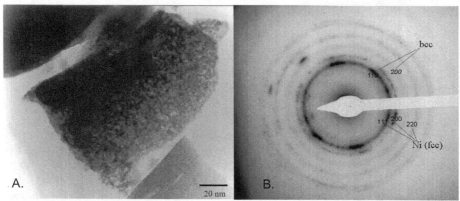

Figure 57.11. The HRTEM of the CA53 activated at 110°C for 120 min in both the transmission mode (A) and the diffraction mode at a wide lens aperture (B) (30).

Figure 57.11 displays the HRTEM of the CA53 that was leached at 110°C for 120 min. One can see in the transmission mode that the catalyst particles have a spongy porous structure that takes on the block-like shape of the original Ni_2Al_3 phase. These particles contain 5 to 15 nm blocks of slightly disordered crystallites and due to the alignment differences of a few degrees, the scattering from the different crystals was not coherent. The average crystal size determined by electron diffraction is ~10 nm and the half widths of the powder lines (FWHM) in the XRD are about 1° which corresponds to an average crystal size of ~9 nm (11). Both of these crystal sizes are in agreement with those we calculated from the chemisorption literature of CO-H titration (46) and H_2 desorption (47) which range from 8.5 to 10.7 nm (assuming 8.9 x 10~-21 g Ni per nm^3, 6.67 x 10^{20} nm^2 per g Ni and a cubic structure). The HRTEM diffraction pattern at the wide lens aperture setting seen in Figure 57.11B shows the presence of both the Ni fcc and Ni/Al bcc phases in the

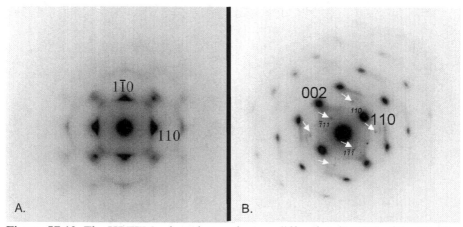

Figure 57.12. The HRTEM selected area electron diffraction (SAED) of the CA53 leached for 45 min at rt showing the presence of the Ni/Al bcc phase (A) and the coexistence of both the Ni-Al bcc phase with the original Ni_2Al_3 alloy (B).

final catalyst and the EDX analysis of the Ni/Al bcc phase showed that it contains 85% ± 5 wt.% Ni with the rest being Al. The SAED patterns seen in Figure 57.12 of the 45 min rt leached CA53 also demonstrated the presence of the Ni/Al bcc phase (with a lattice constant of 0.286 nm) in this catalyst as well as areas where this bcc

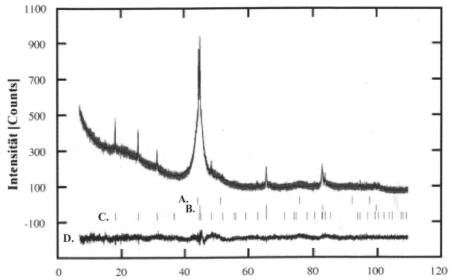

Figure 57.13. The Rietveld XRD refinement for CA53 after being leached 45 min at rt for the determination of the Ni fcc (A), Ni/Al bcc (B) and the Ni_2Al_3 (C) contents. The residual error analysis of the curve fit is displayed in line D.

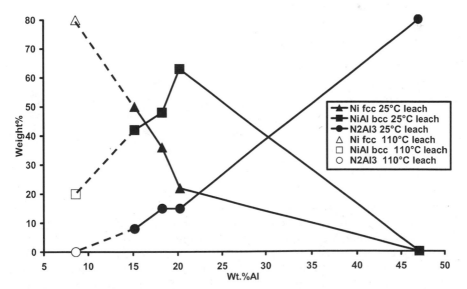

Figure 57.14. The wt.% of Ni fcc, Ni/Al bcc and Ni_2Al_3 as Al was leached from the CA53 as calculated by the Rietveld refinement method.

phase was in coexistence with the Ni_2Al_3 alloy (30). X-ray does not provide any evidence for the bcc phase in the precursor alloy; hence, it must have formed during the leaching process. We used the Rietveld refinement of the XRD powder patterns from the activated CA53 to determine the Ni fcc, Ni/Al bcc and Ni_2Al_3 contents of the catalyst as it was leached to different Al contents (see Figure 57.13). This method performs a curve fit of the standard Ni fcc, Ni/Al bcc and Ni_2Al_3 patterns to what was measured in order to calculate the corresponding concentrations of the components with a relative error of ±10%. Figure 57.14 shows the results of these calculations with respect to the wt.% Al remaining in the catalyst during leaching. One can see that as the Ni_2Al_3 concentration steadily decreases during leaching, the amount of Ni/Al bcc increases to a peak of 63 wt.% at ~20 wt.% Al before it decreases to 20 wt.% at 8.6 wt.% Al. Simultaneously, the amount of the Ni fcc slowly increases as the Al content drops to ~20 wt.% and once the Ni/Al bcc has reached its maximum, the amount of Ni fcc increases even faster with the loss of additional Al. The data strongly suggests that the Ni/Al bcc is an intermediate phase between the Ni_2Al_3 and the Ni fcc during leaching. The CA53 that was leached at 110°C for 120 min was found by SEM EDX to have 8.6 wt.% Al and the Rietveld refinement of this sample's XRD determined that it consists of 80 wt.% Ni fcc and 20 wt.% of Ni/Al bcc. Since the bcc phase contains 15 wt.% Al, that would mean that 3 wt.% of the Al in this catalyst is associated with the Ni/Al bcc and the other 5.6 wt.% Al would have to be in the Ni fcc phase. This is in very good agreement with the literature that states up to 6 wt.% metallic Al should be soluble in Ni according to the NiAl phase diagram (37). This also agrees well with the optimal Al content (from 5 to 6 wt.%) of a Ni cathode for the generation of H_2 (48).

There exists a relationship between the trigonal Ni_2Al_3 structure and the bcc structure (11,13,49,50) as demonstrated in Figure 57.15 by the subdivision of the Ni_2Al_3 structure (bold lines: unit cell) into cubic cells (thin lines). The Ni_2Al_3 structure may be considered as a bcc structure ($a_0 \sim 0.284$ nm) with Ni atoms, Al atoms and voids ordered in a way that breaks the cubic symmetry (11). Thus, the bcc can form from the Ni_2Al_3 phase by the loss of Al atoms followed by the short order rearrangement of Ni atoms for which only small deviations from the original positions are required. There is also a relationship between fcc and bcc structures that is known as the Bain relationship and it has been described in detail for the martensitic transformation of iron alloys from fcc to bcc structure (51). Thus, as the depleted Ni/Al alloy loses more Al atoms, the bcc structure then transforms martensitically into fcc Ni by the short distance rearrangement of atoms. In summary, the leaching process starts with the selective dissolution of Al to form vacancies in the Ni_2Al_3 lattice and Al-deficient regions in the particle. As the Al content decreases further, the Ni_2Al_3 collapses and transforms by a short order rearrangement into an intermediate bcc structure whose composition is 85±5 wt.% Ni with the balance being Al. This process occurs in small nm-sized regions leading to the observed crystallite size. The leaching of more Al from the bcc phase then leads to its martensitic transformation into fcc Ni by another short distance rearrangement of atoms. The finished catalyst consists of ~10 nm fcc Ni crystals with defects having up to ~6 wt.% Al and finely dispersed between them are some bcc Ni/Al phases. Our data strongly suggests that the fcc Ni is more active than the bcc Ni/Al

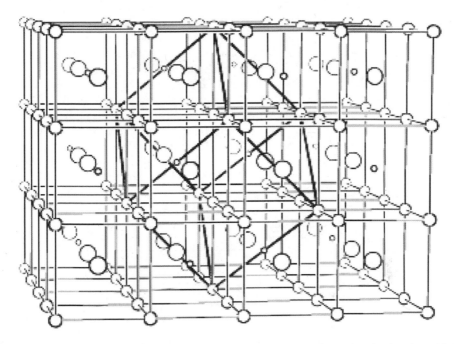

Figure 57.15. The structural correlation between the trigonal Ni_2Al_3 (the thick outline) and the bcc structure of the disordered Ni-Al phase. The large circles are Ni, the medium ones are Al and the small ones are voids (11).

and this transformation is usually accompanied with an increase in activity, as seen with the activation of the CA53 alloy for more than 60 min.

The Influence of the Alloy's Cooling Rate on the Kinetics of Leaching. Figure 57.16 shows the rt leaching of the casted single-phase powders, the casted multiphase block and CA53. The $NiAl_3$ leaches at roughly the same rate for both the single-phase powder and the multiphase block, hence there were no effects on leaching due to sample preparation. The most important Ni_2Al_3 comparisons made here will be between alloys containing 50 to 53 wt.% Ni and these conclusions will be limited to this Ni content. Although the 10 wt.% NaOH used to leach CA53 should be slightly more active than 20 wt.% NaOH due to its lower viscosity, this effect is too weak at rt (21) to explain the measured differences between the leaching of CA53 to that of the other alloys in Figure 57.16. As shown before, the isolated Ni_2Al_3 in the single-phase powder is inert to NaOH at rt and the Ni_2Al_3 of the multiphase block takes 120 min to reach ~11 wt.% Al. The CA53 with 80 vol.% Ni_2Al_3 leached rather rapidly to ~15 wt.% Al after merely 45 min and only $NiAl_3$ leached faster in Figure 57.16. Even though CA53 contained more of the difficult to leach Ni_2Al_3 than the multiphase block (see Table 57.1), its S_v was 43% higher meaning that the average size of its Ni_2Al_3 domains is much smaller than that of the multiphase block. Hence the Ni_2Al_3 should have more interfacial contact to $NiAl_3$ in the CA53 than it has in the multiphase block, even though CA53 has only has ~15 vol.% $NiAl_3$ (with ~5%

Al-eutectic) in comparison to 30 vol.% NiAl₃ for the multiphase block. Since the CA53 activated much faster than the multiphase block, it must be the quantity of the Ni₂Al₃-to-NiAl₃ interface and not the vol.% NiAl₃ that determines the ability of the NiAl₃ to enhance the leaching of Ni₂Al₃ for slowly cooled casted alloys. In contradiction to the theory of Timofeeva (36), this would imply that the greater heat generated by leaching 30 vol.% NiAl₃ in the vicinity of the Ni₂Al₃ for the multiphase block is less important for the NiAl₃ accelerated leaching of Ni₂Al₃ than the increased S_v between them as seen with CA53. The rapidly cooled alloys containing 50 wt.% Al exhibited exactly the opposite relationship where the smaller Ni₂Al₃ domains of WA50 were far more difficult to leach than the larger ones of the NA50 (Table 57.2). Hence increasing the amount of the interfacial contact between Ni₂Al₃ and NiAl₃ for the rapidly cooled alloys does not lead to more NiAl₃ assisted leaching of the Ni₂Al₃ phase as it did with the slowly cooled ones. Therefore, the Ni₂Al₃/NiAl₃ interface of the slowly cooled casted alloys is considerably different than that for the rapidly cooled ones.

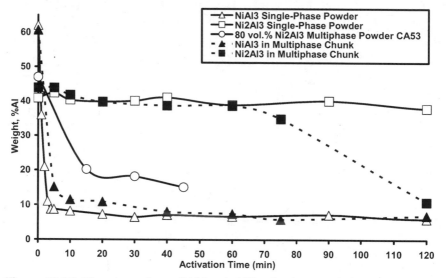

Figure 57.16. The comparison of Al leachability for the casted single-phase alloy powders, the casted multiphase alloy block and CA53. The CA53 was activated at rt with 10 wt.% NaOH and its Al content was determined by x-ray. The other casted alloys were activated at rt with 20 wt.% NaOH and their Al contents were ascertained by SEM EDX.

According to the phase diagram, the slowly cooled CA53 should form NiAl crystals as the primary phase and then react with the melt peritectically to form Ni₂Al₃. After further cooling, the peritectic reaction of Ni₂Al₃ with the melt should result in the formation of the NiAl₃ phase. Fortunately the nucleation of the unleachable (52) NiAl is suppressed during cooling (11). Instead, the solidification of CA53 starts by the primary formation of Ni₂Al₃ crystals that subsequently react with the melt peritectically to form NiAl₃ on its favorable nucleation sites. The peritectic reaction is incomplete leading to a larger solidification interval with some of the melt below the peritectic temperature. Once the eutectic temperature is reached, the melt solidifies eutectically to form eutectic $NiAl_3$ crystals on top of the peritectic $NiAl_3$ leaving behind Al with some Ni in solid solution (11). This slow peritectic growth of $NiAl_3$ on Ni_2Al_3 is drastically different than the development of this interphase with the rapidly cooled alloys. The solid Ni_2Al_3 is initially formed during the rapid solidification of the alloy, quickly thereafter the Ni-deficient melt starts to form $NiAl_3$ and the two phases grow into each other. There may be some peritectic growth of $NiAl_3$ onto Ni_2Al_3 during the rapid cooling of alloys; however, the ratio of this peritectic reaction to that of $NiAl_3$ nucleation in the melt is determined by the cooling rate with higher cooling rates favoring nucleation in the melt over the peritectic reaction. The faster the liquid-solid interface advances, the more the long-range diffusion of atoms will be stifled and the nucleation speed will be faster than the growth rate leading to more nuclei and finer crystallites (53). Hence, the more rapid a phase is formed, the smaller it is and the more it will act as if it is isolated. Thus, the activation level of Ni_2Al_3 and the likelihood that $NiAl_3$ will enhance its leaching will both become lower as the cooling rate increases. In this respect, practically all the sites along the $NiAl_3/Ni_2Al_3$ interface of a rapidly cooled alloy will have similarly slow leaching rates and its overall activation kinetics will follow that of an advancing interface (a.k.a., shrinking core (19)) model (15) like that of single-phase alloys (17,18).

The slowly cooled casted alloys do experience a sizeable acceleration of the Ni_2Al_3 leaching rate when it is in the presence of $NiAl_3$. This is due to the peritectic growth of the $NiAl_3$ phase initially on the preferred nucleation sites of the Ni_2Al_3. This acceleration effect may be from the migration of Al from $NiAl_3$ to Ni_2Al_3 as it peritectically grows on it, due to the ability of Ni_2Al_3 to accept Al via the displacement of Ni in the lattice (54). The ease of Al migration in Ni/Al alloys is well known (44) and such an interface would readily enhance the leaching rate of the Ni_2Al_3 phase. In such a situation, not all of the sites along the $NiAl_3/Ni_2Al_3$ interface of a slowly casted alloy may leach at the same rate and this would explain the site nucleation kinetics of casted alloys as reported in the literature (20,21). The leachability of the Ni_2Al_3 phase for alloys having 50 to 53 wt.% Ni depends on the speed of solidification and the sizes of the phases as summarized in Figure 57.17.

Selection of the Right Alloy and Activation Method. Selecting the right alloy type and activation method for the desired application and its reaction conditions will improve the economics of a chemical process. If a reaction is performed at temperatures lower than 75°C, one may consider an alloy with ~60 wt.% Al due to the higher $NiAl_3$ concentration, which upon leaching results in smaller Ni fcc crystals (5) and more accessible pores (6) than leached Ni_2Al_3. The pores of activated $NiAl_3$ are thought to be more cylindrical than the ink-well type of pores (36,55) formed during Ni_2Al_3 leaching. The 40 wt% Ni / 60 wt.% Al alloy should also be leached at 75°C or lower and its caveats are its sensitivity to sintering (7-9) and mechanical attrition due to the weaker structure of the activated $NiAl_3$ phase. Most industrial applications do not employ such mild conditions and this type of catalyst is very rarely used, if at all. A catalyst made from a slowly cooled casted alloy having ~50

1. <u>Medium</u> Ni₂Al₃ phases having NiAl₃ <u>peritectically</u> grown on them in <u>ground slowly</u> cooled powder alloys leach faster than.....

2. <u>Very large</u> Ni₂Al₃ phases having NiAl₃ <u>peritectically</u> grown on them in <u>slowly</u> cooled alloy chunks Leach very much faster than.....

3. <u>Small</u> Ni₂Al₃ phases with <u>more</u> NiAl₃ <u>nucleation in the melt</u> than peritectic growth in <u>rapidly</u> cooled alloys leach faster than.....

4. <u>Very small</u> Ni₂Al₃ phases with <u>much more</u> NiAl₃ <u>nucleation in the melt</u> than peritectic growth in <u>very rapidly</u> cooled alloys Leach very much faster than.....

5. <u>Isolated</u> Ni₂Al₃ in the <u>absence</u> of NiAl₃

Figure 57.17. The top to bottom ranking of Ni₂Al₃ leachability in Ni/Al alloys containing 50 to 53 wt.% Ni.

to 53 wt.% Ni, that is activated from 75 to 110°C offers the best performance with high activity, selectivity and stability for the overwhelming majority of industrial reactions and conditions. The higher leaching temperatures might be preferred to convert as much of the Ni/Al bcc to the more active fcc Ni as possible. However, the bcc to fcc conversion may be left incomplete at lower leaching temperatures if its occurrence in the reaction medium is an advantage. The high Ni₂Al₃ content in this alloy type also provides a very strong backbone for the leached catalyst due to the Ni₂Al₃ structural memory effect.

The rapidly cooled alloys are being used more in certain niche areas of industry; however, they still have problems with their propensity to sinter. The WA alloys have an additional issue in that their particles are very fragile and their mechanical attrition could cause problems with filtration and the pumping of their slurry. On the other hand, the irregular structures of WA alloys make them perfect for the production of porous fixed bed catalysts (56-68) where the particles are not exposed to the same mechanical sheer and stress as powders in a slurry reactor. The NA alloys are more mechanically stable than the WA ones, but they still are prone to rapid sintering during use and/or if they are activated too aggressively. The rapidly cooled alloys are preferred for applications where it is necessary to avoid the build up of unwanted aluminate-type compounds formed by the further leaching of the catalyst during the reaction. In this situation, using a difficult to leach catalyst from a rapidly cooled alloy can be an advantage. The hydrogenation of dinitrotoluene (DNT) to toluene diamine (TDA) is one example where the reaction conditions will leach out the residual Al of the activated Ni catalyst resulting in the formation of undesired solid aluminate compounds such as takovite (69). These solid aluminate compounds can build up in the reactor and cause difficulties with heat transfer, the separation of the catalyst from the reaction mixture and product filtration. In the worst case, the buildup of these solid aluminates can lead to safety problems. Hence, the use of catalysts from rapidly cooled alloys that are more resistant to leaching, so that the buildup of aluminates can be managed, will be advantageous for DNT hydrogenation. Larson (70) was the first to use catalysts from "water shattered" alloys for the hydrogenation of DNT with commercially satisfactory results and this technology was optimized further by Birkenstock et al. (71) and Wegener et al. (72,73). Regardless of these optimizations, troubles still remain with the mechanical and chemical stability of the catalysts made from rapidly cooled alloys during DNT hydrogenation, and new catalyst technologies have been developed to solve these issues as well as the aluminate problems discussed above (74).

The Formation of Steric Ni Ensembles for the Hydrogenation of Nitriles. While the choice of alloy and activation procedure determines the characteristics of the basic catalyst, the surface active sites can be further customized to improve the yield of a desired reaction. Nitriles can be hydrogenated via the Von Braun mechanism (75) to yield a mixture of 1°, secondary (2°) and tertiary (3°) amines over an activated Ni catalyst, where the 1° amine is usually the desired product. The activity of Ni can be improved by promoting it with Mo. However, the positively charged Mo (76) coordinates strongly with nitriles and the hydrogenation intermediates to lengthen their residence time on the catalyst resulting in more 2° and 3° amines. Adding NH_3 may improve selectivity (57), but further adjustments were necessary to obtain the desired results. The idea that worked the best was to limit the steric Ni ensemble size so that the bulkier 2° and 3° amines will not form. This is readily done by the deposition of CO-like residues onto the catalyst's surface via the rt decomposition of H_2CO in a stirred aqueous catalyst suspension. Additional information about this treatment can be found in the literature (23-27). It was found that H_2CO can disproportionate over Ni (110) surfaces at low temperature (95 K) to form MeOH and CO (77). The freshly formed CO adsorbs strongly on the metal to function as a site blocker and possibly as an electronic modifier of the nearby active sites. These species desorb around 170°C (78-81) and that agrees with our TPO of the formaldehyde treated (FT) catalyst where a measured amount of CO_2 in the range of 200 to 370°C was detected (23,26). In comparison to the TPO of the untreated catalyst, the catalyst's FT only created a small change in the O_2 uptake during the temperature range where CO_2 was produced. Hence, the other structural characteristics of the catalyst were maintained during this treatment (23,26) and these CO-like species are adsorbed strong enough to survive the hydrogenation of nitriles. Other reactions which could take place during this treatment include: (a) the reverse reaction of CO with H_2 to give H_2CO; however, Newton and Dodge (82) found that Ni at even at 200°C is far more likely to decompose H_2CO to CO and H_2 ($K_{decomposition}$ = 1800) than it is to hydrogenate CO to H_2CO ($K_{hydrogenation}$ = 2.3 x 10-5); (b) the in situ generated MeOH is also reabsorbed during the treatment and this may form chemisorbed CO and hydrogen, or even possibly a hydrogen deficient polymeric species on the catalyst. These stable deposits were very effective in "fencing in" the smaller steric Ni ensembles and as expected, the catalyst's average steric Ni ensemble size became smaller as it was treated with more H_2CO.

The Influence of FT on the Hydrogenation of Benzonitrile. Figure 57.18 displays the reaction scheme for the hydrogenation of benzonitrile (BN) to benzylimine (BI), benzylamine (BA), dibenzylimine (DBI) and dibenzylamine (DBA). This reaction proceeds stepwise proceeds stepwise from BN to BI and from BI to BA; however,

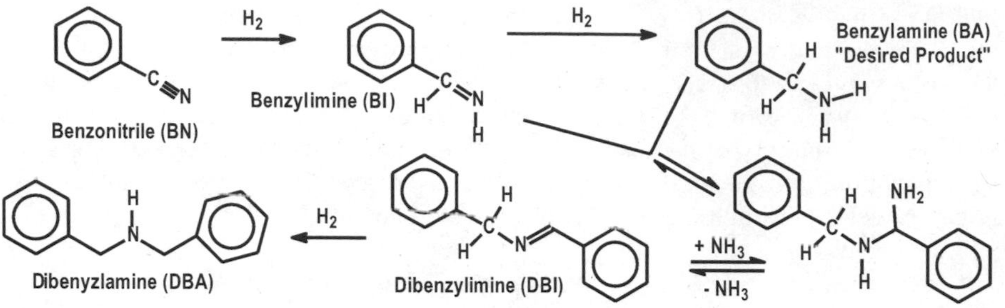

Figure 57.18. The hydrogenation scheme of benzonitrile (BN)

the adsorbed BI can also react with BA to form an α-amino 2° amine that can give off NH₃ to form DBI. The DBI can be irreversibly hydrogenated to DBA or it can react with NH₃ to reproduce the α-amino 2° amine and possibly go back to BA and adsorbed BI. Obviously, the key to this reaction is to understand and control the formation and activity of the adsorbed BI intermediate. Figure 57.19 compares the selectivity patterns of BN hydrogenation in the absence of NH₃ over two different activated Ni catalysts that were treated with various levels of H₂CO. One of the activated Ni catalysts was promoted with Mo for enhanced activity and its particle size distribution (PSD) had a D_{50} of 31 μm. The other catalyst was not promoted with Mo and had a PSD D_{50} of 53 μm. The first observation was that BI could not be found at anytime during the reaction. Unlike the stepwise hydrogenation of phenyl acetylene (PA) to styrene (ST) and eventually ethyl benzene (EB), where it is possible to selectively produce the semi-hydrogenated ST (83), BN does not semi-hydrogenate to produce BI in an isolatable fashion. This is probably due to the influence of the unshared electrons located on the nitrogen atom of the imine group. The two π-clouds of a triple bond are perpendicular to each other and they cannot be hydrogenated at the same time. Instead, one π-cloud is hydrogenated, the molecule desorbs in this semi-hydrogenated state (e.g., ST for PA), rotates 90° and readsorbs so that the final π-cloud can be reduced. In the case of PA, its adsorption strength is much stronger than that of ST so that ST cannot competitively readsorb for its reduction to EB until almost all of the PA is converted to ST. Hence, it is rather easy to selectively semi-hydrogenate PA to ST. Although BN is initially reduced to BI (like PA to ST), the unshared electrons of BI's nitrogen atom can also adsorb to keep it tethered to Ni while rotating 90° for the adsorption and hydrogenation of its remaining C=N π-cloud to BA. Thus, BI does not desorb and it does not have to competitively readsorb against BN before it is reduced to BA or forms the α-amino 2° amine.

Figure 57.19 shows that the selectivity patterns of BN hydrogenation are strongly influenced by the FT level. As the FT level increases the DBA level decreases, the BA yield goes through a maximum and DBI only starts to appear at and after the optimal FT level for BA has been reached. These trends are the same for both catalysts; however, the Mo promoted catalyst requires more H₂CO to reach the optimal FT level for BA formation. This may be due to its smaller PSD, the higher surface area of the Mo promoted catalyst, and/or each surface Mo atom may decompose more H₂CO molecules than Ni. The data suggests that there is an optimal ensemble size that allows for BA formation and impedes DBA generation. One only starts to detect DBI in the reaction mixture once the optimal BA ensemble size is reached (1.33 mmol H_2CO / g of Ni and 2 mmol H_2CO / g of Mo doped Ni) and the DBI level steadily increases as the ensembles become smaller and BA starts to become more difficult to form. These trends suggest that DBA is most readily produced on the largest Ni ensembles, BA is preferentially formed on the moderately sized steric Ni ensembles and DBI can be formed and desorbed from the smallest ones. The formation of DBI over the smallest steric Ni ensembles implies that it is not formed via a Hinshelwood-Langmuir (HL) mechanism between an adsorbed BA and an adsorbed BI to give the α-amino 2° amine intermediate. Instead, the formation of DBI occurs via an Eley-Rideal (ER) mechanism as displayed in Figure 57.20 for a catalyst with small steric Ni ensembles. The formation of DBI starts with an attack by the unshared electrons of the nitrogen atom on a BA molecule in the reaction medium to the carbon side of the imine moiety of an adsorbed BI. This shift of the imine's π electrons to a H on the attacking N so that both of the electrons of the H-N bond can be transferred to N in order to neutralize its positive charge that resulted from the initial attack as the ensuing proton forms and creates a bond with the π electrons and N of the former imine group. This generates the α-amino 2° amine where the previous imine carbon has been converted from a sp^2 to a sp^3 configuration with considerably more rotational freedom. This molecule can then lose NH_3 to form DBI; however, this would require that the carbon with the amino group changes from a sp^3 configuration to the more restrictive sp^2. The DBImolecule is at its lowest energy state when the imine is in resonance with the attached phenyl group. This means that the methylene of the benzyl group, the imine moiety itself and the phenyl group will all be in the same plane as seen in Figure 57.20. This bulky DBI structure is too large for the smaller steric Ni ensembles and the resulting

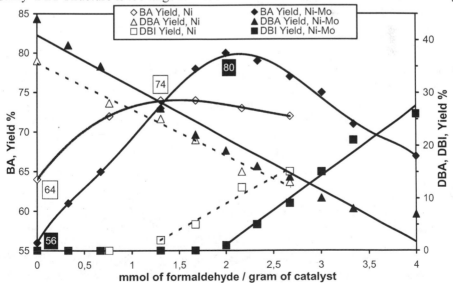

Figure 57.19. The influence of the catalyst's FT level on the yields of benzylamine (BA), dibenzylimine (DBI) and dibenzylamine (DBA) during the hydrogenation of benzonitrile (BN) in the absence of NH_3.

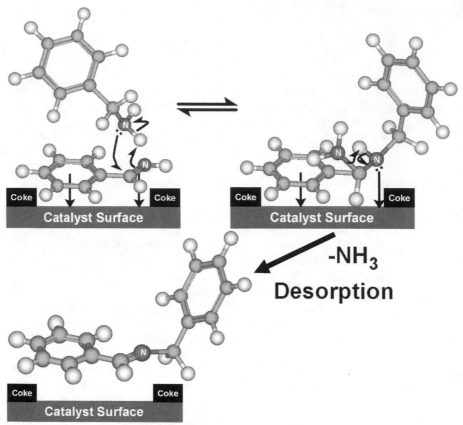

Figure 57.20. DBI formation during BN hydrogenation over small steric Ni ensembles.

steric interactions either inhibit the formation of DBI or force it to desorb without being hydrogenated. Once the DBI is desorbed, it will not be able to readsorb on these small steric Ni ensembles resulting in its buildup in the reaction medium. If the phenyl group is not in resonance with the imine, the spatial arrangement of the imine plane will be even bulkier meaning that this form also will not hydrogenate before desorbing or readsorb for hydrogenation.

Figure 57.21 displays the formation of DBA over large Ni ensembles where the α-amino 2° amine is created via the ER attack of BA from the reaction medium on an adsorbed BI in the same manner as was exhibited in Figure 57.20. The major difference between the reaction mechanism on the small steric Ni ensembles and the larger ones is that the loss of NH_3 from the α-amino 2° amine results in a strongly held DBI on the large Ni ensembles. This strongly adsorbed DBI will either hydrogenate to form DBA or it will react with NH_3 to push the equilibrium towards the α-amino 2° amine, however it will not desorb. Figure 57.22 demonstrates how BN is hydrogenated stepwise to BI and then BA over optimally sized steric Ni ensembles. Even though the 180° bond angle of Ph-C≡N: changes to 60° for

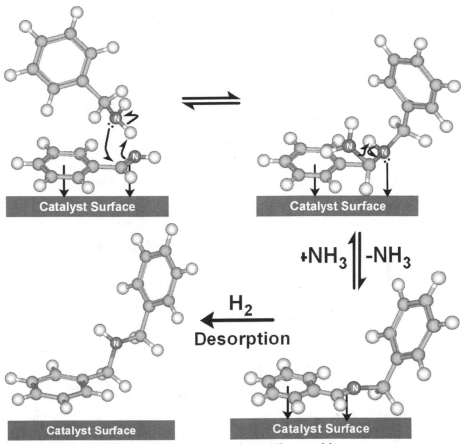

Figure 57.21. The formation of DBA over large Ni ensembles.

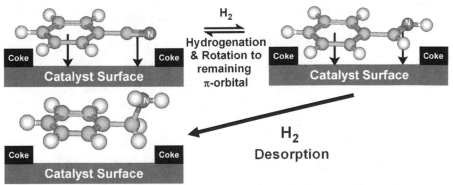

Figure 57.22. The hydrogenation of BN to BI and BA over a catalyst with moderately sized steric Ni ensembles.

Ph-C=N:, this does not force the BI to desorb even on the smallest ensembles and the steric Ni ensemble requirement of both molecules appears to be the same. Since

Design of Activated Nickel Catalysts

hydrogenations occur between adsorbed hydrogen atoms and the adsorbed π-cloud of the unsaturated substrate via a Hinshelwood-Langmuir (HL) mechanism, the logical conclusion would be that HL mechanisms become more difficult as the average steric Ni ensemble size goes from moderate (at optimal BA yield) to small resulting in less BA, a longer residence time of BI on the surface and more DBI via the ER path.

The results of Figure 57.19 illustrate that the Mo doped Ni has a lower BA yield than the unpromoted Ni when they have not underwent FT. However, as the FT level of these catalysts increases, the 1° amine yield of the Mo doped Ni becomes better than that of the unpromoted one. Hence, the positively charged coordinating properties of the surface Mo that are responsible for increased 2° amine formation have been eliminated by the direct decomposition of H_2CO onto Mo. Not only does Mo make H_2CO decomposition more effective, but the resulting coke residue takes on the structure of the distributed Mo over the Ni surface leading to the improved performance of this combination. Samples of the Mo doped Ni were also treated with sodium formate at 90°C to the levels of 1.66, 2.0 and 3.0 mmol of modifier per g of catalyst and the results for BN hydrogenation were the same as with H_2CO at rt for these modifier levels. Hence, other molecules can also be used for this purpose (23).

Figure 57.23 shows how the BN hydrogenation batch time lengthens as the FT level increases. This drop in activity is due to the loss of catalytic surface during the FT and it is to be expected. The higher activity of the Mo promoted Ni is due to the presence of Mo and its smaller PSD, and the rate of its activity drop with respect to the FT level is much greater than it is for the Ni without Mo. Since the surface Mo accelerates the conversion of nitriles, selectively covering it with H_2CO should lead to a faster deactivation rate as the FT level is increased. This is yet a further

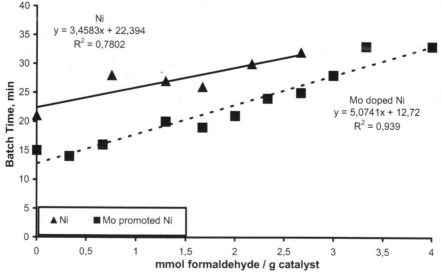

Figure 57.23. The effect of FT level on the BN hydrogenation batch time without NH_3.

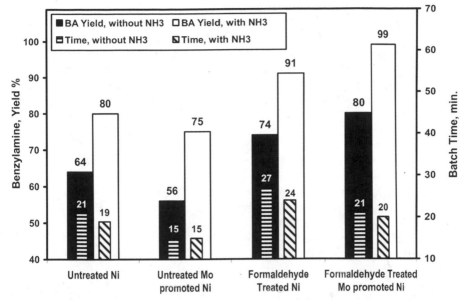

Figure 57.24. The batch times and BA yields of the Ni and Mo doped Ni with and without FT. The FT levels were 1.33 mmol H_2CO per g Ni and 2.0 mmol H_2CO per g of Mo promoted Ni.

indication that Mo is selectively poisoned via H_2CO decomposition during FT. In spite of this, the Mo promoted Ni catalyst treated with 2 mmol H_2CO / g catalyst had the optimal BA yield of 80% (in the absence of NH_3) and a fast batch time of 21 min while the untreated Ni catalyst without Mo had the BA yield of 64% with also a 21 min batch time. Clearly the combination of effects for the FT Mo promoted Ni led to the most selective catalyst in the absence of NH_3 while maintaining the desired batch time. It is interesting to note that the amount of H_2CO needed to modify the Mo promoted Ni to its highest BA yield was an order of magnitude more than the absolute amount of Mo on the catalyst. Hence, even if each Mo atom decomposes more than one H_2CO molecule, the majority of the decomposed H_2CO on the optimally treated catalyst should not be associated intimately with Mo; nonetheless the presence of Mo undoubtedly dictates the performance of this modification. This suggests that the decomposition of the first H_2CO (as with other modifiers as well) molecules is key to the performance of the FT catalyst and that Mo helps with the orientation and kinetics of this decomposition.

Up to now, all of the BN hydrogenation data was in the absence of NH_3 and Figure 57.24 shows the effects of adding NH_3 to the reaction mixture for the Ni and Mo promoted Ni catalysts either with or without FT. The amounts of FT were 1.33 mmol H_2CO per g of Ni and 2.0 mmol H_2CO per g of Mo promoted Ni, because the BA yields for these catalysts were the highest at these FT levels in the absence of NH_3. The amount of NH_3 added to these reactions was 10 ml of a 32 wt.% aqueous NH_3 solution and this is relatively much lower than what is normally added to enhance the 1° amine selectivity of nitrile hydrogenation. The addition of NH$_3$ to the untreated catalysts improves the BA selectivity by driving the equilibrium between the alpha amino 2° amine and BDI towards the alpha amino 2° amine, so that the regeneration of BA and adsorbed BI becomes more likely. Since DBI is also a strongly adsorbed intermediate that can act as a reversible poison to this hydrogenation, its avoidance via NH$_3$ addition has also decreased the batch time for the unpromoted activated Ni catalyst. The BA yield for the Mo promoted Ni was also improved by NH$_3$ addition; however, it was always lower than that of the unpromoted Ni under the same conditions. The coordinating effect of Mo also decreased the batch time of BN hydrogenation to the shortest levels measured here for the untreated Mo doped Ni and the addition of NH$_3$ did not shorten it further, because the influence of Mo was stronger. The combined effect of FT with NH$_3$ addition provided the highest BA yields measured here and the Mo doped Ni provided a better platform than the unpromoted one for the FT regardless if NH$_3$ was added or not. The best reaction combination for the highest BA yield was NH$_3$ addition to a FT Mo promoted Ni. Not only was the BA yield of this combination 99%, but the reaction time was comparable to that of the untreated Ni. This presence of only a small amount of NH$_3$ enhanced the BA yield of FT catalysts considerably. As mentioned earlier for the untreated catalyst, NH$_3$ pushes the equilibrium between DBI and the α-amino 2° amine towards the α-amino 2° amine. However, this would not be very effective for the FT catalysts if the steric Ni ensemble size is so small that the DBI desorbs upon formation and cannot readsorb. In this case, NH$_3$ enhances the 1° amine yield by being a more competitive reactant for BI adsorbed on small steric Ni ensembles (see Figure 57.25) than BA due to its size and because the gem-diamino benzyl intermediate (the aminal compound) can form without any

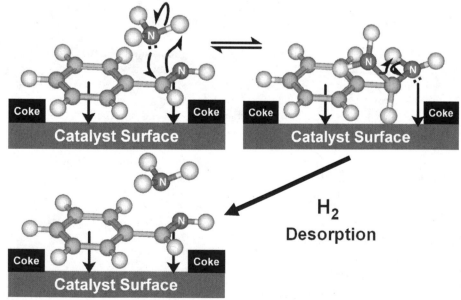

Figure 57.25. The reaction between NH$_3$ and BI adsorbed on a small steric Ni ensemble.

noticeable increase of the steric Ni ensemble requirement as is experienced with BA during the formation of DBI (Figure 57.20). Because the amount of NH_3 needed to enhance the 1° amine yield is greatly reduced for the FT catalysts, this new technology should also reduce the time and costs of the desired nitrile hydrogenation on a commercial scale.

The Selective Hydrogenation of Pynitrile over FT Catalysts. Figure 57.26 displays the hydrogenation scheme of pynitrile to the desired grewe diamine and the unwanted 2° amine. The hydrogenation of pynitrile is an important step in the synthesis of vitamin B1, where the reaction selectivity, hydrogenation rate and avoidance of heavy metal impurities are critical for the commercial success of this process (24,25). Table 57.3 displays the performance of supported Ni/SiO_2, activated Co and activated Ni catalysts either with or without FT for the

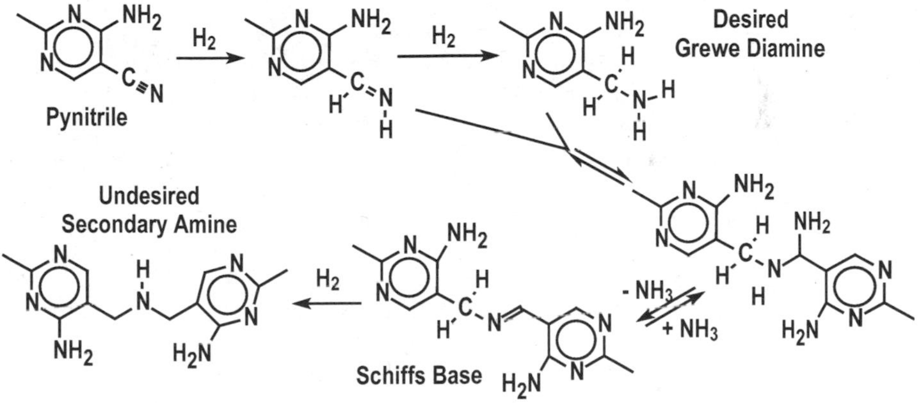

Figure 57.26. The pynitrile hydrogenation scheme.

Table 57.3. The Pynitrile Hydrogenation Data Over Modified and Unmodified Catalysts (24).

Catalyst	Modification Data		Pynitrile Selectivity Data		
	Modifier	mmol mod./g. catalyst	1° amine	2° amine	others
Ni / SiO$_2$	none	---	96.8	1.9	1.3
	H$_2$CO	0.831	98.1	0.1	1.8
Activated Co	none	---	96.8	2.2	1.0
	H$_2$CO	0.840	99.1	0.1	0.8
Activated Ni	none	---	96.4	1.9	1.7
	H$_2$CO[a]	0.840	99.7	0.1	0.2
	Carbon monoxide[a]	0.268	98.8	0.1	0.8
	Acetone	2.296	97.6	1.6	0.8
	Acetaldehyde[b]	0.908	97.3	0.1	2.6

[a] These CO and H$_2$CO modifications lasted 0.5 h and all the others were 1 h.
[b] The modification temperature for acetaldehyde was 60°C and the others were rt.

Table 57.4. The Hydrogenation of Pyridine-3-Carobnitrile (P3C) and Valeronitrile (VN) with Untreated and FT Catalysts (24).

Reactan Structure	mmol H_2CO / g catalyst[a]	Reaction Data			
		time, h	% 1° amine	% 2° amine	% others
(P3C structure) $P3C^b$	---	5	78.1	17.7	4.2
	0.908	3	87.	8.9	4.1
(VN structure) VN^c	---	5	94.4	5.2	0.4
	3.090	3	.96.9	3.1	0

[a] These H_2CO modifications were performed at rt for 1 h.
[b] 110°C, 40 bar, 40 g P3C, 31 g NH_3 and 2000 ml MeOH.
[c] 120°C, 20 bar, 80 g VN, 15 g NH_3 and 2000 ml MeOH.

hydrogenation of pynitrile. Before FT, both the supported Ni/SiO_2 and activated Co catalysts produce the 1° amine in 96.8% yield, while the activated Ni only reached the level of 96.4%. However Co is far more expensive than Ni and the Co impurities in the product are more problematic. Additionally the spent supported Ni/SiO_2 catalyst is much harder to refine leading to far lower Ni recovery rates than with the activated Ni. Thus it would best if one could reach the desired 1° amine yield with the activated Ni catalyst as opposed to the others. The FT improved the performance of all of these catalysts for the hydrogenation of pynitrile and the FT activated Ni yielded 99.7% 1° amine. Although the FT activated Co was not as impressive as the FT activated Ni, it was still much better than the FT Ni/SiO_2. Hence this H_2CO-induced improvement in 1° amine yield is best realized with the catalysts made à la Murray Raney in comparison to the supported one and the FT activated Ni was superior to the FT activated Co. Table 57.3 also displays the performance enhancing capabilities of modifiers other than H_2CO. Although carbon monoxide (CO), acetone and acetaldehyde all improved the 1° amine yield, none of them were as effective as H_2CO. The CO treatment came the closest to FT for 1° amine yield enhancement; however, its mmol modifier to g of catalyst level was much lower than that for FT, because we wanted to avoid the formation of $Ni(CO)_4$ due to its gaseous and poisonous properties. In any case, the 1° amine enhancement capability of CO supports the concept that the best coke residue for the creation of steric Ni ensembles is the CO-like species that was proposed earlier. This may have both steric and electronic reasons since the formation of the Ni=C=O adsorbed species will draw electron density from the Ni atom and its neighbors resulting in nearby electron deficient surface sites that adsorb electron rich substrates (e.g., nitriles) stronger. The lower abilities of the acetone and acetaldehyde treatments to improve the 1° amine yield may have to do with the types of carbonaceous layers they form on Ni as well as their lower likelihood of forming a CO-like species.

The Hydrogenation of Pyridine-3-Carbonitrile and Valeronitrile. Table 57.4 displays the effect of FT on the 1° amine yields and reaction rates of pyridine-3-carbonitrile (P3C) and valeronitrile (VN). Unlike with the hydrogenation of BN, these FT Ni catalysts were able to hydrogenate P3C and VN within 3 hours instead of the 5 hours required for the untreated catalysts. Hence, even the smallest differences in substrate structure and reaction conditions seem to play a role in the performance of these catalysts after FT. The FT also improved the 1° amine selectivities of these reactions, and a greater enhancement was possible for P3C than for VN. This could be due to the differentiating aspects of the larger steric Ni ensemble requirement for adsorbed P3C in comparison to VN. However, the 1° amine selectivity of VN was already very high for the untreated catalyst (94.4%) and it obviously was more difficult to improve further with FT (96.9%) than that of VN (87% with FT and 78.7% without). Since the level of FT was not optimized for these reactions, it is likely that one can improve these results with further experimentation. The higher 1° amine selectivity of VN hydrogenation may be due to its lower residence time on the activated Ni surface, because it doesn't have other conjugating and adsorbing groups like P3C.

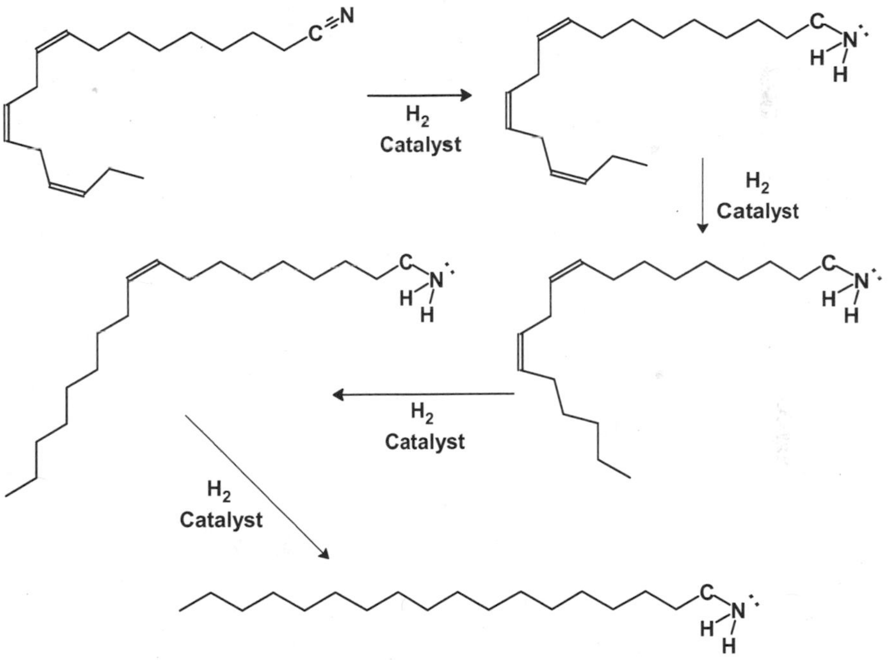

Figure 57.27. The stepwise hydrogenation of a C18:3 fatty nitrile.

The Chemo-Selective Hydrogenation of Tallow Nitriles. Tallow nitriles are a mixture of fatty nitriles having mostly 18 carbons with some 16 carbon molecules and traces of ones with 14 and 20. These molecules might have one to three *cis*-double bonds on the 9th, 12th and/or 15th carbons starting from the nitrile position, and the iodine value (IV) titration method was used to monitor the level of unsaturation. The tallow nitrile mixture used here started with an IV of ~51. One of the molecules present in this mixture is the C18:3 fatty nitrile (18 carbons with 3 double bonds) whose stepwise hydrogenation is displayed in Figure 57.27. Initially the nitrile group is reduced to form the C18:3 amine that can be hydrogenated further to give C18:2

followed by C18:1 and finally C18:0. This stepwise hydrogenation of the olefin moieties makes the molecule more linear as the number of carbon-carbon double bonds decreases leading to improved molecule stacking, a drop in the melting point and the eventual formation of a solid. This undesirable linear structure with its improved stacking and lower melting point can also be achieved by the isomerization of the *cis*-alkenes to *trans* meaning that the adsorption of the alkenes should be avoided altogether. Desirable unsaturated liquid tallow amines are obtained by the complete hydrogenation of the nitriles, as monitored by the total amine value (TAV), without the reduction or the *trans*-isomerization of the olefins. The goal of this work was to design a chemo-selective hydrogenation catalyst with specific steric Ni ensembles of the optimal size. As displayed in Figure 57.28, only a very small

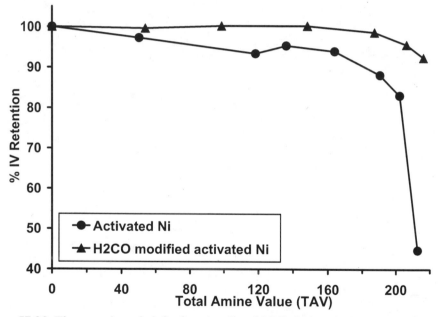

Figure 57.28. The retention of olefin functionality (% IV) during the hydrogention of fatty nitriles to fatty amines over activated Ni with and without H$_2$CO modification.

percentage of the olefins are hydrogenated with the untreated Ni before most of the strongly adsorbed nitriles are reduced. It is only after most of the nitriles were converted to amines (TAV > 165) that the rate of olefin reduction on the untreated Ni increased dramatically as seen by the fast drop in the percentage of IV retention. The initial chemo-selective hydrogenation of nitriles on the untreated catalyst is due to the surface being saturated with the strongly adsorbed nitrile so that the olefins cannot adsorb and hydrogenate. As the nitrile conversion level increases, the availability of adsorption sites for the olefins becomes more prevalent leading to lower IV. This behavior has also been seen in the literature for smaller unsaturated nitriles (84) and these authors claimed that the adsorption of the resulting amine also blocked the adsorption and reduction of the olefins. If that is the case, then the blocking power of the adsorbed amine is not sufficient enough to obtain liquid tallow amines under the typical commercial reaction conditions used here.

Table 57.5. The Tallow Nitrile Hydrogenation Data with FT and Untreated Ni Catalysts at 17 bar NH_3, 23 bar H_2 and 140°C.

Catalyst	Time, h[a]	TAV[b]	% IV retention[c]	2/3 A[d]	Initial activity, mol H_2 / min g catalyst
Activated Ni	300	212.8	45	6.5	1.63
FT Activated Ni[e]	256	216	92	4	1.20

[a] Reaction time determined by hydrogen uptake measurements
[b] TAV = total amine value made by the titration of amines to give nitrile conversion
[c] 2/3 A = the percentage of 2° and 3° amines
[d] This catalyst was modified with 1.33 mmol H_2CO at rt for 1 h

Figure 57.28 also shows that the FT Ni retained 99% of its IV up to a TAV of 187 and when the hydrogenation stopped at the TAV of 216, this mixture still retained 92% of its IV. Table 57.5 makes a complete comparison of the untreated Ni to the FT Ni, and one can see that the FT Ni produced a higher quality unsaturated tallow amine with a lower level of 2° and 3° amines, a higher TAV and a far superior retention of alkene functionality. Although the initial hydrogenation activity of the FT Ni was lower than that of the untreated catalyst, the FT catalyst was able to reach a higher TAV much faster. This is similar to the hydrogenation results of P3C and VN where high levels of NH_3 were also used. However, these results are different than those of BN where NH_3 was either absent or present at a very low level. It could be that the presence of high NH_3 levels reduces or eliminates the drop in overall activity for the FT catalyst in comparison to the untreated ones and/or it may improve it by augmenting the tailor-made steric Ni ensembles. One could also look at this as a regio-selective hydrogenation, where the terminally positioned and strongly adsorbed nitrile group is selectively hydrogenated over the internally located and weaker adsorbed alkenes. In this respect, the nitrile can be preferentially adsorbed and hydrogenated on small steric Ni ensembles without any interference from the long fatty chain, while the alkenes located in the middle of the chain will experience more steric inhibition towards their adsorption on the small steric Ni ensembles. This would also suggest that the alkenes closer to the other end of the fatty chain (e.g., position 15 of a C16 or position 15 of a C18) are more likely to be hydrogenated and they were probably the ones that were reduced as the tallow amine lost 8% of its IV over the FT Ni.

Conclusions

The most suitable alloy for the production of mechanically and chemically stable catalysts is the slowly cooled casted alloy where the $NiAl_3$ phase grows peritectically on the initially solidified Ni_2Al_3. This slowly formed interface led to the $NiAl_3$ enhanced leaching of Ni_2Al_3 during catalyst activation and this leaching process best fits a site nucleation type of kinetics to give a more thoroughly activated catalyst. The leaching of Al out of the trigonal Ni_2Al_3 phase of the casted alloy converts it initially to a bcc phase via the short-order rearrangement of Ni atoms. According to the well-known Bain relationship between the bcc and fcc structures, the further leaching of Al from the bcc phase transforms it martensitically to a fcc structure with ~ 10 nm crystals having a lot of defects and up to 6 wt.% Al. Catalysts with Al contents higher than 6 wt.% have the additional Al in the retained bcc structures dispersed in between the fcc crystals.

The activated rapidly cooled alloys have found a niche market where it is important to avoid the unmanageable formation of aluminate compounds (e.g., takovite formation during DNT hydrogenation) during their use. However, there are new technologies on the horizon that will replace this difficult to leach, readily sintered and mechanically less stable catalyst for the hydrogenation of DNT (74). Perhaps the best use for the WA alloys is the production of fixed bed catalysts where the irregularly shaped particles provide a very accessible pore structure and the particles are not subjected to the same mechanically demanding conditions as a powder catalyst would be in a slurry phase reactor (56,57). During the solidification of a rapidly cooled alloy, the Ni_2Al_3 initially crystallizes and then the Ni-deficient melt starts to form the solid $NiAl_3$ that grows into the advancing solid front of Ni_2Al_3. There may be some peritectic $NiAl_3$ growth on the Ni_2Al_3, but most of it is not and the more rapid the cooling rate is, the lower the level of peritectic $NiAl_3$ growth will be resulting in less $NiAl_3$ enhancement of Ni_2Al_3 leaching. In other words, the more rapidly cooled the alloy is, the more the Ni_2Al_3 will start to act like an isolated phase which decreases its leachability. This is best illustrated by comparing the high Ni_2Al_3 retention during the leaching of the very rapidly cooled WA alloy to the moderate levels of Ni_2Al_3 retention for the rapidly cooled NA alloy and the complete leaching of the Ni_2Al_3 phase in the slowly cooled casted alloys. These difficult to leach, rapidly cooled alloys follow a shrinking core kinetic model during activation and they are prone to sintering when they are leached at 110°C or higher. Another catalyst with very little industrial use is from a $NiAl_3$-rich alloy with ~60 wt.% Al or more, due to its poor mechanical and chemical stability at 110°C or higher.

Activated Ni catalysts can also be made more selective for the hydrogenation of nitriles to 1° amines by the formation of optimally sized steric Ni ensembles via the decomposition of H_2CO or similar molecules on to the catalytic surface. The large Ni ensembles allow for the hydrogenation of 2° imines to 2° amines and the moderately sized steric Ni ensembles are optimal for the formation of 1° amines. The very small steric Ni ensembles inhibit HL hydrogenations as well as the 2° and 1° amines that result from them, while allowing for the formation of 2° imines via an ER mechanism. This concept provided us the ability to design a very selective catalyst for the hydrogenation of pynitrile to the desired 1° amine that is used in vitamin B1 synthesis. This optimization of steric Ni ensembles was also very effective in improving the chemo-selective hydrogenation of unsaturated fatty nitriles to unsaturated fatty amines for the maintenance of clear liquid fatty amine properties.

Acknowledgements

The author would like to thank Professor H.E. Exner and Dr. S. Knies of the Technical University of Darmstadt for the very fruitful collaboration concerning the correlation of alloy type and structure to catalyst performance. A special appreciation goes to M. Berweiler, S. Röder, B. Bender, Dr. R. Olindo (currently at the Technical University of Munchen) and Dr. P. Panster of Evonik Degussa for their technical support and discussions. The author would also like to take this opportunity to thank Professor G.V. Smith of Southern Illinois University at Carbondale, IL, Professor M. Bartók of József Attila Tudományegyetem (József Attila Science University) in Szeged, Hungary and Professor W.M.H. Sachtler of Northwestern University in Evanston, IL for being his mentors and for the motivation to pursue a career in catalysis.

References

1. M. Raney, US Patent 1628191 (1927).
2. A. Knappwürst and K.H. Mader, Naturwissenschaften, **52** (1965) 590.
3. R. Schrötter, *Newer Methods of Preparative Organic Chemistry*, Interscience, New York, 1948, p. 61.
4. R. L. Augustine, *Heterogeneous Catalysis for the Synthetic Chemist*, Marcel Dekker, Inc. New York, 1996, pp. 241–243.
5. R. Wang, Z. Lu and T. Ko, Dianzi Xianwei Xuebao, **16**, 3 (1997) 302-306 [CAN 128:170696].
6. C. De Bellefon and P. Fouilloux; Catal Rev.-Sci. Eng., **36** (1994) 459-506.
7. M.L. Bakker, D.J. Young and M.S. Wainwright, J. Mater. Sci., **23**, 11 (1988) 3921–3926.
8. P. Colin, S. Hama-Thibault and J.C. Joud, J. Mater Sci., **27** (1992) 2326-2334.
9. S.D. Mikahilenko, A.B. Fasman, N.A. Maksimova and E.V. Leongard, Appl. Catal., **12** (1984) 141–150.
10. A.A. Presnyakov, K.T. Chernousova, T. Kabiev, A.B. Fasman and T.T., Bocharova, J. Appl. Chem. USSR (Eng. Translation), **40** (1967) 929-934.
11. S. Knies, G. Miehe, M. Rettenmayr, and D.J. Ostgard, Zeitschrift für Metallkunde, **92** (2001) 596-599.
12. Sassoulas, R., Trambouze, Y., Bulletin Societes Chimique Frances, **5** (1964) 985-988.
13. J. Gros, S. Hama-Thibault, J.C. Joud, Surf. Interface Anal. **11** (1988) 611–616.
14. Reynaud, F., Journal of Applied Crystallography, **9** (1976) 263-268.
15. R, Wang, Z. Lu and T. Ko, Journal of Material Science, **36** (2001) 5649-5657.
16. J. Rothe, C. Hormes, C. Shild and B. Pennemann, J. Catal., **191** (2000) 294–300.
17. M.L. Baker, D.J. Young, and M.S. Wainwright, J. Mater. Sci., **23**, 11 (1988) 3921-3926.
18. F. Delanny, Reactivity of Solids, **2** (1986) 235–243.
19. Levinspiel, *Chemical Reaction Engineering*, Wiley, New York (1962).
20. V.R. Choudhary, S.K. Chaudhari, and A.N. Gokarn, Ind. Eng. Chem. Res. **28** (1989) 33-37.

21. V.R. Choudhary and S.K. Chaudhari, J. Chem. Technol. Biotechnol., **33A** (1983) 339-349.
22. S. Knies, M. Berweiler, P. Panster, H. E. Exner and D.J. Ostgard, *Studies of Surface Science in Catalysis*, A. Corma, F.V. Melo, S. Mendioroz, and J.L.G. Fierro editors, Elsevier, Amsterdam, 2000, **130**, 2249-2254.
23. D. Ostgard, V. Duprez, R. Olindo, S. Röder and M. Berweiler, PCT Patent WO2006050749 (2006).
24. D.J. Ostgard, F. Roessler, R. Karge and T. Tacke, Chemical Industries (CRC Press LLC) **115** (Catal. Org. React.), 227-234 (2007) (and references therein).
25. D.J. Ostgard, Specialty Chemicals Magazine, April (2008) 28–31.
26. D.J. Ostgard, R. Olindo, M. Berweiler, S. Roder and T. Tacke, Catalysis Today, **121** (2007) 106–114.
27. D.J. Ostgard, R. Olindo, M. Berweiler, S. Röder and T. Tacke, in *Science and Technology in Catalysis 2006* (eds. K. Eguchi, M. Machida and I. Yamanaka), Kondansha Elservier, Tokyo, 2007, 537–538.
28. D.J. Ostgard, V. Duprez, M. Berweiler, S. Röder and T. Tacke, Chemical Industries (CRC Press LLC) **115** (Catal. Org. React.), 197-211 (2007).
29. D.J. Ostgard, V. Duprez, M. Berweiler, S. Röder and T. Tacke, in *Science and Technology in Catalysis 2006* (eds. K. Eguchi, M. Machida and I. Yamanaka), Kondansha Elservier, Tokyo, 2007, 185-188.
30. S. Knies, Ph.D. Dissertation, Technische Universität Darmstadt (2001).
31. G.J. Rietveld, Acta Crystallographia, **22** (1967) 151-152.
32. S.R. Scherrer, Nachrichten von der Gesellschaft der Wissenschaften, Göttingen, **2** (1918) 98-100.
33. H.E. Exner and H.P. Hougardy, in *Einführung in die Quantitative Gefügeanalyse*, DGM Informationsgesellschaft, Oberursel/Frankfurt, 1986.
34. D.J. Ostgard, M. Berweiler, P. Panster, R. Müller and F. Roessler, Chemical Industries (Decker) **82** (Catal. Org. React.), 109-115 (2001).
35. J. Petró, Chemical Industries (Decker) **82** (Catal. Org. React.), 1-34 (2001).
36. V. F. Timofeeva and A.B. Fasman, Russian Journal of Physical Chemistry, **52**, 4 (1978) 549.
37. P. Fouilloux, Appl. Catalysis, **8** (1983) 1–42.
38. A. B. Fasman, V.F. Timofeeva, V. N. Rechkin, Y.F. Klyuchnikov and Y.A. Sapukov, Kinetika i Kataliz, **13** (1971) 1513 [CAN 78:76283].
39. H. Lei, Z. Song, X. Bao, X. Mu, B. Zong and E. Min, Surf. Interface Anal., **32** (2001) 210-213.
40. H. Lei, Z. Song, D. Tan, X. Bao, X. Mu, B. Zong and E. Min, Appl. Catal. A:General, **214** (2001) 69–76.
41. B.H. Zeifert, J. Salmones, J.A. Hernández, R. Reynoso, N. Nava, J.G. Cabanas-Moreno and G. Aguilar-Ríos, Catalysis Letters, **63** (1999) 161–165.
42. T. Ronghou, S. Shijie, L. Silin, L. Xuequan, M. Ruzhang, L. Tao., Trans Nonferrous Met. Soc. China, **9**, 3 (1999) 562–565.
43. V.P. Gostikin, L-.G- Nishchenkova, G.V. Golubkova and L.V. Kozlova, Kinetics and Catalysis, **36**, 1 (1995) 107–110.
44. J.C. Klein and D.M. Hercules, Anal. Chem., **53** (1981) 754–758.
45. C. Suryanarayana, Materials Science and Technology (VCH-Verlag Weinheim) **15** (Processing of Metals and Alloys) 57-110 (1991).

46. S.R. Schmidt, Chemical Industries (Decker) **62** (Catal. Org. React.) 45-59 (1995).
47. J. Nicolau and R.B. Anderson, J. Catal., **69** (1981) 339–348.
48. S. Tanaka, N. Hirose and T. Tanaki, Denki Kagaku, **65**, 12 (1997) 1044–1048.
49. J. Gros, S. Hamar-Thibault and J.C. Joud, J. Mater. Sci., **24** (1989) 2987–2998.
50. S. Hamar-Thibault, J. Thibault an J.C. Joud, Z. Metallkd., **83** (1992) 258-265.
51. C.M. Wayman, in Physical Metallurgy, 4th Edition, **2**, R.W. Cahn and P. Haasen eds., North Holland, Amsterdam (1996).
52. R. Sossoulas and Y. Trambouze, Bull. Soc. Chim. Fr., **5** (1964) 985–988.
53. H. Hu, M. Qiao, S. Wang, K. Fan, H. Li, B. Zong, and X. Zhang, J. Catal., **221** (2004) 612–618.
54. A. Taylor and N.J Doyle, J. Appl. Cryst., **5** (1972) 201.
55. A.B. Fasman, Chemical Industries (Decker) **75** (Catal. Org. React.) 151-168 (1998).
56. D.J. Ostgard, M. Berweiler, S. Röder, K. Möbus, A Freund and P. Panster, Chemical Industries (Decker) **82** (Catal. Org. React.) 75-89 (2003).
57. D.J. Ostgard, M. Berweiler, S. Röder and P. Panster, Chemical Industries (Decker) **89** (Catal. Org. React.) 273-294 (2003).
58. D.J. Ostgard, A. Freund, M. Berweiler, B. Bender, K. Möbus and P. Panster, Chemie-Anlyge+Verfahren, **9** (1999) 118-119.
59. D.J. Ostgard, P. Panster, C. Rehren, M. Berweiler, G. Stephani and L. Schneider, US Patent 6573213 (2003).
60. D.J. Ostgard, O. Puisségur, S. Roder and M. Berweiler, PCT Patent WO 2005042153 (2005).
61. D.J. Ostgard, M. Berweiler and S. Röder, PCT Patent WO 02055476 (2002).
62. D.J. Ostgard, M. Berweiler and S. Röder, PCT Patent WO 02055453 (2002).
63. D.J. Ostgard, M. Berweiler and S. Röder, US Patent 6649799 (2003).
64. D.J. Ostgard, M. Berweiler and S. Röder, US Patent 6486366 (2002).
65. D.J. Ostgard, M. Berweiler, S. Röder, J. Sauer, B. Jaeger and N. Finke, US Patent 6437186 (2002).
66. D.J. Ostgard, K. Möbus, M. Beweiler, B. Bender, G. Stein, US Patent 6489521 (2002).
67. D.J. Ostgard, R. Olindo, V. Duprez, S. Röder and M. Berweiler, PCT Patent WO2006050742 (2006).
68. D.J. Ostgard, M. Berweiler, T. Quandt and S. Röder, PCT Patent WO 2007028411 (2007).
69. P. Marion, US Patent 6423872 (2002).
70. F.G. Larson Jr., US Patent 3839011 (1974).
71. U. Birkenstock, J. Scharschmidt, P. Kunert, H. Meinhardt, P. Häusel and P. Meier, US Patent 5090997 (1992).
72. G. Wegener, E. Waldau, B. Pennemann, B. Temme, H. Warlimont and U. Kühn, US Patent 6395934 (2002).
73. G. Wegener, M. Brandt, L. Duda, J. Hofmann, B. Klesczewski, D. Koch, R.J. Kumpf, H. Orzesek, H.G. Pirkl, C. Six, C. Steinlein and M. Weibeck, Appl. Catal. A: General, **221** (2001) 303–335.
74. D.J. Ostgard, M. Berweiler, M. Göttlinger, S. Laporte and M. Schwarz, unpublished results from Evonik Degussa, Patent Pending (2008).

75. J. Von Braun, G. Blessing and F. Zobel, Chem. Ber., **36** (1923) 1988.
76. C. Bellefon and P. Fouilloux, Catal. Rev.-Sci. Eng., **36** (1994) 459.
77. L.J. Richter and W. Ho, J. Chem. Phys, **83**, 5 (1985) 2165-2169.
78. J.T. Yates and D.W. Goodman, J.Chem. Phys., **73**, 10 (1980) 5371-5375.
79. A. Bandara, S. Katano, J. Kubota, K. Onda, A. Wada, K. Domen and C. Hirose, Chem. Phys Lett., **290** (1998) 261-267.
80. A. Bandara, S. Dobashi, J. Kubota, K. Onda, A. Wada, K. Domen, C. Hirose and S.S. Kano, Surf. Sci., **387** (1997) 312-319.
81. A. Bandara, S.S. Kano, K. Onda, S. Katano, J. Kubota, K. Domen, C. Hirose and A. Wada, Bull. Chem. Soc. Jpn., **75** (2002) 1125-1132.
82. R.H. Newton, B.F. Dodge, J. Am. Chem. Soc., **55** (1933) 4747-4759.
83. D.J. Ostgard, K.M. Crucilla and F.P. Daly, Chemical Industries (Marcel) **68** (Catal. Org. React.), 199-212 (1996).
84. P. Kukula and K. Koprivova, J. Catal., **234** (2005) 161–171.

Author Index

Keyword Index